Lecture Notes in Computer Science 9236

Commenced Publication in 1973
Founding and Former Series Editors:
Gerhard Goos, Juris Hartmanis, and Jan van Leeuwen

More information about this series at http://www.springer.com/series/7407

Christian Urban · Xingyuan Zhang (Eds.)

Interactive
Theorem Proving

6th International Conference, ITP 2015
Nanjing, China, August 24–27, 2015
Proceedings

 Springer

Editors
Christian Urban
Department of Informatics
King's College London
London
UK

Xingyuan Zhang
PLA University of Science and Technology
Nanjing
China

ISSN 0302-9743   ISSN 1611-3349 (electronic)
Lecture Notes in Computer Science
ISBN 978-3-319-22101-4   ISBN 978-3-319-22102-1 (eBook)
DOI 10.1007/978-3-319-22102-1

Library of Congress Control Number: 2015944507

LNCS Sublibrary: SL1 – Theoretical Computer Science and General Issues

Springer Cham Heidelberg New York Dordrecht London

Printed on acid-free paper

Springer International Publishing AG Switzerland is part of Springer Science+Business Media
(www.springer.com)

Preface

This volume contains the papers presented at ITP 2015, the 6th International Conference on Interactive Theorem Proving held during August 24–27, 2015, in Nanjing, China.

ITP brings together researchers working in interactive theorem proving and related areas, ranging from theoretical foundations to implementation aspects and applications in program verification, security and formalisation of mathematics. The ITP conference series originated in the TPHOLs conferences and the ACL2 workshops, the former starting in 1991 and the latter in 1999. The first ITP conference was held in Edinburgh in 2010, and after that in Nijmegen, Princeton, Rennes and Vienna. This is the first time that ITP and its predecessor conferences have been hosted in Asia.

This year there were 54 submissions. Each submission was reviewed by at least three Programme Committee members. The committee decided to accept 30 papers, three of which are Rough Diamonds. The programme also includes invited talks and there were two three day tutorials on Isabelle/HOL and Coq. As is the tradition, half a day of ITP is dedicated to an excursion, which this year took the participants to Yangzhou and the Slender West Lake.

We are very grateful for the support received from the PLA University of Science and Technology. We also thank the local support team, especially Chunhan Wu and Jinshang Wang, for their help in staging ITP 2015. Finally we are very grateful to the Programme Committee for their hard work of reviewing and discussing the submissions.

We are also looking forward to the next ITP in 2016 that will be held in Nancy, France, organised by Jasmin Blanchette and Stephan Merz.

June 2015

Christian Urban
Xingyuan Zhang

Organisation

Programme Committee

Andrea Asperti	University of Bologna, Italy
Jesper Bengtson	IT University of Copenhagen, Denmark
Stefan Berghofer	Secunet Security Networks AG, Germany
Yves Bertot	Inria, France
Lars Birkedal	Aarhus University, Denmark
Sandrine Blazy	University of Rennes, France
Bob Constable	Cornell University, USA
Thierry Coquand	University of Gothenburg, Sweden
Xinyu Feng	University of Science and Technology, China
Ruben Gamboa	University of Wyoming, USA
Herman Geuvers	Radboud University Nijmegen, The Netherlands
Mike Gordon	Cambridge University, UK
Elsa Gunter	University of Illinois, Urbana-Champaign, USA
John Harrison	Intel Corporation, USA
Hugo Herbelin	Inria, France
Matt Kaufmann	University of Texas at Austin, USA
Gerwin Klein	NICTA, Australia
Cesar Munoz	NASA Langley Research Center, USA
Tobias Nipkow	TU Munich, Germany
Michael Norrish	NICTA, Australia
Scott Owens	University of Kent, UK
Randy Pollack	Harvard University, USA
Carsten Schuermann	IT University of Copenhagen, Denmark
Konrad Slind	Rockwell Collins, USA
Alwen Tiu	Nanyang Technological University, Singapore
Christian Urban	King's College London, UK
Dimitrios Vytiniotis	Microsoft Research Cambridge, UK
Xingyuan Zhang	PLA University of Science and Technology, China

Contents

Verified Over-Approximation of the Diameter of Propositionally Factored Transition Systems

Mohammad Abdulaziz[1,2]($\boxtimes$), Charles Gretton[1,3], and Michael Norrish[1]

[1] Canberra Research Laboratory, NICTA, Canberra, Australia
mohammad.abdulaziz8@gmail.com
[2] Australian National University, Canberra, Australia
[3] Griffith University, Brisbane, Australia

Abstract. To guarantee the completeness of bounded model checking (BMC) we require a *completeness threshold*. The *diameter* of the Kripke model of the transition system is a valid completeness threshold for BMC of safety properties. The *recurrence diameter* gives us an upper bound on the diameter for use in practice. Transition systems are usually described using (propositionally) factored representations. Bounds for such lifted representations are calculated in a compositional way, by first identifying and bounding atomic subsystems, and then composing those results according to subsystem dependencies to arrive at a bound for the concrete system. Compositional approaches are invalid when using the diameter to bound atomic subsystems, and valid when using the recurrence diameter. We provide a novel overapproximation of the diameter, called the *sublist diameter*, that is tighter than the recurrence diameter. We prove that compositional approaches are valid using it to bound atomic subsystems. Those proofs are mechanised in HOL4. We also describe a novel verified compositional bounding technique which provides tighter overall bounds compared to existing bottom-up approaches.

1   Introduction

Problems in *model checking* and *automated planning* are typically formalised in terms of transition systems. For model checking safety formulae—*i.e.*, globally true formulae of the form $\mathbf{G}p$—one asks: Does every sequence of transitions from the initial state include only states satisfying p? In planning one asks the converse question: Is there a sequence of transitions from a given initial state to a state satisfying a goal condition? In other words, model checking poses a classical planning problem with a goal condition *"p is false"*. For bounded versions of those questions we have an upper bound on the number of transitions that can be taken—*i.e.*, Can the goal be achieved using N transitions? The *diameter* of a system is the length of the longest minimum-length sequence of transitions between any pair of states. Taking "$N = \text{diameter}$" we have that bounded model

M. Abdulaziz, C. Gretton and M. Norrish—NICTA is funded by the Australian Government through the Department of Communications and the Australian Research Council through the ICT Centre of Excellence Program.

© Springer International Publishing Switzerland 2015
C. Urban and X. Zhang (Eds.): ITP 2015, LNCS 9236, pp. 1–16, 2015.
DOI: 10.1007/978-3-319-22102-1_1

checking of safety formulae is complete (Biere *et al.* [3]). In other words, if the goal condition cannot be reached in N transitions, then it cannot be reached irrespective of the number of transitions taken. In this sense, the diameter is equal to a *completeness threshold* for bounded model checking of safety (Kroening and Strichman [11]).

A variety of uses for bounds providing the completeness of bounded model checking have been proposed. If the bound N is small, problems reduce in practice to fixed-horizon variants—*i.e.*, is there a sequence of transitions of length less-than-or-equal-to N whose final state satisfies a desired property? That fixed-horizon problem can be posed as a Boolean SAT(isfiability) problem and solved efficiently (Biere *et al.* [3], Kautz and Selman [10]). Also, SAT-based approaches which implement query strategies to focus search effort at important horizon lengths, as discussed in Rintanen [14], and Streeter and Smith [19], can also benefit from knowing a small bound. Finally, where system models are given by factored propositional representations and the search goal is to prove that no state satisfying a large conjunct is reachable, by additionally considering the *cones of influence* of variable sets one can use such bounds to identify a small subset of goal propositions which can be easily proved unreachable (see, for example, Rintanen and Gretton [15]).

Despite the important applications above, obtaining useful bounds in practice is not a simple matter. Computing the system diameter is intractable. An encoding of the required formula in quantified Boolean logic to test whether the diameter is N is provided in [3], and is not known to be practically useful. Indeed, for practice the proposal in [3] is to use an overapproximation of the system diameter, namely the *recurrence diameter*. The recurrence diameter corresponds to the length of the longest non-looping path in the Kripke model, thus its computation poses the NP-hard longest path problem. A difficulty in practice is that such structures are described using a compact factored problem description. The Kripke structure corresponding to such a description is much (sometimes exponentially) larger than the description itself, and is rarely explicitly represented. Proposals in this setting employ decompositional bounding procedures which exploit independence between sets of state-characterising variables. Bounds for atomic subsystems are first computed in isolation, which gives the advantage of a potential exponential reduction in the cost of computing the bound on the diameter (*e.g.*, if the recurrence diameter is to be computed). Then the bounds on subsystems are combined either multiplicatively or additively depending on subsystem dependencies. Intuitively, the resulting overapproximation is given by an additive expression whose terms correspond to bounds for independent abstract subproblems. The decompositional approach from Baumgartner *et al.* [1] treats design netlists, and in [15] the decomposition treats the *dependency graph* (*causal graph*) associated with a STRIPS-style description. Importantly, decompositional approaches calculate invalid bounds when using the system diameter to bound atomic subproblems. One workaround, as *per* Baumgartner *et al.* is to use the recurrence diameter instead.

Treating the *dependency graph* of STRIPS-style descriptions, we have three primary contributions. First, we develop the concept of *sublist diameter* which is a tighter over-approximation of diameter compared to the recurrence diameter. Second, we develop a novel approach to combining the bounds of atomic subproblems, and show that it can provide relatively tight bounds compared to bottom-up approaches, such as those given in [1,15]. Third, we have proved the correctness of both our sublist diameter, and our approach to exploiting it in problem decompositions. Importantly, our proofs are mechanised in the HOL interactive theorem proving system [18].

2 Definitions

Definition 1 (States and Actions). *A transition system is defined in terms of* states *and* actions: *(i) States are finite maps from variables—i.e., state-characterizing propositions—to Booleans, abbreviated as a* state. *We write* $\mathcal{D}(s)$ *for the domain of s. (ii) An action π is a pair of finite maps over subsets of those variables. The first component of the pair,* $\mathsf{pre}(\pi)$, *is the* precondition *and the second component of the pair,* $\mathsf{eff}(\pi)$, *is the* effect.

We give examples of states and actions using sets of literals. For example, $\{x, \neg y, z\}$ is the state where state variables x and z are (map to) true, and y is false.

Definition 2 (Factored Transition System). *A factored transition system Π contains two components, (i) $\Pi.\mathsf{I}$, a finite map, whose domain is the domain of the system; and (ii) $\Pi.\mathsf{A}$ a set of actions, as above. We write* transition-system *Π when the domains of the system's actions' preconditions and effects are all subsets of the system's domain. We also write $\mathcal{D}(\Pi)$ for the domain of a system Π, and will often take complements of variable sets with respect to it, so that $\overline{vs}$ is the set of variables $\mathcal{D}(\Pi) \setminus vs$. The set of valid states, $\mathbb{U}(\Pi)$, for a factored transition system is $\{s \mid \mathcal{D}(s) = \mathcal{D}(\Pi)\}$. The set of valid action sequences, $\mathbb{A}(\Pi)$, for a factored transition system is $\{\vec{\pi} \mid \mathsf{set}\,\vec{\pi} \subseteq \Pi.A\}$.*

Definition 3 (Action Execution). *When an action $\pi \ (= (p, e))$ is executed at state s, written* $\mathsf{e}(s, \pi)$, *it produces a successor state s'. If p is not a submap of s then $s' = s$.[1] Otherwise s' is a valid state where $e \sqsubseteq s'$ and $s(x) = s'(x)$ $\forall x \in \mathcal{D}(s) \setminus \mathcal{D}(e)$. This operation,* state-succ, *has the following formal definition in HOL:*

$$\mathsf{state\text{-}succ} \quad s \ (p, \ e) \ = \ \mathbf{if} \ p \sqsubseteq s \ \mathbf{then} \ e \uplus s \ \mathbf{else} \ s$$

We lift e *to sequences of executions, taking an action sequence $\vec{\pi}$ as the second argument. So $\mathsf{e}(s, \vec{\pi})$ denotes the state resulting from successively applying each action from $\vec{\pi}$ in turn, starting from s.*

A Key concept in formalising compositional reasoning about transition systems is *projection*.

---

[1] Allowing any action to execute in any state is a deviation from standard practice. By using total functions in our formalism, much of the resulting mathematics is relatively simple.

Definition 4 (Projection). *Projecting an object (a state s, an action π, a sequence of actions $\vec{\pi}$ or a factored transition system Π) on a set of variables vs refers to restricting the domain of the object or its constituents to vs. We denote these operations as $s\lfloor_{vs}$, $\pi\lfloor_{vs}$, $\vec{\pi}\lfloor_{vs}$ and $\Pi\lfloor_{vs}$ for a state, action, action sequence and transition system respectively. HOL provides domain restriction for finite maps in its standard library; projections on composite objects use this operation on the constituents. The one (minor) exception is projection on action sequences, where a projected action is dropped entirely if it has an empty effect:*

$$[]\lfloor_{vs} = []$$
$$((p,\ e){::}\ \vec{\pi})\lfloor_{vs} = \texttt{if}\ \mathcal{D}(e\lfloor_{vs}) \neq \emptyset\ \texttt{then}\ (p\lfloor_{vs},\ e\lfloor_{vs})\ {::}\ \vec{\pi}\lfloor_{vs}\ \texttt{else}\ \vec{\pi}\lfloor_{vs}$$

3   Upper Bounding the Diameter with Decomposition

We define the diameter of a transition system in terms of its factored represen-tation as[2]

Definition 5 (Diameter).

$$d(\Pi) = \mathsf{MAX}\ \{\,\mathsf{MIN}\ (\Pi^d\ (s,\ \vec{\pi},\ \Pi))\ |\ s \in \mathbb{U}(\Pi) \wedge \vec{\pi} \in \mathbb{A}(\Pi)\,\}$$

where Π^d is defined as

$$\Pi^d\ (s,\ \vec{\pi},\ \Pi) = \{\,|\vec{\pi}'|\ |\ \mathsf{e}\ (s,\ \vec{\pi}') = \mathsf{e}\ (s,\ \vec{\pi}) \wedge \vec{\pi}' \in \mathbb{A}(\Pi)\,\}$$

If the transition system under consideration exhibits a modular or hierarchi-cal structure, upper bounding its diameter compositionally would be of great utility in terms of computational cost; *i.e.,* if the whole system's diameter is bounded by a function of its components' diameters. One source of structure in transition systems, especially hierarchical structure, is *dependency* between state variables.

Definition 6 (Dependency). *A variable v_2 is dependent on v_1 in a factored transition system Π iff one of the following statements holds:[3] (i) v_1 is the same as v_2, (ii) For some action π in A such that v_1 is a precondition of π and v_2 is an effect of π, or (iii) There is an action π in A such that both v_1 and v_2 are effects of π. We write $v_1 \rightarrow v_2$ if v_1 is dependent on v_2 in Π. Formally in HOL we define this as[4]:*

$$v_1 \rightarrow v_2 \Longleftrightarrow$$
$$(\exists\ p\ e\ .$$
$$(p, e) \in \Pi.\mathsf{A} \wedge$$
$$(v_1 \in \mathcal{D}(p) \wedge v_2 \in \mathcal{D}(e) \vee v_1 \in \mathcal{D}(e) \wedge v_2 \in \mathcal{D}(e))) \vee$$
$$v_1 = v_2$$

---

[2] With this definition the diameter will be one less than how it was defined in [2].

[3] We are using the standard definition of dependency described in [5,20].

[4] $\rightarrow$ has a Π parameter, but we use it with HOL's *ad hoc* overloading ability.

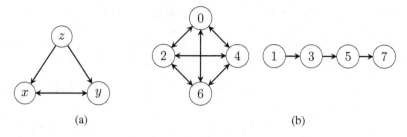

(a) (b)

Fig. 1. (a) the dependency and (b) the state transition graphs of the system in Example 1.

We also lift the concept of dependency to sets of variables. A set of variables vs_2 is dependent on vs_1 in a factored transition system Π (written $vs_1 \to vs_2$) iff all of the following conditions hold: (i) vs_1 is disjoint from vs_2 and (ii) There exist variables $v_1 \in vs_1$ and $v_2 \in vs_2$ such that $v_1 \to v_2$. We define this relation in HOL as[4]:

$$vs_1 \to vs_2 \iff$$
$$\exists\ v_1\ v_2.\ v_1 \in vs_1\ \wedge\ v_2 \in vs_2\ \wedge\ \mathsf{DISJOINT}\ vs_1\ vs_2\ \wedge\ v_1 \to v_2$$

One tool used in analysing dependency structures is the *dependency graph*. The dependency graph of a factored transition system Π is a directed graph, written G, describing variable dependencies. We use it to analyse hierarchical structures in transition systems. This graph was conceived under different guises in [5, 20], and is also commonly referred to as a *causal graph*.

Definition 7 (The Dependency Graph). *The dependency graph has one vertex for each variable in D. An edge from v_1 to v_2 records that $v_1 \to v_2$. When we illustrate a dependency graph we do not draw arcs from a variable to itself although it is dependent on itself.*

We also lift the concept of dependency graphs and refer to lifted dependency graphs (written G_{vs}). Each node in G_{vs}, for a factored transition system Π, represents a member of a partition of $\mathcal{D}(\Pi)$, i.e. all vertices in G_{vs} represent disjoint sets of variables.

The structures we aim to exploit are found via dependency graph analysis, where, for example, the lifted dependency graph exhibits a *parent-child* structure, defined as follows:

Definition 8 (Parent-Child Structure). *A system Π has a parent-child structure if $\mathcal{D}(\Pi)$ is comprised of two disjoint sets of variables vs_1 and vs_2 satisfying $vs_2 \not\to vs_1$. In HOL we model this as follows:*

$$\mathsf{child\text{-}parent\text{-}rel}\ (\Pi,\ vs) \iff vs \not\to \overline{vs}$$

It is intuitive to seek an expression that is no worse than multiplication to combine subsystem diameters, or an upper bound for it, into a bound for the entire system's diameter. For instance, for systems with the parent-child structure, previous work in the literature suggests that $(b(\Pi|_{vs}) + 1)(b(\Pi|_{\overline{vs}}) + 1)$ should bound $b(\Pi)$, for some upper bound on the diameter $b : \Pi \to \mathbb{N}$ to be considered *decomposable* for parent-child structures. The following example shows that the diameter is not a decomposable bound for parent-child structures.

Example 1. *Consider a factored transition system* Π *with the following set of actions*

$$A = \left\{ \begin{array}{l} a = (\{\neg x, \neg y\}, \{x\}), b = (\{x, \neg y\}, \{\neg x, y\}), c = (\{\neg x, y\}, \{x\}), \\ z_1 = (\{\neg z\}, \{x, y\}), z_2 = (\{\neg z\}, \{\neg x, y\}), z_3 = (\{\neg z\}, \{x, \neg y\}), \\ z_4 = (\{\neg z\}, \{\neg x, \neg y\}) \end{array} \right\}.$$

The dependency graph of Π *is shown in Fig. 1a.* $\mathcal{D}(\Pi)$ *is comprised of two sets of variables* $P = \{z\}$, *and the set* $C = \{x, y\}$, *where* $C \not\rightarrow P$ *holds.* $d(\Pi) = 3$, *as this is the length of the longest shortest transition sequence in* Π. *Specifically, that sequence is* $[a; b; c]$ *from* $\{\neg x, \neg y, z\}$ *to* $\{x, y, z\}$. $d(\Pi|_P) = 0$ *because in* $\Pi|_P$ *the state cannot be changed, and* $d(\Pi|_C) = 1$ *because any state in* $\Pi|_C$ *is reachable from any other state via one transition with one of the actions* $\{z_1|_C, z_2|_C, z_3|_C, z_4|_C\}$. *Accordingly, d is not decomposable.*

3.1 Decomposable Diameters

Baumgartner *et al.* [1] show that the *recurrence diameter* gives a decomposable upper bound on diameter.

Definition 9 (Recurrence Diameter). *Following Biere et al. [2], the recurrence diameter is formally defined as follows:*

$$rd(\Pi) = \mathsf{MAX} \ \{ |p| \mid \mathsf{valid\_path} \ \Pi \ p \ \wedge \ \mathsf{ALL\_DISTINCT} \ p \}$$

where

valid_path Π [] $\iff$ T
valid_path Π [s] $\iff$ $s \in \mathsf{U}(\Pi)$
valid_path Π (s_1 :: s_2 :: $rest$) $\iff$
    $s_1 \in \mathsf{U}(\Pi) \ \wedge \ (\exists \ \pi. \ \pi \in \Pi.\mathsf{A} \wedge \mathsf{e} \ (s_1, [\pi]) = \ s_2) \ \wedge$
valid_path Π (s_2 :: $rest$)

We provide a tighter and decomposable diameter: the *sublist diameter*, ℓ.

Definition 10 (Sublist Diameter). *List* l_1 *is a scattered sublist of* l_2 *(written* $l_1 \preceq l_2$*) if all the members of* l_1 *occur in the same order in* l_2. *This is defined in HOL as:*

[] $\preceq \ell_1 \iff$ T
h :: $t \preceq$ [] $\iff$ F
x :: $\ell_1 \preceq y$ :: $\ell_2 \iff$ $x = y \wedge \ell_1 \preceq \ell_2 \vee x$ :: $\ell_1 \preceq \ell_2$

Based on that the sublist diameter is defined as:

$$\ell(\Pi) = \mathsf{MAX} \ \{\mathsf{MIN} \ \Pi^{\preceq}(s, \dot{\pi}) \mid s \in \mathsf{U}(\Pi) \wedge \dot{\pi} \in \mathsf{A}(\Pi)\}$$

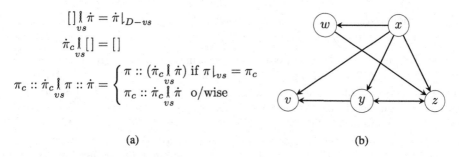

$$[]\,\substack{\l \\ vs}\,\dot{\pi} = \dot{\pi}\lfloor_{D-vs}$$

$$\dot{\pi}_c\,\substack{\l \\ vs}\,[] = []$$

$$\pi_c :: \dot{\pi}_c\,\substack{\l \\ vs}\,\pi :: \dot{\pi} = \begin{cases} \pi :: (\dot{\pi}_c\,\substack{\l \\ vs}\,\dot{\pi}) \text{ if } \pi\lfloor_{vs} = \pi_c \\ \pi_c :: \dot{\pi}_c\,\substack{\l \\ vs}\,\dot{\pi} \ \ \text{o/wise} \end{cases}$$

(a)　　　　　　　　　　　　　　　　(b)

Fig. 2. (a) The definition of the stitching function ($\l$), and (b) is the dependency graph of the system in Example 4.

where $\Pi^{\preceq\cdot}$ *is defined as*

$$\Pi^{\preceq\cdot}(s,\dot{\pi}) = \{\,|\dot{\pi}'| \mid \text{e } (s,\dot{\pi}') = \text{e } (s\ \dot{\pi}) \wedge \dot{\pi}' \preceq\cdot \dot{\pi}\,\}$$

It should be clear that $\ell(\Pi)$ is an upper bound on $d(\Pi)$ and that it is also a lower bound on $rd(\Pi)$, as demonstrated in the following theorem.

$$\vdash \text{ transition-system } \ \Pi \ \Rightarrow \ d(\Pi) \le \ell(\Pi) \wedge \ell(\Pi) \le rd(\Pi)$$

The sublist diameter is tighter than the recurrence diameter because it exploits the factored representation of transitions as actions, as shown in the next example.

Example 2. *Consider a factored transition system Π with the following set of actions*

$$A = \{\, a_1 = (\emptyset, \{x, y\}), a_2 = (\emptyset, \{\neg x, y\}), a_3 = (\emptyset, \{x, \neg y\}), a_4 = (\emptyset, \{\neg x, \neg y\}) \,\}.$$

For this system $d(\Pi) = 1$ because any state is reachable from any state with one action. $rd(\Pi) = 3$ because there are many paths with length 3 with no repeated states, but not any longer than that. Lastly, $\ell(\Pi) = 1$ because for any non empty action sequence $\dot{\pi} \in \mathbb{A}(\Pi)$ the last transition π in $\dot{\pi}$ can reach the same destination as $\dot{\pi}$, and $[\pi]$ is a sublist of $\dot{\pi}$.

The other important property of ℓ is that it is decomposable as demonstrated in the following example.

Example 3. *Consider the factored transition system Π in Example 1. The values of $\ell(\Pi)$ and $\ell(\Pi\lfloor_P)$ are the same as those of $d(\Pi)$ and $d(\Pi\lfloor_P)$, respectively. However, $\ell(\Pi\lfloor_C) = 3$ because, although any state in $\Pi\lfloor_C$ is reachable from any other state via one transition, there is no shorter sublist of the transition sequence $[a\lfloor_C; b\lfloor_C; c\lfloor_C]$ that starts at $\{\neg x, \neg y\}$ and results in $\{x, y\}$.*

Now we prove that ℓ is decomposable.

Theorem 1. *If the domain of Π is comprised of two disjoint sets of variables vs_1 and vs_2 satisfying $vs_2 \not\rightarrow vs_1$, we have:*

$$\ell(\Pi) < (\ell(\Pi\lfloor_{vs_1}) + 1)(\ell(\Pi\lfloor_{vs_2}) + 1)$$

$\vdash$ transition-system Π $\wedge$ child-parent-rel $a(\Pi, vs)$ $\Rightarrow$
$$\ell(\Pi) < (\ell(\Pi\!\restriction_{vs}) + 1) \times (\ell(\Pi\!\restriction_{\overline{vs}}) + 1)$$

Proof. To prove Theorem 1 we use a construction which, given any action sequence $\dot\pi \in \mathbb{A}(\Pi)$ violating the stated bound and a state $s \in \mathbb{U}(\Pi)$, produces a sublist, $\dot\pi'$, of $\dot\pi$ satisfying that bound and $\mathsf{e}(s, \dot\pi) = \mathsf{e}(s, \dot\pi')$. The premise $vs_2 \not\rightarrow vs_1$ implies that actions with variables from vs_2 in their effects never include vs_1 variables in their effects. Hereupon, if for an action π, $\mathsf{eff}(\pi) \subseteq s$ is true, we call it a *s-action*. Because vs_1 and vs_2 capture all variables, the effects of vs_2-actions after projection to the set vs_2 are unchanged. Our construction first considers the abstract action sequence $\dot\pi\!\restriction_{vs_2}$. Definition 10 of ℓ provides a scattered sublist $\dot\pi'_{vs_2} \preccurlyeq \dot\pi\!\restriction_{vs_2}$ satisfying $|\dot\pi'_{vs_2}| \leq \ell(\Pi\!\restriction_{vs_2})$. Moreover, the definition of ℓ can guarantee that $\dot\pi'_{vs_2}$ is equivalent, in terms of the execution outcome, to $\dot\pi\!\restriction_{vs_2}$. The stitching function described in Fig. 2a is then used to remove the vs_2-actions in $\dot\pi$ whose projections on vs_2 are not in $\dot\pi'_{vs_2}$. Thus our construction arrives at the action sequence $\dot\pi'' = \dot\pi'_{vs_2} \wr_{vs_2} \dot\pi$ with at most $\ell(\Pi\!\restriction_{vs_2})$ vs_2-actions.

We are left to address the continuous lists of vs_1-actions in $\dot\pi''$, to ensure that in the constructed action sequence any such list satisfies the bound $\ell(\Pi\!\restriction_{vs_1})$. The method by which we obtain $\dot\pi''$ guarantees that there are at most $\ell(\Pi\!\restriction_{vs_2}) + 1$ such lists to address. The definition of ℓ provides that for any abstract list of actions $\dot\pi\!\restriction_{vs_1}$ in $\Pi\!\restriction_{vs_1}$, there is a list that achieves the same outcome of length at most $\ell(\Pi\!\restriction_{vs_1})$. Our construction is completed by replacing each continuous sequence of vs_1-actions in $\dot\pi''$ with witnesses of appropriate length ($\ell(\Pi\!\restriction_{vs_1})$). $\square$

The above construction can be illustrated using the following example.

Example 4. *Consider a factored transition system with the set of actions*

$$A = \left\{ \begin{array}{l} a = (\emptyset, \{x\}), b = (\{x\}, \{y\}), c = (\{x\}, \{\neg v\}), d = (\{x\}, \{w\}), \\ e = (\{y\}, \{v\}), f = (\{w, y\}, \{z\}), g = (\{\neg x\}, \{y, z\}) \end{array} \right\}$$

whose dependency graph is shown in Fig. 2b. The domain of Π is comprised of the two sets $vs_2 = \{v, y, z\}$ and $vs_1 = \{w, x\}$, where $vs_2 \not\rightarrow vs_1$. In Π, the actions b, c, e, f, g are vs_2-actions, and a, d are vs_1-actions. An action sequence $\dot\pi \in \mathbb{A}(\Pi)$ is $[a; a; b; c; d; d; e; f]$ that reaches the state $\{v, w, x, y, z\}$ from $\{v, \neg w, \neg x, \neg y, \neg z\}$. When $\dot\pi$ is projected on vs_2 it becomes $[b\!\restriction_{vs_2}; c\!\restriction_{vs_2}; e\!\restriction_{vs_2}; f\!\restriction_{vs_2}]$, which is in $\mathbb{A}(\Pi\!\restriction_{vs_2})$. A shorter action sequence, $\dot\pi_c$, achieving the same result as $\dot\pi\!\restriction_{vs_2}$ is $[b\!\restriction_{vs_2}; f\!\restriction_{vs_2}]$. Since $\dot\pi_c$ is a scattered sublist of $\dot\pi\!\restriction_{vs_2}$, we can use the stitching function to obtain a shorter action sequence in $\mathbb{A}(\Pi)$ that reaches the same state as $\dot\pi$. In this case, $\dot\pi_c \wr_{vs_2} \dot\pi$ is $[a; a; b; d; d; f]$. The second step is to contract the pure vs_1 segments which are $[a; a]$ and $[d; d]$, which are contracted to $[a]$ and $[d]$ respectively. The final constructed action sequence is $[a; b; d; f]$, which achieves the same state as $\dot\pi$.

4   Decomposition for Tighter Bounds

So far we have seen how to reason about bounds on underlying transition system sublist diameters by treating sublist diameters of subsystems separately in an example structure (*i.e.* the parent child structure). We now discuss how to exploit a more general dependency structure to compute an upper bound on the sublist diameter of a system, after separately computing subsystems' sublist diameters. We exploit branching one-way variable dependency structures. An example of that type of dependency structure is exhibited in Fig. 3a, where S_i are sets of variables each of which forms a node in the lifted dependency graph. Recall that an edge in this graph from a node S_i to a node S_j means $S_i \to S_j$, which means that there is at least one edge from a variable in S_i to one in S_j. Also, an absence of an edge from a node S_i to a node S_j means that $S_i \not\to S_j$, and which means that is not a variable in S_j that depends on a variable in S_i.

In this section we present a general theorem about the decompositional properties of the sublist diameter to treat the more general structures. Then we provide a verified upper bounding algorithm based on it. Consider the following more general form of the parent child structure:

Definition 11 (Generalised Parent-Child Structure). *For a factored transition system Π and two sets of variables vs_1 and vs_2, the generalised parent-child relation holds between vs_1 and vs_2 iff (i) $vs_2 \not\to vs_1$, (ii) $vs_1 \not\to \overline{(vs_1 \cup vs_2)}$, and (iii) no bidirectional dependencies exist between any variable in vs_2 and $\overline{(vs_1 \cup vs_2)}$. Formally, we define this relation in HOL as follows:*

gen-parent-child　$(\Pi,\ vs,\ vs') \Longleftrightarrow$
　　DISJOINT $vs\ vs' \wedge vs' \not\to vs \wedge vs \not\to \overline{vs \cup vs'} \wedge$
　　$\forall\ v\ v'.v \in vs' \wedge v' \in \overline{vs \cup vs'} \Rightarrow v \not\to v' \vee v' \not\to v$

To prove the more general theorem we need the following lemma:

Lemma 1. *Let $\mathsf{n}(vs, \dot\pi)$ be the number of vs-actions contained within $\dot\pi$. Consider Π, in which the generalised parent-child relation holds between sets of variables p and c. Then, any action sequence $\dot\pi$ has a sublist $\dot\pi'$ that reaches the same state as $\dot\pi$ starting from any state such that: $\mathsf{n}(p, \dot\pi') \le \ell(\Pi\!\downarrow_p)(\mathsf{n}(c, \dot\pi') + 1)$ and $\mathsf{n}(\overline{p}, \dot\pi') \le \mathsf{n}(\overline{p}, \dot\pi)$.*

$\vdash$ transition-system $\Pi \wedge s \in \mathsf{U}(\Pi) \wedge \dot\pi \in \mathsf{A}(\Pi) \wedge$
　　gen-parent-child　$(\Pi, p, c) \Rightarrow$
　　　$\exists\ \dot\pi'.$
　　　　$\mathsf{n}\,(p, \dot\pi') \le \ell(\Pi\!\downarrow_p) \times (\mathsf{n}\,(c, \dot\pi') + 1) \wedge \dot\pi' \preceq \dot\pi \wedge$
　　　　$\mathsf{n}\,(\overline{p}, \dot\pi') \le \mathsf{n}\,(\overline{p}, \dot\pi) \wedge \mathsf{e}\,(s, \dot\pi) = \mathsf{e}\,(s, \dot\pi')$

Proof. The proof of Lemma 1 is a constructive proof. Let $\dot\pi_{\overline{c}}$ be a contiguous fragment of $\dot\pi$ that has no c-actions in it. Then perform the following steps:

- By the definition of ℓ, there must be an action sequence $\dot\pi_p$ such that $\mathsf{e}(s, \dot\pi_p) = \mathsf{e}(s, \dot\pi_{\overline{c}}\!\downarrow_p)$, and satisfies $|\dot\pi_p| \le \ell(\Pi\!\downarrow_p)$ and $\dot\pi_p \preceq \dot\pi_{\overline{c}}\!\downarrow_p$.
- Because $p \not\to \mathcal{D}(\Pi) \backslash p \backslash c$ holds and using the same argument used in the proof of Theorem 1, $\dot\pi'_{\overline{c}}(= \dot\pi_p\downharpoonright_p \dot\pi_{\overline{c}}\!\downarrow_{D\backslash c})$ achieves the same $D \backslash c$ assignment as $\dot\pi_{\overline{c}}$ (*i.e.*, $\mathsf{e}(s, \dot\pi'_{\overline{c}})\!\downarrow_{D\backslash c} = \mathsf{e}(s, \dot\pi_{\overline{c}})\!\downarrow_{D\backslash c})$, and it is a sublist of $\dot\pi_{\overline{c}}$. Also, $\mathsf{n}(p, \dot\pi'_{\overline{c}}) \le \ell(\Pi\!\downarrow_p)$ holds.

– Finally, because $\dot{\pi}_{\overline{c}}$ has no c-actions, no c variables change along the execution of $\dot{\pi}_{\overline{c}}$ and accordingly any c variables in preconditions of actions in $\dot{\pi}_{\overline{c}}$ always have the same assignment. This means that $\dot{\pi}'_{\overline{c}} \restriction_{D \setminus c} \dot{\pi}_{\overline{c}}$ will achieve the same result as $\dot{\pi}_{\overline{c}}$, but with at most $\ell(\Pi\restriction_p)$ p-actions.

Repeating the previous steps for each $\dot{\pi}_{\overline{c}}$ fragment in $\dot{\pi}$ yields an action sequence $\dot{\pi}'$ that has at most $\ell(\Pi\restriction_p)(\mathsf{n}(c,\dot{\pi}) + 1)$ p-actions. Because $\dot{\pi}'$ is the result of consecutive applications of the stitching function, it is a scattered sublist of $\dot{\pi}$. Lastly, because during the previous steps, only p-actions were removed as necessary, the count of the remaining actions in $\dot{\pi}'$ is the same as their number in $\dot{\pi}$. □

Corollary 1. *Let $F(p, c, \dot{\pi})$ be the witness action sequence of Lemma 1. We know then that:*

– $\mathsf{e}(s, F(p, c, \dot{\pi})) = \mathsf{e}(s, \dot{\pi})$,
– $\mathsf{n}(p, F(p, c, \dot{\pi})) \leq \ell(\Pi\restriction_p)(\mathsf{n}(c, \dot{\pi}) + 1)$.
– $F(p, c, \dot{\pi}) \preceq \dot{\pi}$, *and*
– $\mathsf{n}(\overline{p}, F(p, c, \dot{\pi})) \leq \mathsf{n}(\overline{p}, \dot{\pi})$.

Branching one-way variable dependencies are captured when G_{vs} is a directed acyclic graph (DAG).

Definition 12 (DAG Lifted Dependency Graphs). *In HOL we model a G_{vs} that is a DAG with the predicate* top-sorted *that means that G_{vs} is a list of nodes of a lifted dependency graph topologically sorted w.r.t. dependency. This predicate is defined as follows in HOL:*

top-sorted $[] \iff \mathsf{T}$
top-sorted $(vs :: G_{vs}) \iff$
$\quad (\forall\ vs'.\ vs' \in \mathsf{set}\ G_{vs} \Rightarrow vs' \not\rightarrow vs) \wedge$ top-sorted G_{vs}

We also define the concept of children *for a node vs in G_{vs}, written as $\mathcal{C}(vs)$ to denote the set $\{vs_0 \mid vs_0 \in G_{vs} \wedge vs \rightarrow vs_0\}$, which are the children of vs in G_{vs}. In HOL this is modelled as the list:[5]*

$\mathcal{C}(vs) = \mathsf{FILTER}\ (\lambda\ vs'.\ vs \rightarrow vs')\ G_{vs}$

We now use Corollary 1 to prove the following theorem:

Theorem 2. *For a factored transition system Π, and a lifted dependency graph G_{vs} that is a DAG, the sublist diameter $\ell(\Pi)$ satisfies the following inequality:*

$$\ell(\Pi) \leq \Sigma_{vs \in G_{vs}} \mathsf{N}(vs) \tag{1}$$

where $\mathsf{N}(vs) = \ell(\Pi\restriction_{vs})(\Sigma_{C \in \mathcal{C}(vs)} \mathsf{N}(C) + 1)$.

---

[5] top-sorted has a Π parameter and $\mathcal{C}$ has Π and G_{vs} as parameters hidden with overloading.

Alternatively, in the HOL presentation:

$\vdash$ ALL_DISTINCT G_{vs} $\wedge$ ALL_DISJOINT G_{vs} $\Rightarrow$
    $\forall \Pi$.
        transition-system Π $\wedge$ $\mathcal{D}(\Pi.\mathsf{I}) = \bigcup$ (set G_{vs}) $\wedge$
        top-sorted G_{vs} $\Rightarrow$
        $\ell(\Pi) <$ SUM (MAP N G_{vs}) $+ 1$

where N is defined as[6]

    transition-system Π $\wedge$ top-sorted G_{vs} $\Rightarrow$
    N$(vs) = \ell(\Pi\!\restriction_{vs}) \times$ (SUM (MAP N $\mathcal{C}(vs)$) $+ 1$)

Proof. Again, our proof of this theorem follows a constructive approach where we begin by assuming we have an action sequence $\dot{\pi} \in \mathbb{A}(\Pi)$ and a state $s \in \mathbb{U}(\Pi)$. The goal of the proof is to find a witness sublist, $\dot{\pi}'$, of $\dot{\pi}$ such that $\forall$ $vs \in G_{vs}$. n$(vs, \dot{\pi}') \leq$ N(vs) and e$(s, \dot{\pi}) = e(s, \dot{\pi}')$. We proceed by induction on l_{vs}, a topologically sorted list of nodes in G_{vs}. The base case is the empty list $[]$, in which case $\mathcal{D}(\Pi) = \emptyset$ and accordingly $\ell(\Pi) = 0$.

In the step case, we assume the result holds for any system for which l_{vs} is a topologically sorted node list of one of its lifted dependency graphs. We then show that it also holds for Π, a system whose node list is $vs :: l_{vs}$, where vs has no parents (hence its position at the start of the sorted list). Since l_{vs} is a topologically sorted node list of a lifted dependency graph of $\Pi\!\restriction_{\overline{vs}}$, the induction hypothesis applies. Accordingly, there is a $\dot{\pi}_{\overline{vs}}$ for $\Pi\!\restriction_{\overline{vs}}$ such that e$(s, \dot{\pi}_{\overline{vs}}) = e(s, \dot{\pi}\!\restriction_{\overline{vs}})$, $\dot{\pi}_{\overline{vs}} \preceq \cdot$ $\dot{\pi}\!\restriction_{\overline{vs}}$, and $\forall$ $K \in l_{vs}$. n$(K, \dot{\pi}') \leq$ N(K). Since $vs :: l_{vs}$ is topologically sorted, $(\overline{vs}) \not\rightarrow vs$ holds. Let $\dot{\pi}'_{\overline{vs}} \equiv \dot{\pi}_{\overline{vs}}\! \underset{vs}{\wr}\, \dot{\pi}$. Therefore e$(s, \dot{\pi}'_{\overline{vs}}) = e(s, \dot{\pi})$ (using the same argument used in the proof of Theorem 1). Furthermore, $\forall K \in l_{vs}$. n$(K, \dot{\pi}'_{\overline{vs}}) \leq$ N(K) and $\dot{\pi}'_{\overline{vs}} \preceq \dot{\pi}$. The last step in this proof is to show that $F(vs, \bigcup \mathcal{C}(vs), \dot{\pi}'_{\overline{vs}})$ is the required witness, which is justified because the generalised parent-child relation holds for Π, vs and $\bigcup \mathcal{C}(vs)$. From Corollary 1 and because the relations $=$, $\leq$ and $\preceq$ are transitive, we know that

- e$(s, \dot{\pi}) = e(s, F(vs, \bigcup \mathcal{C}(vs), \dot{\pi}'_{\overline{vs}}))$,
- n$(vs, F(vs, \bigcup \mathcal{C}(vs), \dot{\pi}'_{\overline{vs}})) \leq \ell(\Pi\!\restriction_{vs})(\Sigma_{C \in \mathcal{C}(vs)}n(C, \dot{\pi}'_{\overline{vs}}) + 1)$,
- $F(vs, \bigcup \mathcal{C}(vs), \dot{\pi}'_{\overline{vs}}) \preceq \dot{\pi}$, and
- n$(\mathcal{D}(\Pi) \setminus vs, F(vs, \bigcup \mathcal{C}(vs), \dot{\pi}'_{\overline{vs}})) \leq n(\mathcal{D}(\Pi) \setminus vs, \dot{\pi}'_{\overline{vs}})$.

Since $\Sigma_{K \in l_{vs}}$n$(K, \dot{\pi}'_{\overline{vs}}) = n(\mathcal{D}(\Pi) \setminus vs, \dot{\pi}'_{\overline{vs}})$ is true, $\forall K \in l_{vs}$. n$(K, \dot{\pi}'_{\overline{vs}}) \leq$ N(K) is true, and n$(vs, F(vs, \bigcup \mathcal{C}(vs), \dot{\pi}'_{\overline{vs}})) \leq \ell(\Pi\!\restriction_{vs})(\Sigma_{C \in \mathcal{C}(vs)}N(C))$ is true, therefore $F(vs, \bigcup \mathcal{C}(vs), \dot{\pi}'_{\overline{vs}})$ is an action sequence demonstrating the needed bound. $\square$

4.1 A Bounding Algorithm

We now discuss an upper bounding algorithm that we prove is valid. Consider the function N$_b$, defined over the nodes of a lifted dependency DAG as:

$$\text{N}_b(vs) = b(\Pi\!\restriction_{vs})(\Sigma_{C \in \mathcal{C}(vs)} \text{N}_b(C) + 1)$$

---

[6] N has Π and G_{vs} as parameters hidden with overloading.

Note that N_b is a general form of the function N defined in Theorem 2, parameterised over a *base case* function $b : \Pi \to \mathbb{N}$. Viewing $\mathfrak{N}_b = \Sigma_{vs \in G_{vs}} N_b(vs)$ as an algorithm, the following theorem shows that it calculates a valid upper bound for a factored transition system's sublist diameter if the base case calculation is a valid bound.

Theorem 3. *For a base case function* $b : \Pi \to \mathbb{N}$, *if* $\forall \Pi. \ell(\Pi) \leq b(\Pi)$ *then* $\ell(\Pi) \leq \mathfrak{N}_b(vs)$.

$\vdash\ (\forall\, \Pi'.\ \text{transition-system}\ \ \Pi' \Rightarrow \ell(\Pi') \leq b(\Pi')) \Rightarrow$
$\quad \forall\ G_{vs}\,.$
$\qquad \text{ALL\_DISTINCT}\ \ G_{vs} \wedge \text{ALL\_DISJOINT}\ \ G_{vs} \Rightarrow$
$\qquad\quad \forall\, \Pi.$
$\qquad\qquad \text{transition-system}\ \ \Pi \wedge \mathcal{D}(\Pi.\mathsf{I}) = \bigcup\ (\text{set}\ \ G_{vs}) \wedge$
$\qquad\qquad \text{top-sorted}\ \ G_{vs} \Rightarrow$
$\qquad\qquad\quad \ell(\Pi) < \text{SUM}\ (\text{MAP}\ N_b\ \ G_{vs}) + 1$

where N_b is characterised by the following theorem[7] as

$(\forall\, \Pi'.\ b(\Pi'\!\downarrow_\emptyset)\ =\ 0)\ \Rightarrow$
$\quad \forall\, \Pi\ \ vs\ \ G_{vs}\,.$
$\qquad \text{transition-system}\ \ \Pi\ \ \wedge\ \ \text{top-sorted}\ \ G_{vs}\ \Rightarrow$
$\qquad\quad N_b(vs)\ =\ b(\Pi\!\downarrow_{vs})\ \times\ (\text{SUM}\ (\text{MAP}\ \ N_b\ \ \mathcal{C}(vs))\ +\ \ 1)$

Without any detailed analysis we are able to take $b(\Pi) = 2^{|\mathcal{D}(\Pi)|} - 1$. In other words, an admissible base case is one less than the number of states representable by the set of state variables of the system being evaluated. That is a valid upper bound for both the recurrence and sublist diameters.

5   Bounds in Practice

In this section we provide an evaluation of the upper bounds produced by the algorithm from Sect. 4.1. We first compare it to previously suggested compositional bounding approaches in planning benchmarks. We also evaluate its performance on a practical model checking problem.

5.1   Evaluating in Benchmarks from Automated Planning

In this section we compare the bounds computed by $\mathfrak{N}_b$ with the ones computed using the algorithm suggested by Baumgartner *et al.* [1] for treating design netlists. This algorithm, and that in Rintanen and Gretton [15], both traverse the structure of the factored transition system in a bottom-up way. By way of contrast, our algorithm traverses the same structure top-down. Before we model the bottom-up calculation, we need to define the concepts of *ancestors* and *leaves*.

---

[7] N_b has Π and G_{vs} as parameters hidden with overloading.

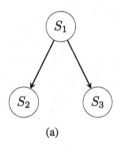

(a)

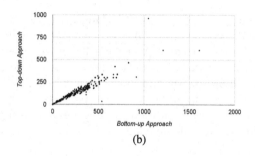

(b)

Fig. 3. (a) A lifted dependency graph, and (b) The bounds computed by $\mathfrak{N}_b$ versus $\mathfrak{M}_b$ with $2^{|D(\pi)|} - 1$ as a base function.

Definition 13 (Leaves and Ancestors). *We define the set of leaves $\mathcal{L}(G_{vs})$ to contain those vertices of G_{vs} from which there are no outgoing edges. We also write $\mathcal{A}(vs)$ to denote the set of ancestors of vs in G_{vs} i.e. the set $\{vs_0 \mid vs_0 \in G \wedge vs_0 \rightarrow vs^+\}$, where $\rightarrow^+$ is the transitive closure of $\rightarrow$.*

To model bottom-up calculation, consider the function

$$M_b(vs) = b(\Pi\!\downarrow_{vs}) + (1 + b(\Pi\!\downarrow_{vs}))\Sigma_{A \in \mathcal{A}(vs)} M_b(\Pi\!\downarrow_A)$$

The bottom-up approach, $\mathfrak{M}_b$, can be described as $\mathfrak{M}_b = \Sigma_{vs \in \mathcal{L}(G_{vs})} M_b(vs)$, where b is a base case function.

Consider the lifted dependency DAG in Fig. 3a. Given a base a function b, and letting $b(\Pi\!\downarrow_{S_i})$ be b_i, the values of $N_b(S_2)$ and $N_b(S_3)$ are b_2 and b_3, respectively. The value of $N_b(S_1)$ is $b_1 + b_1b_2 + b_1b_3$. Accordingly the value of $\mathfrak{N}_b = b_1 + b_1b_2 + b_1b_3 + b_2 + b_3$.

On the other hand, $M_b(S_1) = b_1$, $M_b(S_2) = b_1 + b_1b_2 + b_2$ and $M_b(S_3) = b_1 + b_1b_3 + b_1$. Accordingly $\mathfrak{M}_b = 2b_1 + b_1b_2 + b_1b_3 + b_2 + b_3$. The value of $\mathfrak{M}_b$ has an extra b_1 term over that of $\mathfrak{N}_b$. This extra term is because $\mathfrak{M}_b$ counts every ancestor node in the lifted dependency graph as many times as the size of its posterity, which is a consequence of the bottom-up traversal of the dependency graph. Figure 3b shows the computed bounds of $\mathfrak{N}_b$ versus $\mathfrak{M}_b$ with the function $2^{|\mathcal{D}(\Pi)|} - 1$ as the base function for a 1030 different International Planning Competition benchmarks. That figure shows that $\mathfrak{M}_b$ computes looser bounds as it repeats counting the ancestor nodes unnecessarily.

5.2   Hotel Key Protocol

We consider the verification of a safety property of the hotel key distribution protocol introduced by Jackson [9] and further discussed in [4,13]. To our knowledge this domain has not previously been explored along with decompositional bounding techniques. We model the problem in the Planning Domain Description Language [12]. There are three actions schemas representing the three categories of change in the system's state which are: (i) entering a room with a new key

(enter-new-key), (ii) checking in the hotel (check-in), and (iii) checking out
(check-out). Note that we omit the fourth action which is entering the room
with its current key, because it has no effect on the system's state. We have a
predicate safe, that is true of a room while the safety property of that room is
maintained. All rooms are initially safe, and then the value of "safe" is set in
the effect of enter-new-key, and reset in the effect of check-in. The goal is to
have at least one room that is entered by a guest who does not occupy it and
for which "safe" is true.

Table 1. A table showing the bounds computed by $\mathcal{N}_b$ for the hotel key example.
Rows 1–3 show the bounds with keeping two of the parameters constant (= 5) and the
third ranging from 1–10.

$r = g = 5$	190	662	1397	2532	4067	6002	8337	11072	14207	17742
$k = g = 5$	859	1661	2463	3265	4067	4869	5671	6473	7275	8077
$r = k = 5$	803	1619	2435	3251	4067	4883	5699	6515	7331	8147

In our experiment, we parameterised instances of this problem by the number
of: guests (g), rooms (r) and keys per room (k). The initial state of an instance
asserts: (i) the types of the guests, the rooms and the keys, (ii) which key is
owned by which rooms, and (iii) an ordering of the keys such that the keys
owned by a room form a series. Table 1 shows the output of $\mathcal{N}_b$ on different
instances, with an unverified base case function based on invariants analysis.[8]
We computed the sets of variables that satisfy the invariance condition with
Fast Downward [8]. Each row shows the computed bounds keeping two of the
parameters constant and equal to 5, while the third parameter ranges from 1–10.
The computed upper bound increases linearly with the r and with g.

6   Related Work

The notions of diameter and recurrence diameter were introduced in Biere
et al. [2,3]. In this work they describe how to test whether k is the recurrence
diameter using a SAT formula of size $O(k^2)$. It was later shown by Kroening and
Strichman [11] that this test can be done using a SAT formula of size $O(k \log k)$.
An *inductive* algorithm for computing the recurrence diameter was introduced
by Sheeran et al. [17].

Other work exploits the structure or the type of system being verified for efficient computation of the diameter as well as for obtaining tighter bounds on it. For
example, Ganai et al. [7] show that an upper bound on the completeness threshold
for checking the safety of some software errors—such as array bound violations—
can be computed using a SAT formula of size $O(k)$. Also Konnov et al. [16] show

---

[8] The point of this experiment is to show how the computed upper bound grows with
different parameters, regardless of the base case function used.

how some components in threshold-based distributed algorithms have diameters quadratic in the number of transitions in the component. Most relevant to our work, Baumgartner *et al.* [1,6] show that the recurrence diameter can be used to calculate a bound for the diameter in a decompositional way using design netlists. In 2013, Rintanen and Gretton [15] described a similar method for calculating a transition system's diameter in the context of planning. The algorithms for calculating a bound in both of those works operate in a bottom-up way.

7   Conclusion

We considered computing admissible completeness thresholds for model checking safety properties in transition models with factored representations. We developed the concept of *sublist diameter*, a novel, tighter overapproximation of diameter for the factored case. We also developed a novel procedure for computing tighter bounds for factored systems by exploiting compositionality, and have formally verified dominance and correctness results associated with these.

The insights which led us to develop the sublist diameter followed from attempting to formalise the results in [15]. That effort helped us find a bug in their formal justification of their approach, where they incorrectly theorise that the diameter can be used directly to bound atomic components for compositional algorithms. Errors such as those make a strong case for the utility of mechanical verification.

In future, we hope to find efficient procedures for efficiently computing/tightly-approximating sublist diameters for the atomic subsystems in our compositional approach.

HOL4 Notation and Availability. All statements appearing with a turnstile (⊢) are HOL4 theorems, automatically pretty-printed to LaTeX. All our HOL scripts, experimental code and data are available from https://MohammadAbdulaziz@ bitbucket.org/MohammadAbdulaziz/planning.git.

Acknowledgements. We thank Daniel Jackson for suggesting applying diameter upper bounding on the hotel key protocol verification.

References

1. Baumgartner, J., Kuehlmann, A., Abraham, J.: Property checking via structural analysis. In: Brinksma, E., Larsen, K.G. (eds.) CAV 2002. LNCS, vol. 2404, pp. 151–165. Springer, Heidelberg (2002)
2. Biere, A., Cimatti, A., Clarke, E.M., Strichman, O., Zhu, Y.: Bounded model checking. Adv. Comput. **58**, 117–148 (2003)
3. Biere, A., Cimatti, A., Clarke, E., Zhu, Y.: Symbolic model checking without BDDs. In: Cleaveland, W.R. (ed.) TACAS 1999. LNCS, vol. 1579, pp. 193–207. Springer, Heidelberg (1999)

4. Blanchette, J.C., Nipkow, T.: Nitpick: a counterexample generator for higher-order logic based on a relational model finder. In: Kaufmann, M., Paulson, L.C. (eds.) ITP 2010. LNCS, vol. 6172, pp. 131–146. Springer, Heidelberg (2010)
5. Bylander, T.: The computational complexity of propositional STRIPS planning. Artif. Intell. **69**(1–2), 165–204 (1994)
6. Case, M.L., Mony, H., Baumgartner, J., Kanzelman, R.: Enhanced verification by temporal decomposition. In: Proceedings of 9th International Conference on Formal Methods in Computer-Aided Design, FMCAD 2009, 15–18 November 2009, Austin, Texas, USA, pp. 17–24 (2009)
7. Ganai, M.K., Gupta, A.: Completeness in smt-based BMC for software programs. In: Design, Automation and Test in Europe, DATE 2008, Munich, Germany, 10–14 March 2008, pp. 831–836 (2008)
8. Helmert, M.: The Fast Downward planning system. J. Artif. Intell. Res. **26**, 191–246 (2006)
9. Jackson, D.: Software Abstractions: Logic, Language, and Analysis. MIT Press, Cambridge (2006)
10. Kautz, H.A., Selman, B.: Planning as satisfiability. In: ECAI, pp. 359–363 (1992)
11. Kroning, D., Strichman, O.: Efficient computation of recurrence diameters. In: Zuck, L.D., Attie, P.C., Cortesi, A., Mukhopadhyay, S. (eds.) VMCAI 2003. LNCS, vol. 2575, pp. 298–309. Springer, Heidelberg (2002)
12. Mcdermott, D., Ghallab, M., Howe, A., Knoblock, C., Ram, A., Veloso, M., Weld, D., Wilkins, D.: PDDL: The planning domain definition language. Technical report, CVC TR-98-003/DCS TR-1165, Yale Center for Computational Vision and Control (1998)
13. Nipkow, T.: Verifying a hotel key card system. In: Barkaoui, K., Cavalcanti, A., Cerone, A. (eds.) ICTAC 2006. LNCS, vol. 4281, pp. 1–14. Springer, Heidelberg (2006)
14. Rintanen, J.: Evaluation strategies for planning as satisfiability. In: Proceedings of the 16th European Conference on Artificial Intelligence, pp. 682–687. IOS Press (2004)
15. Rintanen, J., Gretton, C.O.: Computing upper bounds on lengths of transition sequences. In: International Joint Conference on Artificial Intelligence (2013)
16. Sastry, S., Widder, J.: Solvability-based comparison of failure detectors. In: 2014 IEEE 13th International Symposium on Network Computing and Applications, NCA 2014, 21–23 August 2014, Cambridge, MA, USA, pp. 269–276 (2014)
17. Sheeran, M., Singh, S., Stålmarck, G.: Checking safety properties using induction and a SAT-solver. In: Johnson, S.D., Hunt Jr., W.A. (eds.) FMCAD 2000. LNCS, vol. 1954, pp. 108–125. Springer, Heidelberg (2000)
18. Slind, K., Norrish, M.: A brief overview of HOL4. In: Mohamed, O.A., Muñoz, C., Tahar, S. (eds.) TPHOLs 2008. LNCS, vol. 5170, pp. 28–32. Springer, Heidelberg (2008)
19. Streeter, M.J., Smith, S.F.: Using decision procedures efficiently for optimization. In: Proceedings of the 17th International Conference on Automated Planning and Scheduling, pp. 312–319. AAAI Press (2007)
20. Williams, B.C., Nayak, P.P.: A reactive planner for a model-based executive. In: International Joint Conference on Artificial Intelligence, pp. 1178–1185. Morgan Kaufmann Publishers (1997)

Formalization of Error-Correcting Codes: From Hamming to Modern Coding Theory

Reynald Affeldt[1]([✉]) and Jacques Garrigue[2]

[1] National Institute of Advanced Industrial Science and Technology, Tsukuba, Japan
reynald.affeldt@aist.go.jp
[2] Nagoya University, Nagoya, Japan
garrigue@math.nagoya-u.ac.jp

Abstract. By adding redundancy to transmitted data, error-correcting codes (ECCs) make it possible to communicate reliably over noisy channels. Minimizing redundancy and (de)coding time has driven much research, culminating with Low-Density Parity-Check (LDPC) codes. At first sight, ECCs may be considered as a trustful piece of computer systems because classical results are well-understood. But ECCs are also performance-critical so that new hardware calls for new implementations whose testing is always an issue. Moreover, research about ECCs is still flourishing with papers of ever-growing complexity. In order to provide means for implementers to perform verification and for researchers to firmly assess recent advances, we have been developing a formalization of ECCs using the SSReflect extension of the Coq proof-assistant. We report on the formalization of linear ECCs, duly illustrated with a theory about the celebrated Hamming codes and the verification of the sum-product algorithm for decoding LDPC codes.

1 Introduction

Error-correcting codes (ECCs) add redundancy to transmitted data to ensure reliable communication over noisy channels. Low-Density Parity-Check (LDPC) codes are ECCs discovered in 1960 by Gallager; they were not thoroughly studied until they were shown in the nineties to deliver good performance in practice. Since then, LDPC codes have found their way into modern devices such as hard-disk storage, wifi communications, etc. and have motivated a new body of works known as modern coding theory.

Implementations of ECCs cannot be crystallized as a generic library that can be deemed correct because extensively tested. Because ECCs are performance-critical, new implementations are required to take advantage of the latest hardware, so that testing is always an issue. Also, research (in particular about LDPC codes) is so active that correctness guarantees for cutting-edge ECCs are scattered in scientific publications of ever-growing complexity.

A formalization of ECCs could help implementers and researchers. First, it would make possible verification of concrete implementations. Today, an implementer willing to perform formal verification should first provide a formal specification of what ECCs are supposed to achieve. In comparison, this is more

C. Urban and X. Zhang (Eds.): ITP 2015, LNCS 9236, pp. 17–33, 2015.
DOI: 10.1007/978-3-319-22102-1_2

difficult than verification of cryptographic functions whose specification requires little infrastructure when they rely on number theory (e.g., [1]).

However, the formalization of ECCs is a difficult undertaking. They rely on a vast body of mathematics: probabilities, graphs, linear algebra, etc. Teaching material is rarely (if ever) structured as algebra textbooks: prose definitions that look incomplete without the accompanying examples, algorithms written in prose, hypotheses about the model that appear during the course of proofs, etc. Monographs and research papers do not provide details for the non-expert reader. It is therefore no wonder that researchers are seeking for means to firmly assess the correctness of their pencil-and-paper proofs: our work is part of such a project.

Still, there is previous work that we can take advantage of to formalize ECCs. The SSREFLECT/MATHCOMP library [8] provides big operators to formalize combinatorial results, a substantial formalization of linear algebra, and tools to reason about graphs. The formalization of the foundational theorems of information theory [2] provides us with basic definitions about channels and probabilities. Last, we are fortunate enough to have colleagues, expert in ECCs, who provided us with details about linear ECCs and LDPC codes [9, Chaps. 7 and 9].

To the best of our knowledge, our effort is the first attempt at a systematic formalization of ECCs inside a proof-assistant. In Sect. 3, we formalize basic results about linear ECCs. Already at this point, some effort was spent in augmenting textbook definitions with their implicit assumptions. In Sect. 4, we formalize Hamming codes. In particular, we provide a concrete encoder and decoder and express the error rate in terms of a closed formula. In Sect. 5, we formalize the key properties of the sum-product algorithm, the standard algorithm for efficient decoding of LDPC codes. Finally, in Sect. 6, we apply our formalization to the verification of a concrete implementation of the sum-product algorithm, making our work the first formal verification of a decoding algorithm for an advanced class of ECCs.

2   Premises on Information Theory and Probabilities

2.1   Channels and Codes in Information Theory

We first recall basic definitions from [2].

The most generic definition of a code is as a *channel code*: a pair of encoder/decoder functions with a finite type M for the message pieces to be encoded. Encoded message pieces (codewords) are represented by row vectors over a finite alphabet A (denoted by 'rV[A]_n in MATHCOMP). The decoder (that may fail) is fed with the outputs of a noisy channel that are also represented by row vectors (possibly over a different[1] alphabet B):

```
Definition encT := {ffun M → 'rV[A]_n}.
Definition decT := {ffun 'rV[B]_n → option M}.
Record code := mkCode { enc : encT ; dec : decT }.
```

---

[1] A and B are different for example in the case of the *binary erasure channel* that replaces some bits with an *erasure*.

A (discrete) noisy channel is modeled as a stochastic matrix that we formalized as a function from the input alphabet A to probability distributions over the output alphabet B:

```
Notation " 𝒞ℋ₁(A, B)" := (A → dist B).
```

dist is the type of probability distributions. They are essentially functions from some finite type to non-negative reals whose outputs sum to 1 (the big operator $\sum_-$(x in P) f x comes from MATHCOMP):

```
Record dist (A : finType) := mkDist {
  pmf :> A → R+ ; (* "→ R+" is a notation *)
  pmf1 : ∑_(a in A) pmf a = 1 }.
```

Given a distribution P, the probability of an event (represented by a finite set of elements: type {set A} in MATHCOMP) is formalized as follows:

```
Definition Pr P (E : {set A}) := ∑_(a in E) P a.
```

Communication of n characters is thought of as happening over the nth extension of the channel, defined as a function from input vectors to distributions of output vectors ({dist T} is a notation for the type of distributions; it hides a function that checks whether T is a finite type):

```
Notation " 𝒞ℋₙ(A, B)" := ('rV[A]_n → {dist 'rV[B]_n}).
```

In this paper, we deal with *discrete memoryless channels* (DMCs). It means that the output probability of a character does not depend on preceding inputs. In this case, the definition of the nth extension of a channel W boils down to a probability mass function that associates to an input vector x the following distribution of output vectors:

```
Definition f (y : 'rV_n) := ∏_(i < n) W (x /_ i) (y /_ i).
```

where x /_ i represents the ith element of the vector x. The notation W ^ n (y | x) ($W^n(y|x)$ in pencil-and-paper proofs) is the probability for the DMC of W that an input x (of length n) is output as y.

Finally, the quality of a code c for a given channel W is measured by its *error rate* (notation: $\bar{e}_{cha}(W, c)$), that is defined by the average probability of errors:

```
Definition ErrRateCond (W : 𝒞ℋ₁(A, B)) c m :=
  Pr (W ^ n (| enc c m)) [set y | dec c y ≠ Some m].
Definition CodeErrRate (W : 𝒞ℋ₁(A, B)) c :=
  1 / INR #| M | * ∑_(m in M) ErrRateCond W c m.
```

W ^ n (| enc c m) is the distribution of outputs corresponding to the codeword enc c m of a message m sent other the DMC of W. [set y | dec c y ≠ Some m] is the set of outputs that do not decode to m. INR injects naturals into reals.

2.2 Aposteriori Probability

Probabilities are used to specify the correctness of probabilistic decoders such as the sum-product algorithm (see Sect. 5).

We first formalize the notion of *aposteriori probability*: the probability that an input was sent knowing that some output was received. It is defined via the Bayes rule from the probability that an output was received knowing that some input was sent. For an input distribution P and a channel W, the aposteriori probability of an input x given the output y is:

$$P^W(x|y) := \frac{P(x)W^n(y|x)}{\sum_{x' \in A^n} P(x')W^n(y|x')}$$

We formalize aposteriori probabilities with the following probability mass function:

```
Definition den := ∑_(x in 'rV_n) P x * W ^ n (y | x).
Definition f x := P x * W ^ n (y | x) / den.
```

This probability is well-defined when the denominator is not zero. This is more than a technical hindrance: it expresses the natural condition that, since `y` was received, then necessarily a suitable `x` (i.e., such that `P x` $\neq 0$ and `W ^ n (y | x)` $\neq 0$) was sent beforehand. The denominator being non-zero is thus equivalent to the `receivable` condition:

```
Definition receivable y := [∃ x, (P x ≠ 0) ∧ (W ^ n (y | x) ≠
0)].
```

In Coq, we denote aposteriori probabilities by `P '^^ W, H ( x | y )` where `H` is a proof that `y` is `receivable`.

Finally, the probability that the n_0^{th} bit of the input is set to b (0 or 1) given the output y is defined by the *marginal aposteriori probability* (K is chosen so that it is indeed a probability):

$$P^W_{n_0}(b|y) := K \sum_{x \in \mathbb{F}_2^n \ x_{n_0}=b} P^W(x|y)$$

In Coq, we will denote this probability by `P'_ n0 '^^ W, H ( b | y )` where `H` is the proof that `y` is `receivable`. See [3] for complete formal definitions.

3   A Formal Setting for Linear ECCs

Linear ECCs are about bit-vectors, i.e., vectors over $\mathbb{F}_2$ (we use the notation `'F_2` from MATHCOMP). Their properties are mostly discussed in terms of Hamming weight (the number of 1 bits) or of Hamming distance (the number of bits that are different). In Coq, we provide a function `wH` for the Hamming weight, from which we derive the Hamming distance:
```
Definition dH n (x y : 'rV_n) := wH (x − y).
```

3.1   Linear ECCs as Sets of Codewords

The simplest definition of a linear ECC is as a set of codewords closed by addition (`n` is called the *length* of the code):

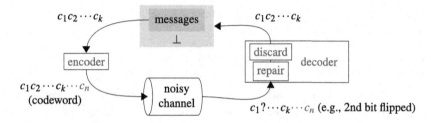

Fig. 1. The setting of error-correcting codes

```
Record lcode0 n := mkLcode0 {
  codewords :> {set 'rV['F_2]_n} ;
  lclosed : addr_closed codewords }.
```

In practice, a linear ECC is defined as the kernel of a *parity check matrix* (PCM), i.e., the matrix whose rows correspond to the checksum equations that codewords fulfill (*m is multiplication and ^T is transposition of matrices):

```
Definition syndrome (H : 'M['F_2]_(m, n)) (y :'rV_n) := H *m y^T.
Definition kernel H := [set c | syndrome H c = 0 ].
```

Since the kernel of the PCM is closed by addition, it defines a linear ECC:

```
Lemma kernel_add H : addr_closed (kernel H). Proof. ... Qed.
Definition lcode0_kernel H := mkLcode0 (kernel_add H).
```

When H is a $m \times n$ matrix, $k = n - m$ is called the *dimension* of the code.

A code is trivial when it is reduced to the singleton with the null vector:

```
Definition not_trivial := ∃ cw, (cw ∈ C) ∧ (cw ≠ 0).
```

When a linear ECC C is not trivial (proof C_not_trivial below), one can define the *minimum distance* between any two codewords, or, equivalently, the minimum weight of non-zero codewords, using SSREFLECT's xchoose and arg_min functions:

```
Definition non_0_codeword := xchoose C_not_trivial.
Definition min_wH_codeword :=
  arg_min non_0_codeword [pred cw in C | wH cw ≠ 0] wH.
Definition d_min := wH min_wH_codeword.
```

The minimum distance $d_{\min}$ defines in particular the number of errors $\lfloor \frac{d_{\min}-1}{2} \rfloor$ that one can correct using minimum distance decoding (see Sect. 3.3):

```
Definition mdd_err_cor := (d_min. - 1)/2.
```

3.2 Linear ECCs with Coding and Decoding Functions

In practice, a linear ECC is not only a set of codewords but also a pair of coding and decoding functions to be used in conjunction with a channel (see Fig. 1 [5, p. 16]). We combine the definition of a linear ECC as a set of codewords (Sect. 3.1) and as a pair of encoding and decoding functions (i.e., a channel code—Sect. 2.1) with the hypotheses that (1) the encoder is injective and (2) its image is a subset of the codewords:

```
Record lcode n k : Type := mkLcode {
  lcode0_of :> lcode0 n ;
  enc_dec :> code 'F_2 'F_2 'rV ['F_2]_k n ;
  enc_inj : injective (enc enc_dec) ;
  enc_img : enc_img_in_code lcode0_of (enc enc_dec) }.
```

enc_img_in_code is the hypothesis that the image of the messages (here, 'rV['F_2]_k) by the encoder (enc enc_dec) is included in the set of codewords (here, lcode0_of). Note that k $\leq$ n can be derived from the injectivity of the encoder.

As indicated by Fig. 1, the decoder is decomposed into (1) a function that repairs the received output and (2) a function that discards the redundancy bits:

```
Record lcode1 n k := mkLcode1 {
  lcode_of :> lcode n k;
  repair : repairT n ; (* 'rV['F_2]_n → option ('rV['F_2]_n) *)
  discard : discardT n k ; (* 'rV['F_2]_n → 'rV['F_2]_k *)
  dec_is_repair_discard :
    dec lcode_of = [ffun y ⇒ omap discard (repair y)];
  enc_discard_is_id : cancel_on lcode_of (enc lcode_of) discard }.
```

enc_discard_is_id is a proof that discard followed by encoding enc is the identity over the domain lcode_of.

Example. The r-repetition code encodes one bit by replicating it r times. It has therefore two codewords: $00\cdots0$ and $11\cdots1$ (r times). The PCM can be defined as $H = A \| 1$ where A is a column vector of $r - 1$ 1's, and the corresponding encoder is the matrix multiplication by $G = 1 \| (-A)^T$. More generally, let A be a $(n - k) \times k$ matrix, $H = A \| 1$ and $G = 1 \| (-A)^T$. Then H is the PCM of a (n, k)-code with the (injective) encoding function $x \mapsto x \times G$. Such a linear ECC is said to be in *systematic form* (details in [3]).

3.3   The Variety of Decoding Procedures

There are various strategies to decode the channel output. *Minimum distance decoding* chooses the closest codeword in terms of Hamming distance. When such a decoder decodes an output y to a message m, then there is no other message m' whose encoding is closer to y:

```
Definition minimum_distance_decoding :=
  ∀ y m, (dec c) y = Some m →
    ∀ m', dH ((enc c) m) y ≤ dH ((enc c) m') y.
```

We can now formalize the first interesting theorem about linear ECCs [11, p. 10] that shows that a minimum distance decoder can correct mdd_err_cor (see Sect. 3.1) errors:

```
Lemma encode_decode m y : (dec C) y ≠ None →
dH ((enc C) m) y ≤ mdd_err_cor C_not_trivial →
(dec C) y = Some m.
```

For example, a repetition code can decode $(r - 1)/2$ errors (with r odd) since minimum distance decoding can be performed by majority vote (see [3] for formal proofs):

```
Definition majority_vote r (s : seq'F_2) : option'F_2 :=
  let cnt := num_occ 1\,s in
  if r/2 < cnt then Some 1
  else if (r/2 = cnt) ∧ ~~ odd r then None
  else Some 0.
```

Maximum Likelihood (ML) Decoding decodes to the message that is the most likely to have been encoded according to the definition of the channel. More precisely, for an encoder f, a ML decoder ϕ is such that $W^n(y|f(\phi(y))) = \max_{m \in M} W^n(y|f(m))$:

```
Definition maximum_likelihood_decoding :=
  support (enc c) → ∀ y, receivable W P y →
  ∃ m, (dec c) y = Some m ∧
    W ^ n (y | (enc c) m) = \rmax_(m' in M) W ^
n (y | (enc c) m').
```

The assumption `receivable W P y` says that we consider outputs with non-zero probability (see Sect. 2.2). The assumption `support (enc c)` says that only codewords can be input. Textbooks do not make these assumptions explicit but they are essential to complete formal proofs.

ML decoding is desirable because it achieves the smallest error rate among all the possible decoders [3, Lemma `ML_smallest_err_rate`]. Still, it is possible to achieve ML decoding via minimum distance decoding. This is for example the case with a binary symmetric channel (that inputs and outputs bits) with error probability $p < \frac{1}{2}$. Formally, for a code c with at least one codeword:

```
Lemma MD_implies_ML : p < 1/2 → minimum_distance_decoding c →
  (∀ y, (dec c) y ≠ None) → maximum_likelihood_decoding W c P.
```

Maximum aposteriori probability (MAP) decoding decodes to messages that maximize the aposteriori probability (see Sect. 2.2). MAP decoding is desirable because it achieves ML decoding [3, lemma `MAP_implies_ML`]. *Maximum posterior marginal (MPM) decoding* is similar to MAP decoding: it decodes to messages such that each bit maximizes the marginal aposteriori probability. The sum-product algorithm of Sect. 5 achieves MPM decoding.

4 Formalization of Hamming Codes and Their Properties

We formalize Hamming codes. In particular, we show that the well-known decoding procedure for Hamming code is actually a minimum distance decoding and that the error rate can be stated as a closed formula.

Formal Definition. Hamming codes are $(n = 2^m - 1, k = 2^m - m - 1)$ linear ECCs, i.e., one adds m extra bits for error checking. The codewords are defined by the PCM whose columns are the binary representations of the $2^m - 1$ non-null words of length m. For a concrete illustration, here follows the PCM of the $(7,4)$-Hamming code:

$$\cdot\ hamH_{7,4} = \begin{pmatrix} 0\ 0\ 0\ 1\ 1\ 1\ 1 \\ 0\ 1\ 1\ 0\ 0\ 1\ 1 \\ 1\ 0\ 1\ 0\ 1\ 0\ 1 \end{pmatrix}$$

Formally, for any m, we define the PCM using a function `nat2bin_cV` that builds column vectors with the binary representation of natural numbers (e.g., for the matrix H above, `nat2bin_cV 3 1` returns the first column vector, `nat2bin_cV 3 2` the second, etc.):

```
Definition hamH := \matrix_(i < m, j < n) (nat2bin_cV m j+1 i 0).
Definition hamC : lcode0 n := lcode0_kernel hamH.
```

Minimum Distance. The minimum distance of Hamming codes is 3, and therefore, by minimum distance decoding, Hamming codes can correct 1-bit errors (by the lemma `encode_decode` of Sect. 3.3). The fact that the minimum distance is 3 is proved by showing that there are no codewords of weights 1 and 2 (by analysis of H) while there is a codeword of weight 3 ($7 \times 2^{n-3} = (1110\cdots0)_2$). Hamming codes are therefore not trivial and their minimum distance is 3:

```
Lemma hamming_not_trivial : not_trivial hamC.
Lemma minimum_distance_is_3 : d_min hamming_not_trivial = 3.
```

Minimum Distance Decoding. The procedure of decoding for Hamming codes is well-known. To decode the output y, compute its syndrome: if it is non-zero, then it is the binary representation of the index of the bit to flip back. The function `ham_detect` computes the index i of the bit to correct and prepare a vector to repair the error. The function `ham_repair` fixes the error by adding this vector:

```
Definition ham_detect y :=
  let i := bin2nat_cV (syndrome hamH y) in
  if i is 0 then 0 else nat2bin_rV n (2 ^ (n - i)).
Definition ham_repair : decT _ _ m := [ffun y =>
  let ret := y + ham_detect y in
  if syndrome hamH ret = 0 then Some ret else None].
```

Let `ham_scode` be a linear ECC using the `ham_repair` function. We can show that it implements minimum distance decoding:

```
Lemma hamming_MD : minimum_distance_decoding ham_scode.
```

It is therefore an ML decoding (by the Lemma `MD_implies_ML` from Sect. 3.3).

The Encoding and Discard Functions. We now complete the formalization of Hamming codes by providing the encoding and discard functions (as in Fig. 1). Modulo permutation of the columns, the PCM of Hamming codes can be transformed into systematic form $sysH = sysA\,\|\,1$ (as explained in the example about repetition codes in Sect. 3.2). This provides us with a generating matrix $sysG = 1\,\|\,(-sysA)^T$. For illustration in the case of the $(7,4)$-Hamming code:

$$sysH_{7,4} = \begin{pmatrix} 0\,1\,1\,1 & 1\,0\,0 \\ 1\,0\,1\,1 & 0\,1\,0 \\ 1\,1\,0\,1 & 0\,0\,1 \end{pmatrix} \qquad sysG_{7,4} = \begin{pmatrix} 1\,0\,0\,0 & 0\,1\,1 \\ 0\,1\,0\,0 & 1\,0\,1 \\ 0\,0\,1\,0 & 1\,1\,0 \\ 0\,0\,0\,1 & 1\,1\,1 \end{pmatrix}$$

Let `sysH_perm` be the column permutation that turns H into sysH. The parity check and generating matrices in systematic form are formalized as follows:

```
Definition sysH :'M['F_2]_(m, n) := col_perm sysH_perm hamH.
Definition sysG :=
  castmx (erefl, subnK (m_len m')) (row_mx 1%:M (-sysA)^T).
```

(`castmx` is a cast that deals with dependent types.) The discard function in systematic form is obvious:

```
Definition sysDiscard :'M['F_2]_(n - m, n) :=
  castmx (erefl, subnK (m_len m')) (row_mx 1%:M 0).
```

Using the column permutation `sysH_perm` the other way around, we can produce the discard function and the generating matrix corresponding to the original *hamH*:

```
Definition ham_discard := col_perm sysH_perm^-1 sysDiscard.
Definition hamG := col_perm sysH_perm^-1 sysG.
```

Coupled with the `ham_repair` function above, `hamG` and `ham_discard` provides us with a complete definition of Hamming encoders and decoders:

```
Definition ham_channel_code := mkCode
[ffun t ⇒ t *m hamG] [ffun x ⇒
omap ham_discard (ham_repair _ x)].
```

Error Rate. Finally, we show, in the case of a binary symmetric channel W, that the error rate (see Sect. 2.1) of Hamming codes can be expressed as a closed formula:

```
Lemma hamming_error_rate : p < 1/2 →
ēcha(W, ham_channel_code) =
1 - ((1 - p) ^ n) - INR n * p * ((1 - p) ^ (n - 1)).
```

The existence of codes with arbitrary small error rates is the main result of Shannon's theorems. But Shannon's proofs are not constructive. Our formalization of Hamming codes with a closed formula for their error rate provides us with a concrete candidate.

5 Formalization of the Properties of Sum-Product Decoding

The sum-product algorithm provides efficient decoding for LDPC codes. It computes for each bit its marginal aposteriori probability by propagating probabilities in a graph corresponding to the PCM. We explain those graphs in Sect. 5.1, the summary operator used to specify the sum-product algorithm in Sect. 5.2, and the main properties of the sum-product algorithm in Sect. 5.3.

5.1   Parity Check Matrix as Tanner Graphs

The vertices of a Tanner graph correspond to the rows and columns of a parity-check matrix H with an edge between m and n when $H_{m,n} = 1$. Rows are called *function nodes* and columns are called *variable nodes*. By construction, a Tanner graph is bipartite.

Sets of successor nodes and *subgraphs* of Tanner graphs appear as indices of big operators in the definitions and proofs of the sum-product algorithm.

Let g be a graph (formalized by a binary relation) and m and n be two connected vertices. The subgraph rooted at the edge m–n is the set of vertices reachable from m without passing through n:

```
Variables (V : finType) (g : rel V).
Definition except n := [rel x y | g x y ∧ (x ≠ n) ∧ (y ≠ n)].
Definition subgraph m n :=
[set v | g n m ∧ connect (except n) m v].
```

For Tanner graphs, we distinguish successors and subgraphs of variable nodes and of function nodes. We denote the successors of the function (resp. variable) node m0 (resp. n0) by 'V m0 (resp. 'F n0). We denote the function nodes of the subgraph rooted at edge m0–n0 by 'F(m0, n0). Similarly, we denote the variable nodes of the subgraph rooted at edge m0–n0 *(to which we add n0)* by 'V(m0, n0). Figure 2 provides an explanatory illustration, see [3] for complete definitions.

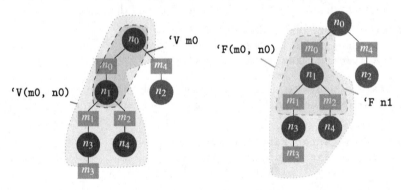

Fig. 2. Successors and subtrees in an acyclic Tanner graph

It will be important to distinguish acyclic Tanner graphs:

```
Definition acyclic g := ∀ l, 2 < size l → ~ path.ucycle g l.
```

Technically, we will need to establish partition properties when proving the properties of the sum-product decoding algorithm (see Sect. 5.3 for a concrete example).

5.2   The Summary Operator

Pencil-and-paper proofs in modern coding theory [12] make use of a special summation called the *summary operator* [10]. It is denoted by $\sum_{\sim s}$ and indicates the variables *not* being summed over. This operator saves the practitioner "from a flood of notation" [12, p. 49], for example by writing steps such as:

$$\prod_{m_0 \in F(n_0)} \sum_{\sim\{n_0\}} \cdots = \sum_{\sim\{n_0\}} \prod_{m_0 \in F(n_0)} \cdots, \tag{1}$$

the reader being trusted to understand that both operators sum over different sets.

We formalize the summary operator as a sum over vectors x such that x /_ i is fixed using a default vector d when i ∉ s and write $\sum$_(x # s, d) instead of $\sum_{\sim s}$:

Definition summary (s : {set ′I_n}) (d x : ′rV[A]_n) :=
  [∀ i, (i ∈ ˜: s) ⇒ (x /_ i = d /_ i)].
Notation "$\sum$ (x ′#′ s ,′ d) e" := ($\sum$ (x | summary s d x) e)

Indeed, $\sum_{\sim s}$ can be understood as a sum over vectors $[x_0; x_1; ...; x_{n-1}]$ such that x_i is fixed when $i \in s$. We found it difficult to recover the terseness of the pencil-and-paper summary operator in a proof-assistant. First, the precise set of varying x_j is implicit; it can be inferred by looking at the x_j appearing below the summation sign but this is difficult to achieve unless one reflects most syntax. Second, it suggests working with vectors x of varying sizes, which can be an issue when the size of vectors appears in dependent types (tuples or row vectors in MATHCOMP). Last, it is not clear about the values of x_i when $i \in s$.

In contrast, our formalization makes clear, for example, that in Eq. (1) the first summary operator sums over $V(m_0, n_0) \backslash \{n_0\}$ while the second one sums over $[1, \dots, n] \backslash \{n_0\}$ (see Sect. 5.3 for the formalization). More importantly, we can benefit from MATHCOMP lemmas about big operators to prove the properties of the sum-product decoding thanks to our encoding (see Sect. 5.3).

Alternatively, our summary operator $\sum$_(x # s , d) e x can also be thought as $\sum_{x_1 \in \mathbb{F}_2} \cdots \sum_{x_{|s|} \in \mathbb{F}_2} e \, d[s_1 := x_1] \cdots [s_{|s|} := x_{|s|}]$ where $d[i := b]$ represents the vector d where index i is updated with b. Put formally (enum s below is the list $[s_1; s_2; \cdots; s_{|s|}]$):

Definition summary_fold (s : {set ′I_n}) d e :=
foldr (fun n0 F t ⇒$\sum$(b in ′F_2) F (t ′[n0 := b])) e (enum s) d.

This is equivalent ($\sum$_(x # s, d) e x = summary_fold s d e) but we found it easier to use summary_fold to prove our implementation of the sum-product algorithm in Sect. 6.

5.3   Properties of the Sum-Product Decoding

Correctness of the Estimation. Let us consider a channel W and a channel output y. With sum-product decoding, we are concerned with evaluating $P_{n_0}^W(b|y)$ where b is the value of the n_0^{th} bit of the input codeword (see Sect. 2.2). In the following, we show that it is proportional to the following quantity:

$$P_{n_0}^W(b|y) \propto W(y_{n_0}|b) \prod_{m_0 \in F(n_0)} \alpha_{m_0, n_0}(b).$$

$\alpha_{m_0,n_0}(b)$ (formal definition below) is the contribution to the marginal aposte-riori probability of the n_0^{th} bit coming from a subtree of the Tanner graph (we assume that the Tanner graph is acyclic).

We now provide the formal statement. Let W be a channel. Let H be a $m \times n$ PCM such that the corresponding Tanner graph is acyclic (hypothesis `acyclic_graph (tanner_rel H)`, where `tanner_rel` turns a PCM into the corre-sponding Tanner graph). Let y be the channel output to decode. We assume that it is receivable (hypothesis Hy below, see Sect. 2.2). Finally, let d be the vector used in the summary operator. Then the aposteriori probability $P_{n_0}^W(b|y)$ can be evaluated by a closed formula:

```
Lemma estimation_correctness (d : 'rV_n) n0 :
  let b := d /_ n0 in let P := 'U C_not_empty in
  P '_ n0 '^^ W, Hy (b | y) =
  Kmpp Hy * Kpp W H y * W b (y /_ n0) * ∏_(m0 in 'F n0) α m0 n0 d .
```

Kmpp and Kpp are normalization constants (see [3]). P is a uniform distribution. The distribution `'U C_not_empty` of codewords has the following probability mass function: $cw \mapsto 1/|C|$ if $cw \in C$ and 0 otherwise. α is the marginal aposteriori probability of the n_0^{th} bit of the input codeword in the modified Tanner graph that includes only function nodes from the subgraph rooted at edge m_0–n_0 and in which the received bit y /_ n0 has been erased. The formal definition relies on the summary operator:

```
Definition α m0 n0 d := ∑_(x # 'V(m0, n0) :\ n0 , d)
  W ^ _ (y # 'V(m0, n0) :\ n0 | x # 'V(m0, n0) :\ n0) *
    ∏_(m1 in 'F(m0, n0)) INR (δ ('V m1) x).
```

δ s x is an indicator function that performs checksum checks:

```
Definition δ n (s : {set 'I_n}) (x : 'rV['F_2]_n) :=
  (\big[+%R/Zp0]_(n0 in s) x /_ n0) = Zp0 .
```

Let us comment about two technical aspects of the proof of `estimation_correctness`. The first one is the need to instrument Tanner graphs with partition lemmas to be able to decompose big sums/prods. See the next paragraph on lemma `recursive_computation` for a concrete example. The second one is the main motivation for using the summary operator. We need to make big sums commute with big prods in equalities like:

```
∏_(m0 in 'F n0) ∑_(x # 'V(m0, n0) :\ n0 , d) ... =
  ∑_(x # setT :\ n0 , d) ∏_(m0 in 'F n0) ...
```

Such steps amount to apply the MATHCOMP lemma `big_distr_big_dep` together with technical reindexing. This is one of our contributions to provide lemmas for such steps.

Recursive Computation of α's. The property above provides a way to evalu-ate $P_{n_0}^W(b|y)$ but not an efficient algorithm because the computation of $\alpha_{m_0,n_0}(b)$ is about the whole subgraph rooted at the edge m_0–n_0. The second property that we formalize introduces β probabilities such that α's (resp. β's) can be computed from neighboring β's (resp. α's). This is illustrated by Fig. 3 whose meaning will be made clearer in Sect. 6. We define β using α as follows:

```
Definition β n0 m0 (d : 'rV_n) :=
  W (d /_ n0) (y /_ n0) * ∏_(m1 in 'F n0 :\ m0) α m1 n0 d.
```

We prove that α's can be computed using β's by the following formula (we assume the same setting as for the Lemma `estimation_correctness`):

```
Lemma recursive_computation m0 n0 d : n0 ∈ 'V m0 →
  α m0 n0 d = ∑_(x # 'V m0 :\ n0 , d)
    INR (δ ('V m0) x) * ∏_(n1 in 'V m0 :\ n0) β n1 m0 x.
```

This proof is technically more involved than the lemma `estimation_correctness` but relies on similar ideas: partitions of Tanner graphs to split big sums/prods and commutations of big sums and big prods using the summary operator. Let us perform the first proof step for illustration. It consists in turning the inner product of α messages $\prod\_($m1 in 'F$($m0, n0$)$:\ m0$)$ INR $(\delta$ ('V m1) x$)$ into:

```
∏_(n1 in 'V m0 :\ n0) ∏_(m1 in 'F n1 :\ m0)
  ∏_(m2 in 'F(m1, n1)) INR (δ ('V m2) x)
```

This is a consequence of the fact that 'F(m0, n0) :\ m0 can be partitioned (when H is acyclic) into smaller 'F(m1, n1) where n1 is a successor of m0 and m1 is a successor of n1, i.e., according to the following partition:

```
Definition Fgraph_part_Fgraph m0 n0 : {set {set 'I_m}} :=
  (fun n1 ⇒ ⋃_(m1 in 'F n1 :\ m0) 'F(m1, n1)) @: (('V m0) :\ n0).
```

Once `Fgraph_part_Fgraph m0 n0` has been shown to cover 'F(m0, n0) :\ n0 with pairwise disjoint sets, this step essentially amounts to use the lemmas `big_trivIset` and `big_imset` from MATHCOMP. See [3, `tanner_partition.v`] for related lemmas.

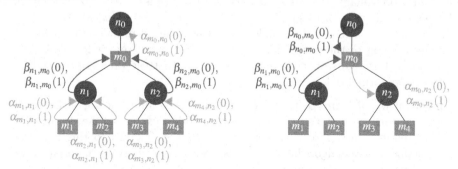

Fig. 3. Illustrations for `sumprod_up` and `sumprod_down`. Left: `sumprod_up` computes the up links from the leaves to the root. Right: `sumprod_down` computes the down link of edge m_0–n_2 using the β's of edges m_0–n_i $(i \neq 2)$.

6   Implementation and Verification of Sum-Product Decoding

An implementation of sum-product decoding takes as input a Tanner graph and an output y, and computes for all variable nodes, each representing a bit of the decoded codeword, its marginal aposteriori probability. One chooses to decode the n_0^{th} bit either as 0 if $P_{n_0}^W(0|y) \geq P_{n_0}^W(1|y)$ or as 1 otherwise, so as to perform MPM decoding.

The algorithm we implement is known in the literature as the forward/backward algorithm and has many applications [10]. It uses the tree view of an acyclic Tanner graph to structure recursive computations. In a first phase it computes α's and β's (see Sect. 5.3) from the leaves towards the root of the tree, and then computes α's and β's in the opposite direction (starting from the root that time). Figure 3 illustrates this.

Concretely, we provide Coq functions to build the tree, compute α's and β's, and extract the estimations, and prove formally that the results indeed agree with the definitions from Sect. 5.3.

Definition of the Tree. Function nodes and variable nodes share the same data structure, and are just distinguished by their kind.

```
Definition R2 := (R * R)%type.
Inductive kind : Set := kf | kv.
Fixpoint negk k := match k with kf ⇒ kv | kv ⇒ kf end.
Inductive tag : kind → Set := Func : tag kf | Var : R2 →
tag kv.
Inductive tn_tree (k : kind) (U D : Type) : Type :=
  Node { node_id : id; node_tag : tag k;
         children : seq (tn_tree (negk k) U D);
         up : U; down : D }.
```

This tree is statically bipartite, thanks to the switching of the kind for the children. Additionally, in each variable node, `node_tag` is expected to contain the channel probabilities for this bit to be 0 or 1, i.e., the pair $(W(y_{n_0}|0), W(y_{n_0}|1))$. The `up` and `down` fields are to be filled with the values of α and β (according to the kind), going to the parent node for `up`, and coming from it for `down`. Here again we will use pairs of the 0 and 1 cases. Note that the values of α's and β's need not be normalized.

Computation of α and β. The function α_β takes as input the tag of the source node, and the α's and β's from neighboring nodes, excluding the destination, and computes either α or β, according to the tag. Thanks to this function, the remainder of the algorithm keeps a perfect symmetry between variable and function nodes.

```
Definition α_op (out inp : R2) :=
   let (o,o') := out in let (i,i') := inp in
   (o*i + o'*i', o*i' + o'*i).
Definition β_op (out inp : R2) :=
   let (o,o') := out in let (i,i') := inp in (o*i, o'*i').
Definition α_β k (t : tag k) : seq R2 → R2 :=
   match t with
   | Func ⇒ foldr α_op (1,0)
   | Var v ⇒ foldl β_op v
   end.
```

The definition for β is clear enough: assuming that v contains the channel probabilities for the corresponding bit, it suffices to compute the product of these probabilities with the incoming α's. For α, starting from the `recursive_computation` lemma, we remark that assuming a bit to be 0 leaves the parity unchanged, while assuming it to be 1 switches the parities. This way, the sum-of-products can be computed as an iterated product, using α_op. This optimization is described in [10, Sect. 5-E]. We will of course need to prove that these definitions compute the same α's and β's as in Sect. 5.3.

Propagation of α and β. sumprod_up and sumprod_down compute respectively the contents of the up and down fields.

```
Fixpoint sumprod_up {k} (n : tn_tree k unit unit)
  : tn_tree k R2 unit :=
  let children' := map sumprod_up (children n) in
  let up' := α_β (node_tag n) (map up children') in
  Node (node_id n) (node_tag n) children' up' tt.
Fixpoint seqs_but1 (a b : seq R2) :=
  if b is h::t then (a++t)::seqs_but1 (rcons a h) t else [::].
Fixpoint sumprod_down {k} (n : tn_tree k R2 unit)
  (from_above : option R2) : tn_tree k R2 R2 :=
  let (arg0, down') :=
    if from_above is Some p then ([::p],p) else ([::],(1,1)) in
  let args := seqs_but1 arg0 (map up (children n)) in
  let funs := map
    (fun n' l ⇒ sumprod_down n' (Some (α_β (node_tag n) l)))
    (children n) in
  let children' := apply_seq funs args in
  Node (node_id n) (node_tag n) children' (up n) down'.
Definition sumprod {k} n := sumprod_down (@sumprod_up k n) None.
```

The from_above argument is None for the root of the tree, or the β coming from the parent node otherwise. apply_seq applies a list of functions to a list of arguments. This is a workaround to allow defining sumprod_down as a Fixpoint.

Building the Tree. A parity-check matrix H and the probability distribution rW for each bit (computed from the output y and the channel W) is turned into a tn_tree, using the function build_tree, and fed to the above sumprod algorithm:

```
Variables (W : 𝒞ℋ₁('F_2, B)) (y : 'rV[B]_n).
Let rW n0 := (W 0 (y /_ n0), W 1 (y /_ n0)).
Let computed_tree := sumprod (build_tree H rW (k := kv) ord0).
```

Extraction of Estimations. We finally recover normalized estimations from the tree:

```
Definition normalize (p : R2) :=
  let (p0, p1) := p in (p0 / (p0 + p1), p1 / (p0 + p1)).
Fixpoint estimation {k} (n : tn_tree k R2 R2) :=
  let l := flatten (map estimation (children n)) in
  if node_tag n is Var _ then
```

```
   (node_id n, normalize (β_op (up n) (down n))) :: 1
 else 1 (* node_tag n is Func *).
```

Correctness. The correctness of the algorithm above consists in showing that the estimations computed are the intended aposteriori probabilities:

```
Let estimations := estimation computed_tree.
Definition esti_spec n0 (x : 'rV_n) :=
  ('U C_not_empty) '_ n0 '^^ W, Hy (x /_ n0 | y).
Definition estimation_spec := uniq (unzip1 estimations) ∧
  ∀ n0, (inr n0, p01 (esti_spec n0) n0) ∈ estimations.
```

where p01 f n0 applies f, to a vector whose n_0^{th} bit is set to 0 and 1.

Theorem `estimation_ok` in [3, `ldpc_algo_proof.v`] provides a proof of this fact. As key steps, it uses the lemmas `recursive_computation` and `estimation_correctness` from Sect. 5.3.

Concrete Codes. All proofs of probabilistic sum-product decoding assume the Tanner graph to be acyclic [10]. In practice codes based on acyclic graphs are rare and not very efficient [7]. We tested our implementation with one of them [3, `sumprod_test.ml`].

In the general case where the Tanner graph contains cycles, one would use an alternative algorithm that computes the α's and β's repeatedly, propagating them in the graph until the corrected word satisfies the parity checks, failing if the result is not reached within a fixed number of iterations [10, Sect. 5]. This works well in practice but there is no proof of correctness, even informal. In place of this iterative approach, one could also build a tree approximating the graph, by unfolding it to a finite depth, and apply our functional algorithm.

7   Related Work

Coding theory has been considered as an application of the interface between the Isabelle proof-assistant and the Sumit computer algebra system [4]. In order to take advantage of the computer algebra system, proofs are restricted to a certain code length. Though the mathematical background about polynomials has been formally verified, results about coding theory are only asserted. In comparison, we formally verify much more (generic) lemmas. Yet, for example when proving that certain bitstrings are codewords, we found ourselves performing formal proofs close to symbolic computation. With this respect, we may be able in a near future to take advantage of extensions of the MATHCOMP library that provide computation [6].

8   Conclusion

In this paper, we have proved the main properties of Hamming codes and sum-product decoding. It is interesting to contrast the two approaches, respectively known as classical and modern coding theory.

For Hamming codes, we could provide an implementation of minimal-distance decoding, and prove that it indeed realizes maximum likelihood decoding, i.e., the best possible form of decoding.

For sum-product decoding, which provides the basis for LDPC codes, one can only prove that the sum-product algorithm allows to implement MPM decoding. However, this is two steps away from maximum likelihood: the proof is only valid for acyclic Tanner graphs, while interesting codes contain cycles, and MPM is an approximation of MAP decoding, with only the latter providing maximum likelihood. Yet, this "extrapolation" methodology does work: sum-product decoding of LDPC codes is empirically close to maximum likelihood, and performs very well in practice.

Acknowledgments. T. Asai, T. Saikawa, K. Sakaguchi, and Y. Takahashi contributed to the formalization. The formalization of modern coding theory is a collaboration with M. Hagiwara, K. Kasai, S. Kuzuoka, and R. Obi. The authors are grateful to the anonymous reviewers for their comments. This work is partially supported by a JSPS Grant-in-Aid for Scientific Research (Project Number: 25289118).

References

1. Affeldt, R., Nowak, D., Yamada, K.: Certifying assembly with formal security proofs: the case of BBS. Sci. Comput. Program. **77**(10–11), 1058–1074 (2012)
2. Affeldt, R., Hagiwara, M., Sénizergues, J.: Formalization of Shannon's theorems. J. Autom. Reason. **53**(1), 63–103 (2014)
3. Affeldt, R., Garrigue, J.: Formalization of error-correcting codes: from Hamming to modern coding theory. Coq scripts. https://staff.aist.go.jp/reynald.affeldt/ecc
4. Ballarin, C., Paulson, L.C.: A pragmatic approach to extending provers by computer algebra–with applications to coding theory. Fundamenta Informaticae **34**(1–2), 1–20 (1999)
5. Betten, A., Braun, M., Fripertinger, H., Kerber, A., Kohnert, A., Wassermann, A.: Error-Correcting Linear Codes-Classification by Isometry and Applications. Springer, Heidelberg (2006)
6. Dénès, M., Mörtberg, A., Siles, V.: A refinement-based approach to computational algebra in Coq. In: Beringer, L., Felty, A. (eds.) ITP 2012. LNCS, vol. 7406, pp. 83–98. Springer, Heidelberg (2012)
7. Etzion, T., Trachtenberg, A., Vardy, A.: Which codes have cycle-free Tanner graphs? IEEE Trans. Inf. Theory **45**(6), 2173–2181 (1999)
8. Gonthier, G., Mahboubi, A., Tassi, E.: A small scale reflection extension for the Coq system. Technical report RR-6455, INRIA (2008). Version 14, March 2014
9. Hagiwara, M.: Coding theory: mathematics for digital communication. Nippon Hyoron Sha (2012) (in Japanese)
10. Kschischang, F.R., Frey, B.J., Loeliger, H.-A.: Factor graphs and the sum-product algorithm. IEEE Trans. Inf. Theory **47**(2), 498–519 (2001)
11. MacWilliams, F.J., Sloane, N.J.A.: The Theory of Error-Correcting Codes. North-Holland, Amsterdam (1977). 7th impression (1992)
12. Richardson, T., Urbanke, R.: Modern Coding Theory. Cambridge University Press, Cambridge (2008)

ROSCoq: Robots Powered by Constructive Reals

Abhishek Anand[✉] and Ross Knepper

Cornell University, Ithaca, NY 14850, USA
abhishek.anand.iitg@gmail.com

Abstract. We present ROSCoq, a framework for developing certified Coq programs for robots. ROSCoq subsystems communicate using messages, as they do in the Robot Operating System (ROS). We extend the logic of events to enable holistic reasoning about the cyber-physical behavior of robotic systems. The behavior of the physical world (e.g. Newton's laws) and associated devices (e.g. sensors, actuators) are specified axiomatically. For reasoning about physics we use and extend CoRN's theory of constructive real analysis. Instead of floating points, our Coq programs use CoRN's exact, yet fast computations on reals, thus enabling accurate reasoning about such computations.

As an application, we specify the behavior of an iRobot Create. Our specification captures many real world imperfections. We write a Coq program which receives requests to navigate to specific positions and computes appropriate commands for the robot. We prove correctness properties about this system. Using the ROSCoq shim, we ran the program on the robot and provide even experimental evidence of correctness.

1 Introduction

Cyber-Physical Systems (CPS) such as ensembles of robots can be thought of as distributed systems where agents might have sensing and/or actuation capabilities. In fact the Robot Operating System (ROS) [15] presents a unified interface to robots where subcomponents of even a single robot are represented as nodes (e.g. sensor, actuator, controller software) that communicate with other nodes using asynchronous message passing. The Logic of Events (LoE) [3] framework has already been successfully used to develop certified functional programs which implement important distributed systems like fault-tolerant replicated databases [19]. Events capture interactions between components and observations rather than internal state. This enables specification and reasoning at higher-levels while integrating easily with more detailed information [22]. CPSs are arguably harder to get right, because of the additional complexity of reasoning about physics and how it interacts with the cyber components. In this work, we show that an event-based semantics is appropriate for reasoning about CPSs too. We extend the LoE framework to enable development of certified Coq programs for CPSs.

There are several challenges in extending LoE to provide a semantic foundation for CPSs, and thus enable holistic reasoning about such systems: (1) One

© Springer International Publishing Switzerland 2015
C. Urban and X. Zhang (Eds.): ITP 2015, LNCS 9236, pp. 34–50, 2015.
DOI: 10.1007/978-3-319-22102-1_3

has to model the physical quantities, e.g. the position, direction and velocity of each robot and also the physical laws relating them. (2) Time is often a key component of safety proofs of a CPS. For example, the software controller of a robot needs to send correct messages (commands) to the motors before it collides with something. (3) The software controller of a robot interacts with devices such as sensors and actuators which measure or influence the physical quantities. The specification of these devices typically involve both cyber and physical aspects. (4) Robotic programs often need to compute with real numbers, which are challenging to reason about accurately.

Our Coq framework addresses each of these challenges. Our running example is that of a robotic system consisting of an iRobot Create[1] and its controller-software. This setup can be represented as a distributed system with 3 agents (a.k.a. nodes in ROS) : (a) the hardware agent which represents the robot along with its ROS drivers/firmware. It receives messages containing angular and linear velocities and adjusts the motors accordingly to achieve those velocities. (b) the software agent which sends appropriate velocity messages to the hardware agent (c) the external agent which sends messages to the software agent telling where the robot should go. The message sequence diagram below shows a sample interaction between the agents. Click here for the corresponding video.

To define a CPS in ROSCoq, one has to first define its physical model and then define each agent independently. The physical model specifies how the relevant physical quantities evolve over time. These are represented as continuous real-valued functions over time, where time is represented as a non-negative real number. In our example, the relevant physical quantities are the position, orientation and velocities (angular and linear) of the robot. Thanks to dependent

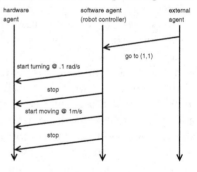

types of Coq, it is easy to express physical constraints such as the fact that the velocity is the derivative of the position w.r.t. time. We extensively used CoRN's [8] rich library of definitions and theorems about derivatives, integrals, continuity, trigonometry etc. to represent and reason about physical components. The assumption of continuity allows us to get around many decidability issues related to constructive reals. During the course of this project, we contributed some generic lemmas about constructive real analysis to CoRN, such as a stronger constructive version of Intermediate Value Theorem which we found more useful while reasoning about CPSs[2].

Events have time-stamps and one can specify assumptions on the time needed by activities like message delivery, sensing, actuation, computation etc. to happen. These will have to be empirically validated; currently one cannot statically reason about the running time of Coq programs.

Agents of a CPS are represented as a relation between the physical model (how physical quantities evolve over time) and the trace of observable events (sending and receiving of messages) generated by the agent. This representation allows incomplete and non-deterministic specifications. For hardware devices such as sensors and actuators, this relation is specified axiomatically. For example, the relation for the hardware agent mentioned above asserts that whenever it receives a message requesting a velocity v, within some time δ the robot attains some velocity close to v. The semantics of software agents (e.g. the middle one in the above figure) can be specified indirectly by providing "message handlers" written as Coq functions. Because Coq is a pure functional language and has no IO facilities, we provide a ROS shim which handles sending and receiving of messages for such Coq programs. Given a received message as input, message handlers compute messages that are supposed to be sent by the shim in response. They can also request the shim to send some messages at a later time. For example, to get a robot to turn by a right angle, one can send a message requesting a positive angular velocity (say 1 rad/s) and send another message requesting an angular velocity of 0 after time $\frac{\pi}{2}$s. While reasoning about the behavior of the system, we assume that the actual time a message is sent is not too different from the requested time.

Clearly robotics programs need to compute with real numbers. In CoRN, real numbers (e.g. π) are represented as Coq's functional programs that can be used to compute arbitrarily close rational ($\mathbb{Q}$) approximations to the represented real number. Most operations on such reals are *exact*, e.g. Field operations, trigonometric functions, integral, etc. Some operations such as equality test are undecidable, hence only approximate. However, the error in such approximations can be made *arbitrarily* small (see Sect. 4.2). We prove a parametric upper bound on how far the robot will be from the position requested by the external agent. The parameters are bounds on physical imperfections, above mentioned computational errors, variations in message-delivery timings, etc. Using our shim, we ran our Coq program on an actual robot. We provide measurements over several runs and videos of the system in action.

Section 2 describes how to specify a physical model in ROSCoq. Section 3 describes the semantics of events and message delivery. Section 4 describes the semantics of agents. Section 5 describes some proof techniques for holistic reasoning about a CPS and the properties proven about our running example. It ends with a description of our experiments. Finally, we discuss related work and conclude. ROSCoq sources and more details are available at the companion website [17].

2   Physics

One of the first steps in developing a CPS using ROSCoq is to accurately specify its physical model. It describes how all of the relevant physical quantities in the system evolve over time. In our running example, these include the position and orientation of the robot and their derivatives. Using dependent types, one can also include the constraints between the physical quantities, e.g. the constraint that velocity is the derivative of position w.r.t. time. Other examples include physical laws such as Newton's laws of motion, laws of thermodynamics. We use 2 of the 3 versions of constructive reals implemented in CoRN. In our programs which are supposed to be executed, we use the faster implementation (CR), while we use the slower (IR) for reasoning. CoRN provides field and order isomorphisms between these 2 versions. To avoid confusion, we use the notation $\mathbb{R}$ for both versions. However, clicking at colored text often jumps to its definition, either in this document or in external web pages.

Time is defined as non-negative reals, where 0 denotes the time when the system starts. For each relevant physical quantity, the physical model determines how it evolves over time. This can be represented as a member (say f) of the function type Time $\to \mathbb{R}$. The intended meaning is that for time t, $f\ t$ denotes the value of the physical quantity at time t. However, physical processes are usually continuous, at least at the scale where classical physics is applicable. For example, a car does not suddenly disappear at some time and appear miles apart at the exact same time. See [9] for a detailed discussion of the importance of continuity in physics. So, we choose to represent evolution of physical quantities as continuous functions. The type TContR is similar to the function type Time $\to \mathbb{R}$, except that it additionally requires that its members be continuous functions. We have proved that TContR is an instance of the Ring typeclass [20], where ring operations on TContR are pointwise versions of the corresponding operations on real numbers. Apart from the proofs of the ring laws, this instance also involves proving that those ring operations result in continuous functions. As a result of this proof, one can use notations like $+$, $*$ on members of TContR, and the **ring** tactic of Coq can automate equational reasoning (about TContR expressions) that follows just from ring laws.

Using records, which are just syntactic sugars for dependent pairs, one can model multiple physical quantities and also the associated physical laws. The record type below defines the physical model in our running example. It represents how the physical state of an iCreate robot evolves over time.

```
Record iCreate : Type := {
    position : Cart2D TContR;
    theta : TContR;        linVel : TContR;        omega : TContR;

    derivRot : isDerivativeOf omega theta;
    derivX : isDerivativeOf (linVel * ( FCos theta)) (X position);
    derivY : isDerivativeOf (linVel * ( FSin theta)) (Y position);

    init1: ({X position} 0) ≡ 0 ∧ ({Y position} 0) ≡ 0;
    init2: ({theta} 0) ≡ 0 ∧ ({linVel} 0) ≡ 0 ∧ ({omega} 0) ≡ 0
}.
```

For any type, A, the type Cart2D A is isomorphic to the product type $A \times A$. X and Y are the corresponding projection functions. The type Polar2D A is similar, except that rad and θ are the projection functions. So, the first field (position) of the record type (iCreate) above is essentially a pair of continuous functions, modeling the evolution of X and Y coordinates over time, respectively. The next line defines 3 fields which respectively model the orientation, linear velocity and angular velocity. The types of remaining fields depend on one or more of the first 4 fields. This dependence is used to capture constraints on the first four fields. The last 2 fields specify the initial conditions. The 3 fields in the middle characterize the derivatives of position and orientation of the robot. The first of those (derivRot) is a constructive proof/evidence that omega is the derivative of theta. The definition of the relation isDerivativeOf is based on CoRN's constructive notion of a derivative, which in turn is based on [4]. The next two are slightly more complicated and involve some trigonometry. FCos denotes the pointwise cosine function of type TContR $\rightarrow$ TContR. So, FCos theta is a function describing how the cosine of the robot's orientation (theta) evolves over time. Recall that here $*$ represents pointwise multiplication of functions. derivX and derivY imply that the linear motion of the robot is constrained to be along the instantaneous orientation of the robot (as defined by theta).

Our definition of a CPS is parametrized by an arbitrary type which is supposed to represent the physical model of the system. In the case of our running example, that type is (iCreate) (defined above). In the future, we plan to consider applications of our framework to systems of multiple robots. For a system of 2 iCreate robots, one could use the type (iCreate) $\times$ (iCreate) to represent the physical model. In Sect. 4, we will see that the semantics of hardware agents of a CPS is specified partly in terms of the physical model of the CPS.

3 Events and Message Delivery

As mentioned in Sect. 1, CPSs such as ensembles of robots can be thought of as distributed systems where agents might have sensing and/or actuation capabilities. The Logic of Events (LoE) framework has already been successfully used to reason about complicated distributed systems like fault-tolerant replicated databases [19]. It is based on seminal work by Lamport and formalizes the notion of message sequence diagrams which are often used in reasoning about the behavior of distributed systems. A distributed system (also a CPS) can be thought of as a collection of agents (components) that communicate via message passing. This is true at several levels of abstraction. In a collection of robots collaborating on a task (e.g. [5]), each robot can be considered as an agent. Moreover, when one looks inside one of those robots, one sees another CPS where the agents are components like software controllers, sensors and actuators. As mentioned before, the Robot Operating System (ROS) [15] presents a unified interface to robots where the subcomponents of even a single robot (e.g. sensors, actuators, controller) are represented as agents (a.k.a. nodes in ROS) that communicate with other agents using message passing. In a message sequence diagram such as the one in Sect. 1, agents are usually represented as vertical lines where the downward direction denotes passing of time. In ROSCoq, one specifies the collection of agents by an arbitrary type (say *Loc*) with decidable equality.

The next and perhaps most central concept in LoE is that of an event. In a message sequence diagram, these are points in the vertical lines usually denoting receipt or sending of messages by an agent. The slant arrows denote flight of messages. We model events by defining an abstract type *Event* which has a bunch of operations, such as:

eLoc : *Event* → *Loc*;   eMesg : *Event* → Message;
causedBy : *Event* → *Event* → Prop;   causalWf : well_founded causedBy

For any event *ev*, eLoc *ev* denotes the agent associated with the event. For receive-events, this is the agent who received the message. For send-events, this is the agent who sent the message. eMesg *ev* is the associated Message. The relation causedBy captures the causal ordering on events. causalWf formalizes the assumption that causal order is well-founded [10]. It allows one to prove properties by induction on causal order.

So far, our definition of an event is a straightforward translation (to Coq) of the corresponding Nuprl definition [19]. For CPSs, we need to associate more information with events. Perhaps the most important of those is a physical (as opposed to logical) notion of time when events happen. For example, the software agent needs to send appropriate messages to the hardware agent before the robot collides with something. One needs to reason about the time needed for activities like sensing, message delivery, computation to happen. So, we add the following operation:

eTime : *Event* → QTime;
globalCausal : ∀ (*e1 e2* : *Event*), causedBy *e1 e2* → (eTime *e1* < eTime *e2*)

For any event *ev*, eTime *ev* denotes the physical (Newtonian) time when it happened. QTime is a type of non-negative rational numbers where 0 represents the time when the system was started[3]. Note that this value of time is merely used for reasoning about the behavior of the system. As we will see later, a software controller cannot use it. This is because there is no way to know the exact time when an event, e.g. receipt of a message happened. For that, one would have to exactly synchronize clocks, which is impossible in general. One could implement provably correct approximate time-synchronization in our framework and then let the software controllers access an approximately correct value of the time when an event happened.

3.1   Message Delivery

Our message delivery semantics formalizes the publish-subscribe pattern used in ROS. The Coq definition of each agent includes a list of topics to which the agent publishes and a list of topics to which it subscribes. In ROSCoq, one can specify the collection of topics by an arbitrary type (say *Topic*) with decidable equality. In addition, one has to specify a function (say *topicType*) of type *Topic* → Type, that specifies the payload type of each topic. The type of messages can then be defined as follows:

---

[3] Section 4.1 explains the difference between Time and QTime.

Definition Message : Type := $\{tp : Topic \times (topicType\ tp)\} \times$ Header .

A message is essentially a 3-tuple containing a topic (tp), a payload corresponding to tp, and a header. Currently the header of a message only has one field (delay) which can be used by software agents (Sect. 4.2) to request the shim to send a message at a later time. For our running example, we use 2 topics:

Definition topicType (t : Topic)

: Type := match t with
| VELOCITY ⇒ Polar2D Q
| TARGETPOS ⇒ Cart2D Q
end.

The topic TARGETPOS is used by the external agent (see Figure in Sect. 1) to send the cartesian coordinates of the target position (relative to the robot's current position) to the software agent. The topic VELOCITY is used by the software agent to send the linear and angular velocity commands to the robot hardware agent. One also provides a ternary relation to specify acceptable message delivery times between any two locations. Finally, we assume that message delivery is ordered.

4   Semantics of Agents

For verification of distributed systems [19], one assumes that each agent is running a functional reactive program. These programs indirectly specify a property about the sequence of events at an agent, namely it should be one that the program could generate. In a CPS, there usually are agents which represent hardware components (along with their ROS drivers) like sensors and actuators. Often, informal specifications about their behavior is available, not their internal design or firmware. Hence, one needs to axiomatically specify the observable behavior (sequence of events) of such devices. Moreover, these hardware devices often depend on (e.g. sensors), or influence (e.g. actuators) the evolution of some physical quantities. A specification of their behavior needs to talk about how the associated physical quantities evolve over time. Hence, an appropriate way of specifying the behavior of agents is to specify them as a *relation* between the physical model (how the physical quantities evolve over time) and the sequence of messages associated with the agent. As we will see below, for hardware agents one can directly specify that relation. For software agents, this relation would only have a vacuous dependence on the physical model and can be specified indirectly as a Coq program, which is often more succinct (Sect. 4.2).

4.1   Hardware Agents

For our running example, the physical model is specified by the type (iCreate) (Sect. 2). The type $\mathbb{N} \rightarrow$ (option *Event*) can be used to represent a possibly finite sequence of events. So, the specification of the behavior of the hardware agent is a relation (HwAgent) of the following type: (iCreate) $\rightarrow$ ($\mathbb{N} \rightarrow$ option *Event*) $\rightarrow$ Prop.

We will first explain it pictorially
and then show the actual Coq definition. iCreate is primarily an actuation
device and this relation asserts how
the angular and linear velocity (see
omega and linVel in the definition of
(iCreate) above) of the robot changes
in response to the received messages.
It is quite close to informal manuals[4]. The iCreate hardware driver only

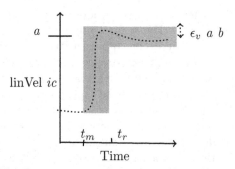

receives messages on the topic VELOCITY. It reacts to such messages by adjusting the speed of the two motors (one on each side) so the robot's linear and
angular velocities are close to the requested values. The figure above illustrates
how the linear velocity of an iCreate is supposed to change in response to a message requesting a linear velocity a and angular velocity b. t_m denotes the time
when the message was received. HwAgent asserts that there must exist a time
t_r by which the linear velocity of the robot becomes close to a. The parameter
reacTime is an upper bound on t_r - t_m. ϵ_v and ϵ_ω are parameters modeling
the actuation accuracy. After time t_r, the linear velocity of the robot remains
at least ϵ_v a b close to the a. Similarly (not shown in the figure) the angular
velocity remains at least ϵ_ω a b close to b. The only assumption we make about
the ϵ is that ϵ_v 0 0 and ϵ_ω 0 0 are 0, i.e. the robot complies exactly (after
a certain amount of reaction time) when asked to both stop turning and stop
moving forward. In particular, we don't assume that ϵ_ω a 0 is 0. When the robot
is asked to move forward at a m/s and not turn at all, it may actually turn a
bit. For a robot to move in a perfect straight line, one will likely have to make
sure that the size of the two wheels are *exactly* the same, the two motors are
getting *exactly* the same amount of current and so on. A consequence of our
realistic assumptions is that some integrals become more complicated to reason
about. For example, unlike in the case for perfect linear motion, the angle in the
derivative of position cannot be treated as a constant. Here is the definition of
HwAgent (mentioned above), which captures the above pictorial intuition:

Definition HwAgent (*ic*: iCreate) (*evs* : nat $\rightarrow$ option *Event*): **Prop** :=
onlyRecvEvts *evs* $\wedge$ $\forall$ *t*: QTime,
   let (*lastCmd*, t_m) := latestVelPayloadAndTime *evs* *t* **in**
   let *a* : Q := rad (*lastCmd*) **in**
   let *b* : Q := θ (*lastCmd*) **in** $\exists$ t_r : QTime, $(t_m \leq t_r \leq t_m$ + *reacTime*)
    $\wedge$ ($\forall$ *t'* : QTime, $(t_m \leq t' \leq t_r)$
       $\rightarrow$ (Min ({linVel *ic*} t_m) $(a - \epsilon_v$ a $b)$
          $\leq$ {linVel *ic*} *t'* $\leq$ Max ({linVel *ic*} t_m) $(a + \epsilon_v$ a $b)))$
    $\wedge$ ($\forall$ *t'* : QTime, $(t_r \leq t' \leq t) \rightarrow$ |{linVel *ic*} *t'* - a | $\leq \epsilon_v$ a b)

The function latestVelPayloadAndTime searches the sequence of events *evs*
to find the latest message of VELOCITY topic received before time *t*. We assume

---

[4] http://pharos.ece.utexas.edu/wiki/index.php/Writing_A_Simple_Node_that_
Moves_the_iRobot_Create_Robot\#Talking_on_Topic_cmd_vel.

that there is a positive lower-bound on the time-gap between any two events at an agent. Hence, one only needs to search a finite prefix of the sequence *evs* to find that event. It returns the payload of that message and the time the corresponding event occurred (e.g. t_m in the figure). If there is no such message, it returns the default payload with 0 as the velocities and 0 as the event time. The last conjunct above captures the part in the above figure after time t_r. The 2nd last conjunct captures the part before t_r where the motors are transitioning to the new velocity. There are 2 more conjuncts not shown above. These express similar properties about angular velocity (omega *ic*), *b* and ϵ_ω *a b*.

Because the semantics of a hardware agent is specified as a relation between the physical model and the sequence of events at the agent, it is equally easy to express the specification of sensing devices where typically the physical model determines the sequence of events. [17] contains a specification of a proximity sensor. Although the external agent in our running example is not exactly a hardware agent, we specify its behavior axiomatically. We assume that there is only one event in its sequence, and that event is a send event on the topic TARGETPOS.

We conclude this subsection with an explanation of some differences between Time and QTime. The former represents non-negative real numbers, while the latter represents non-negative rational numbers. Clearly, there is an injection from QTime to Time. We have declared this injection as an implicit coercion, so one can use a QTime where a Time is expected. Because CoRN's theory of differential calculus is defined for functions over real numbers, we can directly use them for functions over real-valued time (i.e. TContR). However, QTime is often easier to use because comparison relations (equality, less than etc.) on rationals are decidable, unlike on real numbers. For example, if the time of events (eTime) were represented by Time, one could not implement the function latestVelPayloadAndTime mentioned above. Because members of TContR are *continuous* functions over real-valued time, they are totally defined merely by their value on rational numbers, i.e. QTime. For example, the specification above only bounds velocities of the robot at rational values of time. However, it is easy to derive the same bound for all other values of time.

4.2   Software Agents

As mentioned before in this section, the behavior of software agents can be specified indirectly by just specifying the Coq program that is being run by the agent. Following [23], these programs are message handlers which can also maintain some state of an arbitrary type. A software agent which maintains state of type S can be specified as a Coq function of the following type: $S \rightarrow$ Message $\rightarrow (S \times$ list Message$)$. Given the current state and a received message, a message handler computes the next state and a list of messages that are supposed to be sent in response. We provide a ROSJava[5] shim which handles sending and receiving of messages for the above pure functions. It communicates with the Coq

---

[5] http://wiki.ros.org/rosjava.

toplevel (`coqtop`) to invoke message handlers. It also converts received messages to Coq format and converts the messages to be sent to ROSJava format. However, the state is entirely maintained in Coq, i.e. never converted to Java. We define SwSemantics, a specification of how the shim is supposed to respond to received messages. For a message handler, it defines the behavior of the corresponding software (Sw) agent, essentially as a property of the sequence of (send/receive) events at the agent. As mentioned before, the semantic relation of a software agent has vacuous dependence on the physical model. A software agent does not directly depend on or influence physical quantities of a CPS. It does so indirectly by communicating with hardware agents like the one described in the previous subsection.

The definition of our shim (in Java) and SwSemantics (in Coq) can be found at [17]. Here, we explain some key aspects. SwSemantics asserts that whenever a message m is received (at a receive event), the message handler (in Coq) is used to compute the list of messages (say l) that are supposed to be sent. There will be $|l|$ send events which correspond to sending these messages one by one. Let s_i be the time the ith of these send events happened. Recall from Sect. 3.1 that the header of messages contain a delay field. Let d_i be the value of the delay field of the ith message in the list l. Let t be the time the computation of l finished. The shim ensures that s_0 is close to $t + d_0$. It also ensures that $\forall$ appropriate i, s_{i+1} is close to $d_{i+1} + s_i$. The current state is updated with the new state computed along with l. SwSemantics also asserts that there are no other send events; each send event must be associated to a receive event in the manner explained above.

In our running example, the software agent receives a target position for the robot on the topic TARGETPOS and sends velocity-control messages to the motor on the topic VELOCITY. Recall from Sect. 3.1 that the payload type for the former and latter topics are Cart2D Q and Polar2D Q respectively. So the software agent reacts to data of the former type and produces data of the latter type. Its program can be represented as the following pure function:

Definition robotPureProgam (*target* : Cart2D Q) : list (Q × Polar2D Q) :=
   let *polarTarget* : Polar2D ℝ := Cart2Polar *target* **in**
   let *rotDuration* : ℝ := | θ *polarTarget* | / *rotspeed* **in**
   let *translDuration* : ℝ := (rad *polarTarget*) / *speed* **in**
   [(0,{| rad:= 0 ; θ := (*polarθSign target*) * *rotspeed* |})
     ; (tapprox *rotDuration delRes delEps* , {| rad := 0 ; θ := 0 |})
     ; (*delay* , {| rad := *speed* ; θ := 0 |})
     ; (tapprox *translDuration delRes delEps* , {| rad := 0 ; θ := 0 |})].

For any type A and a and b of type A, {| X := a ; Y := b |} denotes a member of type Cart2D A. {| rad := a ; θ := b |} denotes a member of type Polar2D A. The program produces a list of 4 pairs, each corresponding to one of the 4 messages that the software agent will send to the hardware agent (see Figure in Sect. 1). One can compose this program with ROSCoq utility functions to lift it to a message handler. The first component of each pair denotes the delay field of the message's header. The second component corresponds to the payload of the message. Recall that a payload {| rad := a ; θ := b |} represents a request to set the linear velocity to a and the angular velocity to b. In the above program, *polarTarget* represents the result of converting the input to polar coordinates. Note that even though the coordinates in *target* are rational numbers, those in *polarTarget* are real numbers. For example, converting $\{\mid X := 1 ; Y := 1 \mid\}$ corresponds to irrational polar coordinates: $\{\mid \text{rad} := \sqrt{2} ; \theta := \frac{\pi}{4} \mid\}$.

The program first instructs the robot to turn so that its orientation is close to θ *polarTarget*, i.e., in the direction of *target*. *speed*, *rotspeed*, *delEps*, *delRes*, *delay* are parameters in the program. These are arbitrary positive rationals. The robot will turn at speed *rotspeed*, but it can turn in either direction : clockwise or counter-clockwise, depending on the sign of θ *polarTarget*. However, the problem of finding the sign of a real number is undecidable in general. Fortunately, because θ *polarTarget* was computed from rational coordinates (*target*), one can look at *target* and indirectly determine what the sign of θ *polarTarget* would be. We have proved that *polarθSign* does exactly that. *polarθSign target* will either be $+1$ or -1 (in the first message).

The 2nd message which requests the robot to stop (turning) should ideally be sent after a delay of *rotDuration*, which is defined above as $\mid \theta$ *polarTarget* $\mid$ / *rotspeed*. It is a real number because θ *polarTarget* is so. However our Java shim currently uses *java.util.Timer*[6] and only accepts delay requests of integral number of milliseconds[7]. It might be possible to use a better hardware/shim to accept delay requests of integral number of microseconds or nanoseconds. So we use an arbitrary parameter *delRes* which is a positive integer such that $\frac{1}{delRes}$ represents the resolution of delay provided by the shim. For our current shim, one will instantiate *delRes* to 1000. So, we should approximate *rotDuration* by the closest rational number whose denominator is *delRes*. In classical mathematics, one can prove that there "exists" a rational number that is at most $\frac{1}{2*delRes}$ away from *rotDuration*. However, *finding* such a rational number is an undecidable problem in general. Fortunately, one can arbitrarily minimize the suboptimality in this step. We have proved that for *any* positive rational number *delEps*, tapprox *rotDuration delRes delEps* is a rational number whose denominator is *delEps* and is at most $\frac{1+2*delEps}{2*delRes}$ (denoted as R2QPrec) away from *rotDuration*.

tapprox was easy to define because CoRN's real numbers of the type CR are functional programs which approximate the represented real number to arbitrarily close rationals. Note however that cartesian to polar conversion was exact. Unlike with floating points, most operations on real numbers are exact : field operations, trigonometric functions, integrals, etc. One does not have to worry about errors at each step of computation. Instead, one can directly specify the desired accuracy for the final discrete result. So we think constructive reals are ideal for robotic programs written with the intent of rigorous verification.

The 3rd message sets linear velocity to *speed*. The parameter *delay* is the delay value for this message. We assume *delay* is large enough (w.r.t other parameters, e.g. *reacTime*) to ensure that motors have fully stopped in response to the previous message by the time this message arrives. The final message asks the robot to stop. Again it is sent after a nearly right amount of delay.

---

[6] http://docs.oracle.com/javase/7/docs/api/java/util/Timer.html.
[7] Also, recall that the shim is only required to approximately respect these requests.

5 Reasoning About the System

After all the agents of a CPS have been specified, one can reason about how the overall system will behave. For local reasoning about an agent's behavior, one can use natural induction on its sequence of events. For global behavior, one can use induction on the causal order of messages. In our running example, we are interested in how close the robot will be to the target position. In the previous section we already saw that there might be some error in approximating real numbers to certain rational values of time which the shim can deal with. However, that was just one source among myriad other sources of errors : messages cannot be delivered at exact times, actuation devices are not perfect (infinitely precise), and so on. Our goal is to derive parametric bounds on how far the robot can be from the ideal target position, in terms of bounds on the above error sources. Below, we consider an arbitrary run of the system. The external agent asks the robot to go to some position *target* of type Cart2D Q. *ic* of type (iCreate) denotes how the physical quantities evolve in this run.

The first step is to prove that the sequence of events at each agent looks exactly like the figure in Sect. 1. In particular, we prove that there are exactly 4 events at the hardware agent and those events correspond to the four messages (in order) computed by the program described in the previous section. These proofs mostly involve using the properties about topic-subscriptions and guarantees provided by the messaging layer, such as guaranteed and ordered delivery of messages. We use mt0, ..., mt3 to refer respectively to the time of occurrences of those 4 events. The remaining proofs mostly involve using the specification of motor to characterize the position, orientation and velocities of the robot at each of those times. The specification of motor provides us bounds on velocities. We then use CoRN's lemmas on differential calculus, such as the FTC to characterize the position and orientation of the robots.

Because we assumed that ϵ_v 0 0 and ϵ_ω 0 0 are both 0 (Sect. 4.1), and initial velocities (linear and angular) are 0, the velocities will remain *exactly* 0 till mt0. So the position and orientation of the robot at mt0 is exactly the same as that in the initial state (initial conditions are specified in the definition of iCreate). At mt0, the robot receives a message requesting a non-zero angular velocity (say w). Recall that w is either *rotspeed* or - *rotspeed*. Between mt0 and mt1, the robot turns towards the target position. At mt1, it receives a message to stop, however it might take some time to totally stop. At mt2, it receives a message to start moving forward. Ideally, it should be oriented towards the target position by mt2. However, that might not be the case because of several sources of imperfections. The following lemma characterizes how imperfect the orientation of the robot can be at mt2.

Definition $ideal\theta$: $\mathbb{R}$:= θ (Cart2Polar *target*).
Definition $\theta ErrTurn$: $\mathbb{R}$:= *rotspeed* $*$ (timeErr + 2 $*$ *reacTime*)
    + (ϵ_ω 0 w) $*$ (timeErr + *reacTime*) +((ϵ_ω 0 w) / *rotspeed*) $*$ | $ideal\theta$ | .
Lemma ThetaAtEV2 : | {theta *ic*} mt2 - $ideal\theta$ | $\leq \theta ErrTurn$.

Because Sin and Cos are periodic, there are several ways to define θ (Cart2Polar *target*). Our choice enabled us to prove that it is in the range $[-\pi, \pi]$. It minimizes the turning that the robot has to do (vs., e.g. $[0, 2\pi]$). It also enables us to replace $|\ ideal\theta\ |$ by π in the above upper bound. The three terms in the definition $\theta ErrTurn$ correspond to errors that are respectively proportional, independent and inversely proportional to *rotspeed*. Recall (Sect. 4.1) that *reacTime* is the upper bound on the amount of time the robot takes to attain the requested velocity. timeErr has been proved to be an upper bound on the error of the value mt1 – mt0 w.r.t. its ideal value. It is an addition of terms bounding the inaccuracy of sending times, variance of message delivery times, R2QPrec which bounds the inaccuracy introduced when we approximated the ideal real-valued delay by a rational value (Sect. 4.2). A higher value of *rotspeed* means that the error in the duration of turn will lead to more errors in the final angle. A lower value of *rotspeed* increases the duration for which the robot has to turn, thus accumulating more errors because of imperfect actuation of angular velocity (as modeled by ϵ_w 0 w). However, if ϵ_w 0 w is directly proportional to the absolute value of w, which is *rotspeed*, the division in the last term will cancel out. In such a case, a lower value of *rotspeed* will always result in a lower upper bound on turn error.

From mt2 to mt3, the robot will move towards the target. To analyse this motion, we find it convenient to rotate the axis so that the new X axis (X') points towards the target position (shown as a circle in the RHS figure). We characterize the derivative of the position of the robot in the rotated coordinate frame:

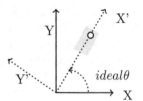

Definition Y'Deriv : TContR :=
  (linVel *ic*) $*$ (FSin (theta *ic* – FConst *ideal*θ)).
Definition X'Deriv := (linVel *ic*) $*$ (FCos (theta *ic* – FConst *ideal*θ)).

The advantage of rotating axes is that unlike theta *ic* which could be any value (depending on *target*), (theta *ic* – FConst *ideal*θ) is a small angle: $\forall\ t$, mt2 $\leq t \leq$ mt3 $\rightarrow$ |{theta *ic*} t – *ideal*θ| $\leq \theta ErrTrans + \theta ErrTurn$.

As explained above, $\theta ErrTurn$ is a bound on the error (w.r.t. *ideal*θ), at mt2. Even though the robot is supposed to move straight towards the goal from time mt2 to mt3, it might turn a little bit due to imperfect actuation (as discussed in Sect. 4.1). We proved that $\theta ErrTrans$ is an upper bound on that. $\theta ErrTrans$ is proportional to mt3 – mt2, which in turn is proportional to the distance of the target position from the origin. For the remaining proofs, we assume $\theta ErrTrans + \theta ErrTurn \leq \frac{\pi}{2}$, which is a reasonable assumption unless the target position is too far away or the actuation is very imprecise. In other words, we are assuming that there cannot be more than a difference of 90 degrees between the direction the robot thinks it is going and the actual direction. For robots that are supposed to move for prolonged periods of time, one usually needs a localization mechanism such as a GPS and/or a compass. In the future, we plan to consider such closed-loop setups by adding another hardware agent in our CPS which will periodically send much more accurate estimates of the robot's position and orientation as messages to the software agent.

Using the above assumption, it is easy to bound the derivatives of the robot's position in the rotated coordinate frame. For example, between times mt2 and mt3, the value of (FSin (theta *ic* − FConst *idealθ*)) will be bounded above by the constant Sin ($\theta ErrTrans + \theta ErrTurn$). We prove that at mt3, the robot will be inside a rectangle which is aligned to the rotated axes. In the above figure, such a rectangle is shown in gray. Recall that mt3 is the final event in the system, where the robot receives a message requesting it to stop. The following defines the upper bound we proved on the distance of the X' aligned sides of the rectangle from the X' axis. In other words, it is a bound on the Y' coordinate of the robot in the rotated coordinate frame. Ideally, this value should be zero.

Definition ErrY': $\mathbb{R}$:= (ϵ_v 0 w) * (*reacTime* + Ev01TimeGapUB)
+ (Sin ($\theta ErrTrans + \theta ErrTurn$)) * (| *target* | + *speed*∗timeErr
                                        + Ev23TimeGapUB * (ϵ_v *speed* 0)) .

The first line of the above definition corresponds to the error accrued in the position while turning (between mt0 and a little after mt1 when turning totally stops). The second and third lines denotes the error accrued after mt2 when the robot moves towards the target position. Similarly, we proved bounds on the distance of the Y' aligned sides from the Y' axis (see [17]).

We also considered the case of a *hypothetical* train traveling back and forth repeatedly between two stations. This CPS has 3 hardware agents : a proximity sensor at each end of the train and a motor at the base for 1D motion. The software controller uses messages generated by the proximity sensor to reverse the direction of motion when it comes close to an endpoint. We proved that it will never collide with an endpoint [17]. We haven't physically implemented it.

5.1   Experiments

Using our shim, we were able to use the Coq program in Sect. 4.2 to actually drive an iCreate robot to the position requested by a human via a GUI. While a detailed estimation of parameters in the model of hardware, message delivery, etc. is beyond the scope of this paper, we did some

Target		Actual		Video link
X	Y	X	Y	
−1	1	−1.06	0.94	vid1
−1	−1	−1.02	−0.99	vid2
1	1	1.05	0.94	vid3

experiments to make sure that the robot is in the right ball park. The table above shows some measurements (in meters) from the experiments.

6   Related Work

Hybrid automata [2] is one of the earliest formalisms to simultaneously model and reason about both the cyber (usually discrete dynamics) and physical (usually continuous dynamics) components of a CPS. Several tools have been developed for approximate reachability analysis, especially for certain sub-classes of hybrid automata (see [1] for a survey). However, hybrid automata provide little structure to implement complicated CPSs in a modular way. Also for CPSs with several communicating agents, it is rather non-trivial to come up with a hybrid automata model which accounts for all possible interactions in such distributed systems. In ROSCoq, we independently specify the agents of a distributed CPS and explicitly reason about all possible interactions.

The KeYmaera [14] tool takes a step towards more structural descriptions of CPSs. It has a non-deterministic imperative language to specify hybrid programs. It also comes with a dynamic-logic style proof theory for reasoning about such programs [13]. Unlike ROSCoq, the semantics of KeYmaera's programming language pretends that one can exactly compare two real numbers, which is impossible in general. When one uses floating point numbers to implement such programs, the numerical errors can add up and cause the system to violate the formally proven properties [11]. In contrast, the use of constructive reals forces us to explicitly account for inexactness of certain operations (like comparison) on real numbers and hence there is one less potential cause of runtime errors. In [21], they consider 2D dynamics similar to ours. They don't consider the possibility of the robot turning a little when asked to go straight. Finally, the semantics of their system assumes that all the robots are executing a synchronized control loop. Our asynchronous message passing based model is more realistic for distributed robotic systems.

Unlike the above tools, our focus is on correct-by-construction, i.e. we intend to prove properties of the actual software controller and not a simplified model of it. Some tools [16,18] automatically synthesize robot-controllers from high-level LTL specifications. However, these fully automatic approaches do not yet scale up to complicated robotic systems. Also, the specifications of these controllers are at a very high level (they discretize the continuous space) and do not yet account for imperfections in sensing, actuation, message delivery, etc.

Unlike the above formalisms, in ROSCoq one uses Coq's rich programming language to specify their hybrid programs and its powerful higher order logic to succinctly express the desirable properties. Coq's dependent types allow one to reuse code and enforce modularity by building interfaces that seamlessly specify not only the supported operations but also the logical properties of the operations. To trust our proofs, one only needs to trust Coq's type checker. Typical reasoning in KeYmaera relies on quantifier elimination procedures implemented using Mathematica, a huge tool with several known inconsistencies [6]. Our framework did not require adding any axiom to Coq's core theory. This is mainly because CoRN's real numbers are actual computable functions of Coq, unlike the axiomatic theory of reals in Coq's standard library. Interactive theorem provers have been previously used to verify certain aspects of hybrid systems [7,12]. Like ROSCoq, [7] uses constructive reals and accounts for numerical errors. However, it only supports reasoning about hybrid systems expressed as hybrid automata. [12] is primarily concerned about checking absence of collisions in completely specified flight trajectories.

7   Conclusion and Future Work

We presented a Coq framework for developing certified robotic systems. It extends the LoE framework to enable holistic reasoning about both the cyber and physical aspects of such systems. We showed that the constructive theory of analysis originally developed by Bishop and later made efficient in the CoRN project is powerful enough for reasoning about physical aspects of practical systems. Constructivity is a significant advantage here because the real numbers in this theory have a well defined computational meaning, which we exploit in our robot programs. Our reasoning is very detailed as it considers physical imperfections and computational imperfections while computing with real numbers.

We plan to use our framework to certify more complicated systems involving collaboration between several robots [5]. Also, we plan to develop tactics to automate as much of the reasoning as possible. We thank Jean-Baptiste Jeannin, Mark Bickford, Vincent Rahli, David Bindel and Gunjan Aggarwal for helpful discussions, Bas Spitters and Robbert Krebbers for help with CoRN, and Liran Gazit for providing the robot used in the experiments.

References

1. Alur, R.: Formal verification of hybrid systems. In: EMSOFT, pp. 273–278. IEEE (2011)
2. Alur, R., Courcoubetis, C., Henzinger, T.A., Ho, P.-H.: Hybrid automata: An algorithmic approach to the specification and verification of hybrid systems. In: Grossman, R.L., Nerode, A., Ravn, A.P., Rischel, H. (eds.) HS 1993. LNCS, vol. 736, pp. 209–229. Springer, Heidelberg (1993)
3. Bickford, M., Constable, R.L., Eaton, R., Guaspari, D., Rahli V.: Introduction to EventML (2012). www.nuprl.org/software/eventml/IntroductionToEventML.pdf
4. Bishop, E., Bridges, D.: Constructive Analysis, p. 490. Springer Science and Business Media, New york (1985)
5. Dogar, M., Knepper, R.A., Spielberg, A., Choi, C., Christensen, H.I., Rus, D.: Towards coordinated precision assembly with robot teams. In: ISER (2014)
6. Duráan, A.J., Péerez, M., Varona, J.L.: the misfortunes of a trio of mathematicians using computer algebra systems. Can we trust in them? In: AMS Notices 61.10, p. 1249, November 1 2014
7. Geuvers, H., Koprowski, A., Synek, D., van der Weegen, E.: Automated machine-checked hybrid system safety proofs. In: Kaufmann, M., Paulson, L.C. (eds.) ITP 2010. LNCS, vol. 6172, pp. 259–274. Springer, Heidelberg (2010)
8. Krebbers, R., Spitters, B.: Type classes for efficient exact real arithmetic in Coq. In: LMCS 9.1, February 14 2013
9. Lamport, L.: Buridan's principle. In: Foundations of Physics 42.8, pp. 1056–1066, August 1 2012
10. Lamport, L.: Time, clocks, and the ordering of events in a distributed system. Commun. ACM **21**(7), 558–565 (1978)
11. Mitsch, S., Platzer, A.: ModelPlex: Verified runtime validation of verified cyber-physical system models. In: Bonakdarpour, B., Smolka, S.A. (eds.) RV 2014. LNCS, vol. 8734, pp. 199–214. Springer, Heidelberg (2014)

12. Narkawicz, A., Munoz, C.A.: Formal verification of con ict detection algorithms for arbitrary trajectories. In: Reliable Computing, this issue (2012)
13. Platzer, A.: Logics of dynamical systems. In: LICS 2012, pp. 13–24 (2012)
14. Platzer, A., Quesel, J.-D.: KeYmaera: A hybrid theorem prover for hybrid systems (system description). In: AR, pp. 171–178. Springer (2008)
15. Quigley, M., Conley, K., Gerkey, B., Faust, J., Foote, T., Leibs, J., Wheeler, R., Ng, A.Y.: ROS: an open-source robot operating system. In: ICRA Workshop on Open Source Software. vol. 3, p. 5 (2009)
16. Raman, V. Kress-Gazit, H.: Synthesis for multi-robot controllers with inter- leaved motion. In: ICRA, pp. 4316–4321, May 2014
17. ROSCoq online reference. http://www.cs.cornell.edu/~aa755/ROSCoq
18. Sarid, S., Xu, B., Kress-Gazit, H.: Guaranteeing high-level behaviors while exploring partially known maps. In: RSS, p. 377, Sydney July 2012
19. Schiper, N., Rahli, V., Renesse, R.V., Bickford, M., Constable, R.L.: Developing correctly replicated databases using formal tools. In: DSN, pp. 395–406. IEEE (2014)
20. Spitters, B., Van Der Weegen, E.: Type classes for mathematics in type theory. MSCS **21**(4), 795–825 (2011)
21. Mitsch, S., Ghorbal, K., Platzer, A.: On provably safe obstacle avoidance for autonomous robotic ground vehicles. In: RSS (2013)
22. Talcott, C.: Cyber-physical systems and events. In: Wirsing, M., Banâtre, J.-P., Hölzl, M., Rauschmayer, A. (eds.) Soft-Ware Intensive Systems. LNCS, vol. 5380, pp. 101–115. Springer, Heidelberg (2008)
23. Wilcox, J.R., Woos, D., Panchekha, P., Tatlock, Z., Wang, X., Ernst, M.D., Anderson, T.: Verdi: a framework for implementing and formally verifying distributed systems. In: PLDI, ACM (2015)

Asynchronous Processing of Coq Documents: From the Kernel up to the User Interface

Bruno Barras, Carst Tankink, and Enrico Tassi[⊠]

Inria, Paris, France
{bruno.barras,carst.tankink,enrico.tassi}@inria.fr

Abstract. The work described in this paper improves the reactivity of the Coq system by completely redesigning the way it processes a formal document. By subdividing such work into independent tasks the system can give precedence to the ones of immediate interest for the user and postpone the others. On the user side, a modern interface based on the PIDE middleware aggregates and presents in a consistent way the output of the prover. Finally postponed tasks are processed exploiting modern, parallel, hardware to offer better scalability.

1 Introduction

In recent years Interactive Theorem Provers (ITPs) have been successfully used to give an ultimate degree of reliability to both complex software systems, like the L4 micro kernel [7] and the CompCert C compiler [8], and mathematical theorems, like the Odd Order Theorem [5]. These large formalization projects have pushed interactive provers to their limits, making their deficiencies apparent. The one we deal with in this work is *reactivity*: how long it takes for the system to react to a user change and give her feedback on her point of interest. For example if one takes the full proof of the Odd Order Theorem, makes a change in the first file and asks the Coq prover for any kind of feedback on the last file she has to wait approximately two hours before receiving any feedback.

To find a solution to this problem it is important to understand how formal documents are edited by the user and checked by the prover. Historically ITPs have come with a simple text based Read Eval Print Loop (REPL) interface: the user inputs a command, the system runs it and prints a report. It is up to the user to stock the right sequence of commands, called script, in a file. The natural evolution of REPL user interfaces adds a very basic form of script management on top of a text editor: the user writes his commands inside a text buffer and tells the User Interface (UI) to send them one by one to the same REPL interface. This design is so cheap, in terms of coding work required on the prover side, and so generic that its best incarnation, Proof General [2], has served as the reference UI for many provers, Coq included, for over a decade.

The simplicity of REPL comes at a price: commands must be executed in a linear way, one after the other. For example the prover must tell the UI if

C. Urban and X. Zhang (Eds.): ITP 2015, LNCS 9236, pp. 51–66, 2015.
DOI: 10.1007/978-3-319-22102-1_4

a command fails or succeeds before the following command can be sent to the system. Under such constraint to achieve better reactivity one needs to speed up the execution of each and every command composing the formal document. Today one would probably do that by taking advantage of modern, parallel hardware. Unfortunately it is very hard to take an existing system coded in a imperative style and parallelize its code at the fine grained level of single commands. Even more if the programming language it is written in does not provide light weight execution threads. Both conditions apply to Coq.

If we drop the constraint imposed by the REPL, a different, complementary, way to achieve better reactivity becomes an option: process the formal document *out-of-order*,[1] giving precedence to the parts the user is really interested in and postpone the others. In this view, even if commands do not execute faster, the system needs to execute fewer of them in order to give feedback to the user. In addition to that, when the parts the document is split in happen to be independent, the system can even process them in parallel.

Three ingredients are crucial to process a formal document out-of-order.

First, the UI must not impose to the prover to run commands linearly. A solution to this problem has been studied by Wenzel in [13] for the Isabelle system. His approach consists in replacing the REPL with an *asynchronous interaction loop*: each command is marked with a unique identifier and each report generated by the prover carries the identifier of the command to which it applies. The user interface sends the whole document to the prover and uses these unique identifiers to present the prover outputs coherently to the user. The asynchronous interaction loop developed by Wenzel is part of a generic middleware called PIDE (for Prover IDE) that we extend with a Coq specific back end.

Second, the prover must be able to perform a *static analysis of the document* in order to organize it into coarse grained tasks and take scheduling decisions driven by the user's point of interest. Typically a task is composed of many consecutive commands. In the case of Coq a task corresponds to the sequence of tactics, proof commands, that builds an entire proof. In other words the text between the `Proof` and `Qed` keywords.

Third, the system must feature an *execution model* that allows the reordering of tasks. In particular we model tasks as pure computations and we analyze the role their output plays in the checking of the document. Finally we implement the machinery required in order to execute them in parallel by using the coarse grained concurrency provided by operating system processes.

In this work we completely redesigned from the ground up the way Coq processes a formal document in order to obtain the three ingredients above. The result is a more reactive system that also performs better at checking documents in batch mode. Benchmarks show that the obtained system is ten times more reactive when processing the full proof of the Odd Order Theorem and that it scales better when used on parallel hardware. In particular fully checking such proof on a twelve core machine is now four times faster than before. Finally,

---

[1] In analogy with the "out-of-order execution" paradigm used in most high-performance microprocessors.

by plugging Coq in the PIDE middleware we get a modern user interface based on the jEdit editor that follows the "spell checker paradigm" made popular by tools like Eclipse or Visual Studio: the user freely edits the document and the prover constantly annotates it with its reports, like underlining in red problematic sentences. The work of Wenzel [11–14] on the Isabelle system has laid the foundations for our design, and has been a great inspiration for this work. The design and implementation work spanned over three years and the results are part of version 8.5 of the Coq system. All authors were supported by the Paral-ITP ANR-11-INSE-001 project.

The paper is organized as follows. Section 2 describes the main data structure used by the prover in order to statically analyze the document and represent the resulting tasks. Section 3 describes how asynchronous, parallel, computations are modeled in the logical kernel of the system and how they are implemented in the OCaml programming language. Section 4 describes how the document is represented on the UI side, how data is aggregated and presented to the user. Section 5 presents a reactivity benchmark of the redesigned system on the full proof of the Odd Order Theorem. Section 6 concludes.

2   Processing the Formal Document Out-of-Order

Unlike REPL in the asynchronously interaction model promoted by PIDE [13] the prover is made aware of the whole document and it is expected to process it giving precedence to the portion of the text the user is looking at. To do so the system must identify the parts of the document that are not relevant to the user and postpone their processing. The most favorable case is when large portions of the document are completely independent from each other, and hence one has complete freedom on the order in which they must be processed. In the specific case of Coq, *opaque proofs* have this property. Opaque proofs are the part of the document where the user builds a proof evidence (a proof term in Coq's terminology) by writing a sequence of tactics and that ends with the Qed keyword (lines four to seven in Fig. 1). The generated proof term is said to be opaque because it is verified by the kernel of the system and stored on disk, but the term is never used, only its corresponding statement (its type) is. The user can ask the system to print such term or to translate it to OCaml code, but from the viewpoint of the logic of Coq (type theory) the system commits not to use the proof term while checking other proofs.[2]

The notion of proof opacity was introduced a long ago in Coq version 5.10.5 (released in May 1994) and is crucial for us: given that the proof term is not used, we can postpone the processing of the piece of the document that builds it as much as we want. All we need is its type, that is the statement that is spelled out explicitly by the user. Proofs are also lengthy and usually the time spent in processing them dominates the overall time needed in order to check the entire document. For example in the case of the Odd Order Theorem proofs amount

---

[2] In the Curry-Howard correspondence Coq builds upon lemmas and definitions are the same: a term of a given type. An opaque lemma is a definition one cannot unfold.

```
1 (* global *) Definition decidable (P : Prop) := P \/ ~ P.
2
3 (* branch *) Theorem dec_False : decidable False.
4 (* tactic *) Proof.
5 (* tactic *) unfold decidable, not.
6 (* tactic *) auto.
7 (* merge  *) Qed.
```

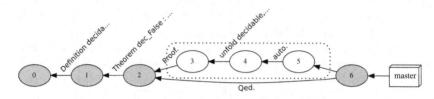

Fig. 1. Coq document and its internal representation

to 60 % of the non-blanks (3.175 KB over 5.262 KB) and Coq spends 90 % of its time on them. This means that, by properly scheduling the processing of opaque proofs, we can increase the reactivity of the system of a factor of ten. In addition to that, the independence of each proof from the others makes it possible to process many proofs at the same time, in parallel, giving a pretty good occasion to exploit modern parallel hardware.

In the light of that, it is crucial to build an internal representation for the document that makes it easy to identify opaque proofs. Prior to this work Coq had no internal representation of the document at all. To implement the Undo facility, Coq has a notion of *system state* that can can be saved and restored on demand. But the system used to keep no trace of how a particular system state was obtained; which sequence of commands results in a particular system state.

The most natural data structure for keeping track of how a system state is obtained is a Directed Acyclic Graph (DAG), where nodes are system states and edges are labeled with commands. The general idea is that in order to obtain a system state from another one, one has to process all the commands on the edges linking the two states. The software component in charge of building and main-taining such data structure is called State Transaction Machine (STM), where the word transaction is chosen to stress that commands need to be executed atomically: there is no representation for a partially executed command.

2.1   The STM and the Static Analysis of the Document

The goal of the static analysis of the document is to build a DAG in which proofs are explicitly separated from the rest of the document. We first parse each sen-tence obtaining the abstract syntax tree of the corresponding command in order to classify it. Each command belongs to only one of the following categories: commands that start a proof (*branch*), commands that end a proof (*merge*), proof commands (*tactic*), and commands that have a global effect (*global*). The

DAG is built in the very same way one builds the history of a software project in a version control system. One starts with an initial state and a default branch, called master, and proceeds by adding new commits for each command. A commit is a new node and an edge pointing to the previous node. Global commands add the commit on the main branch; branching commands start a new branch; tactic commands add a commit to the current branch; merging commands close the current branch by merging it into master. If we apply these simple rules to the document in Fig. 1 we obtain a DAG where the commands composing the proof of dec_False have been isolated. Each node is equipped with a unique identifier, here positive numbers. The edge from the state six to state five has no label, it plays the role of making the end of the proof easily accessible but has not to be "followed" when generating state six.

Indeed to compute state six, or anything that follows it, Coq starts from the initial state, zero, and then executes the labeled transactions until it reaches state six, namely "Definition...", "Theorem..." and Qed. The nodes in gray (1, 2 and 6) are the ones whose corresponding system state has been computed. The nodes and transactions in the dotted region compose the proof that is processed asynchronously. As a consequence of that the implementation of the merging command has to be able to accommodate the situation where the proof term has not been produced yet. This is the subject of Sect. 3. For now the only relevant characteristic of the asynchronous processing of opaque proofs is that such process is a pure computation. The result of a pure computation does depend only on its input. It does not matter when it is run nor in which environment and it has no visible global effect. If we are not immediately interested in its result we can execute it lazily, in parallel or even remotely via the network.

Tactic commands are not allowed to appear outside a proof; on the contrary global commands can, an in practice do, appear in the middle of proofs, as in Fig. 2. Mixing tactics with global commands is not a recommend style for finished proof scripts, but it is a extremely common practice while one writes the script and it must be supported by the system. The current semantics of Coq documents makes the effect of global commands appearing in the middle of proofs persist outside proof blocks (hence the very similar documents in Figs. 1 and 2 have a different semantics). This is what naturally happens if the system has a single, global, imperative state that is updated by the execution of commands. In our scenario the commands belonging to opaque proofs may even be executed remotely, hence a global side effect is lost. To preserve the current semantics of documents, when a global command is found in the middle of a proof its corresponding transaction is placed in the DAG twice: once in the proof branch at its original position, and another time on the master branch. This duplication is truly necessary: only the transaction belonging to the master branch will retain its global effect; the one in the proof branch, begin part of a pure computation, will have a local effect only. In other words the effect of the "Hint..." transaction from state 3 to 4 is limited to states 4 and 5.

Branches identifying opaque proofs also define compartments from which errors do not escape. If a proof contains an error one can replace such proof by

```
1 (* global *) Definition decidable (P : Prop) := P \/ ~ P.
2
3 (* branch *) Theorem dec_False : decidable False.
4 (* tactic *) Proof.
5 (* global *) Hint Extern 1 => unfold decidable, not.
6 (* tactic *) auto.
7 (* merge  *) Qed.
```

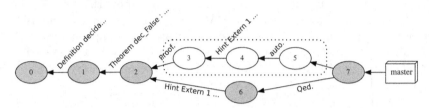

Fig. 2. Coq document and its internal representation

another one, possibly a correct one, without impacting the rest of the document. As long as the statement of an opaque proof does not change, altering the proof does not require re-checking what follows it in the document.

3 Modelling Asynchronously Computed Proofs

The logical environment of a prover contains the statements of the theorems which proof has already been checked. In some provers, like Coq, it also contains the proof evidence, a proof term. A theorem enters the logical environment only if its proof evidence has been checked. In our scenario the proof evidences may be produced asynchronously, hence this condition has to be relaxed, allowing a theorem to be part of the environment before its proof is checked. Wenzel introduced the very general concept of proof promise [11] in the core logic of Isabelle for this precise purpose.

3.1 Proof Promises in Coq

Given that only opaque proofs are processed asynchronously we can avoid introducing in the theory of Coq the generic notion of a sub-term asynchronously produced. The special status of opaque proofs spares us to change the syntax of the terms and lets us just act on the logical environment and the Well Founded (WF) judgement. The relevant rules for the WF judgements, in the presentation style of [1], are the following ones. We took the freedom to omit some details, like T being a well formed type, that play no role in the current discussion.

$$\frac{E \vdash \mathrm{WF} \quad E \vdash b : T \quad d \text{ fresh in } E}{E \cup (\mathbf{definition}\ d : T := b) \vdash \mathrm{WF}} \qquad \frac{E \vdash \mathrm{WF} \quad E \vdash b : T \quad d \text{ fresh in } E}{E \cup (\mathbf{opaque}\ d : T \mid b) \vdash \mathrm{WF}}$$

Note that the proof evidence b for opaque proofs is checked but not stored in the environment as the body of non-opaque definitions. After an opaque proof enters the environment, it shall behave exactly like an axiom.

$$\frac{E \vdash \text{WF} \quad d \text{ fresh in } E}{E \cup (\textbf{axiom } d : T) \vdash \text{WF}}$$

We rephrase the WF judgement as the combination of two new judgements: Asynchronous and Synchronous Well Founded (AWF and SWF respectively). The former is used while the user interacts with the system. The latter complements the former when the complete checking of the document is required.

$$\frac{E \vdash \text{AWF} \quad E \vdash b : T \quad d \text{ fresh in } E}{E \cup (\textbf{definition } d : T := b) \vdash \text{AWF}} \qquad \frac{E \vdash \text{AWF} \quad d \text{ fresh in } E}{E \cup (\textbf{opaque } d : T \mid [f]_E) \vdash \text{AWF}}$$

$$\frac{E \vdash \text{SWF}}{E \cup (\textbf{definition } d : T := b) \vdash \text{SWF}} \qquad \frac{E \vdash \text{SWF} \quad b = \text{run } f \text{ in } E \quad E \vdash b : T}{E \cup (\textbf{opaque } d : T \mid [f]_E) \vdash \text{SWF}}$$

$$\frac{E \vdash \text{AWF} \quad E \vdash \text{SWF}}{E \vdash \text{WF}}$$

The notation $[f]_E$ represents a pure computation in the environment E that eventually produces a proof evidence b. The implementation of the two new judgements amounts to replace the type of opaque proofs (`term`) with a function type (`environment -> term`) and impose that such computation is pure. In practice the commitment of Coq to not use the proof terms of theorems terminated with `Qed` makes it possible to run these computations when it is more profitable. In our implementation this typically happens in the background.

As anticipated the Coq system is coded using the imperative features provided by the OCaml language quite pervasively. As a result we cannot simply rely on the regular function type (`environment -> term`) to model pure computations, since the code producing proof terms is not necessarily pure. Luckily the `Undo` facility lets one backup the state of the system and restore it, and we can use this feature to craft our own, heavyweight, type of pure computations. A pure computation c_0 pairs a function f with the system state it should run in s_0 (that includes the logical environment). When c_0 is executed, the current system state s_c is saved, then s_0 in installed and f is run. Finally the resulting value v and resulting state s_1 are paired in the resulting computation c_1, and the original state s_c is restored. We need to save s_1 only if the computation c_1 needs to be chained with additional impure code. A computation producing a proof is the result of chaining few impure functions with a final call to the kernel's type checker that is a *purely functional* piece of code. Hence the final system state is always discarded, only the associated value (a proof term) is retained.

The changes described in this section enlarged the size of the trusted code base of Coq by less than 300 lines (circa 2 % of its previous size).

3.2   Parallelism in OCaml

The OCaml runtime has no support for shared memory and parallel thread execution, but provides inter process communication facilities like sockets and data marshaling. Hence the most natural way to exploit parallel hardware is to split the work among different worker processes.

While this approach imposes a clean message passing discipline, it may clash with the way the state of the system is represented and stocked. Coq's global state is unstructured and fragmented: each software module can hold some private data and must register a pair of functions to take a snapshot and restore an old backup to a central facility implementing the Undo command. Getting a snapshot of the system state is hence possible, but marshalling it and sending it to a worker process is still troublesome. In particular the system state can contain a lot of unnecessary data, or worse data that cannot be sent trough a channel (like a file descriptor) or even data one does not want to send to a precise worker, like the description of the tasks he is not supposed to perform.

Our solution to this problem is to extrude from the system state the unwanted data, put a unique key in place of it and finally associate via a separate table the data to the keys. When a system state is sent to a worker process the keys it contains lose validity, hence preventing the worker process to access the unwanted data.

The only remaining problem is that, while when a system state becomes unreferenced it is collected by the OCaml runtime, the key-value table still holds references to a part of that system state. Of course we want the OCaml runtime to also collect such data. This can be achieved by making the table "key-weak", in the sense that the references it holds to its keys do not prevent the garbage collector from collecting them and that when this happens the corresponding data has also to be collected. Even if the OCaml runtime does not provide such notion of weak table natively, one can easily code an equivalent finalization mechanism known as *ephemeron* (key-value) pairs [6] by attaching to the keys a custom finalization function (Fig. 3).

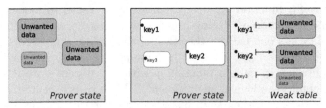

Fig. 3. Reorganization of the prover state

3.3   The Asynchronous Task Queue and the Quick
Compilation chain

The STM maintains a structured representation of the document the prover is processing, identifies independent tasks and delegates them to worker processes. Here we focus on the interaction between Coq and the pool of worker processes.

The main kind of tasks identified by the STM is the one of opaque proofs we discussed before, but two other tasks are also supported. The first one is queries: commands having no effect but for producing a human readable output. The other one is tactics terminating a goal applied to a set of goals via the "par:" goal selector.[3] In all cases the programming API for handling a pool of worker processes is the `AsyncTaskQueue`, depicted Fig. 4. Such data structure, generic enough to accommodate the three aforementioned kind of tasks, provides a priority queue in which tasks of the same kind can be enqueued. Tasks will eventually be picked up by a worker manager (a cooperative thread in charge of a specific worker process), turned into requests and sent to the corresponding worker.

```
module type Task = sig
  type task
  type request
  type response
  val request_of_task : [ 'Fresh | 'Old ] -> task -> request option
  val use_response : task -> response -> [ 'Stay | 'Reset ]
  val perform : request -> response
end
module MakeQueue(T : Task) : sig
  val init : max_workers:int -> unit
  val priority_rel : (T.task -> T.task -> int) -> unit
  val enqueue_task : T.task -> cancel_switch:bool ref -> unit
  val dump : unit -> request list
end
module MakeWorker(T : Task) : sig
  val main_loop : unit -> unit
end
```

Fig. 4. Programming API for asynchronous task queues

By providing a description of the task, i.e. a module implementing the `Task` interface, one can obtain at once the corresponding queue and the main function for the corresponding workers. While the `task` datatype can, in principle, be anything, `request` and `response` must be marshalable since they will be exchanged between the master and a worker processes. `request_of_task` translates a `task` into a `request` for a given worker. If the worker is an 'Old one, the representation of the request can be lightweight, since the worker can save the context in which the tasks take place and reuse it (for example the proof context is the same for each goal to which the `par:` goal selector applies). Also a `task`, while waiting in the queue, can become obsolete, hence the option type. The worker manager calls `use_response` when a response is received, and decides whether the worker has to be reset or not. The `perform` function is the code that is run in the worker. Note that when a task is enqueued a handle to cancel the execution if it becomes outdated is given. The STM will flip that bit eventually, and the worker manager will stop the corresponding worker.

---

[3] The "par:" goal selector is a new feature of Coq 8.5 made possible by this work.

This abstraction is not only valuable because it hides to the programmer all the communication code (socket, marshalling, error handling) for the three kinds of tasks. But also because it enables a new form of separate *quick compilation* (batch document checking) that splits such job into two steps. A first and quick one that essentially amounts at checking everything but the opaque proofs and generates an (incomplete) compiled file, and a second one that completes such file. An incomplete file, extension .vio, can already be used: as we pointed out before the logic of Coq commits not to use opaque proofs, no matter if they belong to the current file or to another one. Incomplete .vio files can be completed into the usual .vo files later on, by resuming the list of requests also saved in the .vio file.

The trick is to set the number of workers to zero when the queue is initialized. This means that no task is processed at all when the document is checked. One can then dump the contents of the queue in terms of a list of requests (a marshalable piece of data) and stock it in the .vio files. Such lists represents all the tasks still to be performed in order to check the opaque proofs of the document. The performances of the quick compilation chain are assessed in Sect. 5.

4   The User Side

A prover can execute commands out-of-order only if the user can interact with it asynchronously. We need an interface that systematically gives to the prover the document the user is working on and declares where the user is looking at to help the system take scheduling decisions. This way the UI also frees the user from the burden of explicitly sending a portion of its text buffer to the prover. Finally this interface needs to be able to interpret the reports the prover generates in no predictable order, and display them to the user.

An UI like this one makes the user experience in editing a formal document way closer to the one he has when editing a regular document using a mainstream word processor: he freely edits the text while the system performs "spell checking" in the background, highlighting in red misspelled words. In our case it will mark in red illegal proof steps. Contextual information, like the current goal being proved, is displayed according to the position of the cursor. It is also natural to have the UI aggregate diverse kinds of feedbacks on the same piece of text. The emblematic example is the full names of identifiers that are displayed as an hyperlink pointing to the place where the identifier is bound.

4.1   PIDE and Its Document Model

To integrate user interfaces with asynchronous provers in a uniform way, Wenzel developed the PIDE middleware [13]. This middleware consists of a number of API functions in a frontend written in the Scala programming language, that have an effect on a prover backend. The front-, and backend together maintain a shared data structured, PIDE's notion of a *document*. Figure 5 depicts how the front and backend collaborate on the document, and its various incarnations. The

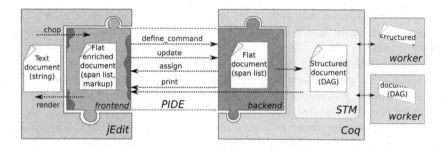

Fig. 5. PIDE sitting between the prover and the UI (Color figure online)

portions of the figure in dark red represent parts of PIDE that were implemented to integrate Coq into the PIDE middleware.

This document has a very simple, prover-independent, structure: it is a flat list of spans. A span is the smallest portion of text in which the document is chopped by a prover specific piece of code. In the case of Coq it corresponds to a full command. The frontend exposes an *update* method on the document, which allows an interface to notify the prover of changes. The second functionality of the document on the frontend is the ability to query the document for any markup attached to the command spans. This markup is provided by the prover, and is the basis for the rich feedback we describe in Sect. 4.4.

The simple structure of the document keeps the frontend "dumb", which makes it easy to integrate a new prover in PIDE: for Coq, this only required implementing a parser that can recognize, chop, command spans (around 120 lines of Scala code) and a way to start the prover and exchange data with it (10 more lines). The prover side of this communication is further described in Sect. 4.2, which also describes the representation of the document text in the PIDE backend for Coq.

The way an interface uses the PIDE document model is by updating it with new text, and then reacting to new data coming from the prover, transforming it in an appropriate way. For example, Coq can report the abstract syntax tree (AST) of each command in the proof document, which the interface can use to provide syntax highlighting which is not based on error-prone regular expressions. Beyond the regular interface services, the information can also be used to support novel interactions with a proof document. In the case of an AST, for example, the tree can be used to support refactoring operations robustly.

4.2   Pidetop

Pidetop is the toplevel loop that runs on the Coq side and translates PIDE's protocol messages into instructions for the STM. Pidetop maintains an internal representation of the shared document. When its update function is invoked, the toplevel updates the internal document representation, inserting and deleting command phrases where appropriate. The commands that are considered as new by pidetop are not only those that have changed but also those that follow: because the client is dumb, it does not know what the impact of a change is. Instead it relies on the statical analysis the STM performs to determine this.

Finally PIDE instructs the STM to start checking the updated document. This instruction can include a *perspective*: the set of spans that frontend is currently showing to the user. This perspective is used by the STM to prioritize processing certain portions of the document.

Processing the document produces messages to update PIDE of new information. During the computation, PIDE can interrupt the STM at any moment, to update the document following the user's changes.

4.3   Metadata Collection

The information Coq needs to report to the UI can be classified according to its nature: persistent or volatile. *Persistent information* is part of the system state, and can be immediately accessed. For this kind of information we follow the model of asynchronous prints introduced by Wenzel in [14]. When a system state is ready it is reported to (any number of) asynchronous printers, which process the state, reporting some information from it as marked up messages for the frontend. Some of these printers, like those reporting the goal state at each span, are always executed, while others can be printed on-demand. An example of the latter are Coq's queries, which are scheduled for execution whenever the user requests it. *Volatile information* can only be reported during the execution of a command, since it is not retained in the system state. An example is the resolution of an identifier resulting in an hyper link: the relation between the input text and the resulting term is not stored, hence localized reports concerning sub terms can only be generated while processing a command. Volatile information can also be produced optionally and on demand by processing commands a second time, for example in response to the user request of more details about a term. From the UI perspective both kind of reports are asynchronously generated and handled homogeneously.

4.4   Rich Feedback

A user interface building on PIDE's frontend can use the marked up metadata produced by the prover to provide a much richer editing experience than was previously possible: the accessible information is no longer restricted to the output of the last executed command. For example jEdit, the reference frontend based on PIDE, uses the spell checker idiom to report problems in the entire proof document: instead of refusing to proceed beyond a line containing an error the interface underlines all the offending parts in red, and the user can inspect each of them.

This modeless interaction model also allows the interface to take different directions than the write-and-execute one imposed by REPL based UI. For example it is now easy to decorrelate the user's point of observation and the place where she is editing the document. A paradigmatic example concerns the

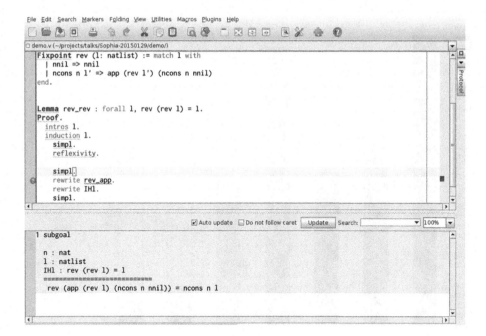

Fig. 6. jEdit using PIDE

linguistic construct of postfix bookkeeping of variables and hypotheses (the **as** intro pattern in Coq). The user is encouraged to decorate commands like the one that starts an induction with the names of the new variables and hypotheses that are available in each subgoal. This decoration is usually done in a sloppy way: when the user starts to work on the third subgoal she goes back to the induction command in order to write the intro pattern part concerning the third goal. In the REPL model she would lose the display of the goal she is actually working on because the only goal displayed would be the one immediately before the induction command. By using a PIDE based interface to Coq the user can now leave the observation point on the third goal and see how it changes *while* she is editing the intro pattern (note the "Do not follow caret" check box in Fig. 6).

These new ways of interaction are the first experiments with the PIDE model for Coq, and we believe that even more drastically different interaction idioms can be provided, for example allowing the user to write more structured proofs, where the cases of a proof are gradually refined, possibly in a different order than the one in which they appear in the finished document.

The PIDE/jEdit based interface for Coq 8.5 is distributed at the following URL: http://coqpide.bitbucket.org/.

5   Assessment of the Quick Compilation Chain

As described in Sect. 3.3 our design enables Coq to postpone the processing of opaque proofs even when used as a batch compiler. To have a picture of the

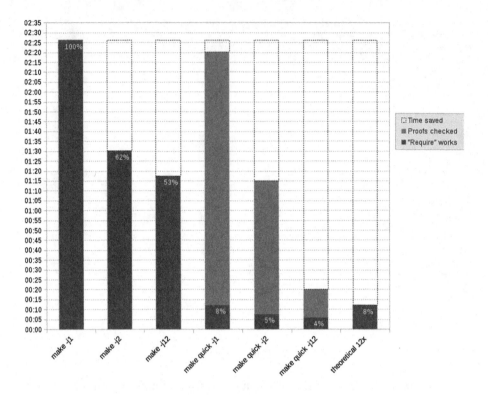

Fig. 7. Benchmarks

improvement in the reactivity of the system we consider the time that one needs
to wait (we call that latency) in order to be able to use (`Require` in Coq's
terminology) the last file of the proof of the Odd Order Theorem after a change
in the first file of such proof. With Coq 8.4 there is nothing one can do but
to wait for the full compilation of the whole formalization. This takes a bit less
than two and a half hours on a fairly recent Xeon 2.3 GHz machine using a single
core (first column). Using two cores, and compiling at most two files at a time,
one can cut that time to ninety minutes (second column). Unfortunately batch
compilation has to honour the dependency among files and follow a topological
order, hence by adding extra 10 cores one can cut only twelve minutes (third
column). Hence, in Coq 8.4 the best latency is of the order of one hour.

Thanks to the static analysis of the document described in Sect. 2.1, the
work of checking definitions and statements (in blue) can be separated by the
checking of opaque proofs (in red), and one can use a (partially) compiled Coq
file even if its opaque proofs are not checked yet. In this scenario, the same
hardware gives a latency of twelve minutes using a single core (fourth column),
eight minutes using two cores (fifth column) and seven minutes using twelve cores
(sixth column). After that time one can use the entire formalization. When one
then decides to complete the compilation he can exploit the fact that each proof

is independent to obtain a good degree of parallelism. For example checking all proofs using twelve cores requires thirteen extra minutes after the initial analysis, for a total of twenty minutes. This is 166 % of the theoretical optimum one could get (seventh column). Still with 12 cores the latency is only six minutes, on par with the theoretical optimum for 24 cores (Fig. 7).

The reader may notice that the quick compilation chain using 1 core (4th column) is slightly faster than the standard compilation chain. This phenomenon concerns only the largest and most complicated files of the development. To process these files Coq requires a few gigabytes of memory and it stresses the OCaml garbage collector quite a bit (where it spends more than 20 % of the time). The separation of the two compilation phases passes trough marshalling to (a fast) disk and un-marshaling to an empty process space. This operation trades (non blocking, since the size of files fits the memory of the computer) disk I/O for a more compact and less fragmented memory layout that makes the OCaml runtime slightly faster.

6   Concluding Remarks and Future Directions

This paper describes the redesign Coq underwent in order to provide a better user experience, especially when used to edit large formal developments. The system is now able to better exploit parallel hardware when used in batch mode, and is more reactive when used interactively. In particular it can now talk with user interfaces that use the PIDE middleware, among which we find the one of Isabelle [9] that is based on jEdit and the Coqoon [3,4] one based on Eclipse.

There are many ways this work can be improved. The most interesting paths seem to be the following ones.

First, one could make the prover generate, on demand, more metadata for the user consumption. A typical example is the type of sub expressions to ease the reading of the document. Another example is the precise list of theorems or local assumptions used by automatic tactics like `auto`. This extra metadata could be at the base of assisted refactoring functionalities the UI could provide.

Another interesting direction is to refine the static analysis of the document to split proofs into smaller, independent, parts. In a complementary way one could make such structure easier to detect in the proof languages supported by the system. The extra structure could be used in at least two ways. First, one could give more accurate feedback on broken proofs. Today the system stops at the first error it encounters, but a more precise structure would enable the system to backup by considering the error confined to the sub-proof in which it occurs. Second, one could increase the degree of parallelism we can exploit.

Finally one could take full profit of the PIDE middleware by adapting to Coq interesting user interfaces based on it. For example clide [10] builds on top of PIDE and provides a web based, collaborative, user interface for the Isabelle prover. The cost of adapting it to work with Coq seems now affordable.

References

1. Asperti, A., Ricciotti, W., Sacerdoti Coen, C., Tassi, E.: A compact kernel for the calculus of inductive constructions. Sadhana **34**(1), 71–144 (2009)
2. Aspinall, D.: Proof general: a generic tool for proof development. In: Graf, S. (ed.) TACAS 2000. LNCS, vol. 1785, pp. 38–42. Springer, Heidelberg (2000)
3. Bengtson, J., Mehnert, H., Faithfull, A.: Coqoon: eclipse plugin providing a feature-complete development environment for Coq (2015). Homepage: https://itu.dk/research/tomeso/coqoon/
4. Faithfull, A., Tankink, C., Bengtson, J.: Coqoon - an IDE for interactive proof development in Coq. Submitted to CAV 2015
5. Gonthier, G., et al.: A machine-checked proof of the odd order theorem. In: Blazy, S., Paulin-Mohring, C., Pichardie, D. (eds.) ITP 2013. LNCS, vol. 7998, pp. 163–179. Springer, Heidelberg (2013)
6. Hayes, B.: Ephemerons: a new finalization mechanism. In: Proceedings of OOPSLA, pp. 176–183. ACM (1997)
7. Klein, G., Andronick, J., Elphinstone, K., Heiser, G., Cock, D., Derrin, P., Elkaduwe, D., Engelhardt, K., Kolanski, R., Norrish, M., Sewell, T., Tuch, H., Winwood, S.: seL4: formal verification of an operating-system kernel. Commun. ACM **53**(6), 107–115 (2010)
8. Leroy, X.: Formal verification of a realistic compiler. Commun. ACM **52**(7), 107–115 (2009)
9. Nipkow, T., Wenzel, M., Paulson, L.C.: Isabelle/HOL: A Proof Assistant for Higher-order Logic. Springer, Heidelberg (2002)
10. Ring, M., Lüth, C.: Collaborative interactive theorem proving with clide. In: Klein, G., Gamboa, R. (eds.) ITP 2014. LNCS, vol. 8558, pp. 467–482. Springer, Heidelberg (2014)
11. Wenzel, M.: Parallel proof checking in Isabelle/Isar. In: Proceedings of PLMMS, pp. 13–21 (2009)
12. Wenzel, M.: Isabelle as document-oriented proof assistant. In: Davenport, J.H., Farmer, W.M., Urban, J., Rabe, F. (eds.) Calculemus/MKM 2011. LNCS, vol. 6824, pp. 244–259. Springer, Heidelberg (2011)
13. Wenzel, M.: READ-EVAL-PRINT in parallel and asynchronous proof-checking. In: Proceedings of UITP. EPTCS, vol. 118, pp. 57–71 (2013)
14. Wenzel, M.: Asynchronous user interaction and tool integration in Isabelle/PIDE. In: Klein, G., Gamboa, R. (eds.) ITP 2014. LNCS, vol. 8558, pp. 515–530. Springer, Heidelberg (2014)

A Concrete Memory Model for CompCert

Frédéric Besson[1], Sandrine Blazy[2], and Pierre Wilke[2(✉)]

[1] Inria, Rennes, France
[2] Université Rennes 1 - IRISA, Rennes, France
pierre.wilke@irisa.fr

Abstract. Semantics preserving compilation of low-level C programs is challenging because their semantics is implementation defined according to the C standard. This paper presents the proof of an enhanced and more concrete memory model for the CompCert C compiler which assigns a definite meaning to more C programs. In our new formally verified memory model, pointers are still abstract but are nonetheless mapped to concrete 32-bit integers. Hence, the memory is finite and it is possible to reason about the binary encoding of pointers. We prove that the existing memory model is an abstraction of our more concrete model thus validating formally the soundness of CompCert's abstract semantics of pointers. We also show how to adapt the front-end of CompCert thus demonstrating that it should be feasible to port the whole compiler to our novel memory model.

1   Introduction

Formal verification of programs is usually performed at source level. Yet, a theorem about the source code of a safety critical software is not sufficient. Eventually, what we really value is a guarantee about the run-time behaviour of the compiled program running on a physical machine. The CompCert compiler [17] fills this verification gap: its semantics preservation theorem ensures that when the source program has a defined semantics, program invariants proved at source level still hold for the compiled code. For the C language the rules governing so called *undefined behaviours* are subtle and the absence of undefined behaviours is in general undecidable. As a corollary, for a given C program, it is undecidable whether the semantic preservation applies or not.

To alleviate the problem, the semantics of CompCert C is executable and it is therefore possible to check that a given program execution has a defined semantics. Jourdan *et al.* [12] propose a more comprehensive and ambitious approach: they formalise and verify a precise C static analyser for CompCert capable of ruling out undefined behaviours for a wide range of programs. Yet, these approaches are, by essence, limited by the formal semantics of CompCert C: programs exhibiting undefined behaviours cannot benefit from any semantic preservation guarantee. This is unfortunate as real programs do have behaviours

This work was partially supported by the French ANR-14-CE28-0014 AnaStaSec.

C. Urban and X. Zhang (Eds.): ITP 2015, LNCS 9236, pp. 67–83, 2015.
DOI: 10.1007/978-3-319-22102-1_5

that are undefined according to the formal semantics of CompCert C[1]. This can be a programming mistake but sometimes this is a design feature. In the past, serious security flaws have been introduced by optimising compilers aggressively exploiting the latitude provided by undefined behaviours [6,22]. The existing workaround is not satisfactory and consists in disabling optimisations known to be triggered by undefined behaviours.

In previous work [3], we proposed a more concrete and defined semantics for CompCert C able to give a semantics to low-level C idioms. This semantics relies on symbolic expressions stored in memory that are normalised into genuine values when needed by the semantics. It handles low-level C idioms that exploit the concrete encoding of pointers (e.g. alignment constraints) or access partially undefined data structures (e.g. bit-fields). Such properties cannot be reasoned about using the existing CompCert memory model [18,19].

The memory model of CompCert consists of two parts: standard operations on memory (e.g. alloc, store) that are used in the semantics of the languages of CompCert and their properties (that are required to prove the semantic preservation of the compiler), together with generic transformations operating over memory. Indeed, certain passes of the compiler perform non-trivial transformations on memory allocations and accesses: for instance, in the front-end, C local variables initially mapped to individually-allocated memory blocks are later on mapped to sub-blocks of a single stack-allocated activation record. Proving the semantic preservation of these transformations requires extensive reasoning over memory states, using memory invariants relating memory states during program execution, that are also defined in the memory model.

In this paper, we extend the memory model of CompCert with symbolic expressions [3] and tackle the challenge of porting memory transformations and CompCert's proofs to our memory model with symbolic expressions. The complete Coq development is available online [1]. Among others, a difficulty is that we drop the implicit assumption of an infinite memory. This has the consequence that allocation can fail. Hence, the compiler has to ensure that the compiled program is using less memory than the source program.

This paper presents a milestone towards a CompCert compiler adapted with our semantics; it makes the following contributions.

- We present a formal verification of our memory model within CompCert.
- We prove that the existing memory model of CompCert is an abstraction of our model thus validating the soundness of the existing semantics.
- We extend the notion of memory injection, the main generic notion of memory transformation.
- We adapt the proof of CompCert's front-end passes, from CompCert C until Cminor, thus demonstrating the feasibility of our endeavour.

The paper is organised as follows. Section 2 recalls the main features of the existing CompCert memory model and our proposed extension. Section 3 explains how to adapt the operations of the existing CompCert memory model

---

[1] The official C standard is in general even stricter.

to comply with the new requirements of our memory model. Section 4 shows that the existing memory model is, in a provable way, an abstraction of our new memory model. Section 5 presents our re-design of the notion of memory injection that is the cornerstone of compiler passes modifying the memory layout. Section 6 details the modifications for the proofs for the compiler front-end passes. Related work is presented in Sect. 7; Sect. 8 concludes.

2   A More Concrete Memory Model for CompCert

In previous work [3], we propose an enhanced memory model (with symbolic expressions) for CompCert. The model is implemented and evaluated over a representative set of C programs. We empirically verify, using the reference interpreter of CompCert, that our extension is sound with respect to the existing semantics and that it captures low-level C idioms out of reach of the existing memory model. This section first recalls the main features of the current CompCert memory model and then explains our extension to this memory model.

2.1   CompCert's Memory Model

Leroy *et al.* [18] give a thorough presentation of the existing memory model of CompCert, that is shared by all the languages of the compiler. We give a brief overview of its design in order to highlight the differences with our own model.

Abstract values used in the semantics of the CompCert languages (see [19]) are the disjoint union of 32-bit integers (written as int(i)), 32-bit floating-point numbers (written as float(f)), locations (written as ptr(l)), and the special value undef representing an arbitrary bit pattern, such as the value of an uninitialised variable. The abstract memory is viewed as a collection of separated blocks. A location l is a pair (b,i) where b is a block identifier (i.e. an abstract address) and i is an integer offset within this block. Pointer arithmetic modifies the offset part of a location, keeping its block identifier part unchanged. A pointer ptr(b,i) is valid for a memory M (written valid_pointer(M,b,i)) if the offset i is within the two bounds of the block b.

Abstract values are loaded from (resp. stored into) memory using the load (resp. store) memory operation. Memory chunks appear in these operations, to describe concisely the size, type and signedness of the value being stored. These operations return option types: we write $\emptyset$ for failure and $\lfloor x \rfloor$ for a successful return of a value x. The free operation may also fail (e.g. when the locations to be freed have been freed already). The memory operation alloc never fails, as the size of the memory is unbounded.

In the memory model, the byte-level, in-memory representation of integers and floats is exposed, while pointers are kept abstract [18]. The concrete memory is modelled as a map associating to each location a concrete value cv that is a byte-sized quantity describing the current content of a memory cell. It can be either a concrete 8-bit integer (written as bytev(b)) representing a part of an

```
struct {                          1   struct {  unsigned char __bf1;} bf;
  int a0 : 1; int a1 : 1;         2
} bf;                             3   int main(){
                                 4     bf.__bf1 = (bf.__bf1 & ~0x2U) |
int main() {                     5               ((unsigned int) 1 << 1U & 0x2U);
  bf.a1 = 1; return bf.a1;}      6     return (int) (bf.__bf1 << 30) >> 31;}
```

(a) Bitfield in C   (b) Bitfield in CompCert C

Fig. 1. Emulation of bitfields in CompCert

integer or a float, $\mathrm{ptrv}(l, i)$ to represent the i-th byte ($i \in \{1, 2, 3, 4\}$) of the location l, or **undefv** to model uninitialised memory.

2.2   Motivation for an Enhanced Memory Model

Our memory model with symbolic expressions [3] gives a precise semantics to low-level C idioms which cannot be modelled by the existing memory model. The reason is that those idioms either exploit the binary representation of pointers as integers or reason about partially uninitialised data. For instance, it is common for system calls, e.g. mmap or sbrk, to return -1 (instead of a pointer) to indicate that there is no memory available. Intuitively, -1 refers to the last memory address 0xFFFFFFFF and this cannot be a valid address because mmap returns pointers that are aligned – their trailing bits are necessarily 0s. Other examples are robust implementations of malloc: for the sake of checking the integrity of pointers, their trailing bits store a checksum. This is possible because those pointers are also aligned and therefore the trailing bits are necessarily 0s.

Another motivation is illustrated by the current handling of bitfields in CompCert: they are emulated in terms of bit-level operations by an elaboration pass preceding the formally verified front-end. Figure 1 gives an example of such a transformation. The program defines a bitfield bf such that a0 and a1 are 1 bit long. The main function sets the field a1 of bf to 1 and then returns this value. The expected semantics is therefore that the program returns 1. The transformed code (Fig. 1b) is not very readable but the gist of it is that field accesses are encoded using bitwise and shift operators. The transformation is correct and the target code generated by CompCert correctly returns 1. However, using the existing memory model, the semantics is undefined. Indeed, the program starts by reading the field __bf1 of the uninitialised structure bf. The value is therefore undef. Moreover, shift and bitwise operators are strict in undef and therefore return undef. As a result, the program returns undef. As we show in the next section, our semantics is able to model partially undefined values and therefore gives a semantics to bitfields. Even though this case could be easily solved by modifying the pre-processing step, C programmers might themselves write such low-level code with reads of undefined memory and expect it to behave correctly.

```
Record compat(cm, m) : Prop := {
  addr_space: ∀ b o, [o] ∈ bound(m, b) → 0 < [cm(b)+o] < Int.max_unsigned;
  overlap : ∀ b b' o o',
     b ≠ b' → [o] ∈ bound(m,b)  → [o'] ∈ bound(m, b') → [cm(b)+o] ≠ [cm(b')+o'];
  alignment: ∀ b,  cm(b)  & alignt(m, b)  = cm(b)  }.
```

Fig. 2. Compatibility relation betw. a memory mapping and an abstract memory

2.3   A Memory Model with Symbolic Expressions

To give a semantics to the previous idioms, a direct approach is to have a fully concrete memory model where a pointer is a genuine integer and the memory is an array of bytes. However, this model lacks an essential property of CompCert's semantics: determinism. For instance, with a fully concrete memory model, allocating a memory chunk returns a non-deterministic pointer – one of the many that does not overlap with an already allocated chunk. In CompCert, the allocation returns a block that is computed in a deterministic fashion. Determinism is instrumental for the simulation proofs of the compiler passes and its absence is a show stopper.

Our approach to increase the semantics coverage of CompCert consists in delaying the evaluation of expressions which, for the time being, may have a non-deterministic evaluation. In memory, we therefore consider side-effect free symbolic expressions of the following form: $sv ::= v \mid op_1^\tau \; sv \mid sv \; {}^\tau op_2{}^\tau \; sv$, where v is a value and op_1^τ (resp. ${}^\tau op_2^\tau$) is a unary (resp. binary) arithmetic operator.

Our memory model is still made of a collection of blocks with the difference that (1) each block possesses an explicit alignment, and (2) memory blocks contain symbolic byte expressions. The alignment of a block b (written alignt(m,b)) states that the address of b has a binary encoding such that its trailing alignt(m,b) bits are zeros. Memory loads and stores cannot be performed on symbolic expressions which therefore need to be normalised beforehand to a genuine pointer value. Another place where normalisation is needed is before a conditional jump to ensure determinism of the jump target. The normalisation gives a semantics to expressions in terms of a concrete mapping from blocks to addresses. Formally, we define in Fig. 2 a compatibility relation stating that:[2]

- a valid location is in the range [0x00000001; 0xFFFFFFFE] (see addr_space);
- valid locations from distinct blocks do not overlap (see overlap);
- a block is mapped to an address abiding to the alignment constraints.

The normalisation is such that normalise $m \; e \; \tau$ returns a value v of type τ if and only if the side-effect free expression e evaluates to v for every concrete mapping cm: $block \rightarrow \mathbb{B}_{32}$ of blocks to concrete 32 bits addresses which are compatible with the block-based memory m (written compat(cm,m)).

---

[2] The notation [i] denotes the machine integer i when interpreted as unsigned.

We define the evaluation of expressions as the function $\llbracket \cdot \rrbracket_{cm}$, parametrised by the concrete mapping cm. Pointers are turned into their concrete value, as dictated by cm. For example, the evaluation of $\text{ptr}(b, o)$ results in cm (b)$+o$. Then, symbolic operations are mapped to the corresponding value operations.

Consider the code of Fig. 1b. Unlike the existing semantics, operators are not strict in undef but construct symbolic expressions. Hence, in line 4, we store in bf._bf1 the symbolic expression e defined by (undef &~0x2U)|(1 << 1 U&0x2U) and therefore return the normalisation of the expression (e << 30) >> 31. The value of the expression is 1 whatever the value of undef and therefore the normalisation succeeds and returns, as expected, the value 1.

3   Proving the Operations of the Memory Model

CompCert's memory model exports an interface summarising all the properties of the memory operations necessary to prove the compiler passes. This section details how the properties and the proofs need to be adapted to accommodate for symbolic expressions. First, we have to refine our handling of undefined values. Second, we introduce an equivalence relation between symbolic expressions.

3.1   Precise Handling of Undefined Values

Symbolic expressions (as presented in Sect. 2.3) feature a unique undef token. This is a shortcoming that we have identified during the proof. With a single undef, we do not capture the fact that different occurrences of undef may represent the same unknown value, or different ones. For instance, consider two uninitialised char variables x and y. Expressions x - x and x - y both construct the symbolic expression undef - undef, which does not normalise. However we would like x - x to normalise to 0, since whatever the value stored in memory for x, say v, the result of $v - v$ should always be 0. To overcome this problem, each byte of a newly allocated memory chunk is initialised with a fresh undef value. After the allocation of block b, the value stored at location (b, o) is $\text{undef}(b, o)$. This value is fresh because block identifiers are never reused. Hence, x - x constructs the symbolic expression $\text{undef}(b, o)$ - $\text{undef}(b, o)$ for some b and o which obviously normalises to 0, because $\text{undef}(b, o)$ now represents a unique value rather than the set of all values. Our symbolic expressions do not use the existing CompCert's values but the following type sval ::= undef(l) | int(i) | float(f) | ptr(l).

3.2   Memory Allocation

CompCert's alloc operation always allocates a memory chunk of the requested size and returns a fresh block to the newly allocated memory (i.e. it models an infinite memory). In our model, memory consumption needs to be precisely monitored and a memory m is a dependent record which, in addition to a CompCert block-based memory, provides guarantees about the concrete memories compatible (see Fig. 2) with the CompCert block-based memory. Hence, our `alloc` operation is partial and returns $\emptyset$ if it fails to construct a memory object.

The first guarantee is that for every memory m there exists at least a concrete memory compatible with the abstract CompCert block-based memory.

Lemma mem_compat: $\forall$ m, $\exists$ cm, compat(cm, m).

To get this property, the `alloc` function runs a greedy algorithm constructing a compatible cm mapping. Given a memory m, `size_mem`(m) returns the size of the constructed memory (i.e. the first fresh address as computed by the allocation). The algorithm makes the pessimistic assumption that the allocated blocks are maximally aligned – for CompCert, this maximum is 3 bits (addresses are divisible by 2^3). It places the allocated block at the first concrete address that is free and compliant with a 3-bit alignment. Allocation fails if no such address can be found. The rationale for the pessimistic alignment assumption is discussed in Sect. 5: it is essential to ensure that certain memory re-arrangements are always feasible (i.e. would not exhaust memory).

Without the `mem_compat` property, a symbolic expression e could have a normalisation v before allocation and be undefined after allocation. The existence of a concrete compatible memory ensures that the normalisation of expressions is strictly more defined after allocation.

Lemma normalise_alloc : $\forall$ m lo hi m' b e τ v,
  alloc m lo hi = $\lfloor$(m', b)$\rfloor$ $\wedge$ normalise m e τ = v $\wedge$ v $\neq$ undef $\rightarrow$
  normalise m' e τ = v.

3.3 Good Variable Properties

In CompCert, the so-called *good variable properties* axiomatise the behaviour of the memory operations. For example, the property `load_store_same` states that, starting from memory m_1, if we store value v at location l with some chunk κ, then when we read at l with κ, we get back the value v. During a `store`, a symbolic expression is split into symbolic bytes using the function `extr(sv,i)` which extracts the ith byte of a symbolic expression `sv`. The reverse operation is the concatenation of a symbolic expression sv_1 with a symbolic expression sv_2 representing a byte. These operations can be defined as `extr(sv,i)` = (sv shr (8*i)) & 0xFF and concat(sv_1, sv_2) = sv_1 shl 8 + sv_2.

As a result, the axiom `load_store_same` needs to be generalised because the stored value v is not syntactically equal to the loaded value v_1 but is equal modulo normalisation (written $v_1 \equiv v_2$). This equivalence relation is a key insight for generalising the properties and the proofs of the memory model.

Axiom load_store_same: $\forall$ κ m_1 b ofs v m_2, store κ m_1 b ofs v = $\lfloor m_2 \rfloor$ $\rightarrow$
  $\exists$ v_1, load chunk m_2 b ofs = $\lfloor v_1 \rfloor$ $\wedge$ v_1 $\equiv$ Val.load_result κ v.

We have generalised and proved the axioms of the memory model using the same principle. The proof effort is non-negligible as the memory model exports more than 150 lemmas. Moreover, if the structure of the proofs is similar, our proofs are complicated by the fact that we reason modulo normalisation of expressions.

4   Cross-Validation of Memory Models

The semantics of the CompCert C language is part of the *trusted computing base* of the compiler. Any modelling error can be responsible for a buggy, though formally verified, compiler. To detect a glitch in the semantics, a first approach consists in running tests and verifying that the CompCert C interpreter computes the expected value. With this respect, the CompCert C semantics successfully run hundreds of random test programs generated by CSmith [23]. Another indirect but original approach consists in relating formally different semantics for the same language. For instance, when designing the CompCert C semantics, several equivalences between alternate semantics were proved to validate this semantics [4]. Our memory model is a new and interesting opportunity to apply this methodology and perform a cross-validation of the C operators which are the building blocks of the semantics.

Our memory model aims at giving a semantics to more operations (e.g. low-level pointer operations). However, when for instance a C binary operation $v_1 \; op \; v_2$ evaluates to a defined value v', using the existing memory model, the symbolic expression $v_1 \; op \; v_2$ should normalise to the same value v'. For pointer values, this property is given in Fig. 3, where `sem_binary_operation_expr` is our extension to symbolic expressions of the function `sem_binary_operation`.

During the proof, we have uncovered several issues. The first issue is that our normalisation only returns a location within the bounds of the block. This is not the case for CompCert C that allows, for instance, to increment a pointer with an arbitrary offset. If the resulting offset is outside the bounds, our normalisation returns `undef`. For us, this is a natural requirement because we only apply normalisation when valid pointers are needed (i.e. before a memory read or write). To cope with this discrepancy, we add in lemma `expr_binop_ok_ptr` the precondition `valid_pointer(m,b,o)`. Another issue was a mistake in our syntactic construction of symbolic expressions: a particular cast operator was mapped to the wrong syntactic constructor. After the easy fix, we found two interesting semantics discrepancies with the current semantics of CompCert C.

One issue is related to *weakly valid* pointers [16] which model a subtlety of C stating that *pointers one past the end of an array object* can be compared. As a result, in CompCert C, if (b, o) is a valid location, then $(b, o) < (b, o+1)$ always returns true. In our model, if $(b, o+1)$ wraps around (because of an integer overflow) it may return 0 and therefore the property does not hold. To avoid this corner situation, we state that a valid address of our model excludes `Int.max_unsigned` (see Fig. 2). This is sufficient to prevent the offset from wrapping around and to be compatible with the semantics of CompCert C.

The last issue is related to the comparison with the `null` pointer. In CompCert, this is the only pointer which is not represented by a location (b, i) but by the integer 0. The semantics therefore assumes that a genuine location can never be equal to the `null` pointer. In our semantics, a location (b, i) can evaluate to 0 in case of wrap around. This is a glitch in the CompCert semantics that is illustrated by the code snippet of Fig. 4. This program initialises a pointer `p` to the address of the variable `i`. In the loop, `p` is incremented until it equals 0 in which case the loop exits and the program returns 1. With this program, the executable semantics of CompCert C returns 0 because `p==0` is always false whatever the value of `p`. However, when running the compiled program, the pointer is a mere integer, the integer eventually overflows; wraps around and becomes 0. Hence, the test holds and the program returns 1. We might wonder how the CompCert semantic preservation can hold in the presence of such a contradiction. Actually, the pointers are kept logical all the way through to the assembly level, and the comparison with the `null` pointer is treated the same during all the compilation process, thus even the assembly program in CompCert returns 0. The inconsistency only appears when the assembly program is compiled into binary and run on a physical machine.

The fix consists in defining the semantics of the comparison with the `null` pointer only if the pointer is *weakly valid*. This causes the program to have undefined semantics at the C level as soon as we increment the pointer beyond its bounds. The issue was reported and the fix was incorporated in the trunk release of CompCert. After adjusting both memory models, we are able to prove that both semantics agree when the existing CompCert C semantics is defined thus cross-validating the semantics of operators.

5   Redesign of Memory Injections

Memory injections are instrumental for proving the correctness of several compiler passes of CompCert. A memory injection defines a mapping between memories; it is a versatile tool to explain how passes reorganise the memory (e.g. construct an activation record from local variables). This section explains how to generalise this concept for symbolic expressions. It requires a careful handling of undefined values $\texttt{undef}(l)$ which are absent from the existing memory model.

5.1   Memory Injections in CompCert

In CompCert, a memory injection is a relation between two memories m_1 and m_2 parameterised by an injection function `f:block` $\rightarrow$ `option location` mapping blocks in m_1 to locations in m_2. The injection relation is defined over values

```
Lemma expr_binop_ok_ptr: ∀ op v₁ t₁ v₂ t₂ m b o,
  sem_binary_operation op v₁ t₁ v₂ t₂ m = ⌊ptr(b,o)⌋ → valid_ptr(m, b, o) →
  ∃ v', sem_binary_operation_expr op v₁ t₁ v₂ t₂ m = ⌊v'⌋ ∧
       normalise m v' Ptr = ptr(b, o).
```

Fig. 3. Example of cross-validation of binary C operators.

```
int main(){ int i=0, *p = &i;
            for(i=0; i < INT_MAX; i++)  if (p++ == 0) return 1;
            return 0; }
```

Fig. 4. A `null` pointer comparison glitch

(and called `val_inject`) and then lifted to memories (and called `inject`). The `val_inject` relation distinguishes three cases:

1. For concrete values (i.e. integers or floating-point numbers), the relation is reflexive: e.g. `int(i)` is in relation with `int(i)`;
2. `ptr(b, i)` is in relation with `ptr(b', i + \delta)` when $f(b) = \lfloor (b', \delta) \rfloor$;
3. `undef` is in relation with any value (including `undef`).

The purpose of the injection is twofold: it establishes a relation between pointers using the function `f` but it can also specialise `undef` by a defined value.

In CompCert, so-called generic memory injections state that every valid location in memory m_1 is mapped by function `f` into a valid location in memory m_2; the corresponding location in m_2 must be properly aligned with respect to the size of the block; and the values stored at corresponding locations must be in injection. Among other conditions, we have that if several blocks in m_1 are mapped to the same block in m_2, the mapping ensures the absence of overlapping.

5.2 Memory Injection with Symbolic Expressions

The Injection of Symbolic Expressions demands a generalisation because `undef` is now parameterised by a location `l`. The function `f` is still present and serves the same purpose. However, the injection must also be applied to undefined values. Moreover, our generalised injection requires an explicit specialisation function `spe: location → option byte`. Our injection `expr_inject` is therefore defined as the composition of the function `apply_spe spe` which specialises `undef(l)` into concrete bytes, and the function `apply_inj f` which injects locations. Both `spe` and `f` are partial functions. If `spe(l)=∅`, the undefined location is not specialised. If `f(b)=∅` and `b` appears in the expression, it cannot be injected.

Definition expr_inject spe f e_1 e_2 := apply_inj f (apply_spe spe e_1) = $\lfloor e_2 \rfloor$.

Example 1. Consider the injection `f` and the specialisation functions `spe` and `spe'` defined by: $f(b_1) = \lfloor (b_0, 1) \rfloor$, $spe(b_1, 0) = \lfloor 0 \rfloor$ and $spe'(b_1, 0) = \lfloor 1 \rfloor$. The `val_inject` relation (left column) between values becomes in our memory model the following `expr_inject` relation (right column).

val_inject f undef undef	expr_inject spe f undef(b_1,1) undef(b_0,2)
val_inject f undef int(0)	expr_inject spe f undef(b_1,0) int(0)
val_inject f undef int(1)	expr_inject spe' f undef(b_1,0) int(1)
val_inject f ptr(b_1,1) ptr(b_0,2)	expr_inject spe f ptr(b_1,1) ptr(b_0,2)

```
Record inject spe f m₁ m₂ : Prop := { ...
   mi_align: ∀ b b' z, f b = ⌊(b', z)⌋ →
             alignt(m₁, b) ≤ alignt(m₂, b') ∧ 2^[alignt(m₁,b)] | z;
   mi_size_mem:   size_mem m₂ ≤ size_mem m₁; }
```

Fig. 5. Memory injection: extra constraints

Injections of Memories. Like in the existing CompCert, the injection of values is then lifted to memories. With our memory model, the properties of injections need to be adapted to accommodate for symbolic expressions.

Alignment constraints are modelled in the existing CompCert as a property of offsets. Roughly speaking, a value of size s bytes can be stored at a location (b, o) such that the offset o is a multiple of s. For instance, an integer int(i) could be stored at offsets 0, 4, 8, and so on. This model makes the implicit assumption that memory blocks are always sufficiently aligned. In our model, blocks are given an explicit alignment. As a result, we can precisely state that an injection preserves alignement and is given by the mi_align property of Fig. 5. Note that the weaker formulation $2^{\text{alignt}(m_1,b)} | 2^{\text{alignt}(m_2,b')} + z$ is sound. However, the chosen formulation has the advantage of being backward compatible with the existing properties of offsets in CompCert.

The size constraint is evaluated using the size_mem function that is the algorithm used by the allocation function (see Sect. 3.2). This constraint ensures that an injection is compatible with allocation as stated by the following lemma. The hypothesis size_mem m_2 ≤ size_mem m_1 (called mi_size_mem in Fig. 5) ensures that the block b_2 can be allocated in memory m_2.

```
Theorem alloc_parallel_inject : ∀ spe f m₁ m₂ lo hi m₁' b₁,
  0 ≤ lo ≤ hi → inject spe f m₁ m₂ → alloc m₁ lo hi = ⌊(m₁', b₁)⌋ →
  ∃ m₂', ∃ b₂, alloc m₂ lo hi=⌊(m₂',b₂)⌋ ∧ inject spe f[b₁ ↦ ⌊(b₂,0)⌋] m₁' m₂'.
```

Absence of offset overflows. The existing formalisation of inject has a property mi_representable which states that the offset $o+\delta$ obtained after injection does not overflow. With our concrete memory model, this property is not necessary anymore as it can be proved for any injection.

5.3 Memory Injection and Normalisation

Our normalisation is defined w.r.t. all the concrete memories compatible with the CompCert block-based memory (see Sect. 2.3). Theorem norm_inject shows that under the condition that all blocks are injected, if e and e' are in injection, then their normalisations are in injection too. Thus, the normalisation can only get more defined after injection. This is expected as the injection can merge blocks and therefore makes pointer arithmetic more defined. The condition that all blocks need to be injected is necessary. Without it, there could exist a concrete memory cm' in m' without counterpart in m. The normalisation could therefore fail when the expression would evaluate differently in cm'. A consequence of this theorem is that the compiler is not allowed to reduce the memory usage.

```
Theorem norm_inject: ∀ spe f m m' e e' τ,
  all_blocks_injected f m → inject spe f m m'→ expr_inject spe f e e'→
  val_inject f (normalise m e τ) (normalise m' e' τ).
```

6   Proving the Front-End of the CompCert Compiler

The architecture of the front-end of CompCert is given in Fig. 6. The front-end compiles CompCert C programs down to Cminor programs. Later compiler passes are architecture dependent and are therefore part of the back-end. This section explains how to adapt the semantics preservation proofs of the front-end to our memory model with symbolic expressions.

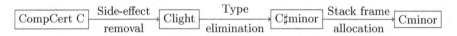

Fig. 6. Architecture of CompCert's front-end

6.1   CompCert Front-End with Symbolic Expressions

The semantics of all intermediate languages need to be modified in order to account for symbolic expressions. In principle, the transformation consists in replacing values by symbolic expressions *everywhere* and introducing the normalisation function when accessing memory. In reality, the transformation is more subtle because, for instance, certain intermediate semantic functions explicitly require locations represented as pairs (b, o). In such situations, a naive solution consists in introducing a normalisation. This solution proves wrong and breaks semantics preservation proofs because introduced normalisations may be absent in subsequent intermediate languages. The right approach consists in delaying normalisation as much as possible. Normalisations are therefore introduced before memory accesses. They are also introduced when evaluating the condition of if statements and to model the lazy evaluation of && and || operators. Using this strategy we have adapted the semantics (with built-in functions as only external functions) of the 4 languages of the front-end.

In our experience, the difficulty of the original semantics preservation proofs is not correlated with the difficulty of adapting the proofs to our memory with symbolic expressions. For instance, the compilation pass from CompCert C to Clight is arguably the most complex pass to prove; the proof is almost identical with symbolic expressions. In the following, we focus on the two other passes which stress different features of our memory model.

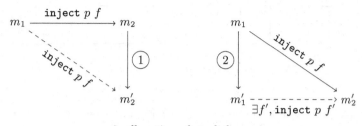

1. allocation of stack frame
2. allocation of local variables

Fig. 7. Structure of `match_callstack_alloc_variables`'s proof in CompCert

6.2   From Clight to C♯minor

The compilation from Clight to C♯minor translates loops and `switch` statements into simpler control structures. This pass does not transform the memory and therefore the existing proof can be reused. The pass also performs type-directed transformations and removes redundant casts. For example, it translates the expression `p + 1` with `p` of type `int *` into the expression `p + sizeof(int)`. For the existing memory model, both expressions compute exactly the same value. However, with symbolic expressions, syntactic equality is a too strong requirement that needs to be relaxed to a weaker equivalence relation. A natural candidate is the equality of the normalisation. However, this relation is too weak and fails to pass the induction step. Indeed, when expressions e_1 and e_2 have the same normalisation ($v_1 \equiv v_2$), it is not the case that $op_1^\tau(v) \equiv op_1^\tau(v')$ when the normalisations are **undef**. A stronger relation is the equality of the evaluation of symbolic expressions in any concrete memory (compatible or not).

We lift the equivalence relation to memories in the obvious way. To carry out the proof, we also extend the interface of the memory model and prove that the memory operations are morphisms for the equivalence relation. With these modifications, the compiler pass can be proved semantics preserving using the existing proof structure.

6.3   From C♯minor to Cminor

The compilation from C♯minor to Cminor allocates the stack frame, thus transforming significantly the memory. The stack frame is a single block and local variables are accessed via offsets in this block. The proof introduces a memory injection stating how the blocks representing local variables in C♯minor are mapped into the single block representing the stack frame in Cminor.

The existing proof can be adapted with our generalised notion of injection (see Sect. 5) with the notable exceptions of two intermediate lemmas whose proofs need to be completely re-engineered. The problem is related with the preservation of the memory injection when allocating and de-allocating the variables in C♯minor and the stack frame in Cminor. The structure of the original

Fig. 8. Variables on the left; stack frame on the right.

proof is depicted in Fig. 7 where plain arrows represent hypotheses and the dotted arrow the conclusion. The existing proof first allocates the stack frame in memory m_2 to obtain the memory m_2'. It then establishes that the existing injection between the intial memories m_1 and m_2 still holds with the memory m_2'. In a second step, the memory m_1' is obtained by allocating variables in memory m_1 and the proof constructs an injection thus concluding the proof.

With our memory model, memory injections need to reduce the memory usage – this is needed to ensure that allocations cannot fail. Here, this is obviously not the case because the memory m_2' contains a stack frame whereas the corresponding variables are not yet allocated in m_1. Our modified proof is directly by induction over the number of allocated variables. In this case, we prove that if the variables do fit into memory, then so does the stack frame. Note that to accommodate for alignment and padding the stack frame might allocate more bytes than the variables. However, our allocation algorithm makes a worst-case assumption about alignment and padding and therefore ensures that there is enough room for allocating the stack frame. We therefore conclude that the memories m_1' and m_2' are in injection.

At function exit, the variables and the stack frame are freed from memory. As before, the arguments of the original proof do not hold with our memory model. Once again, we adapt the two-step proof with a direct induction over the number of variables. To carry out this proof and establish an injection we have to reason about the relative sizes of the memories. We already discussed how the allocation algorithm rules out the possibility for the stack frame not to fit into the memory. Here, we have to deal with the opposite situation where the stack frame could use less memory than the variables.

To avoid this situation, and facilitate the proof, our stack frame currently allocates all the padding introduced when allocating the variables. This pessimistic construction is depicted in Fig. 8 where thick lines identify block boundaries and padding is identified by grey rectangles. It shows our injection of two 8-bits character variables and a 32-bit integer variable into a stack frame.

We are currently investigating on how to give a finer account of the necessary padding to avoid allocating too much memory for the stack frame. Yet, using the strategy described above, we are able to complete the proof of the front-end while reusing as much as possible the architecture of the existing proof.

7   Related Work

Examples of low-level memory models include Norrish's HOL semantics for C [20] and the work of Tuch *et al.* [21]. There, memory is essentially a mapping from addresses to bytes, and memory operations are axiomatised in these terms. Reasoning about program transformations is more difficult than with a block-based model; Tuch *et al.* use separation logic to alleviate these difficulties.

Memory models have been proposed to ease the reasoning about low-level code. VCC [7] generates verification conditions using an abstract typed memory model [8] where the memory is a mapping from typed pointers to structured C values. This memory model is not formally verified. Using Isabelle/HOL, Autocorres [10,11] constructs provably correct abstractions of C programs. The memory models of VCC [8] and Autocorres [11] ensure separation properties of pointers for high-level code and are complete w.r.t. the concrete memory model. For the CompCert model [18], separation properties of pointers are for free because pointers are modelled as abstract locations. For our symbolic extension, the completeness (and correctness) of the normalisation is defined w.r.t. a concrete memory model and therefore allows for reasoning about low-level idioms.

Several formal semantics of C are defined over a block-based memory model (e.g. [9,15,17]). The different models differ upon their precise interpretation of the ISO C standard. The CompCert C semantics [5] provides the specification for the correctness of the CompCert compiler [17]. CompCert is used to compile safety critical embedded systems [2] and the semantics departs from the ISO C standard to capture existing practices. Our semantics extends the existing CompCert semantics and benefits from its infrastructure.

Krebbers *et al.* also extend the CompCert semantics but aim at being as close as possible to the C standard [16]; he formalises sequence points in non-deterministic programs [15] and strict aliasing restrictions in union types of C11 [14]. This is orthogonal to the focus of our semantics which gives a meaning to implementation defined low-level pointer arithmetic and models bit-fields. Most recently, Kang *et al.* [13] propose a formal memory model for a C-like language which allows optimisations in the presence of integer-pointer casts. Pointers are kept logical until they are cast to integers, then a concrete address is non-deterministically assigned to the block of the pointer. Their semantics of C features non-determinism while determinism is a crucial feature of our model.

8   Conclusion

This work is a milestone towards a CompCert compiler proved correct with respect to a more concrete memory model. Our formal development adds about 10000 lines of Coq to the existing CompCert memory model. A side-product of our work is that we have uncovered and fixed a problem in the existing semantics of the comparison with the null pointer. We are very confident that this is the very last remaining bug that can be found at this semantics level. We also prove that the front-end of CompCert can be adapted to our refined memory model. The proof effort is non-negligible: the proof script for our new memory model is twice as big as the existing proof script. The modifications of the front-end are less invasive because the proof of compiler passes heavily rely on the interface of the memory model.

As future work, we shall study how to adapt the back-end of CompCert. We are confident that program optimisations based on static analyses will not be problematic. We expect the transformations to still be sound with the caveat that static analyses might require minor adjustments to accommodate for our more defined semantics. A remaining challenge is register allocation which may allocate additional memory during the spilling phase. An approach to solve this issue is to use the extra-memory that is available due to our pessimistic construction of stack frames. Withstanding the remaining difficulties, we believe that the full CompCert compiler can be ported to our novel memory model. This would improve further the confidence in the generated code.

References

1. Companion website. URL: http://www.irisa.fr/celtique/ext/new-mem
2. França, R.B., Blazy, S., Favre-Felix, D., Leroy, X., Pantel, M., Souyris, J.: Formally verified optimizing compilation in ACG-based flight control software. In: ERTS2 (2012)
3. Besson, F., Blazy, S., Wilke, P.: A precise and abstract memory model for C using symbolic values. In: Garrigue, J. (ed.) APLAS 2014. LNCS, vol. 8858, pp. 449–468. Springer, Heidelberg (2014)
4. Blazy, S.: Experiments in validating formal semantics for C. In: C/C++ Verification Workshop. Raboud University Nijmegen report ICIS-R07015 (2007)
5. Blazy, S., Leroy, X.: Mechanized Semantics for the Clight Subset of the C Language. J. Autom. Reasoning, 43(3), 263–288 (2009)
6. Clements, A.T., Kaashoek, M.F., Zeldovich, N., Morris, R.T., Kohler, E.: The scalable commutativity rule: designing scalable software for multicore processors. In: SOSP. ACM (2013)
7. Cohen, E., Dahlweid, M., Hillebrand, M., Leinenbach, D., Moskal, M., Santen, T., Schulte, W., Tobies, S.: VCC: a practical system for verifying concurrent C. In: Berghofer, S., Nipkow, T., Urban, C., Wenzel, M. (eds.) TPHOLs 2009. LNCS, vol. 5674, pp. 23–42. Springer, Heidelberg (2009)
8. Cohen, E., Moskal, M., Tobies, S., Schulte, W.: A precise yet efficient memory model for C. ENTCS **254**, 85–103 (2009)
9. Ellison, C., Roşu, G.: An executable formal semantics of C with applications. In: POPL. ACM (2012)
10. Greenaway, D., Andronick, J., Klein, G.: Bridging the gap: automatic verified abstraction of C. In: Beringer, L., Felty, A. (eds.) ITP 2012. LNCS, vol. 7406, pp. 99–115. Springer, Heidelberg (2012)
11. Greenaway, D., Lim, J., Andronick, J., Klein, G.: Don't sweat the small stuff: formal verification of C code without the pain. In: PLDI. ACM (2014)
12. Jourdan, J., Laporte, V., Blazy, S., Leroy, X., Pichardie, D.: A formally-verified C static analyzer. In: POPL (2015)
13. Kang, J., Hur, C.-K., Mansky, W., Garbuzov, D., Zdancewic, S., Vafeiadis, V.: A formal C memory model supporting integer-pointer casts. In: PLDI. ACM (2015)
14. Krebbers, R.: Aliasing restrictions of C11 formalized in Coq. In: Gonthier, G., Norrish, M. (eds.) CPP 2013. LNCS, vol. 8307, pp. 50–65. Springer, Heidelberg (2013)
15. Krebbers, R.: An operational and axiomatic semantics for non-determinism and sequence points in C. In: POPL. ACM (2014)

16. Krebbers, R., Leroy, X., Wiedijk, F.: Formal C semantics: "CompCert and the C standard. In: Klein, G., Gamboa, R. (eds.) ITP 2014. LNCS, vol. 8558, pp. 543–548. Springer, Heidelberg (2014)
17. Leroy, X.: Formal verification of a realistic compiler. C. ACM **52**(7), 107–115 (2009)
18. Leroy, X., Appel, A.W., Blazy, S., Stewart, G.: The CompCert memory model. In: Program Logics for Certified Compilers. Cambridge University Press (2014)
19. Leroy, X., Blazy, S.: Formal verification of a C-like memory model and its uses for verifying program transformations. J. Autom. Reasoning **41**(1), 1–31 (2008)
20. Norrish, M.: C formalised in HOL. Ph.D. thesis, University of Cambridge (1998)
21. Tuch, H., Klein, G., Norrish, M.: Types, bytes, and separation logic. In: POPL. ACM (2007)
22. Wang, X., Chen, H., Cheung, A., Jia, Z., Zeldovich, N., Kaashoek, M.: Undefined behavior: What happened to my code? In: APSYS 2012 (2012)
23. Yang, X., Chen, Y., Eide, E., Regehr, J.: Finding and understanding bugs in C compilers. In: PLDI. ACM (2011)

Validating Dominator Trees for a Fast, Verified Dominance Test

Sandrine Blazy[1]([✉]), Delphine Demange[1]([✉]), and David Pichardie[2]([✉])

[1] Université Rennes 1 - IRISA - Inria, Rennes, France
{sandrine.blazy,delphine.demange}@irisa.fr
[2] ENS Rennes - IRISA - Inria, Rennes, France
david.pichardie@irisa.fr

Abstract. The problem of computing dominators in a control flow graph is central to numerous modern compiler optimizations. Many efficient algorithms have been proposed in the literature, but mechanizing the correctness of the most sophisticated algorithms is still considered as too hard problems, and to this date, verified compilers use less optimized implementations. In contrast, production compilers, like GCC or LLVM, implement the classic, efficient Lengauer-Tarjan algorithm [12], to compute dominator trees. And subsequent optimization phases can then determine whether a CFG node dominates another node in constant time by using their respective depth-first search numbers in the dominator tree. In this work, we aim at integrating such techniques in verified compilers. We present a formally verified validator of untrusted dominator trees, on top of which we implement and prove correct a fast dominance test following these principles. We conduct our formal development in the Coq proof assistant, and integrate it in the middle-end of the CompCertSSA verified compiler. We also provide experimental results showing performance improvement over previous formalizations.

1 Introduction and Related Work

Given a control flow graph (CFG) with a single entry node, computing dominators consists in determining, for each node in the graph, the set of nodes that dominate it. Informally, a node d dominates another node n if d belongs to every path from the entry node to n. The problem of computing dominators is ubiquitous in computer science, and occurs in applications ranging from program optimization, to circuit testing, analysis of component systems, and worst-case execution time estimation.

Since 1972, this problem has been extensively studied. Many algorithms have been proposed, trading-off ease of implementation and efficiency. The natural formulation of the problem as data-flow equations is due to Allen and Cocke [1]. It can be directly implemented using an iterative Kildall algorithm, but suffers, in this case, from a quadratic asymptotic complexity. Cooper et al. [4] present

This work was supported by Agence Nationale de la Recherche, grant number ANR-14-CE28-0004 DISCOVER.

C. Urban and X. Zhang (Eds.): ITP 2015, LNCS 9236, pp. 84–99, 2015.
DOI: 10.1007/978-3-319-22102-1_6

another iterative solution for this equation system, based on a more compact representation of dominator sets (only the immediate dominator, i.e. the closest dominator, is computed for each node), and a careful implementation, leading to better performance in practice, despite the same worst-case bound time as [1]. To date, the most popular algorithm remains the one by Lengauer-Tarjan [12], which, as Cooper et al. algorithm, computes a compact representation of the dominance relation (namely the *dominator tree*). But this sophisticated algorithm relies on depth-first search (DFS) spanning tree of the CFG with elaborate path compression and tree balancing techniques to achieve a stunning near-linear complexity. We refer the interested reader to [16] for a more complete survey of the numerous algorithms proposed so far in the literature, and to [10] for a thorough experimental study comparing the leading algorithms.

We consider the problem of dominators in the specific context of compilation, where dominators allow, for instance, the implementation of a variety of powerful and efficient program optimizations (e.g. loops optimization or global code motion), and the construction of the SSA form [5], an intermediate representation of code that is specially tailored towards program optimization. Production compilers, like GCC or LLVM, implement the classic, efficient Lengauer-Tarjan algorithm [12], to compute dominator trees. Subsequent optimization phases can then determine whether a node dominates another node in constant time by using their respective DFS traversal numbers in the dominator tree.

Specifically, the present work is part of a compiler verification effort, where an (optimizing) compiler must be formally proved to preserve the program behaviors along the compilation chain, i.e. the generated code behaves as prescribed by the semantics of the source program, if any. In this context, correctly implementing a time- and space-efficient dominator algorithm is not sufficient; one has to formally prove its correctness. We are not aware of any formal verification of the dominator problem outside of the field of compiler verification. Further, faced with this technical difficulty, existing verified compilers either ignore dominators, or implement simplified and under-optimized versions of dominator algorithms.

For instance, the CompCert C compiler [13,14] is not based on any SSA form for performing code optimization, and no global optimization uses explicit dominance information. The CompCertSSA project extends the CompCert compiler with an SSA-based middle-end. The SSA generation algorithm [2] is proved by *a posteriori* validation of an external checker. Although we prove that the checker ensures the strictness of the SSA generated function (that is, each variable use is dominated by its definition), the checker implementation (a simple, non-iterative, CFG traversal) and soundness proof do not rely on the computation of dominators. The only phase of the CompCertSSA middle-end that depends on such a computation (that we would like to be efficient) is a common sub-expression elimination (CSE) optimization based on Global Value Numbering (GVN). It discovers equivalence classes between program variables, where variables belonging to the same class are supposed to evaluate to the same value. Its implementation, presented in [7], closely follows the choices made in production compilers, and performs some dominance test requests to make sure that the chosen representative of a variable class dominates the definition point of that variable. To date, this dominance test was implemented (and proved directly) with a simple Allen and Cocke algorithm, using a Kildall workset algorithm, thus impacting the performance of our middle-end.

Another SSA-based verified compiler is Vellvm. Zhao et al. [18,20] formalize the LLVM SSA intermediate form and its generation algorithm in Coq. Their work follows closely the LLVM design and their verified transformation can be run inside the LLVM platform itself. Zhao et al. [19] formalize in Coq a fast dominance computation based on the Cooper et al. algorithm [4], but their algorithm is, for verification purposes, a simplified version of the initial algorithm. This is a non-trivial formalization work that also proves in Coq the completeness of the dominance relation computation, an interesting and difficult problem in itself. However, this work does not focus on compilation time. Other CPS or ANF-based verified compilers for functional languages [3,6] implement simple optimizations that do not require dominance information, although their (unverified) peers, like MLton, benefit from dominators for, e.g. contification [8] for inter-procedural optimization.

Facing the conceptual complexity of the most clever variants for computing dominators, there has been a growing interest in proposing ways of checking their results. Georgiadis et al. [11] propose a linear-time checker of the dominator tree, based on the notions of headers and loop nesting forests. Georgiadis et al. [9] propose a linear-time certifying algorithm, producing a certificate (a preorder of the vertices of the dominator tree, with a so-called property *low-high*), that helps simplifying the checking process. Despite that checking the low-high property on the certificate is straightforward, and easily implemented in linear time, linking the low-high property back to the immediate dominance relation (via the concepts of strongly independent spanning trees) remains quite involved. As a matter of fact, to date, these two recent, sophisticated algorithms are still out of the reach of mechanized developments.

In our context of verified compilation, we need two things: compute efficiently the dominance relation, and represent this relation in a compact way, so that the dominance test can be implemented efficiently. Note however that mechanically verifying (or validating) the dominator tree remains, for the time being, unessential. Hence, we believe that the technique used in GCC and LLVM, i.e. computing a dominator tree using Lengauer-Tarjan's algorithm, and then fast-checking dominance with an ancestor test in the DFS numbering of the dominator tree, provides a, perhaps more modest, yet viable, trade-off between efficiency and verifiability. We argue that this technique can also be applied to verified compilers, by relying on an *a posteriori* validation approach. We present a formally verified validator of untrusted dominator trees, on top of which we implement and prove a fast dominance test that follows these principles.

Contributions. After recalling the technical background on dominators and the main algorithms (Sect. 2), we present the following contributions.

- A new, simple and verified validator for the dominance relation (Sect. 3), which leads to a formally verified implementation of a dominance test technique

used in production compilers. The heart of the validator algorithm is our own contribution but it is mixed with well-known graph algorithms for fast ancestor checking. This paper presents, to our knowledge, the first verification of these kinds of techniques.
- Empirical evidence that this technique allows, in practice, a non-negligible performance gain, even in the context of verified compilers (Sect. 4).
- The integration of this dominance computation and dominance test within the CompCertSSA verified compiler. Our formal development and experiments are available online at http://www.irisa.fr/celtique/ext/ssa_dom/.

2 Technical Background and Overview of Algorithms

In this section, we recall the technical background on dominance, together with standard techniques to compute this relation. We also present how dominance and dominance testing is implemented in modern compilers.

2.1 Definitions

A control flow graph $G = (N, E, e)$ is defined as an oriented graph, i.e. a set of nodes N, a set of edges E, and a distinguished entry node $e \in N$ (that is not the successor of any other node). In the following, we depict an edge connecting node $i \in N$ to $j \in N$ by $i \rightarrow j$.

Definition 1 (Dominance relation). *A node d dominates a node n if n is reachable from the graph entry node and if any path from the entry point to n contains d. If $d \neq n$, the dominance is said to be strict.*

For every node n (except the entry e), the set $sdom(n)$ of nodes that strictly dominate n, contains a node $idom(n)$ that is dominated by every other nodes in $sdom(n)$ [12]: the *immediate dominator*. As an extra important corollary, the immediate dominance relation can be represented as a tree [12].

Definition 2 (Dominator tree). *The dominator tree of a CFG is a tree whose nodes are the nodes of the CFG, and where the children of a node are all the nodes that it immediately dominates.*

Figure 1 shows an example of a CFG and its dominator tree. For instance, in the CFG, node 10 is dominated by node 15, since all paths from the entry node 17 to node 10 must go through 15. Hence, in the dominator tree, node 10 must have node 15 as an ancestor. However, 15 is not the immediate dominator of node 10, it is node 13: indeed, in the set $sdom(10) = \{17, 16, 15, 14, 13\}$, node 13 is the one dominated by every other node in the set. Hence, node 10 is a child of node 13 in the dominator tree.

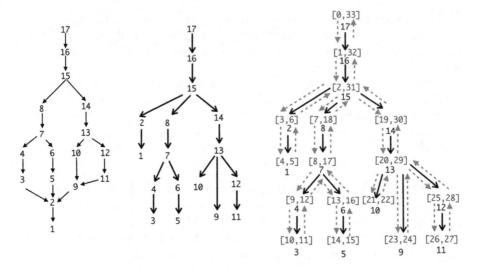

Fig. 1. Left: example CFG, with entry point 17, and nodes ordered in reverse post-order (on the left). Center: its dominator tree (where, if i has j as a child, then i is the immediate dominator of j). Right: dominator tree (solid arrows), annotated with a DFS traversal (dotted arrows), and its corresponding DFS intervals (see Sect. 2.3).

2.2   Standard Techniques for Computing Dominance

Allen-Cocke (AC) Standard Data-Flow Analysis [1]. The AC algorithm is based on the following fixpoint characterization of dominance.

$$dom(n) = \{n\} \cup \bigcap_{n' \to n} dom(n') \tag{1}$$

Intuitively, it captures that every strict dominator of a node n must also dominate every n's predecessors in the CFG. Such fixpoint equation can be solved using a workset fixpoint iteration *à la* Kildall. As is typical for forward data-flow problems, the fixpoint resolution is speeded up if at each workset iteration we choose the node with the lowest rank with respect to a reverse postorder ordering on the CFG (a node is visited before any of its successor nodes has been visited, except when the successor is reached by a back edge). A direct implementation is quadratic in the number of nodes, or $O(|N|^2)$.

Cooper-Harvey-Kennedy (CHK) Algorithm [4]. The CHK dominance computation improves Allen-Cocke data-flow approach using the following properties. First, dominator sets can be characterized by the immediate dominator table.

$$\forall n, \exists k, dom(n) = \{n, idom(n), \ldots, idom^k(n)\} \tag{2}$$

CHK can be understood as a variation of the previous approach where dominance sets are implicitly represented by the immediate dominator tree. Using reverse postorder ordering, by noticing that $\forall n, n \prec_{\text{rpo}} idom(n)$, set intersection can be performed in an efficient way because if $dom(a) \cap dom(b) \neq \emptyset$, then the resulting set is a prefix of both $dom(a)$ and $dom(b)$ [4]. This algorithm performs better in practice than AC, but follows the same $O(|N|^2)$ asymptotic time complexity.

Lengauer-Tarjan (LT) Algorithm [12]. Modern compilers implement dominance using the LT algorithm. It uses depth-first search and union-find data structures to achieve an asymptotic complexity of $O(|N|log|N| + |E|)$. It relies on the subtle notion of *semi-dominator* which provides a convenient intermediate step in the dominators computation. An amortized quasi-linear complexity can be obtained using path compression but it does not seem to be implemented in practice.

2.3   Modern Implementation of Dominance Test in Compilers

As explained above, modern compilers such as GCC or LLVM implement dominance following the LT algorithm. Once they obtain a dominator tree (as shown in Fig. 1), they pre-process it to obtain a constant-time dominance test. The dominance between two nodes d and n can be determined by testing if the node d is an ancestor of n in the dominator tree. For instance, in Fig. 1, node 15 dominates node 10 because there is an upward path from 10 to 15 in the tree.

This test can be performed in constant time thanks to a linear pre-computation (on the $|N| - 1$ edges of the dominator tree). For each node, one computes a *depth-first search interval* $I(n) = [d(n), f(n)]$ where $d(n)$ is the discovery time of node n during the traversal (the first time n exists in the DFS stack) and $f(n)$ is the finishing time (the time where all sons of n have been processed) [17]. In a direct acyclic graph, d is an ancestor of n if and only if $I(n) \subseteq I(d)$. Figure 1 shows, on the right, the results of such an interval computation: intervals bounds are determined according to the starting and ending time clocks when depth-first traversing the tree. There, the fact that 15 dominates node 10 is obtained by observing that interval $I(10) = [21, 22]$ is included in $I(15) = [2, 31]$.

As a result of this pre-computation with complexity $O(|N|log|N| + |E|)$, a constant time dominance can be obtained by storing the intervals information in adequate data structures.

3   Validator and Proof of Dominance Test

Our formalization is done on top of an abstract notion of CFG. Such a graph is defined as follows by an entry node and a set of edges.

```
Variable entry : node.
Variable cfg : node → node → Prop.
```

In the sequel, `reached : node→Prop` is a predicate characterizing the set of nodes that are reachable, via `cfg`, from the node `entry`, and `dom : node→node→Prop` denotes the dominance relation that is defined using a standard definition of CFG paths.

In this section, we assume that an external tool computes a list `dt_edges` that contains the reversed edges of the candidate dominator tree (i.e. the pair `(i,j)` represents that, in the candidate dominator tree, `i` is a child of `j`, or that `j` immediately dominates `i`).

```
Variable dt_edges : list (node * node).
```

We then validate this list and build a dominance test, implemented by the function `test_dom : node→node→bool` that satisfies the following theorem:

```
forall i j, reached j → test_dom i j = true → dom i j.
```

In the rest of this section, we proceed in three steps. First, we give a dominator map `D : node→node` (extracted from `dt_edges`), a specification that entails dominance. Then, we provide an efficient procedure to test whether a node is a descendant of another in the dominator tree (encoded morally in `D`). This procedure is used twice: for checking that `D` meets its specification, and in the final implementation of the dominance test, `test_dom`.

3.1 Validation of Dominator Tree

In this section, we assume a dominator map `D:node→node` that provides an (immediate) dominator candidate for each node. We will explain in Sect. 3.4 how we build `D` from the list `dt_edges`. We provide a formal specification for `D` and prove it entails dominance. Note that we do not prove that it implies immediate domination, as this is not required in our final soundness theorem[1].

The specification, inspired from Eqs. (1) and (2), is defined as follows.

```
Record D_spec := { D_entry : D entry = entry;
                   D_cfg   : forall i j,  cfg i j → Dstar i (D j) }.
```

where `(Dstar i j)` holds whenever `j` is of the form $D^k(i)$, for some k. Formally:

```
Inductive Dstar : node → node → Prop :=
| D_refl : forall i, Dstar i i
| D_trans: forall i j, Dstar i j → Dstar i (D j).
```

We then prove, quite straightforwardly, that `D_spec` implies dominance by induction on the definition of predicate `reached`.

```
Theorem D_spec_correct : D_spec →
   forall i j, reached i → Dstar i j → dom j i.
```

Hence, we can validate the map `D` if we manage to check that it satisfies the specification `D_spec`. Interestingly, we need an executable version of the `Dstar` relation for two distinct usages. First we want to validate `D_spec` on `D`. Second, we want to implement a dominance test using `Dstar`.

---

[1] Such a property would be required to prove *completeness*: if a node d dominates a node n then the dominance test on (d, n) should succeed. To our experience in verified compilation, we never make usage of such a completeness property. The property holds, but we do not need to prove it in Coq.

3.2   Ancestor Test in the Dominator Tree

In this section we assume an acyclic oriented graph[2], defined by an entry node and a map, `sons`, from nodes to the list of their successors[3]. We will later relate this graph with our dominator tree.

```
Variable entry: node.              (* entry node *)
Variable sons : PTree.t (list node).   (* adjacency map *)
```

As outlined in Sect. 2.3, the ancestor test consists in performing a depth-first traversal of the graph, starting from `entry`, and using a traversal clock, that increases each time a node is encountered (by visiting it or by marking it). We compute for each node n, an interval $I(n) = [d(n), f(n)]$ where $d(n)$ is the value of the clock when node n was first encountered, and $f(n)$ is the value of the clock when all successors of n have been processed. If the graph is acyclic and each node is reachable from `entry`, we can use these intervals to perform efficient ancestor tests [17]: there exists a path from n to m in the graph if and only if $I(m) \subseteq I(n)$. For our purpose, we only need to prove that this condition is sufficient. We define intervals, intervals inclusion and our efficient ancestor test in an interval map as follows.[4]

```
Record itv := { pre: Z; post: Z }.
Definition itv_Incl (i1 i2:itv) : Prop :=
  i2.(pre) <= i1.(pre) /\ i1.(post) <= i2.(post).
Definition is_ancestor (itvm: PTree.t itv) (n1 n2:node) : bool :=
  match itvm!n1, itvm!n2 with
  | Some i1, Some i2 ⇒ itv_Incl i2 i1
  | _, _ ⇒ false
  end.
```

Now, to state the correctness of our interval computation, we specify a notion of ancestor called `InSubTree`. A node r is an ancestor of n (or equivalently n belongs to a subtree whose root is r) if $n = r$ or there exists a successor s of n such that s is an ancestor of n.

```
Inductive InSubTree (r:node) : node → Prop :=
|InSubTree_root: InSubTree r r
|InSubTree_sons: forall n s, InSubTree s n→In s (sons r)→InSubTree r n.
```

The interval map is computed by the function `build_itv` that performs the recursive DFS traversal of the graph, accumulating in a record of type `state`, the current interval map, and the current time clock.

```
Record state := { itvm: PTree.t itv;   (* the interval map *)
                  next: Z              (* the current time *) }.
```

---

[2] Not to be confused with the control flow graph here.

[3] PTree is a dictionary implementation using Patricia trees provided in CompCert. Type (Ptree.t a) denotes an associative, partial map with keys of type `positive` – binary encoding of strictly positive integers – with associated data of type **a**. In this paper, types `node` and `positive` are synonyms.

[4] We write `m!n` the lookup of a key `n` in a map `m`.

Note that, to ensure termination of `build_itv_rec`, we use a fuel auxiliary argument, i.e. a natural number counter decreasing at each recursive call. The fuel argument is useful not only to avoid proving termination, but also, and more crucially, to get a useful induction principle on the next inductive predicate.

```
Definition build_itv (fuel:nat) : option state :=  build_itv_rec
    entry    (* start traversing the graph at entry node *)
    {| itvm := PTree.empty _; next := 0 |} (* initial state *)
    fuel.   (* initial fuel *)
Fixpoint build_itv_rec (n:node) (st:state) (fuel:nat) : option state :=
  match fuel with
    | 0 ⇒ None                    (* no more fuel, abort computation *)
    | S fuel ⇒
      let pre_n := st.(next) in   (* current time when we reach node n *)
      match fold_left (fun ost s ⇒   (* we process each successor *)
                      match ost with
                        | None ⇒ None
                        | Some st ⇒ build_itv_rec s st fuel
                      end)
                  (sons n)
                  (Some {| itvm := st.(itvm); next := st.(next)+1 |})
      with
      | None ⇒ None
      | Some st ⇒ (* if no fuel error occurred, we extract st.(next) *)
                  (* to build the last component of n's interval   *)
        let itv_n := {| pre := pre_n; post := st.(next) |} in
      Some {| itvm := PTree.set n itv_n st.(itvm); next := st.(next)+1 |}
      end
  end.
```

The correctness theorem of `build_itv` states that, in the resulting interval map `st.itvm`, interval inclusion implies an ancestor relationship in the tree.

```
Theorem build_itv_correct : forall fuel,
  NoRepetTreeN entry (S fuel) →
  forall st, build_itv fuel = Some st →
  forall n1 n2 itv1 itv2,
    st.(itvm)!n1 = Some itv1 → st.(itvm)!n2 = Some itv2 →
    itv_Incl itv1 itv2 → InSubTree n2 n1.
```

As can be seen, this theorem is proved under the hypothesis that the graph is well-formed, namely that it does not contain duplicates or crossing edges, as expressed by predicate `NoRepetTreeN`, whose formal definition is the following.

```
Inductive NoRepetTreeN (r:node) : nat → Prop :=
| NoRepetTreeN0: NoRepetTreeN r 0
| NoRepetTreeN_sons: forall k,
  (forall s,                             (* sons are well formed *)
        In s (sons r) → NoRepetTreeN s k) →
  (forall s,              (* r does not appear in any of its subtrees *)
        In s (sons r) → ¬ InSubTree s r) →
  (forall s1 s2 n,               (* r's subtrees do not intersect *)
```

```
          In s1 (sons r) → InSubTree s1 n →
          In s2 (sons r) → InSubTree s2 n → s1=s2) →
    (list_norepet (sons r)) →          (* r's sons don't have duplicates *)
    NoRepetTreeN r (S k).
```

The definition of NoRepetTreeN is *staged*, i.e. indexed by a natural number. This level in the definition (that coincides with the height of the tree under consideration) provides a nice induction principle when combined with the fuel argument of function build_itv. Without such a trick, Coq does not generate a useful induction principle.

We prove build_itv_rec correctness using several auxiliary invariants, notably that the clocks are monotonic, that computed intervals are never empty, and that in a given subtree, computed intervals are included in the interval of the root of the subtree.

3.3   Well-Formed Graph Construction

This section explains how we relate the list dt_edges that contains the edges of the dominator tree, with the immediate dominator map D we use in Sect. 3.1, and the graph representation used in Sect. 3.2. We not only build a map of successors, but also check sufficient conditions enforcing the NoRepetTreeN property presented previously.

Starting from the list dt_edges, we straightforwardly build a map D from nodes to their immediate dominator candidate with the function make_D_fun of type make_D_fun (dt_edges:list (node*node)) : node → node. If a node is not in the association list dt_edges, its (correct) immediate dominator is set to itself.

In a similar way to the construction of the candidate dominator tree from dt_edges, we also define the function build_succs of type

```
build_succs (dt_edges:list (node * node)): option (PTree.t (list node))
```

that performs a reverse topological sort to build a map that associates to each node the (candidate) list of immediately dominated nodes. Function build_succs somewhat builds the inverse of the map D. Its correctness theorem states that the output successor tree, if any, is well-formed.

```
Theorem build_succs_no_repet : forall dt_edges sn,
  build_succs dt_edges = Some sn →
  forall fuel, NoRepetTreeN (sons sn) entry (S fuel).
```

This theorem follows from the checks performed during the computation of build_succs. Indeed, in its signature, the option type of the result represents a validation failure.

```
Theorem build_succs_correct_tree : forall dt_edges sn,
  make_D_fun dt_edges = D → build_succs dt_edges = Some sn →
  forall i j, In j (sons sn(i)) → D j = i.
```

During the traversal of dt_edges, we check that it contains no edge of the form (n,n), and that, when processing an edge (n,d), i.e. adding n to the list

```
Definition compute_test_dom (f: function) :
                                   option (node → node → bool) :=
  (* external computation in OCaml, using Lengauer-Tarjan algorithm *)
  let dt_edges : list (node*node) := extern_d f in
  (* immediate dominator map computation from dt_edges *)
  let D : node → node := make_D_fun dt_edges in
  (* successors tree computation from dt_edges *)
  match build_succs (entry f) l with
  | Some sn ⇒
    (* interval computation from successor tree *)
   match build_itv_map (entry f) sn fuel with
     (* same as (build_itv(entry f)(sons sn) fuel) *)
       | Some itvm ⇒
         (* fast ancestor test using interval, see section 3.2  *)
         let td := (fun i j ⇒ is_ancestor itvm (D j) i) in
         (* immediate dominator tree validation using ancestor test *)
         if (check_D_eq f D td) then Some (is_ancestor itvm) else None
       | None ⇒ None
     end
  | None ⇒ None
end.
```

Fig. 2. Dominance test construction

of successors of node d, node d was already seen (i.e. is already a key in the tree), and that node n has not yet been seen. Hence, to be accepted by the validator, the provided list dt_edges should be topologically sorted, and by the same validation, we ensure there is no loop in the graph. For further details, we refer the reader to the formal development available online.

3.4   Final Construction

The final dominance test computation is given in Fig. 2. It takes as input a program represented by its CFG (more precisely any function of this program) and combines the various functions presented earlier. It is proved correct with the following theorem.

```
Theorem dom_test_correct : forall f test_dom,
  compute_test_dom f = Some test_dom →
  forall i j, reached f j → test_dom i j = true → dom f i j.
```

We now discuss its asymptotic complexity. If N denotes the number of nodes in the CFG, and E the number of edges, then the asymptotic complexity of this computation is as follows.

- The list dt_edges has length $N-1$ (every node, except the entry, has a unique immediate dominator).
- The map make_D_fun dt_edges is computed with 1 traversal of dt_edges and 1 map update is performed at each step. The overall complexity is $O(N \log N)$.

- `build_succs` is computed with one traversal of `dt_edges` and several set and map updates are performed at each step. The overall complexity is $O(N \log N)$.
- Intervals are built with a traversal of a graph with N nodes and $N-1$ edges (this is a tree). At each step, some map updates are performed. The overall complexity is $O(N \log N)$.
- One ancestor test requires two map lookups and some integer comparisons. Each integer[5] is between 0 and $2N-1$. The overall complexity of an ancestor test is $O(\log(N))$.
- Dominance tree validation requires, for each edge in the CFG, one ancestor test and some map lookups. The overall complexity of this step is $O(E \log(N))$.

Overall, the dominance tree pre-computation follows an asymptotic complexity of $3O(N \log(N)) + O(E \log(N)) = O(E \log(N))$ ($N \leq E$ as all nodes are reachable from the entry) and the generated dominance test requires $O(\log(N))$ computations.

As will be explained in the next section, we also provide a *native* version of the implementation, that uses native integers for graph nodes and interval bounds. It does not improve the asymptotic time complexity of the whole dominance test construction, but it enables a constant time dominance test since interval lookup is as fast as an array access and interval test inclusion requires four comparisons between native integers.

4   Experimental Results

Implementations. We compare experimentally the following dominance tests:

I-CHK. This is the implementation of the CHK algorithm available from the Vellvm project [19]. To be able to plug it inside our middle-end, we have performed the slightest adaptation possible (essentially by-passing the abstract data-type of `atoms` and making them be bare `positive`; these are used to define program points in the CFG). We have kept the choices of data-structures used in their available development.

I-ACZ. This is the implementation of the AC algorithm available from the Vellvm project [19]. We performed the same adaptations as in I-CHK.

I-AC. This is the implementation of the AC algorithm initially available in CompCertSSA [2,7]. The implementation uses a classical Kildall workset algorithm for solving the data-flow equations, and its correctness is proved directly (no *a posteriori* validator). The CFG is stored in a `PTree` mapping to every node in the graph, the list of its successors. The Kildall solver, taken directly from CompCert, uses a Coq implementation of a heap data structure (splay tree). Dominator sets are implemented using `PTree` while I-ACZ [19] uses unsorted lists.

I-DT. This is the implementation presented in the previous section of this paper. Its correctness is partly validated, partly verified. The CFG and dominance test computation use `PTree`. The external computation of the dominator tree is done in OCaml, using the LT algorithm on arrays, for more efficiency.

---

[5] Recall that we rely on the standard implementation of positive integers, `positive`.

Table 1. Benchmarks description (lines of codes and categories of function sizes)

Program	LoC	Program	LoC	Program	LoC	Program	LoC	Program	LoC
bzip2	7007	mcf	2697	nsichneu	4033	lzss	3063	oggenc	58463
hmmer	13458	raytracer	2867	papabench	4763	lzw	2629	spass	82103

Cat. ID	#nodes	#func	Cat. ID	#nodes	#func	Cat. ID	#nodes	#func
(1)	$[500; 1k[$	229	(4)	$[4k; 8k[$	49	(7)	$[40k; 50k[$	13
(2)	$[1k; 2.5k[$	155	(5)	$[8k; 22k[$	42	(8)	$[50k; max]$	8
(3)	$[2.5k; 4k[$	66	(6)	$[22k; 40k[$	20			

I-DT-NAT. Same as I-DT, but in native mode. The CFG of the function and the dominance test computation use Patricia trees [15] on OCaml integers, and the pre-computed dominance test is stored in an OCaml array to allow for a constant time access. Dominator test is constant time.

I-LLVM. This is the OCaml-only version of the algorithm presented in the previous section, without *a posteriori* validation. The CFG is stored in a `PTree`. The dominator tree is computed using mutable OCaml data-structures (arrays and stacks), for more efficiency. Interval are stored in a mutable array but since graph nodes are encoded as Coq binary numbers (`positive`), dominator test is $O(\log |N|)$ instead of constant time.

I-LLVM-NAT. Same as I-LLVM, except that the CFG of the function and the dominance test computation use arrays. Dominator test is constant time. This provides a lower bound of the performance we could aim to achieve.

I-CHK, I-ACZ, and I-AC represent the state-of-the-art of mechanically verified dominance test (with a direct correctness proof and no extra validator).

Benchmarks. We plug each implementation in our SSA middle-end by extracting its Coq implementation into OCaml code, and running it on some realistic C program benchmarks, described in Table 1, taken from the CompCert test suite, the SPEC2006 benchmarks and WCET-related reference benchmarks. These represent around 192,600 lines of C code, each program ranging from thousands of lines of C code, to tens of thousands.

To evaluate the scalability of the implementations in extreme conditions, we force the compiler to always inline functions with a CFG size below 1000 nodes. We classify some of our results by categories of function size, i.e. the number of CFG nodes of its SSA form (see Table 1). We also present some global results, categorized by programs. Experiments were run on a MacBook OSX 10.8.5, 2.3 GHz Intel Core i7, 8 GB 1600 MHz DDR3.

Measures and Results. To evaluate and compare the dominance test implementations, we measure both the building time of the dominance test, and its practical cost in time, when using it.[6] Results are presented in Figs. 3 and 4.

---

[6] The impact of building and using the dominance test is currently negligible compared to the whole compilation time, as, currently, certain compiler passes (such as the SSA deconstruction) would need performance improvement.

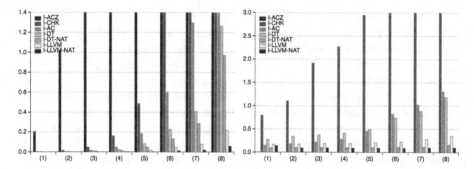

Fig. 3. Building (left) and using (right) times, by function size (the average time for each category). For I-ACZ, we set a time-out of 2 s for building the dominance test of one program function. Because of these time-outs, we do not show using times for I-ACZ (in practice, they are similar to, or higher than the ones for I-CHK).

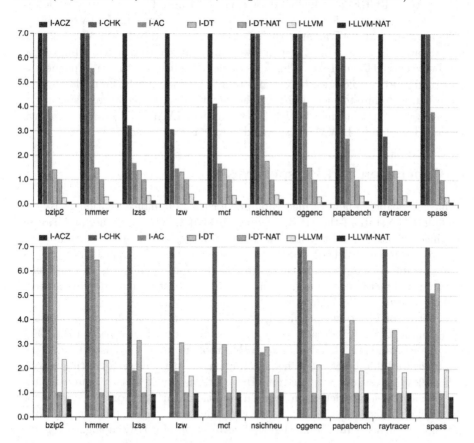

Fig. 4. Total building (top) and using (bottom) time overheads, relatively to the I-DT-NATIVE implementation, classified by programs.

The building time of the dominance test is the time, in seconds, required to compute the function `test_dom: node → node → bool`. For I-DT, and I-DT-NAT, this includes the validation time. As for the using time of dominance tests, we measure for each function the time, in micro-seconds, spent in executing dominance test requests. To avoid glitches in the measures of so small values, we performed 5 times the measures for all tests, and kept the lowest value. The collection of dominance tests is the same for all implementations, these are the ones required by our GVN-based CSE optimization (see Sect. 1). It is worth noting that our GVN-CSE performs the exact same set of requests to the dominance test, independently of the implementation that is used. Additionally, on this set of dominance test requests, we have checked that the various implementations were returning the same verdict (thus establishing their relative correctness and completeness one to each other).

In Figs. 3 and 4, we observe that I-DT performs significantly better than I-AC, I-ACZ, I-CHK for the building time. Setting I-DT to native mode (I-DT-NAT) improves performance. Comparing building times for I-LLVM against I-DT, and I-LLVM-NAT against I-DT-NAT gives a good estimate of the cost of the validator. In terms of using time, we do not observe so much a big difference between I-DT and I-AC. This is as expected, given that both implementations require `PTree` accesses. I-ACZ and I-CHK provide much slower dominance test because dominator sets are represented by lists. In Fig. 3, for using time, we observe a constant time for I-DT-NAT and I-LLVM-NAT thanks to arrays accesses. I-DT and I-LLVM are slower due to `PTrees` accesses.

5   Conclusion and Perspectives

We have described a new verified validator for the dominance relation. It is able to validate the state-of-the-art dominance construction by Lengauer and Tarjan combined with an ancestor test in the dominator tree candidate. This technique, borrowed from (un-verified) production compilers like GCC and LLVM, brings an important speedup compared to previous verified algorithms [2,19]. Using native data-structures after extraction, it builds a constant-time dominance test similar, in terms of efficiency, to the non-verified test.

In terms of program optimization, this dominance test already provides a strong support (i.e. we are able to perform efficient dominance test on the CFG on demand), and we already leverage this tool in our GVN-based CSE. This important building block could help us implement other powerful optimizations such as loop invariant code motion. However, the most efficient implementation of natural loops detection rely on iteration startegies on the dominator tree itself. In this case, the dominance checking is no longer sufficient, and one may have to investigate the mechanized verification of certifying algorithms for the dominator tree, such as the linear-time certifying algorithm by Georgiadis et al. [9].

References

1. Allen, F.E., Cocke, J.: Graph theoretic constructs for program control flow analysis. Technical report, IBM T.J. Watson Research Center (1972)
2. Barthe, G., Demange, D., Pichardie, D.: Formal verification of an SSA-based middle-end for CompCert. ACM TOPLAS **36**(1), 4:1–4:35 (2014)
3. Chlipala, A.: A verified compiler for an impure functional language. In: POPL 2010, pp. 93–106. ACM (2010)
4. Cooper, K.D., Harvey, T.J., Kennedy, K.: A simple, fast dominance algorithm. Technical report, Rice University (2006)
5. Cytron, R., Ferrante, J., Rosen, B.K., Wegman, M.N., Zadeck, F.K.: Efficiently computing static single assignment form and the control dependence graph. ACM TOPLAS **13**(4), 451–490 (1991)
6. Dargaye, Z., Leroy, X.: Mechanized verification of CPS transformations. In: Dershowitz, N., Voronkov, A. (eds.) LPAR 2007. LNCS (LNAI), vol. 4790, pp. 211–225. Springer, Heidelberg (2007)
7. Demange, D., Pichardie, D., Stefanesco, L.: Verifying fast and sparse SSA-based optimizations in Coq. In: Franke, B. (ed.) CC 2015. LNCS, vol. 9031, pp. 233–252. Springer, Heidelberg (2015)
8. Fluet, M., Weeks, S.: Contification using dominators. In: Proceedings of ICFP 2001, pp. 2–13. ACM (2001)
9. Georgiadis, L., Laura, L., Parotsidis, N., Tarjan, R.E.: Dominator certification and independent spanning trees: an experimental study. In: Demetrescu, C., Marchetti-Spaccamela, A., Bonifaci, V. (eds.) SEA 2013. LNCS, vol. 7933, pp. 284–295. Springer, Heidelberg (2013)
10. Georgiadis, L., Tarjan, R.E., Werneck, R.F.: Finding dominators in practice. J. Graph Algorithms Appl. **10**(1), 69–94 (2006)
11. Georgiadis, L., Tarjan, R.E.: Dominator tree verification and vertex-disjoint paths. In: Proceedings of SODA 2005, pp. 433–442. ACM (2005)
12. Lengauer, T., Tarjan, R.E.: A fast algorithm for finding dominators in a flowgraph. ACM TOPLAS **1**(1), 121–141 (1979)
13. Leroy, X.: Formal verification of a realistic compiler. Commun. ACM **52**(7), 107–115 (2009)
14. Leroy, X.: A formally verified compiler back-end. JAR **43**(4), 363–446 (2009)
15. Okasaki, C., Gill, A.: Fast mergeable integer maps. In: Workshop on ML, pp. 77–86 (1998)
16. Parotsidis, N., Georgiadis, L.: Dominators in directed graphs: a survey of recent results, applications, and open problems. In: 2nd International Symposium on Computing in Informatics and Mathematics (ISCIM 2013), vol. 1, pp. 15–20. Epoka University (2013)
17. Rivest, R.L., Cormen, T.H., Leiserson, C.E., Stein, C.: Introduction to Algorithms, 3rd edn. MIT Press, Cambridge (2009)
18. Zhao, J., Nagarakatte, S., Martin, M., Zdancewic, S.: Formal verification of SSA-based optimizations for LLVM. In: PLDI 2013, pp. 175–186. ACM (2013)
19. Zhao, J., Zdancewic, S.: Mechanized verification of computing dominators for formalizing compilers. In: Hawblitzel, C., Miller, D. (eds.) CPP 2012. LNCS, vol. 7679, pp. 27–42. Springer, Heidelberg (2012)
20. Zhao, J., Zdancewic, S., Nagarakatte, S., Martin, M.: Formalizing the LLVM intermediate representation for verified program transformation. In: POPL 2012, pp. 427–440. ACM (2012)

Refinement to Certify Abstract Interpretations, Illustrated on Linearization for Polyhedra

Sylvain Boulmé[(✉)] and Alexandre Maréchal

VERIMAG, Université Grenoble-Alpes, 38000 Grenoble, France
{sylvain.boulme,alex.marechal}@imag.fr

Abstract. Our concern is the modular development of a certified static analyzer in CoQ: we extend a certified abstract domain of convex polyhedra with a linearization procedure approximating polynomial expressions. In order to help such a development, we propose a proof framework, embedded in CoQ, that implements a refinement calculus.

1 Introduction

This paper presents two contributions: first, a certified linearization for an abstract domain of convex polyhedra, approximating polynomials by affine constraints; second, a refinement calculus, helping us to mechanize this proof in CoQ [1]. We detail below the context and the features of these two contributions.

1.1 A Certified Linearization for the Abstract Domain of Polyhedra

We consider the certification of an *abstract interpreter*, which aims to ensure absence of undefined behaviors such as division by zero or invalid memory access in an input source program. This analyzer computes for each program point an *invariant*: a property that the state at that point must satisfy in all executions. Such invariants belong to datatypes called *abstract domains* [2] which are syntactic classes of properties on memory states. For instance, in the abstract domain of *convex polyhedra* [3], invariants are conjunctions of affine constraints written $\sum_i a_i x_i \leq b$ where $a_i, b \in \mathbb{Q}$ are scalar values and x_i are integer program variables. Hence, this domain cannot deal directly with non-linear invariants, *e.g.* $x^2 - y^2 \leq x \times y$. Thus, linearization techniques, such as intervalization [4], are necessary to analyze programs with non-linear arithmetic.

Indeed, intervalization replaces some variables in a non-linear product by intervals of constants. For instance in Example 1, x is replaced by $[0, 10]$ in assignment $r := x.(y - z) + 10.z$. The interval is then eliminated by analyzing the sign of $y - z$, leading to affine constraints usable by the polyhedra domain.

This work was partially supported by ANR project "VERASCO" (INS 2011) and by the European Research Council under the European Union's Seventh Framework Programme (FP/2007-2013)/ERC Grant Agreement nr. 306595 "STATOR".

© Springer International Publishing Switzerland 2015
C. Urban and X. Zhang (Eds.): ITP 2015, LNCS 9236, pp. 100–116, 2015.
DOI: 10.1007/978-3-319-22102-1_7

Example 1 (Intervalization Using a Sign-Analysis).
In any state where $x \in [0, 10]$, assignment "$r := x.(y-z) +$
$10.z$" is approximated by the affine program on the right
hand-side. Here operator $:\in$ performs a non-deterministic
assignment.

if $y - z \geq 0$ then
$\quad r :\in [10.z, 10.y]$
else
$\quad r :\in [10.y, 10.z]$

Our certified linearization procedure is now part of the VERIMAG Polyhedra Library (VPL) [5,6], which provides a certified polyhedra domain to
VERASCO [7], a certified abstract interpreter for COMPCERT C [8]. Following a design proposed in [9], the VPL is organized as a two-tier architecture:
an untrusted oracle, combining OCAML and C code, performs most complex
computations and outputs a Farkas certificate used by a certified front-end to
build a correct-by-construction result. As oracles may have side-effects and bugs,
they are viewed in COQ as non-deterministic computations of an axiomatized
monad [6].

Built on a similar design, our linearization procedure invokes an untrusted
oracle that selects strategies for linearizing an arithmetic expression and produces a certificate which is checked by the certified part of the procedure. It leads
to a correct-by-construction over-approximation of the expression. It is convenient to see such strategies as program transformations, because their correctness
is independent from the implementation of the underlying abstract domain and
is naturally expressed using concrete semantics of programs [10]. Indeed, a linearization is correct if, in the current context of the analysis, any postcondition
satisfied by the output program is also satisfied on the input one (see Example 1).
In such a case, we say that the input program *refines* the output one. This paper
aims to explain how refinement helps to develop certified procedures on abstract
domains, and in particular our linearization algorithm.

1.2 Certifying Computations on Abstract Domains by Refinement

Program refinement [11] consists in decomposing proofs of complex programs
by stepwise applications of correctness-preserving transformations. We provide
a framework in COQ to apply this methodology for certifying the correctness of
computations combining operators of an existing abstract domain. Our framework provides a Guarded Command Language (GCL) called $^\dagger \mathbb{K}$ that contains
these operators. A computation $^\dagger K$ in $^\dagger \mathbb{K}$ comes with two types of semantics: an
abstract and a concrete one. Concrete semantics of $^\dagger K$ is a transformation on
memory states. Abstract semantics of $^\dagger K$ is a transformation on *abstract states*,
i.e. on values of the abstract domain. A $^\dagger \mathbb{K}$ computation also embeds a proof
that abstract semantics is correct *w.r.t.* concrete one: each $^\dagger \mathbb{K}$ operator thus
preserves correctness by definition. Moreover, an OCAML function is extracted
from abstract semantics which is certified to be correct *w.r.t.* concrete semantics. Hence, concrete semantics of $^\dagger K$ acts as a *specification* which is *implemented*
by its abstract semantics. In the following, a transformation on abstract (resp.
memory) states is called an abstract (resp. concrete) computation.

Taking a piece of code as input, our linearization procedure outputs a $^{\dagger}\mathbb{K}$ computation, and its correctness is ensured by proving that concrete semantics of its input refines concrete semantics of its output. It means that the output does not forget any behaviour of the input. Our procedure being developed in a modular way from small intermediate functions, its proof reduces itself to small refinement steps. Each of these refinement steps involves only concrete semantics. Our framework provides a tactic simplifying such refinement proofs by computational reflection of weakest-preconditions. The correctness of abstract semantics *w.r.t.* concrete semantics is ensured by construction of $^{\dagger}\mathbb{K}$ operators.

Our framework supports *impure* abstract computations, *i.e.* abstract computations that invoke imperative oracles whose results are certified *a posteriori*. It also allows to reason conveniently about higher-order abstract computations. In particular, our linearization procedure uses a Continuation-Passing-Style (CPS) [12] in order to partition its analyzes according to the sign of affine sub-expressions. For instance in Example 1, the approximation of the non-linear assignment depends on the sign of $y - z$. In our procedure, CPS is a higher-order programming style which avoids introducing an explicit datatype handling partitions: this simplifies both the implementation and its proof. This also illustrates the expressive power of our framework, since a simple Hoare logic does not suffice to reason about such higher-order imperative programs.

Our refinement calculus could have applications beyond the correctness of linearization strategies. In particular, the top-level interpreter of the analyzer could also be proved correct in this way. Indeed, the interpreter invokes operations on abstract domains in order to over-approximate any execution of the program, but its correctness does not depend on abstract domains implementations (as soon as these implementations are themselves correct). We illustrate this claim on a toy analyzer, also implemented in COQ.

The mathematics involved in our refinement calculus, relating operational semantics to the lattice structure of monotone predicate transformers, are well-known in abstract interpretation theory [13]. In parallel of our work, the idea to use a refinement calculus in formal proofs of abstract interpreters was proposed in [14]. Hence, our contribution is more practical than theoretical. On the theoretical side, we propose a refinement calculus dedicated to certification of *impure* abstract computations. On the practical side, we show how to get a concise implementation of such a refinement in COQ and how it helps on a realistic case study: a linearization technique within the abstract interpreter VERASCO.

1.3   Overview of the Paper

Our refinement calculus is implemented in only 350 lines of COQ (proof scripts included), by a shallow-embedding of our GCL $^{\dagger}\mathbb{K}$ which combines computational reflection of weakest-preconditions [15] with monads [16]. However, it can be more simply understood in classical set theory, using binary relations instead. However, it can be understood in a much simpler setting using binary relations instead of monads and weakest-preconditions, and classical set theory instead of COQ. Section 2 introduces our refinement calculus in this simplified setting, where computations are represented as binary relations. Section 3 presents our certified linearization procedure and how its proof benefits from our refinement calculus. Section 4 explains how we mechanize this refinement calculus in CoQ by using smart encodings of binary relations introduced in Sect. 2.

2   A Refinement Calculus for Abstract Interpretation

We consider an analyzer correct if and only if it rejects all programs that may lead to an *error state*: due to lack of precision, it may also reject safe programs. Let us begin by defining the notion of error state and semantics of concrete computations, combining big-steps operational semantics with Hoare Logic.

Notations on Relations. The whole paper abusively uses classical set theory, whereas our formalization is in the intuitionistic type theory of CoQ without axioms. In particular, it identifies type $A \to \mathtt{Prop}$ of predicates on A with set $\mathcal{P}(A)$. Hence, we note $\mathcal{R}(A, B) \triangleq \mathcal{P}(A \times B)$ the set of binary relations on $A \times B$. Given R of $\mathcal{R}(A, B)$, we note $x \xrightarrow{R} y$ instead of $(x, y) \in R$. We use operators on $\mathcal{R}(A, A)$ inspired from regular expressions: ε is *the identity relation* on A, $R_1 \cdot R_2$ means "*relation R_2 composed with R_1*" (i.e. $x \xrightarrow{R_1 \cdot R_2} z \triangleq \exists y, x \xrightarrow{R_1} y \wedge y \xrightarrow{R_2} z$) and R^* is the reflexive and transitive closure of R. In all the paper, $A \to B$ is a type of *total* functions.

2.1   Stepwise Refinement of Concrete Computations

Given a domain D representing the type of memory states, we add a distinguished element $\notz$ to D in order to represent the error state: we define $D_\notz \triangleq D \uplus \{\notz\}$.

Specifying Concrete Computations with Runtime Errors. We define the set of concrete computations as $\mathbb{K} \triangleq \mathcal{R}(D, D_\notz)$. Hence, an element K of $\mathbb{K}$ corresponds to a (possibly) non-deterministic or non-terminating computation from an *input state* of type D to an *output state* of type $D_\notz$. Typically, the empty relation represents a computation that loops infinitely for any input. It also represents unreachable code (dead code).

   We denote by $\downarrow K$ the normalization of the computation K that returns any output in case of error. It is defined by $d \xrightarrow{\downarrow K} d' \triangleq (d \xrightarrow{K} d' \vee d \xrightarrow{K} \notz)$.

Refinement Pre-order and Hoare Specifications. We equip $\mathbb{K}$ with a *refinement pre-order* $\sqsubseteq$ such that $K_1 \sqsubseteq K_2$ iff $K_1 \subseteq \downarrow K_2$ (or equivalently, $\downarrow K_1 \subseteq \downarrow K_2$). Informally, an abstract analysis correct for K_2 is also correct for K_1. The equivalence relation $\equiv$ associated with this pre-order is given by $K_1 \equiv K_2$ iff $\downarrow K_1 = \downarrow K_2$.

   Hoare logic is a standard and convenient framework to reason about imperative programs. Let us explain how computations in $\mathbb{K}$ are equivalent to specifications of Hoare logic. A computation K corresponds to a Hoare specification $(\mathfrak{p}_K, \mathfrak{q}_K)$ of $\mathcal{P}(D) \times \mathcal{R}(D, D)$, where $\mathfrak{p}_K$ is a precondition ensuring the absence of error, and $\mathfrak{q}_K$ a postcondition relating the input state to a non-error output state,[1] defined by $\mathfrak{p}_K \triangleq D \setminus \{d \mid d \overset{K}{\to} \sharp\}$ and $\mathfrak{q}_K \triangleq K \cap (D \times D)$. Conversely, any Hoare specification (P, Q) corresponds to a computation $\vdash P;Q$ – defined below – such that $K \equiv \vdash \mathfrak{p}_K;\mathfrak{q}_K$. Moreover, the refinement pre-order $K_1 \sqsubseteq K_2$ is equivalent to $\mathfrak{p}_{K_2} \subseteq \mathfrak{p}_{K_1} \wedge \mathfrak{q}_{K_1} \cap (\mathfrak{p}_{K_2} \times D) \subseteq \mathfrak{q}_{K_2}$. Thus, it is equivalent to the usual refinement of specifications in Hoare logic.

Algebra of Guarded Commands. We now equip $\mathbb{K}$ with an algebra of guarded commands inspired by [11].[2] It combines a *complete* lattice structure with operators inspired from regular expressions. Here, we present this algebra in the case where $\mathbb{K}$ is represented as $\mathcal{R}(D, D_\sharp)$. In our CoQ implementation (given in Sect. 4), this representation is changed in order to mechanize refinement proofs.

First, the complete lattice structure of $\mathbb{K}$ (for pre-order $\sqsubseteq$) is given by operator $\sqcap$ defined as "$\cap$ after normalization" (e.g. $\bigsqcap_i K_i \triangleq \bigcap_i \downarrow K_i$) and by operator $\sqcup$ simply defined as $\cup$. In our context, $\sqcup$ represents alternatives that may non-deterministically happen at runtime: the analyzer must consider that each of them may happen. Symmetrically, $\sqcap$ represents some choice left to the analyzer. Empty relation $\emptyset$ is the bottom element and is noted $\bot$. Relation $D \times \{\sharp\}$ is the top element. Given $d \in D_\sharp$, we implicitly coerce it as the constant relation $D \times \{d\}$. Hence, the top element of the $\mathbb{K}$ lattice is simply noted $\sharp$. The notation $\uparrow f$ explicitly lifts function f of $D \to D$ in $\mathbb{K}$.   •

Given a relation $K \in \mathcal{R}(D, D_\sharp)$, we define its lifting $\upharpoonright K$ in $\mathcal{R}(D_\sharp, D_\sharp)$ by $\upharpoonright K \triangleq K \cup \{(\sharp, \sharp)\}$. This allows us to define the sequence of computations by $K_1;K_2 \triangleq K_1 \cdot \upharpoonright K_2$, and the unbounded iteration of this sequence (*i.e.* a loop with a runtime-chosen number of iterations) by $K^* \triangleq (\upharpoonright K)^* \cap (D \times D_\sharp)$.

Given a predicate $P \in \mathcal{P}(D)$, we define the notion of *assumption* (or *guard*) as $\dashv P \triangleq (P \times D) \sqcap \varepsilon$. Informally, if P is satisfied on the current state then $\dashv P$ skips like ε. Otherwise, $\dashv P$ produces no output like $\bot$. We also define the dual notion of *assertion* as $\vdash P \triangleq (\dashv \neg P;\sharp) \sqcup \varepsilon$. If P is not satisfied on the current state, then $\vdash P$ produces an error. Otherwise, it skips.

Hence, $\mathbb{K}$ provides a convenient language to express specifications: any Hoare specification (P, Q) of $\mathcal{P}(D) \times \mathcal{R}(D, D)$ is expressed as the computation $\vdash P;Q$. Moreover, refinement allows to express usual Verification Conditions (VC) of Hoare Logic. For our toy analyzer – described later – we use the usual partial correctness VC of unbounded iteration: K^* is equivalent to produce an output satisfying *every* inductive invariant I of K.

$$K^* \equiv \bigsqcap\nolimits_{I \in \{I \in \mathcal{P}(D) \mid K \sqsubseteq \vdash I; D \times I\}} \vdash I; D \times I$$

---

[1] A postcondition is thus in $\mathcal{P}(D \times D)$ instead of the original $\mathcal{P}(D)$: this standard generalization avoids introducing "auxiliary variables" to represent the input state.

[2] However, in our algebra, $\sqsubseteq$ corresponds to "*refines*", whereas in standard refinement calculus it dually corresponds to "*is refined by*". Actually, our convention follows lattice notations of abstract interpretation.

In this equivalence, the $\sqsubseteq$-way corresponds to the soundness of the VC, whereas the $\sqsupseteq$-way corresponds to its completeness. In our context, such a soundness proof typically ensures that the specification of an abstract computation is refined by concrete semantics of the analyzed code. It guarantees that the analysis is correct *w.r.t.* semantics of the analyzed code.

Example on a Toy Language. Let t stands for an arithmetic term and c be a condition over numerical variables, whose syntax is $c ::= t_1 \bowtie t_2 \mid \neg c \mid c_1 \wedge c_2 \mid c_1 \vee c_2$ with $\bowtie \in \{=, \neq, \leq, \geq, <, >\}$. Semantics $[\![t]\!]$ of t and $[\![c]\!]$ of c work with a domain of integer memories $D \triangleq \mathbb{V} \to \mathbb{Z}$ where $\mathbb{V}$ is the type of variables. Hence, $[\![t]\!] \in D \to \mathbb{Z}$ and $[\![c]\!] \in \mathcal{P}(D)$. We omit their definition here.

Let us now introduce a small imperative programming language named $\mathbb{S}$ for which we will describe a toy analyzer in Sect. 2.2. The syntax of a $\mathbb{S}$ program s is described on Fig. 1 together with its big-steps semantics $[\![s]\!]$ defined as an element of $\mathbb{K}$. This semantics is defined recursively on the syntax of s using guarded commands derived from $\mathbb{K}$. First, we define $\dashv c \triangleq \dashv [\![c]\!]$ and $\vdash c \triangleq \vdash [\![c]\!]$. We also use command "$x := t$" defined as $\uparrow \lambda d.d[x := [\![t]\!](d)]$, where the memory assignment noted "$d[x := n]$" – for $d \in D$, $x \in \mathbb{V}$ and $n \in \mathbb{Z}$ – is defined as the function $\lambda x' : \mathbb{V}, \text{if } x' = x \text{ then } n \text{ else } d(x')$.

s	$\mathbf{assert}(c)$	$x \leftarrow t$	$s_1 ; s_2$	$\mathbf{if}(c)\{s_1\}\mathbf{else}\{s_2\}$	$\mathbf{while}(c)\{s\}$
$[\![s]\!]$	$\vdash c$	$x := t$	$[\![s_1]\!] ; [\![s_2]\!]$	$\dashv c ; [\![s_1]\!]$ $\sqcup$ $\dashv \neg c ; [\![s_2]\!]$	$(\dashv c ; [\![s]\!])^* ; \dashv \neg c$

Fig. 1. Syntax and concrete semantics of $\mathbb{S}$

At this point, we have defined an algebra $\mathbb{K}$ of concrete computations: a language that we use to express specifications – for instance, in the form of Hoare specifications – on abstract computations. This algebra also provides denotations for defining big-steps semantics (like in Fig. 1). Hence, $\mathbb{K}$ is aimed at providing an intermediate level between operational semantics of programs and their abstract interpretations. The next section defines how we certify correctness of abstract computations with respect to $\mathbb{K}$ computations.

2.2 Composing Diagrams to Certify Abstract Computations

Rice's theorem states that the property $d \xrightarrow{K} d'$ is undecidable. In the theory of abstract interpretation, we approximate K by a *computable (terminating) function* $^\sharp K$ working on an approximation $^\sharp D$ of $\mathcal{P}(D)$. Set $^\sharp D$ is called an *abstract domain* and it is related to $\mathcal{P}(D)$ by a concretization function $\gamma : {}^\sharp D \to \mathcal{P}(D)$. Function $^\sharp K$ is called an *abstract interpretation* (or *abstract computation*) of K. This paper considers two abstract domains, intervals and convex polyhedra, associated with the concrete domain $D \triangleq \mathbb{V} \to \mathbb{Z}$ involved in Fig. 1.

1. Given $\mathbb{Z}_\infty \triangleq \mathbb{Z} \uplus \{-\infty, +\infty\}$, an abstract memory $^\sharp d$ of the interval domain is a finite map associating each variable x with an interval $[a_x, b_x]$ of $\mathbb{Z}_\infty \times \mathbb{Z}_\infty$. Its concretization is the set of concrete memory states satisfying the constraints of $^\sharp d$, *i.e.* $\gamma(^\sharp d) \triangleq \{d \in D \mid \forall x, a_x \leq d(x) \leq b_x\}$.

2. The concretization of a convex polyhedron $^\sharp d = \bigwedge_i \sum_j a_{ij}.x_j \leq b_i$, where a_{ij} and b_i are rational constants, and x_j are integer program variables is $\gamma(^\sharp d) \triangleq \{d \in D \mid \bigwedge_i \sum_j a_{ij}.d(x_j) \leq b_i\}$.

Correctness Diagrams of Impure Abstract Computations. Our framework only deals with partial correctness: we do not prove that abstract computations terminate, but only that they are a sound over-approximation of their corresponding concrete computation. Moreover, abstract computations may invoke untrusted oracles, whose results are verified by a certified checker. A bug in those oracles may make the whole computation non-deterministic or divergent. Thus, it is potentially unsound to consider abstract computations as pure functions. In this simplified presentation of our framework, we define abstract computations as relations in $\mathcal{R}(^\sharp D, ^\sharp D)$. In order to extract abstract computations from CoQ to OCAML functions, we will improve this representation of abstract computations in Sect. 4.

We express correctness of abstract computations through commutative diagrams represented on the right hand side and defined as follows.

Definition 1 (Correctness of Abstract Computations). *An abstract computation* $^\sharp K \in \mathcal{R}(^\sharp D, ^\sharp D)$ *is correct w.r.t.a concrete computation* $K \in \mathcal{R}(D, D_\xi)$ *iff*

$$\forall^\sharp d, ^\sharp d' \in {}^\sharp D, \quad \forall d \in D, \forall d' \in D_\xi,$$

$$^\sharp d \xrightarrow{^\sharp K} {}^\sharp d' \wedge d \xrightarrow{K} d' \wedge d \in \gamma(^\sharp d) \quad \Rightarrow \quad d' \in \gamma(^\sharp d')$$

Note that $d' \in \gamma(^\sharp d')$ implies itself that $d' \neq \xi$.

Such a diagram thus corresponds to a pair of an abstract and a concrete computation, with a proof that the abstract one is correct *w.r.t.* the concrete one. As illustrated on the example below, these diagrams allow to build *compositional proofs* that an abstract computation, composed of several simpler parts, is correct *w.r.t.* a concrete computation. Diagrams are indeed preserved by several composition operators, and also by refinement of concrete computations.

As an example, consider two abstract computations $^\sharp K_1$ and $^\sharp K_2$ which are correct *w.r.t.* concrete K_1 and K_2. In order to show that the sequential composition $^\sharp K_1 \cdot {}^\sharp K_2$ is correct *w.r.t.* concrete K, it suffices to prove that $K \sqsubseteq K_1 \,; K_2$, as illustrated on the right hand side scheme.

In the following, we introduce a datatype noted $\mathbb{K}$ to represent these diagrams: a diagram $^\dagger K \in \mathbb{K}$ represents an abstract computation $^\sharp K$ which is correct *w.r.t.* its associated concrete computation K. The core of our approach is to lift guarded-commands on $\mathbb{K}$ involved in Fig. 1 as guarded-commands on $^\sharp\mathbb{K}$. For instance, our toy analyzer $^\sharp[\![s]\!]$ for s in $\mathbb{S}$ is defined similarly to $[\![s]\!]$ of Fig. 1, but from $^\sharp\mathbb{K}$ operators instead of $\mathbb{K}$ ones. For a given diagram $^\dagger K$, we can prove the correctness of an abstract computation $^\sharp K$ w.r.t. a concrete computation K' simply by proving that $K' \sqsubseteq K$. In practice, such refinement proofs are simplified using a weakest-liberal-precondition calculus (see Sect. 4).

Our Interface of Abstract Domains. We derive our guarded-commands on $^\sharp\mathbb{K}$ in a generic way from the VPL interface [6] of abstract domains, reformulated here on Fig. 2. Besides its concretization function γ, an abstract domain $^\sharp D$ provides constants $^\sharp\top$ and $^\sharp\bot$, representing respectively predicate true and false. It also provides abstract computations $^\sharp\dashv c$ and $x^\sharp{:=}t$ of $\mathcal{R}(^\sharp D, {}^\sharp D)$, which are respectively correct w.r.t. concrete computations $\dashv c$ and $x := t$. It provides operator $^\sharp\sqcup$ of $\mathcal{R}(^\sharp D{\times}^\sharp D, {}^\sharp D)$, which over-approximates the binary union on $\mathcal{P}(D)$. At last, it provides inclusion test $^\sharp\sqsubseteq$ of $\mathcal{R}(^\sharp D{\times}^\sharp D, \texttt{bool})$.

$$D \subseteq \gamma(^\sharp\top) \qquad \gamma(^\sharp\bot) \subseteq \emptyset \qquad {}^\sharp d \xrightarrow{^\sharp\dashv c} {}^\sharp d' \;\Rightarrow\; \gamma(^\sharp d) \cap [\![c]\!] \subseteq \gamma(^\sharp d')$$

$$^\sharp d \xrightarrow{x^\sharp{:=}t} {}^\sharp d' \wedge d \in \gamma(^\sharp d) \Rightarrow d[x := [\![t]\!](d)] \in \gamma(^\sharp d')$$

$$(^\sharp d_1, {}^\sharp d_2) \xrightarrow{^\sharp\sqcup} {}^\sharp d' \Rightarrow \gamma(^\sharp d_1) \cup \gamma(^\sharp d_2) \subseteq \gamma(^\sharp d') \qquad (^\sharp d_1, {}^\sharp d_2) \xrightarrow{^\sharp\sqsubseteq} \texttt{true} \Rightarrow \gamma(^\sharp d_1) \subseteq \gamma(^\sharp d_2)$$

Fig. 2. Correctness specifications of our abstract domains

Abstract Computations of Guarded-commands. We now lift each $\mathbb{K}$ guarded-command of Fig. 1 into a $^\sharp\mathbb{K}$ guarded-command. A $^\sharp\mathbb{K}$ operator has the same notation than its corresponding $\mathbb{K}$ operator. Below, we associate each concrete operator of Fig. 1 with an abstract computation. The diagrammatic proof relating them is straightforward from correctness specifications given on Fig. 2.

Concrete commands $\dashv c$ and $x := t$ are associated with $^\sharp\dashv c$ and $x^\sharp{:=}t$. Concrete command $K_1 ; K_2$ is associated with $^\sharp K_1 \cdot {}^\sharp K_2$ – where $^\sharp K_2$ returns $^\sharp\bot$ if the current abstract state is included in $^\sharp\bot$, or runs $^\sharp K_2$ otherwise. Concrete $K_1 \sqcup K_2$ is lifted by applying operator $^\sharp\sqcup$ to the results of $^\sharp K_1$ and $^\sharp K_2$.

Concrete assertion $\vdash c$ is associated with checking that the result of $^\sharp\dashv \neg c$ is *included* in $^\sharp\bot$: otherwise, the abstract computation fails.[3] Hence, concrete $\frac{\iota}{\ell}$ is associated with abstract computation $\emptyset$ (concrete $\bot$ is associated with $^\sharp\bot$).

At last, concrete K^* is associated with an abstract computation which invokes an untrusted oracle proposing an inductive invariant of $^\sharp K$ for the current abstract state. Thus, using inclusion tests, $^\sharp(K^*)$ checks that the invariant proposed by the oracle is actually an inductive invariant (otherwise, it fails), before returning this invariant as the output abstract state.

---

[3] In practice, it may raise an alarm for the user, see our handling of alarms in the extended version of this paper [17].

2.3 Higher-Order Programming with Correctness Diagrams

Our linearization procedure detailed in Sect. 3.2 illustrates how we use GCL $^\dagger\mathbb{K}$ as a programming language for abstract computations. GCL $\mathbb{K}$ is our specification language. Each program $^\dagger K$ of $^\dagger\mathbb{K}$ is associated with a specification K of $\mathbb{K}$ syntactically derived from its code, meaning that each $^\dagger\mathbb{K}$ operator is syntactically associated with the $\mathbb{K}$ operator from which it is lifted in the above paragraph.

Our linearization procedure invokes two other operators of $^\dagger\mathbb{K}$. First, an operator which *casts* $^\dagger K$ to a given specification K': it requires $K' \sqsubseteq K$ in order to produce a new valid $^\dagger\mathbb{K}$ diagram. This cast operator thus leads to a modular design of the certified development since it allows stepwise refinement of $^\dagger\mathbb{K}$ diagrams. Second, given a computation π of $\mathcal{R}(^\sharp D, A)$ where A is a given type, it invokes an operator binding the results of π to a function $^\dagger g$ of $A \to {}^\dagger\mathbb{K}$. This operator requires a *concrete* postcondition Q of $A \to \mathcal{P}(D)$ on the results of π. In other words, under the condition $\forall^\sharp d, \forall x \in A, {}^\sharp d \xrightarrow{\pi} x \Rightarrow \gamma(^\sharp d) \subseteq Q\,x$, we define the diagram $\pi^\dagger\ggeq_Q {}^\dagger g$ as the abstract computation $\{(^\sharp d_1, {}^\sharp d_2) \mid \exists x, {}^\sharp d_1 \xrightarrow{\pi} x \wedge {}^\sharp d_1 \xrightarrow{^\sharp g\,x} {}^\sharp d_2\}$ specified by $\bigsqcap_x \vdash Q\,x\,;g\,x$.

Actually, Sect. 3.2 applies our refinement calculus to certify higher-order abstract computations. Indeed, our linearization procedure *partitions* abstract states in order to increase precision. Continuation-Passing-Style (CPS) [12] is a higher-order pattern which provides a lightweight and modular style to program and certify simple partitioning strategies. Let us now detail this idea.

Typically, given an abstract state $^\sharp d$, our linearization procedure invokes a sub-procedure $^\sharp f$ that splits $^\sharp d$ into a partition $(^\sharp d_i)_{i \in I}$ and computes a value r_i (of a given type A) for each cell $^\sharp d_i$. Then, the linearization procedure *continues* the computation from each cell $(r_i, {}^\sharp d_i)$ to finally return the join of all cells. In other words, from $^\sharp d$, $^\sharp f$ computes $(r_i, {}^\sharp d_i)_{i \in I}$. The main procedure finally computes $^\sharp\bigsqcup_{i \in I}(^\sharp g\ r_i\ {}^\sharp d_i)$ – where $^\sharp g$ is a given function of $A \to \mathcal{R}(^\sharp D, {}^\sharp D)$. In order to avoid explicit handling of partitions, we make $^\sharp g$ a parameter of $^\sharp f$ to perform the join inside $^\sharp f$. In this style, $^\sharp f$ is of type $(A \to \mathcal{R}(^\sharp D, {}^\sharp D)) \to \mathcal{R}(^\sharp D, {}^\sharp D)$ and the parameter $^\sharp g$ of $^\sharp f$ is called its *continuation*.

However, specifying directly the correctness of computations that use CPS is not obvious because of the higher-order parameter. Actually, we define $^\dagger f$ of type $(A \to {}^\dagger\mathbb{K}) \to {}^\dagger\mathbb{K}$ and work with a continuation $^\dagger g$ of type $A \to {}^\dagger\mathbb{K}$, therefore keeping implicit the notion of partition, both in specification and in implementation. This CPS technique could also be applied to simple strategies of trace-partitioning without a trace-partitioning domain [17].

3 Interval-Based Linearization Strategies for Polyhedra

As described in [6], VPL works with affine terms given by the abstract syntax $t ::= n \mid x \mid t_1 + t_2 \mid n.t$ where x is a variable and n a constant of $\mathbb{Z}$. We now extend VPL operators of Fig. 2 to support polynomial terms, where the product "$n.t$" is generalized into "$t_1 \times t_2$".

The VPL derives assignment operator $\overset{\sharp}{=}$ from guard $\overset{\sharp}{\dashv}$ and two low-level operators: projection and renaming. It also derives the guard operator from a restricted one where conditions have the form $0 \bowtie t$ where $\bowtie \in \{\leq, =, \neq\}$. Hence, we only need to linearize the restricted guard $\overset{\sharp}{\dashv} 0 \bowtie t$, where t is a polynomial. Below, we use letter p for polynomials and only keep letter t for affine terms.

Roughly speaking, we approximate a guard $\overset{\sharp}{\dashv} 0 \bowtie p$ by guards $\overset{\sharp}{\dashv} 0 \bowtie [t_1, t_2]$ – where t_1 and t_2 are affine or infinite bounds – such that, in the current abstract state, $p \in [t_1, t_2]$. Approximated guards $\overset{\sharp}{\dashv} 0 \bowtie [t_1, t_2]$ are defined by cases on $\bowtie$:

$\bowtie$	$\leq$	$=$	$\neq$
$\overset{\sharp}{\dashv} 0 \bowtie [t_1, t_2]$	$\overset{\sharp}{\dashv} 0 \leq t_2$	$\overset{\sharp}{\dashv} 0 \leq t_2 \wedge t_1 \leq 0$	$\overset{\sharp}{\dashv} 0 < t_2 \vee t_1 < 0$

Affine intervals are computed using heuristics inspired from [4], except that in order to increase precision, we dynamically partition the abstract state according to the sign of some affine subterms. This process will be detailed further. More complex and precise linearization methods exist, implying more advanced mathematics such as Bernstein's basis [18] or Handelman representation of polynomials [19]. Intervalization is clearly faster than others [10], and its precision-versus-efficiency trade-off may be controlled by several heuristics.

Our certified linearization is built on a two-tier architecture: an untrusted oracle uses heuristics to select linearization strategies and a certified procedure applies them to build a correct-by-construction result. Let us now list these strategies and their effect on the precision-versus-efficiency trade-off.

3.1   Our List of Interval-Based Strategies

Constant Intervalization. Our fastest strategy applies an intervalization operator of the abstract domain. Given a polynomial p, this operator, written $\overset{\sharp}{\pi}(p)$, over-approximates p by an interval where affine terms are reduced to constants. More formally, $\overset{\sharp}{\pi}(p)$ is a computation of $\mathcal{R}(\overset{\sharp}{D}, \mathbb{Z}_\infty^2)$ such that if $\overset{\sharp}{d} \xrightarrow{\overset{\sharp}{\pi}(p)} [n_1, n_2]$, then $\gamma(\overset{\sharp}{d}) \subseteq \{d \mid n_1 \leq [\![p]\!]d \leq n_2\}$. It uses a naive interval domain, built on the top of the polyhedral domain. Arithmetic operations $+$ and $\times$ are approximated by corresponding operations on intervals: $[n_1, n_2] + [n_3, n_4] \triangleq [n_1 + n_3, n_2 + n_4]$ and $[n_1, n_2] \times [n_3, n_4] \triangleq [\min(E), \max(E)]$ where $E = \{n_1.n_3, n_1.n_4, n_2.n_3, n_2.n_4\}$.

Ring Rewriting. A weakness of operator $\overset{\sharp}{\pi}$ is its sensitivity to ring rewriting. For instance, consider a polynomial p_1 such that $\overset{\sharp}{\pi}(p_1)$ returns $[0, n]$, $n \in \mathbb{N}^+$. Then $\overset{\sharp}{\pi}(p_1 - p_1)$ returns $[-n, n]$ instead of the precise result 0. Such imprecision occurs in barycentric computations such as $p_2 \triangleq p_1 \times t_1 + (n - p_1) \times t_2$ where affine terms t_1, t_2 are bounded by $[n_1, n_2]$. Indeed $\overset{\sharp}{\pi}(p_2)$ returns $2n.[n_1, n_2]$ instead of $n.[n_1, n_2]$. Moreover, if we rewrite p_2 into an equivalent polynomial $p_2' \triangleq p_1 \times (t_1 - t_2) + n.t_2$, then $\overset{\sharp}{\pi}(p_2')$ returns $n.[2.n_1 - n_2, 2.n_2 - n_1]$. If $n_1 > 0$ or $n_2 < 0$, then $\overset{\sharp}{\pi}(p_2')$ is strictly more precise than $\overset{\sharp}{\pi}(p_2)$. The situation is reversed otherwise. Consequently, our oracle begins by simplifying the polynomial before trying to factorize it conveniently. But as illustrated above, it is difficult to find a factorization minimizing $\overset{\sharp}{\pi}$ results. We give more details on the ring rewriting heuristics of our oracle in the following.

Sign Partitioning. In order to find more precise bounds of a polynomial p than those given by $\sharp\pi(p)$, we may split the current abstract state $\sharp d$ into a partition $(\sharp d_i)_{i\in I}$ according to the sign of some affine subterms of p, such that each cell $\sharp d_i$ may lead to a distinct affine interval $[t_{i,1}, t_{i,2}]$. Finally, $\sharp\!\!\sqcup 0 \bowtie p$ is over-approximated by computing the join of all $\sharp\!\!\sqcup 0 \bowtie [t_{i,1}, t_{i,2}]$.

For example, given an affine term t and a polynomial p' such that $\sharp\pi(p')$ returns $[n'_1, n'_2]$, if $0 \le t$ then $p' \times t \in [n'_1.t, n'_2.t]$, otherwise $p' \times t \in [n'_2.t, n'_1.t]$. When the sign of t is known, sign partitioning allows to discard one of these two cases and thus gives a fast affine approximation of $p' \times t$. The main drawback of sign partitioning is a worst-case exponential blow-up if applied systematically.

Let us illustrate sign partitioning for the previous barycentric-like computation of p'_2. By convention, our certified procedure partitions the sign of right affine subterms (here, the sign of $t_1 - t_2$). Hence, it founds $p'_2 \in [n.t_2, n.t_1]$ in cell $0 \le t_1 - t_2$, and $p'_2 \in [n.t_1, n.t_2]$ in cell $t_1 - t_2 < 0$. When it joins the two cells, $\sharp\!\!\sqcup 0 \bowtie p'_2$ is computed as $\sharp\!\!\sqcup 0 \bowtie n.[n_1, n_2]$ as we can expect for such a barycentre. Remark that sign partitioning is also sensitive to ring rewriting. In particular, the oracle may rewrite a product of affine terms $t_1 \times t_2$ into $t_2 \times t_1$, in order to discard t_1 instead of t_2 by sign-partitioning.

Focusing. Focusing is a ring rewriting heuristic which may increase the precision of sign partitioning. Given a product $p \triangleq t_1 \times t_2$, we define the *focusing* of t_2 in center n as the rewriting of p into $p' \triangleq n.t_1 + t_1 \times (t_2 - n)$. Thanks to this focusing, the affine term $n.t_1$ appears whereas t_1 would otherwise be discarded by sign partitioning. Let us simply illustrate the effect of this rewriting when $0 \le n \le n'_1$ with t_1 (resp. t_2) bounded by $[n_1, n_2]$ (resp. $[n'_1, n'_2]$). Sign partitioning bounds p in affine interval $[n_1.t_2, n_2.t_2]$ whereas p' is bounded by interval $[n_1.t_2 + n.(t_1 - n_1), n_2.t_2 - n.(n_2 - t_1)]$. The former contains the latter since $t_1 - n_1$ and $n_2 - t_1$ are non-negative. Under these assumptions, the precision is maximal when $n = n'_1$.

Applied carelessly, focusing may also decrease the precision. Consequently, on products $p'' \times t_2$, our oracle uses the following heuristic which can not decrease the precision: if $0 \le n'_1$, then focus t_2 in center n'_1; if $n'_2 \le 0$, then focus t_2 in center n'_2; otherwise, do not try to change the focus of t_2.

Static vs Dynamic Intervalization During Partitioning. Computing the constant bounds of an affine term inside a given polyhedron invokes a costly linear programming procedure. Hence, for a given polynomial p to approximate, we start by computing an environment σ that associates each variable of p with a constant interval: as detailed later, this environment is indeed used by heuristics of our oracle. By default, operator $\sharp\pi$ is called in *dynamic* mode, meaning that each bound is computed dynamically in the current cell – generated from sign partitioning – using linear programming. If one wants a faster use of operator $\sharp\pi$, he may invoke it in *static* mode, where bounds are computed using σ.

For instance, let us consider the sign-partitioning of $p \triangleq t_1 \times t_2$ in the context $0 < n_1, n_2$ and $-n_1 \le t_2 \le t_1 \le n_2$. In cell $0 \le t_2$, static mode bounds p by $[-n_1.t_2, n_2.t_2]$, whereas dynamic mode bounds p by $[0, n_2.t_2]$. In cell $t_2 < 0$, both

Given $^t\pi\, p$ of $(\mathbb{Z}^2_\infty \to {}^t\mathbb{K}) \to {}^t\mathbb{K}$ defined by $^t\pi\, p\, {}^t g_0 \triangleq {}^{\sharp}\pi(p) \, {}^t\!\!\gg=_{\lambda[n_1,n_2],\{d\,|\,n_1\leq \llbracket p \rrbracket d \leq n_2\}} {}^t g_0$
the $^t\mathbb{K}$ program on the right-hand side satisfies the specification below:

$$\bigsqcap_{[t_1,t_2]} \vdash \{d \,|\, t_1 \leq \llbracket p \times t \rrbracket d \leq t_2\}\,; g[t_1,t_2]$$

---

if static then
$\quad ^t\pi\ p\ (\lambda[n_1,n_2],\ (\neg 0 \leq t\,; {}^t g[n_1.t, n_2.t])$
$\qquad\qquad\qquad \sqcup\ (\neg t < 0\,; {}^t g[n_2.t, n_1.t]))$
else
$\qquad (\neg 0 \leq t\,; {}^t\pi\ p\ \lambda[n_1,n_2],\ {}^t g[n_1.t, n_2.t])$
$\qquad \sqcup\ (\neg t < 0\,; {}^t\pi\ p\ \lambda[n_1,n_2],\ {}^t g[n_2.t, n_1.t])$

Fig. 3. Sign-partitioning for $p \times t$ with continuation $^t g$

modes bound p by $[n_2.t_2, -n_1.t_2]$. On the join of these cells, both modes give the same upper bound. But the lower bound is $-n_1.n_2$ for static mode, whereas it is $\frac{n_1.n_2}{n_1+n_2}(t_2 + n_1) - n_1.n_2$ for dynamic mode, which is strictly more precise.

3.2   Design of Our Implementation

For a guard $^{\sharp}\!\neg 0 \bowtie p$, our certified procedure first rewrites p into $p' + t$ where t is an affine term and p' a polynomial. This may keep the non-affine part p' small compared with the affine one t. Typically, if p' is syntactically equal to zero, we simply apply the standard affine guard $^{\sharp}\!\neg 0 \bowtie t$. Otherwise, we compute environment σ for p' variables and invoke our external oracle on p' and σ. This oracle returns a polynomial p'' enriched with tags on subexpressions. We handle three tags to direct the intervalization: AFFINE expresses that the subexpression is affine; STATIC expresses that the subexpression has to be intervalized in static mode; INTERV expresses that intervalization is done using only $^{\sharp}\pi$ (instead of sign-partitioning). Our certified procedure checks that $p' = p''$ using a normalization procedure defined in the standard distribution of CoQ (see [20]). If $p' \neq p''$, our procedure simply raises an error. If $p' = p''$, it invokes a CPS affine intervalization of p'' for continuation $\lambda[t_1,t_2],\neg 0 \bowtie [t_1 + t, t_2 + t]$. The next paragraphs detail this certified CPS intervalization and then, our external oracle.

Certified CPS Affine Intervalization. We implement and prove our affine intervalization using the CPS technique described in Sect. 2.3. On polynomial p'' and continuation $^t g$, the specification of our CPS intervalization is

$$\varepsilon \sqcap \bigsqcap_{[t_1,t_2]} \vdash \{d \,|\, t_1 \leq \llbracket p'' \rrbracket d \leq t_2\}\,; g[t_1,t_2]$$

The ε case corresponds to a failure of our procedure: typically, a subexpression is not affine as claimed by the external oracle. In case of success, the procedure selects non-deterministically some affine intervals $[t_1,t_2]$ bounding p'' before merging continuations on them. The procedure is implemented recursively over the syntax of p''. Figure 3 sketches the implementation and the specification of the sign-partitioning subprocedure. The figure deals with a particular case where p'' is a polynomial written $p \times t$ with t affine. In the implementation part, boolean `static` indicates the mode of $^\sharp\pi$. In static mode, we indeed factorize the computation of $^\sharp\pi$ on both cells of the partition.

Our linearization procedure is written in around 2000 Coq lines, proofs included. Among them, the CPS procedure and its subprocedures take only 200 lines. The bigger part – around 1000 lines – is thus taken by arithmetic operators on interval domains (constant and affine intervals).

Design of Our External Oracle. Only fast strategies may be tractable on big polynomials. Therefore, our external oracle may select systematically static constant intervalization on big polynomials. Otherwise, it ranks variables according to their priority to be discarded by sign-partitioning. Then, it factorizes variables with the highest priority. The priority rank is mainly computed from the size of intervals in the precomputed environment σ: unbounded variables must not be discarded whereas variables bounded by a singleton are always discarded by static intervalization. Our oracle also tries to minimize the number of distinct variables that are discarded: variables appearing in many monomials have a higher priority. The oracle also interleaves factorization with focusing. Our oracle is written in 1300 lines of OCAML code.

4   A Lightweight Refinement Calculus in Coq

Our implementation in Coq reformulates Sect. 2 with a more computational representation of binary relations. We only sketch here these representations. A more detailed description and our Coq sources are available online [17].

4.1   Representation of Abstract Computations

A relation R of $\mathcal{R}(A, B)$ can be equivalently seen as the function of $A \to \mathcal{P}(B)$ given by $\lambda x, \{y \mid x \xrightarrow{R} y\}$. This curryfied representation is the basis of our representations for abstract computations. Indeed, we need to provide a Coq representation of $\mathcal{R}(^\sharp D, ^\sharp D)$ that can be turned into an OCAML type $^\sharp D \to \,^\sharp D$ at extraction. This is achieved by axiomatizing in Coq the type "$\mathcal{P}(^\sharp D)$" as "$?^\sharp D$" where "$?$" is the type transformer of a given may-return monad – defined below. More generally, impure abstract computations of $\mathcal{R}(A, B)$ in Fig. 2 are actually expressed in our Coq development as functions of $A \to ?B$.

Definition 2 (May-Return Monad). *For any type A, type $?A$ represents impure computations returning values of type A. Type transformer "$?$" is equipped with a monad [16] providing a may-return relation [6]*

- *Operator $\gg=_{A,B}: ?A \to (A \to ?B) \to ?B$ encodes OCAML "$\mathtt{let}\ x = k_1\ \mathtt{in}\ k_2$" as "$k_1 \gg= \lambda x, k_2$".*
- *Operator $\varepsilon_A : A \to ?A$ lifts a pure computation as an impure one.*
- *Relation $\equiv_A: ?A \to ?A \to \mathrm{Prop}$ is a congruence (w.r.t. $\gg=$) which represents equivalence of semantics between impure computations. Moreover, operator $\gg=$ is associative and admits ε as neutral element (w.r.t. $\equiv$).*

- *Relation $\leadsto_A$: $?A \to A \to$ Prop, where "$k \leadsto a$" means that "k may return a". This relation must be compatible with $\equiv_A$ and satisfies the axioms*

$$(\varepsilon\, a_1) \leadsto a_2 \;\Rightarrow\; a_1 = a_2 \qquad\qquad (k_1 \ggeq k_2) \leadsto b \;\Rightarrow\; \exists a, k_1 \leadsto a \wedge k_2\, a \leadsto b$$

A computation k of $A \to ?B$ represents a relation of $\mathcal{R}(A, B)$ defined by $d \overset{k}{\to} d' \triangleq k\, d \leadsto d'$. Given k_1 and k_2 in $^\sharp D \to {}^\sharp_? D$, then "$\lambda x, ((k_1\, x) \ggeq k_2)$" corresponds to a subrelation of "$k_1 \cdot k_2$". Formally, our COQ implementation departs from Sect. 2 when we compose abstract computations, because we use operator "$\ggeq$" instead of the less precise "$\cdot$", but this does not differ a lot in practice.

The VPL is parametrized by a *core* may-return monad which axiomatizes external computations. This monad avoids a potential unsoundness by expressing that external oracles are not pure functions but relations. It is instantiated at extraction by providing the identity monad. Of course, the implementation of the core monad remains opaque for our COQ proofs: therefore they are valid for any instance of a may-return monad. Actually, our COQ proofs are sound if we admit that there exists a may-return monad able to denote any typesafe OCAML computation.

4.2   Representation of Concrete Computations

Like in standard refinement calculus [11], we simplify refinement proofs by computations of weakest-preconditions [15]. Precisely, we use weakest-*liberal*-preconditions (WLP) because they appear naturally in correctness diagrams of abstract computations. This will be illustrated in Sect. 4.3.

Given $K \in \mathcal{R}(D, D_?)$, the WLP of K is the function $[K] : \mathcal{P}(D) \to \mathcal{P}(D)$ defined by $[K]P \triangleq \{d \in D \mid \forall d' \in D_?, d \overset{K}{\to} d' \Rightarrow d' \in P\}$. The benefit of WLP is to propagate function computations through sequences of relations. Indeed, WLP transforms a sequence into a function composition: $[K_1; K_2]P = [K_1]([K_2]P)$. Hence, it avoids existential quantifiers in relation composition defining $x \overset{K_1; K_2}{\longrightarrow} z$ as $\exists y, x \overset{K_1}{\longrightarrow} y \wedge y \overset{?K_2}{\longrightarrow} z$, which is tedious to handle in proofs. For functions f of type $D \to D$, $[\uparrow f]P = \{d \mid f(d) \in P\}$. This allows for instance to compute $[\uparrow f_1; \uparrow f_2]P$ as $\{d \mid f_2(f_1(d)) \in P\}$. We embed WLP computations in refinement proofs using the equivalence between $K_1 \sqsubseteq K_2$ and $\forall P, [K_2]P \subseteq [K_1]P$.

Our definition of $\mathbb{K}$ in COQ is based on a shallow embedding. This means that we avoid introducing abstract syntax trees for $\mathbb{K}$ computations, which would induce many difficulties because of binders in $\bigsqcup$ and $\bigsqcap$ operators. Instead, we represent $\mathbb{K}$ computations directly as monotone predicate transformers. Hence, our COQ definition for $\mathbb{K}$ is isomorphic to the following type[4]:

```
Record K: Type := {
  wlp: (D → Prop) → D → Prop;
  wlp_mono: ∀ P Q, (∀ d, P d→ Q d) → ∀ d, wlp P d→ wlp Q d}.
```

---

[4] Actually, our definition of $\mathbb{K}$ derives from a more general structure, see [17].

In other words, our syntax for $\mathbb{K}$ guarded commands is directly provided by a given set of COQ operators on monotone predicate transformers (corresponding to some WLP computations). This shallow embedding is sufficient because we do not need to establish in COQ properties of $\mathbb{K}$ as an algebra. Thus, the COQ code for $\mathbb{K}$ operators is only 150 lines long, proofs and comments included. On refinement goals, we let COQ computes weakest-preconditions and simply solve the remaining goal using standard COQ tactics. In practice, this gives us well-automated proof scripts.

4.3 Definition of Correctness Diagrams

The COQ pseudo-code below defines values of $^\dagger\mathbb{K}$ as triples with a field `impl` being an abstract computation, a field `spec` being a concrete computation and a field `impl_correct` being a proof that `impl` is correct *w.r.t.* `spec`. Such proofs are simplified by applying together the WLP embedded in `spec` and the WLP already designed by [6] which simplifies reasonings with relation $\rightsquigarrow$.

```
Record †𝕂: Type := {
  impl:♯D → ?♯D;  spec:𝕂;
  impl_correct: ∀♯d♯d',(impl ♯d) ⤳♯d' → ∀d,d ∈ γ(♯d) → (wlp spec γ(♯d') d)}.
```

Because `impl` is the only informative field of record $^\dagger\mathbb{K}$, type $^\dagger\mathbb{K}$ is exactly extracted as type $^\sharp D \rightarrow ?^\sharp D$. A $^\dagger\mathbb{K}$ command is also exactly extracted in OCAML as its underlying abstract computation. Here again, the COQ code for $^\dagger\mathbb{K}$ operators, diagrammatic proofs included, is small, around 200 lines.

5 Conclusion and Perspectives

We extended the VPL with certified handling of non-linear multiplication by a modular and novel design. Our computations are performed by an untrusted oracle delivering a certificate to a certified front-end. Our proofs use diagrammatic constructs based on stepwise refinement calculus. Refinement proofs are finally made clear and concise by the computations of Weakest-Liberal-Preconditions.

Our linearization procedure is able to give a fast over-approximation of polynomials thanks to variable intervalization. The precision is increased by domain partitioning (implicitly done with a Continuation-Passing-Style design) and the dynamic computation of bounding affine terms, allowing to finely tune the precision-versus-efficiency trade-off in the oracle.

Because floating arithmetic would make us explicitly handle error terms at each operation, the VPL is for now limited to integers, and so is our linearization. Our implementation also lacks other operators such as division or modulo. For these reasons, it is hard to evaluate our method on real-life programs. Currently, our tests are limited to small handmade examples focusing on classes of mathematical problems, such as parabola or barycentric approximations. On these cases, our oracle is able to give much more precise approximations than the VERASCO interval domain.

We certified a toy analyzer from big-steps semantics by simply interpreting the operators of concrete semantics in abstract semantics. Our approach should scale up on a complex language like COMPCERT C, even if it does not use simple big-steps semantics. Although its semantics allows to distinguish programs (for instance diverging ones invoking system calls or not) that are equivalent for our concrete semantics, such features do not seem necessary to the VERASCO analysis correctness. Hence, our approach would introduce an abstraction over COMPCERT semantics which should even ease the proof of the analyzer.

Acknowledgements. We would like to thank Alexis Fouilhé, Michaël Périn and David Monniaux for their continuous feedback all along this work. We also thank the members of the VERASCO project for their motivating interaction.

References

1. The Coq Development Team: The Coq proof assistant reference manual - version 8.4. INRIA (2012–2014)
2. Cousot, P., Cousot, R.: Abstract interpretation: a unified lattice model for static analysis of programs by construction or approximation of fixpoints. In: POPL. ACM (1977)
3. Cousot, P., Halbwachs, N.: Automatic discovery of linear restraints among variables of a program. In: POPL. ACM (1978)
4. Miné, A.: Symbolic methods to enhance the precision of numerical abstract domains. In: Emerson, E.A., Namjoshi, K.S. (eds.) VMCAI 2006. LNCS, vol. 3855, pp. 348–363. Springer, Heidelberg (2006)
5. Fouilhe, A., Monniaux, D., Périn, M.: Efficient generation of correctness certificates for the abstract domain of polyhedra. In: Logozzo, F., Fähndrich, M. (eds.) Static Analysis. LNCS, vol. 7935, pp. 345–365. Springer, Heidelberg (2013)
6. Fouilhe, A., Boulmé, S.: A certifying frontend for (sub) polyhedral abstract domains. In: Giannakopoulou, D., Kroening, D. (eds.) VSTTE 2014. LNCS, vol. 8471, pp. 200–215. Springer, Heidelberg (2014)
7. Jourdan, J.H., Laporte, V., Blazy, S., Leroy, X., Pichardie, D.: A formally-verified C static analyzer. In: POPL. ACM (2015)
8. Leroy, X.: Formal verification of a realistic compiler. Commun. ACM **52**, 107–115 (2009)
9. Besson, F., Jensen, T., Pichardie, D., Turpin, T.: Certified result checking for polyhedral analysis of bytecode programs. In: Wirsing, M., Hofmann, M., Rauschmayer, A. (eds.) TGC 2010, LNCS, vol. 6084, pp. 253–267. Springer, Heidelberg (2010)
10. Maréchal, A., Périn, M.: Three linearization techniques for multivariate polynomials in static analysis using convex polyhedra. Technical report TR-2014-7, Verimag Research Report (2014)
11. Back, R.J., von Wright, J.: Refinement Calculus - A Systematic Introduction. Graduate Texts in Computer Science. Springer, Heidelberg (1999)
12. Reynolds, J.C.: The discoveries of continuations. Lisp Symb. Comput. **6**, 233–247 (1993)
13. Cousot, P.: Constructive design of a hierarchy of semantics of a transition system by abstract interpretation. TCS **277**, 47–103 (2002)

14. Spiwack, A.: Abstract interpretation as anti-refinement. CoRR abs/1310.4283 (2013)
15. Dijkstra, E.W.: Guarded commands, nondeterminacy and formal derivation of programs. Commun. ACM **18**, 453–457 (1975)
16. Wadler, P.: Monads for functional programming. In: Jeuring, J., Meijer, E. (eds.) AFP 1995. LNCS, vol. 925. Springer, Heidelberg (1995)
17. Boulmé, S., Maréchal, A.: A refinement calculus to certify impure abstract computations of the verimag polyhedra library - documentation and Coq+OCaml sources (2015). http://www-verimag.imag.fr/~boulme/vpl201503
18. Farouki, R.T.: The Bernstein polynomial basis: a centennial retrospective. Comput. Aided Geom. Des. **29**, 379–419 (2012)
19. Handelman, D.: Representing polynomials by positive linear functions on compact convex polyhedra. Pac. J. Math. **132**, 35–62 (1988)
20. Grégoire, B., Mahboubi, A.: Proving equalities in a commutative ring done right in Coq. In: Hurd, J., Melham, T. (eds.) TPHOLs 2005. LNCS, vol. 3603, pp. 98–113. Springer, Heidelberg (2005)

Mechanisation of AKS Algorithm:
Part 1 – The Main Theorem

Hing-Lun Chan[1]([✉]) and Michael Norrish[2]

[1] Australian National University, Canberra, Australia
`joseph.chan@anu.edu.au`
[2] Canberra Research Laboratory, NICTA, Australian National University,
Canberra, Australia
`Michael.Norrish@nicta.com.au`

Abstract. The AKS algorithm (by Agrawal, Kayal and Saxena) is a significant theoretical result proving "PRIMES in P", as well as a brilliant application of ideas from finite fields. This paper describes the first step towards the goal of a full mechanisation of this result: a mechanisation of the AKS Main Theorem, which justifies the correctness (but not the complexity) of the AKS algorithm.

1 Introduction

The AKS algorithm is a decision procedure for primality testing. That is, given a number n, it returns "true" if n is prime and "false" otherwise. As *per* the title of AKS paper [3], "PRIMES is in P", the significance of their work is that the number of steps for the verification is bounded by some polynomial function of the size of n, measured by $\log_2 n$.

There have been several attempts to formalize the AKS Main Theorem (see Sect. 6), but so far none is complete. In this paper, we describe the first complete mechanization of this result. In subsequent work, we aim to demonstrate that the algorithm built on top of this result does indeed compute its answer in polynomial time.

1.1 Overview

A number n is a perfect-power of another number m if there exists an exponent e such that $n = m^e$, and n is power-free if it is a trivial perfect power, *i.e.*, if $n = m^e$ then $e = 1$ and $m = n$. Given a number n, the smallest positive exponent j such that $n^j \equiv 1 \,(\text{mod } k)$ is denoted by $\text{order}_k(n)$. Computation in $(\text{mod } n, \, \mathsf{X}^k - 1)$ means that all numerical as well as polynomial computational results are reduced to remainders after divisions by n and by $\mathsf{X}^k - 1$. The constant polynomial arising from constant c is denoted by boldface $\mathbf{c}$. More notation will be covered in Sect. 1.2. Here is a peek at our HOL4 result.

M. Norrish—NICTA is funded by the Australian Government through the Department of Communications and the Australian Research Council through the ICT Centre of Excellence Program.

© Springer International Publishing Switzerland 2015
C. Urban and X. Zhang (Eds.): ITP 2015, LNCS 9236, pp. 117–136, 2015.
DOI: 10.1007/978-3-319-22102-1_8

Theorem 1. *The AKS Main Theorem.*

$\vdash$ prime n $\iff$
$\quad$ $1 < n \wedge$ power_free $n \wedge$
$\quad$ $\exists k.$
$\qquad$ prime $k \wedge (2 (\log n + 1))^2 \le$ order$_k(n) \wedge$
$\qquad$ $(\forall j.\ 0 < j \wedge j \le k \wedge j < n \Rightarrow \gcd(n, j) = 1) \wedge$
$\qquad$ $(k < n \Rightarrow$
$\qquad\quad$ $\forall c.$
$\qquad\qquad$ $0 < c \wedge c \le 2\sqrt{k} (\log n + 1) \Rightarrow$
$\qquad\qquad$ $(\mathsf{X} + c)^n \equiv (\mathsf{X}^n + c) (\mathrm{mod}\quad n,\ \mathsf{X}^k - 1))$

This theorem then justifies the following algorithm[1] for primality testing:

Input: integer $n > 1$.

1. If ($n = b^m$ for some base b with $m > 1$), return COMPOSITE.
2. Search for a prime k satisfying order$_k(n) \ge (2 (\log n + 1))^2$.
3. For each ($j = 1$ to k) if ($j = n$) break, else if ($\gcd(j, n) \ne 1$), return COMPOSITE.
4. If ($k >= n$), return PRIME.
5. For each ($c = 1$ to $2\sqrt{k} (\log n + 1)$) if $(\mathsf{X} + c)^n \not\equiv (\mathsf{X}^n + c) (\mathrm{mod}$ $n, \mathsf{X}^k - 1)$, return COMPOSITE.
6. return PRIME.

Given a number n, this version of the AKS algorithm requires a search for another prime k in Step 2. Step 4 suggests that it is not always true that $k < n$. Nevertheless, the theorem can still be viewed as a well-founded recursive definition because it turns out that k is roughly bounded by $(\log n)^5$ [3]. So, for sufficiently large values of n, there will always be a $k < n$. For smaller n (effectively the base cases of recursion), a look-up table might be used.

The rest of this paper is devoted to explaining the mechanised proof of this result. Section 2 covers some necessary background. Sections 3 and 4 describe the proof of the AKS Main Theorem. Section 5 discusses our mechanisation experience. Section 6 compares our work with others. Finally, we conclude in Sect. 7.

1.2   Notation

All statements starting with a turnstile ($\vdash$) are HOL4 theorems, automatically pretty-printed to LATEX from the relevant theory in the HOL4 development. Generally, our notation allows an appealing combination of quantifiers

---

[1] The constants involved in this algorithm are based on [10, Algorithm 8.2.1]. They are slightly different from those in the AKS papers [2,3], but such variations do not affect the conclusion of "PRIMES is in P".

($\forall, \exists$), set notation ($\in, \cup$ and comprehensions such as $\{x \mid x < 6\}$), and functional programming (λ for function abstraction, and juxtaposition for function application).

The cardinality of a set s is written $|s|$; the image of set s under the mapping f is written $f(\!(s)\!)$; we write $f : s \hookrightarrow t$ to mean that function f is injective from set s to set t.

Number-Theoretic Notation. With n a natural number, $\sqrt{n}$ is its square-root, and $\log n$ its logarithm to base 2. Both logarithm and square-root are integer functions, being the floor value of the corresponding real functions. We use $\varphi(n)$ to denote the Euler φ-function of n, the count of coprime numbers less than n. We write $n \mid m$ when n divides m.

For the AKS algorithm, we shall use n for the input number, p for its prime factor, k for the prime (existentially quantified) "parameter" of the Main Theorem, and $\ell = 2\sqrt{k}(\log n + 1)$ for a computed parameter (the limit for a range of constants). Note that $\mathsf{order}_k(n)$ is nonzero whenever $\gcd(k, n) = 1$.

Ring, Field, and Polynomial Notation. A ring $\mathcal{R}$ has carrier set R, with **1** and **0** its one and zero. The characteristic of a ring $\mathcal{R}$ is written as $\chi(\mathcal{R})$, often abbreviated to χ for a generic ring. A ring homomorphism from a ring $\mathcal{R}$ to another ring $\mathcal{S}$ *via* a map f is denoted by $f : \mathcal{R} \mapsto_r \mathcal{S}$.

We write $\mathcal{R}[\mathsf{X}]$ to denote the ring of polynomials with coefficients drawn from the underlying ring $\mathcal{R}$. Similarly, the ring $\mathcal{F}[\mathsf{X}]$ has polynomials with coefficients from a field $\mathcal{F}$. Polynomials from those rings are written with the sans-serif font, *e.g.*, p, q, h. The constant polynomial c (in bold) is derived from adding **1** repeatedly c times. The degree of p is written deg p, its leading coefficient is lead p, and monic p means its leading coefficient is **1**. The polynomial X is the monomial (monic of degree 1) with zero constant. The polynomial field quotiented by modulus with an irreducible polynomial h is $\mathcal{F}_\mathsf{h}[\mathsf{X}]$, with multiplicative group $\mathcal{F}_\mathsf{h}^*[\mathsf{X}]$.

Arithmetic (addition, subtraction, multiplication, remainders) on polynomials is written with usual symbols ($+$, $-$, $*$, mod *etc.*), *e.g.*, $\mathsf{X}^k - \mathbf{1}$ is the unity polynomial of degree k. Here we can see HOL4's overloading facilities at work: constant polynomials one and zero are **1** and **0**, the same as those for a ring. More aggressively, we use overloading to conceal "implicit" parameters such as the underlying ring $\mathcal{R}$ in terms such as p $*$ q (polynomial product).

We write p$[\![$q$]\!]$ to denote the substitution of q for every X in p. We use roots p for the set of p's roots, and ipoly p to mean that p is an irreducible polynomial, both with respect to its underlying ring $\mathcal{R}$. The quotient ring formed by $\mathcal{R}[\mathsf{X}]$ and irreducible polynomial h is denoted by $\mathcal{R}_\mathsf{h}[\mathsf{X}]$, which can be shown to be a field. Its multiplicative group is $\mathcal{R}_\mathsf{h}^*[\mathsf{X}]$. The order of an element in this group, *e.g.*, X, is denoted by $\mathsf{order}_\mathsf{h}(\mathsf{X})$.

HOL4 Sources. All our proof scripts can be found at http://bitbucket.org/jhlchan/hol/src/aks/theories.

2 Background

A glance at the algorithm in Sect. 1.1 shows its most prominent feature: polynomial identity tests in modulo $X^k - 1$. To understand this we need to get a feel for the motivation behind the AKS algorithm.

2.1 Finite Fields

The AKS Main Theorem has a setting in finite fields, since the characteristic of a finite field is always prime. A field is also a ring, and a ring with prime characteristic enjoys some wonderful properties.

Theorem 2. *The Freshman-Fermat Theorem*

$\vdash$ Ring $\mathcal{R}$ $\wedge$ prime χ $\Rightarrow$ $\forall c.$ $(X + c)^\chi = X^\chi + c$

Proof. This follows directly from two theorems, (a) Freshman's Theorem:

$\vdash$ Ring $\mathcal{R}$ $\wedge$ prime χ $\Rightarrow$ $\forall p\ q.$ poly p $\wedge$ poly q $\Rightarrow$ $(p + q)^\chi = p^\chi + q^\chi$

and (b) Fermat's Little Theorem for polynomials:

$\vdash$ Ring $\mathcal{R}$ $\wedge$ prime χ $\Rightarrow$ $\forall c.$ $c^\chi = c$

Both theorems (a) and (b) have been mechanized in a previous paper by the same authors [6]. $\square$

The converse, suitably formulated, is also true:

Theorem 3. *A ring has its characteristic prime iff a Freshman-Fermat identity holds for any constant coprime with the characteristic.*

$\vdash$ Ring $\mathcal{R}$ $\Rightarrow$ $\forall c.$ $\gcd(c, \chi) = 1$ $\Rightarrow$ (prime χ $\iff$ $1 < \chi \wedge (X + c)^\chi = X^\chi + c$)

Given a number $n > 1$, we can identify $\mathcal{R}$ with $\mathbb{Z}_n$, and $\chi(\mathbb{Z}_n) = n$. Since $\gcd(1, n) = 1$ always, pick $c = 1$, then this theorem applies, and whether n is prime is just one check of a Freshman-Fermat polynomial identity in $\mathbb{Z}_n$, *i.e.*, (mod n).

Therefore, this theorem amounts to a deterministic primality test. But there is a catch: the left-side, upon expansion, contains $n + 1$ terms. Thus this is an impractical primality test for large values of n.

The AKS idea begins with checking such Freshman-Fermat identities, with two twists:

– Instead of just checking in (mod n), perform the computations in (mod n, $X^k - 1$) for a suitably chosen parameter k. Since results are always the remainder after division by $X^k - 1$, the degree of intermediate polynomials (which determines the number of terms) never exceed k—presumably k is much smaller than n.
– Instead of checking just one coprime value, check for a range of coprime values c, for $0 < c \leq \ell$, up to some maximum limit ℓ—presumably ℓ is related to k, and small compared to n.

Of course, the big question is whether after such modifications there is still a primality test. The AKS team answered this in the affirmative—there exist parameters k and ℓ, bounded by some polynomial function of the size of input number n, *i.e.*, $\log n$, giving a polynomial-time deterministic primality test.

2.2   Introspective Relation

Recall from Sect. 1.1 that AKS computations are done in (mod n, $\mathsf{X}^k - 1$). This double modulo notation is clumsy. Let us work in a generic ring $\mathcal{R}$, later to be identified with instances such as $\mathbb{Z}_n$. The first (mod n) equivalence becomes equality in the ring $\mathcal{R}$ (*i.e.*, $x \equiv y \,(\mathrm{mod}\ n)$ means $x = y$ in $\mathbb{Z}_n$); leaving the symbol ($\equiv$) to indicate the polynomial modulo equivalence in $\mathcal{R}[\mathsf{X}]$.

In this context, of a general ring $\mathcal{R}$, the polynomial identity checks in Theorem 1 take the form:

$$(\mathsf{X} + c)^n \equiv \mathsf{X}^n + c \ (\mathrm{mod}\ \mathsf{X}^k - 1)$$

They look like Freshman-Fermat identities of Theorem 2, only now under modulo by a polynomial. Rewriting with a polynomial substitution, the left and right sides are strikingly similar:

$$(\mathsf{X} + c)^n [\![\mathsf{X}]\!] \equiv (\mathsf{X} + c)[\![\mathsf{X}^n]\!] \ (\mathrm{mod}\ \mathsf{X}^k - 1)$$

The rewrites are trivial, since for any polynomial p, we have $\mathsf{p}[\![\mathsf{X}]\!] = \mathsf{p}$ and $(\mathsf{X} + c)[\![\mathsf{p}]\!] = \mathsf{p} + c$. Superficially, the left-hand side is transformed into the right-hand side simply by shifting of the exponent n. Following the terminology of AKS team, we say n is *introspective* to polynomial p, denoted by $n \bowtie \mathsf{p}$, when:

$$\vdash\ n \bowtie \mathsf{p} \iff \mathsf{poly}\ \mathsf{p} \wedge 0 < k \wedge \mathsf{p}^n \equiv \mathsf{p}[\![\mathsf{X}^n]\!] \,(\mathrm{mod}\ \mathsf{X}^k - 1)$$

Note that the symbol for introspective relation ($\bowtie$) hides the polynomial modulus $\mathsf{X}^k - 1$, and the underlying ring $\mathcal{R}$. We shall include a subscript when the underlying ring is of significance, *e.g.*, $\bowtie_{\mathbb{Z}_n}$.

Therefore, the AKS algorithm verifies, for the input number n, the identities $n \bowtie \mathsf{X} + c$ in $\mathbb{Z}_n$ for $0 < c \le \ell$ up to some maximum ℓ. Moreover, Freshman-Fermat (Theorem 2) can be restated as:

Theorem 4. *For a ring with prime characteristic, its characteristic is introspective to any monomial.*

$$\vdash\ \mathsf{Ring}\ \mathcal{R} \wedge 1 \ne 0 \wedge \mathsf{prime}\ \chi \Rightarrow \forall k.\ 0 < k \Rightarrow \forall c.\ \chi \bowtie \mathsf{X} + c$$

Proof. By introspective definition, this is to show: $(\mathsf{X} + c)^\chi \equiv (\mathsf{X} + c)[\![\mathsf{X}^\chi]\!]$ (mod $\mathsf{X}^k - 1$). Transforming the right side by substitution, $(\mathsf{X} + c)[\![\mathsf{X}^\chi]\!] = \mathsf{X}^\chi + c$. Then both sides are equal by Freshman-Fermat (Theorem 2), hence they are equivalent under modulo by $\mathsf{X}^k - 1$.   □

The fundamental properties of introspective relation are:

Theorem 5. *Introspective relation is multiplicative for exponents.*

$$\vdash\ \mathsf{Ring}\ \mathcal{R} \wedge 1 \ne 0 \Rightarrow \forall k\ \mathsf{p}\ n\ m.\ n \bowtie \mathsf{p} \wedge m \bowtie \mathsf{p} \Rightarrow n\,m \bowtie \mathsf{p}$$

Proof. Working in (mod $\mathsf{X}^k - 1$), we have $\mathsf{p}^n \equiv \mathsf{p}[\![\mathsf{X}^n]\!]$ by $n \bowtie \mathsf{p}$, and $\mathsf{p}^m \equiv \mathsf{p}[\![\mathsf{X}^m]\!]$ by $m \bowtie \mathsf{p}$. The latter means $\mathsf{p}[\![\mathsf{X}]\!]^m - \mathsf{p}[\![\mathsf{X}^m]\!]$ is divisible by $\mathsf{X}^k - 1$. Substitute every X of the previous statement by X^n, and noting $\mathsf{X}^k - 1 \mid (\mathsf{X}^n)^k - 1$ by divisibility of unity polynomials, $\mathsf{p}[\![\mathsf{X}^n]\!]^m \equiv \mathsf{p}[\![(\mathsf{X}^n)^m]\!]$. Therefore, $\mathsf{p}^{n\,m} = (\mathsf{p}^n)^m \equiv \mathsf{p}[\![\mathsf{X}^n]\!]^m \equiv \mathsf{p}[\![(\mathsf{X}^n)^m]\!] = \mathsf{p}[\![\mathsf{X}^{n\,m}]\!]$, or $n\,m \bowtie \mathsf{p}$.   □

Theorem 6. *Introspective relation is multiplicative for polynomials.*

⊢ Ring $\mathcal{R}$ ∧ $1 \neq 0$ ⇒ ∀k p q n. n ⋈ p ∧ n ⋈ q ⇒ n ⋈ p∗q

Proof. Working in (mod $X^k - 1$), we have $p^n \equiv p[\![X^n]\!]$ by n ⋈ p, and $q^n \equiv q[\![X^n]\!]$ by n ⋈ q. Therefore, $(p * q)^n = p^n * q^n \equiv p[\![X^n]\!] * q[\![X^n]\!] = (p*q)[\![X^n]\!]$, or n ⋈ p∗q.   □

3   Main Theorem

We can now restate the AKS Main Theorem (Theorem 1) in terms of the introspective relation.

Theorem 7. *A number is prime iff it satisfies all the AKS checks.*

⊢ prime n ⟺
    $1 < n$ ∧ power_free n ∧
    ∃ k .
        prime k ∧ $(2 (\log n + 1))^2 \leq$ order$_k(n)$ ∧
        (∀j. $0 < j$ ∧ $j \leq k$ ∧ $j < n$ ⇒ $\gcd(n,j) = 1$) ∧
        ($k < n$ ⇒
            ∀c. $0 < c$ ∧ $c \leq 2\sqrt{k} (\log n + 1)$ ⇒ n ⋈$_{\mathbb{Z}_n}$ X + c)

Note how the symbol ⋈$_{\mathbb{Z}_n}$ encapsulates the introspective relation (*i.e.*, mod $X^k - 1$) within $\mathbb{Z}_n$ (*i.e.*, mod n), the double modulo in the AKS computations. We prove this logical equivalence in two parts.

3.1   Easy Part (⇒)

Theorem 8. *The if-part of AKS Main Theorem (Theorem 7).*

⊢ prime n ⇒
    $1 < n$ ∧ power_free n ∧
    ∃ k .
        prime k ∧ $(2 (\log n + 1))^2 \leq$ order$_k(n)$ ∧
        (∀j. $0 < j$ ∧ $j \leq k$ ∧ $j < n$ ⇒ $\gcd(n,j) = 1$) ∧
        ($k < n$ ⇒ ∀c. $0 < c$ ∧ $c \leq 2\sqrt{k} (\log n + 1)$ ⇒ n ⋈$_{\mathbb{Z}_n}$ X + c)

Proof. The first two goals, $1 < n$ and power_free n, are trivial for a prime n. For the third goal, let $m = (2 (\log n + 1))^2$, then parameter k exists by Theorem 25 in Sect. 4.6:

⊢ $1 < n$ ∧ $0 < m$ ⇒ ∃k. prime k ∧ $\gcd(k,n) = 1$ ∧ $m \leq$ order$_k(n)$

If $k \geq n$, the coprime checks are subsumed by ∀j.$0 < j \wedge j < n \Rightarrow \gcd(n,j) = 1$. Otherwise $k < n$, and the coprime checks are subsumed by ∀j.$0 < j \wedge j \leq k \Rightarrow \gcd(n,j) = 1$. Either way this is true since a prime n is coprime with all values less than itself. When $k < n$, the last check is established by Theorem 4, since a prime n gives a field $\mathbb{Z}_n$, with $\chi(\mathbb{Z}_n) = n$.   □

A close equivalent of this Theorem 8 was mechanised by de Moura and Tadeu [9] in Coq, and by Campos *et al.* [5] in ACL2.

3.2 Hard Part ($\Leftarrow$)

Theorem 9. *The only-if part of AKS Main Theorem (Theorem 7).*

$\vdash 1 < n \land$ power_free $n \land$
$(\exists k.$
    prime $k \land (2 (\log n + 1))^2 \leq \text{order}_k(n) \land$
    $(\forall j. \ 0 < j \land j \leq k \land j < n \Rightarrow \gcd(n,j) = 1) \land$
    $(k < n \Rightarrow \forall c. \ 0 < c \land c \leq 2\sqrt{k} (\log n + 1) \Rightarrow n \bowtie_{\mathbb{Z}_n} X + c)) \Rightarrow$
    prime n

Proof. Based on the given parameter k, let $\ell = 2\sqrt{k} (\log n + 1)$. If $k \geq n$ the coprime checks will verify $\forall j. 0 < j \land j < n \Rightarrow \gcd(n,j) = 1$, thus n will be prime since it has no proper factor. Otherwise $k < n$, the coprime checks are $\forall j. 0 < j \land j \leq k \Rightarrow \gcd(n,j) = 1$. In Sect. 3.3 we shall establish:

Theorem 10. *The AKS Main Theorem in $\mathbb{Z}_n$.*

$\vdash 1 < n \Rightarrow$
    $\forall k \ \ell.$
        prime $k \land (2 (\log n + 1))^2 \leq \text{order}_k(n) \land \ell = 2\sqrt{k} (\log n + 1) \land$
        $(\forall j. \ 0 < j \land j \leq k \Rightarrow \gcd(n,j) = 1) \land$
        $(\forall c. \ 0 < c \land c \leq \ell \Rightarrow n \bowtie_{\mathbb{Z}_n} X + c) \Rightarrow$
        $\exists p. \ \text{prime } p \land \text{perfect\_power } n \ p$

Applying this theorem, $n = p^e$ for some prime p and exponent e by definition of a perfect power. But n is assumed power-free, so $e = 1$ and $n = p$, making n a prime. $\qquad\square$

3.3 Shifting Playgrounds

The AKS verifications take polynomials with coefficients from $\mathbb{Z}_n$, a ring for general n. Polynomials with coefficients from a ring can have more roots than their degree, due to the possible existence of zero divisors in a ring.[2] A field has no zero divisors, and polynomials with coefficients from a field have this nice property:

$\vdash$ Field $\mathcal{F} \Rightarrow \forall p. \ \text{poly } p \land p \neq 0 \Rightarrow |\text{roots } p| \leq \deg p$

As we shall see (Sects. 4.3 and 4.4), there will be two important injective maps involved in the AKS proof. To establish the injective property, this restriction on the number of polynomial roots by its degree is of utmost importance.

But where to find a field $\mathcal{F}$ to work with, given that we start in the ring $\mathbb{Z}_n$?

When the number n is not 1, it must have a prime factor p. This leads to the field $\mathbb{Z}_p$. If relationships between monomials $X + c$ are carried over unaffected from $\mathbb{Z}_n[X]$ to $\mathbb{Z}_p[X]$, we are in a better place to investigate the nature of n. This shifting of playgrounds is essential in the proof of Theorem 10:

---

[2] For example, in $\mathbb{Z}_6$, $2 \times 3 = 0$, hence $(X - 2)(X - 3) = X^2 - 5X = X(X - 5)$, which shows a polynomial of degree 2 can have more than 2 roots.

Proof (of Theorem 10). If n is prime, take $p = n$. Otherwise, n has a proper prime factor p such that $p < n$ and $p \mid n$. Introduce two rings, $\mathbb{Z}_n$ and $\mathbb{Z}_p$. The latter ring $\mathbb{Z}_p$ is also a field, in fact a finite field. This is because all nonzero elements are coprime to the prime modulus p, hence they have inverses.

There is a homomorphism between these two rings due to that fact that p divides n:

$$\vdash 0 < n \wedge 0 < p \wedge p \mid n \Rightarrow (\lambda x. \ x \bmod p) : \mathbb{Z}_n \mapsto_r \mathbb{Z}_p$$

This ring homomorphism will preserve monomials $X + c$ if a condition on limit ℓ is met:

$$\vdash 0 < n \wedge 1 < p \wedge \ell < p \Rightarrow$$
$$\forall c. \ 0 < c \wedge c \leq \ell \Rightarrow \forall f. \ f : \mathbb{Z}_n \mapsto_r \mathbb{Z}_p \Rightarrow f(X + c) = X + c$$

Here $f(\mathsf{p})$ denotes applying the ring homomorphism f to each coefficient of a polynomial p. We shall show in Sect. 4.6 that $\ell \leq k$ (Theorem 27). To meet the condition $\ell < p$, we need only to show $k < p$. Note that the given coprime checks on k are (from the statement of Theorem 10):

$$\forall j. \ 0 < j \wedge j \leq k \Rightarrow \gcd(n, j) = 1$$

Taking $j = k$, we conclude $\gcd(n, k) = 1$. This will be useful later. Apply the following theorems:

$$\vdash 1 < n \wedge \mathsf{prime} \ p \wedge p \mid n \Rightarrow \forall j. \ \gcd(n, j) = 1 \Rightarrow \gcd(p, j) = 1$$
$$\vdash 1 < p \Rightarrow \forall k. \ (\forall j. \ 0 < j \wedge j \leq k \Rightarrow \gcd(p, j) = 1) \Rightarrow k < p$$

Tracing the transformation of gcd's gives $k < p$, hence $\ell < p$.

Therefore the monomials are preserved by homomorphism, together with the introspective relation:

$$\vdash 0 < n \wedge 1 < p \wedge p \mid n \wedge 0 < k \wedge \ell < p \Rightarrow$$
$$\forall m \ c. \ 0 < c \wedge c \leq \ell \wedge m \bowtie_{\mathbb{Z}_n} X + c \Rightarrow m \bowtie_{\mathbb{Z}_p} X + c$$

Thus the AKS checks in $\mathbb{Z}_n$ are equivalent to checks in $\mathbb{Z}_p$, a finite field, where p is a prime factor of n. Generalising to arbitrary finite fields, in Sect. 4.5 we will prove:

Theorem 11. *AKS Main Theorem in finite fields.*

$$\vdash \mathsf{FiniteField} \ \mathcal{F} \wedge \mathsf{prime} \ k \wedge k < \chi \Rightarrow$$
$$\forall n.$$
$$1 < n \wedge \chi \mid n \wedge \gcd(n, k) = 1 \wedge (2 \, (\log n + 1))^2 \leq \mathsf{order}_k(n) \wedge$$
$$\ell = 2 \sqrt{k} \, (\log n + 1) \wedge (\forall c. \ 0 < c \wedge c \leq \ell \Rightarrow n \bowtie X + c) \Rightarrow$$
$$\mathsf{perfect\_power} \ n \ \chi$$

We then identify $\mathcal{F}$ with $\mathbb{Z}_p$, noting $\chi(\mathbb{Z}_p) = p$, with $k < p$. Knowing $\gcd(n, k) = 1$ from the gcd checks above, we conclude that n must be a perfect power of its prime factor p, as required. $\qquad\square$

4   Introspective Game

There are two useful facts when working in the context of a finite field $\mathcal{F}$, where χ is necessarily prime:

- We get, for free in $\mathcal{F}[\mathsf{X}]$, the result: $\chi \bowtie \mathsf{X} + c$, by Theorem 4, since a field is a non-trivial ring.
- The modulus polynomial $\mathsf{X}^k - 1$, now in $\mathcal{F}[\mathsf{X}]$, will have a monic irreducible factor $\mathsf{h} \neq \mathsf{X} - 1$.

Both will play significant roles in the proof of Theorem 11. Here are the highlights:

- The finite field $\mathcal{F}$ will enrich the introspective relation, through the interplay between prime χ and n.
- This will give rise to some interesting sets, among them are two finite sets $\widehat{\mathcal{N}}$ and $\mathcal{M}$ (Sect. 4.2).
- The conditions on parameters k and ℓ will establish an injective map from $\widehat{\mathcal{N}}$ to $\mathcal{M}$.
- If n were not a perfect power of χ, then we would have $|\widehat{\mathcal{N}}| > |\mathcal{M}|$, contradicting the Pigeonhole Principle.

Summary of the AKS Proof (Theorem 11)

Our strategy for the AKS proof can be described as a game between two players (see Fig. 1). The introspective relations of n and p, a prime factor of n, give rise to two sets $\mathcal{N}$ and $\mathcal{P}$ (Sect. 4.1). Taking modulo by k (an AKS parameter) and by h (an irreducible factor of $\mathsf{X}^k - 1$), the sets $\mathcal{N}$ and $\mathcal{P}$ map (straight arrows), respectively, to two finite sets $\mathcal{M}$ and $\mathcal{Q}_{\mathsf{h}}$ (Sect. 4.2). Two finite subsets of $\mathcal{N}$ and $\mathcal{P}$, shown as $\widehat{\mathcal{N}}$ and $\widehat{\mathcal{P}}$, can be crafted in such a way that injective maps (curve arrows) between the finite sets can be constructed, if k and ℓ (another AKS parameter) are suitably chosen to satisfy the "if" conditions (Sects. 4.3 and 4.4). The construction of injective maps involves interactions (dashed arrows) between the two players, based on properties of the introspective relation and polynomials in $\mathcal{F}_{\mathsf{h}}[\mathsf{X}]$. Once these are all in place, if n were not a perfect power of p, the grey set $\widehat{\mathcal{N}}$ will have more than $|\mathcal{M}|$ elements, where $\mathcal{M}$ is the target of the left injective map. This contradicts the Pigeonhole Principle (Sect. 4.5). Hence n must be a perfect power of its prime factor p.

4.1   Introspective Sets

As noted above, after shifting to a finite field $\mathcal{F}$ where $p = \chi$ is prime, for the constants $0 < c \leq \ell$, besides the given $n \bowtie \mathsf{X} + c$, we also have $p \bowtie \mathsf{X} + c$ by Theorem 4.

In view of this, we define the following two sets:

$$\vdash \mathcal{N} = \{m \mid \gcd(m, k) = 1 \wedge \forall c.\ 0 < c \wedge c \leq \ell \Rightarrow m \bowtie \mathsf{X} + c\}$$
$$\vdash \mathcal{P} = \{\mathsf{p} \mid \mathsf{poly}\ \mathsf{p} \wedge \forall m.\ m \in \mathcal{N} \Rightarrow m \bowtie \mathsf{p}\}$$

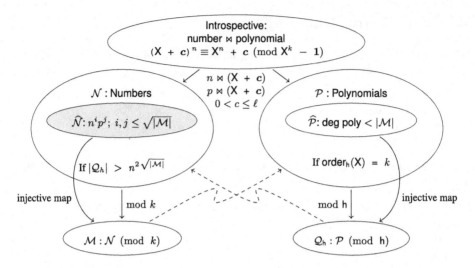

Fig. 1. The AKS proof as a game between numbers and polynomials *via* introspective relation. Refer to summary above for an explanation.

The set $\mathcal{N}$ captures the introspective exponents. Observe that $n \in \mathcal{N}$, $p \in \mathcal{N}$, and trivially, $1 \in \mathcal{N}$. They are all coprime to k, since the coprime checks in Sect. 3.3 give $\gcd(n, k) = 1$ and $k < p$. For a prime p, $k < p$ gives $\gcd(p, k) = 1$.

The set $\mathcal{P}$ captures the introspective polynomials, those with introspective exponents in $\mathcal{N}$. Certainly $\forall c.\ 0 < c \wedge c \le \ell \Rightarrow \mathsf{X} + c \in \mathcal{P}$, and trivially, $1 \in \mathcal{P}$.

Recall the fundamental properties of introspective relation: there will be multiplicative exponents for $\mathcal{N}$ (Theorem 5) and multiplicative polynomials for $\mathcal{P}$ (Theorem 6). Together they imply that the sets $\mathcal{N}$ and $\mathcal{P}$ will be infinitely large. Our contradiction from the Pigeonhole Principle comes when we have derived some related, and finite sets.

4.2 Modulo Sets

One way to get a finite counterpart from an infinite set is by looking at remainders after division, or image of the set under some modulus. For the exponents set $\mathcal{N}$, the parameter k provides a modulus:

$$\vdash \mathcal{M} = (\lambda m.\ m \bmod k)(\!|\mathcal{N}|\!)$$

It is easy to estimate the cardinality of $\mathcal{M}$:

Theorem 12. *The cardinality of set $\mathcal{M}$ is bounded by k and $\mathrm{order}_k(n)$.*

$$\vdash \mathsf{Ring}\ \mathcal{R} \wedge \mathbf{1} \ne \mathbf{0} \wedge 1 < k \Rightarrow \forall n.\ n \in \mathcal{N} \Rightarrow \mathrm{order}_k(n) \le |\mathcal{M}| \wedge |\mathcal{M}| < k$$

Proof. Since there are k remainders under modulo k, $|\mathcal{M}| \le k$. But multiples of k (those n with $n \bmod k = 0$) are not in $\mathcal{N}$, as all elements of $\mathcal{N}$ are coprime to k and $k \ne 1$. Therefore $0 \notin \mathcal{M}$, making $|\mathcal{M}| < k$. Given $n \in \mathcal{N}$, so are all its powers:

$\forall j.\ n^j \in \mathcal{N}$ by Theorem 5. Hence all the remainders $n^j \bmod k$ are in $\mathcal{M}$. Since $\mathrm{order}_k(n)$ is the minimal exponent j before such remainders wrap around to 1, there are at least $\mathrm{order}_k(n)$ distinct remainders. Thus $\mathrm{order}_k(n) \le |\mathcal{M}|$.   □

For the polynomials set $\mathcal{P}$, the irreducible factor h of $\mathsf{X}^k - \mathbf{1}$ provides a modulus:

$$\vdash\ \mathcal{Q}_\mathsf{h}\ =\ (\lambda\mathsf{p}.\ \mathsf{p}\ \bmod\ \mathsf{h})\,(|\mathcal{P}|)$$

For the cardinality of $\mathcal{Q}_\mathsf{h}$, estimation requires more work, due to the change of modulus to h. Let $\mathsf{z} = \mathsf{X}^k - \mathbf{1}$, then monic z and $\deg\mathsf{z} = k$. Note that $\mathsf{z} \equiv \mathbf{0}\,(\bmod\,\mathsf{h})$, since h divides z by being a factor. These facts ensure that polynomial equivalences in $(\bmod\,\mathsf{z})$ are preserved to $(\bmod\,\mathsf{h})$:

Theorem 13. *Polynomial modulo equivalence holds for modulus factor.*

$\vdash$ Ring $\mathcal{R} \wedge$ monic $\mathsf{z} \wedge 0 < \deg\mathsf{z} \wedge$ monic $\mathsf{h} \wedge 0 < \deg\mathsf{h} \wedge \mathsf{z} \equiv \mathbf{0}\,(\bmod\ \mathsf{h}) \Rightarrow$
    $\forall\mathsf{p}\ \mathsf{q}.\ \mathrm{poly}\ \mathsf{p} \wedge \mathrm{poly}\ \mathsf{q} \wedge \mathsf{p} \equiv \mathsf{q}\,(\bmod\ \mathsf{z}) \Rightarrow \mathsf{p} \equiv \mathsf{q}\,(\bmod\ \mathsf{h})$

Proof. When $(\mathsf{p} - \mathsf{q})$ is divisible by z (due to $\mathsf{p} \equiv \mathsf{q}\,(\bmod\,\mathsf{z})$), and z is divisible by h (due to $\mathsf{z} \equiv \mathbf{0}\,(\bmod\,\mathsf{h})$), the difference $(\mathsf{p} - \mathsf{q})$ is also divisible by h due to transitivity of division.   □

An irreducible polynomial h gives a polynomial modulo field $\mathcal{F}_\mathsf{h}[\mathsf{X}]$, and nonzero elements of a field form a multiplicative group. Since $\mathsf{X} \neq \mathbf{0}$, it has a nonzero $\mathrm{order}_\mathsf{h}(\mathsf{X})$, with the following features.

Theorem 14. *When X is a root of unity, order of X is maximal when unity exponent is prime.*

$\vdash$ FiniteField $\mathcal{F} \wedge$ monic $\mathsf{h} \wedge$ ipoly $\mathsf{h} \wedge \mathsf{h} \neq \mathsf{X} - \mathbf{1} \Rightarrow$
    $\forall k.\ \mathrm{prime}\ k \wedge \mathsf{X}^k \equiv \mathbf{1}\,(\bmod\ \mathsf{h}) \Rightarrow \mathrm{order}_\mathsf{h}(\mathsf{X}) = k$

Proof. Let $t = \mathrm{order}_\mathsf{h}(\mathsf{X})$. By definition of order, $\mathsf{X}^t \equiv \mathbf{1}\,(\bmod\ \mathsf{h})$, and given $\mathsf{X}^k \equiv \mathbf{1}\,(\bmod\ \mathsf{h})$. Since t is minimal, t divides k. Given prime k, $t = 1$ or $t = k$. Only $\mathbf{1}$ has order 1, but $\mathsf{X} \not\equiv \mathbf{1}\,(\bmod\ \mathsf{h})$ by assumption. Therefore $\mathrm{order}_\mathsf{h}(\mathsf{X}) = t = k$.   □

Theorem 15. *In the polynomial field $\mathcal{F}_\mathsf{h}[\mathsf{X}]$, powers of X are distinct for exponents less than $\mathrm{order}_\mathsf{h}(\mathsf{X})$.*

$\vdash$ FiniteField $\mathcal{F} \wedge$ monic $\mathsf{h} \wedge$ ipoly $\mathsf{h} \wedge \mathsf{h} \neq \mathsf{X} \Rightarrow$
    $\forall m\ n.\ m < \mathrm{order}_\mathsf{h}(\mathsf{X}) \wedge n < \mathrm{order}_\mathsf{h}(\mathsf{X}) \Rightarrow (\mathsf{X}^m \equiv \mathsf{X}^n\,(\bmod\ \mathsf{h}) \iff m = n)$

Proof. Since $\mathcal{F}_\mathsf{h}[\mathsf{X}]$ is a finite field, its multiplicative group is a finite group. By the given assumption, $\mathsf{X} \not\equiv \mathbf{0}\,(\bmod\,\mathsf{h})$, thus X is an element in this group. Its order is the minimal exponent for the powers of X to wrap around to $\mathbf{1}$. Given the exponents are less than its order, such powers of X are distinct.   □

We shall see how the distinct powers of X lead to a lower bound for $\mathcal{Q}_\mathsf{h}$. This simple result is helpful:

Theorem 16. *Powers of X are equivalent in $\mathsf{X}^k - \mathbf{1}$ if exponents are equivalent in $\mathbb{Z}_k$.*

$\vdash$ Ring $\mathcal{R} \wedge \mathbf{1} \neq \mathbf{0} \Rightarrow \forall k.\ 0 < k \Rightarrow \forall m.\ \mathsf{X}^m \equiv \mathsf{X}^{m\ \bmod\ k}\,(\bmod\ \mathsf{X}^k - \mathbf{1})$

Proof. Since $m = (m\ \mathrm{div}\ k)k + m\ \bmod\ k$ and $\mathsf{X}^k \equiv \mathbf{1}\,(\bmod\ \mathsf{X}^k - \mathbf{1})$, the result follows.   □

4.3 Reduced Polynomials

Referring to Fig. 1, we shall see eventually that the right injective map is essential to give a lower bound for Q_h, and this lower bound is essential to provide the left injective map. These two injective maps are critical in the AKS proof.

To obtain a lower bound for Q_h, we need another way to get something finite from the infinite set $\mathcal{P}$, by taking a reduced subset of $\mathcal{P}$:

$$\vdash \widehat{\mathcal{P}} = \{ p \mid p \in \mathcal{P} \wedge \deg p < |\mathcal{M}| \}$$

This is a finite subset of $\mathcal{P}$ due to the polynomial degree cut-off. We shall prove that there is an injective map from $\widehat{\mathcal{P}}$ to Q_h, hence a lower bound on $|\widehat{\mathcal{P}}|$ will also be a lower bound for $|Q_h|$.

First, note an interesting interaction from $\mathcal{M}$ to $\mathcal{P}$, which is relevant to $\widehat{\mathcal{P}}$ since $\widehat{\mathcal{P}} \subseteq \mathcal{P}$. We know that $\mathcal{P}$ has a lot of elements (Sect. 4.1), so pick two polynomials $p \in \mathcal{P}$ and $q \in \mathcal{P}$. While it is almost impossible to identify any roots for p or q in $\mathcal{F}_h[X]$, it turns out that introspective relation helps to identify some interesting roots of their difference $(p - q)$, from the elements of $\mathcal{M}$.

Theorem 17. *Each element in $\mathcal{M}$ gives a root for any difference polynomial from $\mathcal{P}$ in $\mathcal{F}_h[X]$.*

$\vdash$ Field $\mathcal{F} \wedge$ monic $h \wedge 0 < \deg h \wedge X^k - 1 \equiv 0 \,(\mathrm{mod}\ \ h) \Rightarrow$
$\quad \forall p\ q.$
$\qquad p \in \mathcal{P} \wedge q \in \mathcal{P} \wedge p \equiv q \,(\mathrm{mod}\ \ h) \Rightarrow$
$\qquad\quad \forall n.\ n \in \mathcal{M} \Rightarrow (p - q)[\![X^n]\!] \equiv 0 \,(\mathrm{mod}\ \ h)$

Proof. Given $n \in \mathcal{M}$, there is $m \in \mathcal{N}$ such that $n = m \bmod k$. Therefore $m \bowtie p$ and $m \bowtie q$ by definition of $\mathcal{P}$. Let $z = X^k - 1$. Note that $z \equiv 0 \,(\mathrm{mod}\ h)$ by assumption. We can proceed:

	$p^m \equiv p[\![X^m]\!] \,(\mathrm{mod}\ z)$	by $m \bowtie p$
and	$p[\![X^m]\!] \equiv p[\![X^n]\!] \,(\mathrm{mod}\ z)$	by Theorem 16
so	$p^m \equiv p[\![X^n]\!] \,(\mathrm{mod}\ z)$	by transitivity
or for p , by $z \equiv 0 \,(\mathrm{mod}\ h)$	$p^m \equiv p[\![X^n]\!] \,(\mathrm{mod}\ h)$	by Theorem 13—[1]
Repeat the same steps for q	$q^m \equiv q[\![X^n]\!] \,(\mathrm{mod}\ h)$	by $m \bowtie q$ *etc.*—[2]
Since	$p^m \equiv q^m \,(\mathrm{mod}\ h)$	by $p \equiv q \,(\mathrm{mod}\ h)$ given
so	$p[\![X^n]\!] \equiv q[\![X^n]\!] \,(\mathrm{mod}\ h)$	by [1] and [2] above
or	$(p - q)[\![X^n]\!] \equiv 0 \,(\mathrm{mod}\ h)$	as claimed. $\square$

Due to this, an injective map between the two finite sets derived from $\mathcal{P}$ is possible:

Theorem 18. *There is an injective map from reduced set of $\mathcal{P}$ to modulo set of $\mathcal{P}$.*

$\vdash$ FiniteField $\mathcal{F} \wedge 0 < k \wedge$ monic $h \wedge$ ipoly $h \wedge X^k - 1 \equiv 0 \,(\mathrm{mod}\ \ h) \wedge$
$\quad k = \mathrm{order}_h(X) \Rightarrow$
$\qquad (\lambda p.\ p \bmod h) : \widehat{\mathcal{P}} \hookrightarrow Q_h.$

Proof. Let p, $q \in \widehat{\mathcal{P}}$, with $p \equiv q \,(\mathrm{mod}\ h)$ in $\mathcal{Q}_h$. For our map to be injective, we need to show $p = q$. Since $\widehat{\mathcal{P}} \subseteq \mathcal{P}$, p, $q \in \mathcal{P}$. Theorem 17 applies: each $n \in \mathcal{M}$ gives a root X^n for $(p - q)$. Now $h \neq X$ because, by assumption, $h \mid X^k - 1$, but $X \nmid X^k - 1$, and $n < k$ since $n \in \mathcal{M}$ means n is a remainder in $(\mathrm{mod}\ k)$. By assumption, $k = \mathrm{order}_h(X)$, hence these roots are distinct by Theorem 15. Thus there are at least $|\mathcal{M}|$ distinct roots for $(p - q)$.

But $\deg p < |\mathcal{M}|$ and $\deg q < |\mathcal{M}|$ since p, $q \in \widehat{\mathcal{P}}$, hence $\deg (p - q) < |\mathcal{M}|$. There are more roots than its degree for the difference $(p - q)$ with coefficients from a finite field $\mathcal{F}$. This is possible only when the difference is 0, *i.e.*, $p = q$. $\qquad\square$

This injective map leads to a lower bound for the cardinality of $\mathcal{Q}_h$.

Theorem 19. *The modulo set of $\mathcal{P}$ has a nice lower bound.*

$\vdash \ \mathsf{FiniteField}\ \mathcal{F} \ \wedge \ 1 \ < \ k \ \wedge \ k \ = \ \mathrm{order}_h(X) \ \wedge \ \ell \ < \ \chi \ \wedge \ \mathsf{monic}\ h \ \wedge \ \mathsf{ipoly}\ h \ \wedge$
$X^k \ - \ 1 \ \equiv \ 0 \,(\mathrm{mod}\ h) \ \Rightarrow$
$\qquad 2^{\min(\ell, |\mathcal{M}|)} \ \leq \ |\mathcal{Q}_h|$

Proof. Applying Theorem 18, there is an injective map from $\widehat{\mathcal{P}}$ to $\mathcal{Q}_h$. As both sets are finite, $|\widehat{\mathcal{P}}| \leq |\mathcal{Q}_h|$. We shall estimate $|\widehat{\mathcal{P}}|$, by counting how many polynomials $p \in \mathcal{P}$ have $\deg p < |\mathcal{M}|$.

Note that $1 < |\mathcal{M}|$, since $\mathrm{order}_k(n) \leq |\mathcal{M}|$ by Theorem 12, and $1 < \mathrm{order}_k(n)$ since $n \neq 1$. A simple estimate for $|\widehat{\mathcal{P}}|$ proceeds as follows:

- For $0 < c \leq \ell$, $X + c \in \widehat{\mathcal{P}}$, since each monomial is in $\mathcal{P}$, and each has a degree equal to 1.
- Given $\ell < \chi$, these monomials are distinct, as χ is the least additive wrap-around of 1 in field $\mathcal{F}$.[3]
- By Theorem 6, any product of these monomials will be in $\widehat{\mathcal{P}}$, if the product has a degree less than $|\mathcal{M}|$.
- If $\ell < |\mathcal{M}|$, there are less than $|\mathcal{M}|$ such monomials. Therefore any product drawn from a subset of $\{X + c \mid 0 < c \leq \ell\}$ will have a degree less than $|\mathcal{M}|$. There are 2^ℓ such products.
- If $|\mathcal{M}| \leq \ell$, reduce the constants range to $0 < c \leq |\mathcal{M}|$. Any product drawn from a subset of $\{X + c \mid 0 < c \leq |\mathcal{M}|\}$ will have a degree less than $|\mathcal{M}|$, almost—the product of all such monomials must be excluded. However, $1 \in \widehat{\mathcal{P}}$, but 1 is not a monomial product. There are still $2^{|\mathcal{M}|}$ products.

Considering both cases, we conclude that $2^{\min(\ell, |\mathcal{M}|)} \leq |\mathcal{Q}_h|$. $\qquad\square$

4.4   Reduced Exponents

It turns out that an injective map to $\mathcal{M}$ is possible based on the following set of reduced exponents:

$\vdash \widehat{\mathcal{N}}\ p\ n\ m \ = \ \{p^i\, n^j \mid i \leq m \wedge j \leq m\}$

---

[3] The characteristic χ of a ring $\mathcal{R}$ is defined as the order of 1 in the additive group of $\mathcal{R}$, *i.e.*, $\chi 1 = 0$.

This is generated by the two known elements $n, p \in \mathcal{N}$ (Sect. 4.1), with cut-off m in their exponents. By multiplicative closure of introspective exponents (Theorem 5), $\widehat{\mathcal{N}} \subseteq \mathcal{N}$. Observe the following property:

Theorem 20. *Upper bound of an element in $\widehat{\mathcal{N}}$ p n m.*

$$\vdash 1 < p \wedge p \le n \Rightarrow \forall e \; m. \; e \in \widehat{\mathcal{N}} \; p \; n \; m \Rightarrow e \le n^{2m}$$

Proof. Each $e \in \widehat{\mathcal{N}} \; p \; n \; m$ has the form $p^i n^j$, where $i, j \le m$. Given $p \le n$, we can deduce $e = p^i n^j \le n^i n^j \le n^m n^m = n^{2m}$. □

Note another interesting interaction from $\mathcal{Q}_h$ to $\mathcal{N}$, which is relevant to $\widehat{\mathcal{N}}$ since $\widehat{\mathcal{N}} \; n \; p \; m \subseteq \mathcal{N}$. Pick two exponents $n \in \mathcal{N}$ and $m \in \mathcal{N}$. Consider the special polynomial $X^n - X^m$. It turns out that the introspective relation helps to identify some interesting roots of this special polynomial, from the elements of $\mathcal{Q}_h$.

Theorem 21. *Each element in $\mathcal{Q}_h$ gives a root for a special polynomial from $\mathcal{N}$ in $\mathcal{F}_h[X]$.*

$$\vdash \text{Field } \mathcal{F} \wedge \text{ monic } h \wedge \text{ ipoly } h \wedge X^k - 1 \equiv 0 \,(\text{mod } h) \Rightarrow$$
$$\forall n \; m.$$
$$n \in \mathcal{N} \wedge m \in \mathcal{N} \wedge n \equiv m \,(\text{mod } k) \Rightarrow$$
$$\forall p. \; p \in \mathcal{Q}_h \Rightarrow (X^n - X^m)[\![p]\!] \equiv 0 \,(\text{mod } h)$$

Proof. Given $p \in \mathcal{Q}_h$, there is $q \in \mathcal{P}$ such that $p = q \bmod h$. Therefore $n \bowtie q$ and $m \bowtie q$ by definition of $\mathcal{P}$. Let $z = X^k - 1$. Note that $z \equiv 0 \,(\text{mod } h)$ by given. We can proceed:

	$q^n \equiv q[\![X^n]\!] \,(\text{mod } z)$	by $n \bowtie q$ — [1]
	$q^m \equiv q[\![X^m]\!] \,(\text{mod } z)$	by $m \bowtie q$ — [2]
and	$q[\![X^m]\!] \equiv q[\![X^n]\!] \,(\text{mod } z)$	by Theorem 16
so	$q^n \equiv q^m \,(\text{mod } z)$	by [1], [2], transitivity
Therefore	$q^n - q^m \equiv 0 \,(\text{mod } z)$	by subtraction
by $z \equiv 0 \,(\text{mod } h)$	$q^n - q^m \equiv 0 \,(\text{mod } h)$	by Theorem 13—[3]
Since	$(X^n - X^m)[\![p]\!] \equiv (X^n - X^m)[\![q]\!] \,(\text{mod } h)$	by $p = q \bmod h$ —[4]
and the right-side	$(X^n - X^m)[\![q]\!] = q^n - q^m$	by substitution of q —[5]
Combine [4],[5],[3]	$(X^n - X^m)[\![p]\!] \equiv 0 \,(\text{mod } h)$	as claimed. □

Due to this, an injective map between the two finite sets derived from $\mathcal{N}$ is possible:

Theorem 22. *There is an injective map from reduced set of $\mathcal{N}$ to modulo set of $\mathcal{N}$.*

$$\vdash \text{FiniteField } \mathcal{F} \wedge \text{ monic } h \wedge \text{ ipoly } h \wedge X^k - 1 \equiv 0 \,(\text{mod } h) \Rightarrow$$
$$\forall n \; p.$$
$$1 < p \wedge p < n \wedge n \in \mathcal{N} \wedge p \in \mathcal{N} \wedge n^{2\sqrt{|\mathcal{M}|}} < |\mathcal{Q}_h| \Rightarrow$$
$$(\lambda m. \; m \bmod k) : \widehat{\mathcal{N}} \; p \; n \; \sqrt{|\mathcal{M}|} \hookrightarrow \mathcal{M}$$

Proof. Let $i,\ j\ \in\ \hat{\mathcal{N}}\ n\ p\sqrt{|\mathcal{M}|}$, with $i\ \equiv\ j\,(\mathrm{mod}\ k)$ in $\mathcal{M}$. If the map is to be injective, we need $i=j$. Since $\hat{\mathcal{N}}\ n\ p\sqrt{|\mathcal{M}|}\subseteq\mathcal{N}$, both $i,j\in\mathcal{N}$. Theorem 21 applies: every $\mathsf{p}\in\mathcal{Q}_h$ is a root of $\mathsf{X}^i-\mathsf{X}^j$. Hence there are at least $|\mathcal{Q}_h|$ roots.

By Theorem 20, both $i,\ j$ are bounded by $n^2\sqrt{|\mathcal{M}|}$, hence $\deg\,(\mathsf{X}^i-\mathsf{X}^j)\le n^2\sqrt{|\mathcal{M}|}$. Given $n^2|\mathcal{M}|<|\mathcal{Q}_h|$, there are more roots than its degree for the polynomial $(\mathsf{X}^i-\mathsf{X}^j)$ with coefficients from a finite field $\mathcal{F}$. This is not possible, unless it is $\mathbf{0}$, which means $i=j$.   □

4.5   Punch Line

Given a prime p that divides n, if $n^x=p^y$ for some exponents $x,\ y$ with $x>0$, what can we conclude?

Theorem 23. *A condition that implies a number is a perfect power of prime.*

$\vdash\ 0\ <\ n\ \wedge\ \mathsf{prime}\ p\ \wedge\ p\ |\ n\ \wedge\ (\exists x\ y.\ \ 0<x\wedge p^x\ =\ n^y)\ \Rightarrow\ \mathsf{perfect\_power}\ n\ p$

Proof. Since $p\mid n$, divide n by p as many times as possible, and express $n=p^m q$ where m is the maximum possible, and $p\nmid q$. The equation $p^x=n^y$ becomes $p^x=(p^m q)^y=p^{my}q^y$. By unique factorisation, with prime p and $p\nmid q$ and $x\neq 0$, it must be that $y\neq 0$, and $q^y=1$, *i.e.*, $q=1$.   □

When its generators have a special property, the cardinality of $\hat{\mathcal{N}}\ p\ n\ m$ is simple to express:

Theorem 24. *Cardinality of $\hat{\mathcal{N}}$ when generators n and prime divisor p are not related by perfect power.*

$\vdash\ \mathsf{Ring}\ \mathcal{R}\ \wedge\ 1\ \neq\mathbf{0}\ \wedge\ 1\ <\ k\ \Rightarrow$
$\qquad\forall n\ p\ m.$
$\qquad\qquad n\in\mathcal{N}\ \wedge\ p\in\ \mathcal{N}\ \wedge\ \mathsf{prime}\ p\ \wedge\ p\ |\ n\ \wedge\ \neg\mathsf{perfect\_power}\ n\ p\ \Rightarrow$
$\qquad\qquad|\hat{\mathcal{N}}\ p\ n\ m|\ =\ (m+1)^2$

Proof. Let $f=(\lambda\,(i,j).\ p^i\,n^j)$, $t=\{\,j\mid j\le m\,\}$. From its definition, it is simple to verify that $\hat{\mathcal{N}}\ p\ n\ m=\{\ p^i\,n^j\mid i\le m\wedge j\le m\ \}=f(\!(t\times t)\!)$. More interesting is that the conditions will imply $f:t\times t\ \hookrightarrow\ \hat{\mathcal{N}}\ p\ q\ n$. Once this is proved, being the image of an injective map gives $|\hat{\mathcal{N}}\ p\ q\ n|=|t\times t|\ =\ |t|^2=(m+1)^2$.

To show that the map is injective, assume $p^i\,n^j=p^u\,n^v$ for some i,j and u,v. We need to show $i=u$ and $j=v$. This comes down to analysis by cases.

If $i<u$, only the case $j>v$ is interesting, with $n^{j-v}=p^{u-i}$. As $j-v\neq 0$, Theorem 23 applies, giving $\mathsf{perfect\_power}\ n\ p$, which contradicts the assumption. By the symmetric roles of i,j and u,v, the case $i>u$ leads to the same contradiction. The only possible case is $i=u$, giving $j=v$.   □

This property is crucial in order to complete the proof of AKS Main Theorem (Theorem 11).

Proof. (of Theorem 11). AKS Main Theorem in finite fields

$\vdash$ FiniteField $\mathcal{F}$ $\wedge$ prime k $\wedge$ $k < \chi$ $\Rightarrow$
  $\forall\, n$.
    $1 < n \wedge \chi \mid n \wedge \gcd(n, k) = 1 \wedge (2\,(\log n + 1))^2 \leq \mathrm{order}_k(n) \wedge$
    $\ell = 2\sqrt{k}\,(\log n + 1) \wedge (\forall\, c.\ 0 < c \wedge c \leq \ell \Rightarrow n \bowtie X + c) \Rightarrow$
    perfect_power n χ

Let $p = \chi$. By assumption, $p \mid n$, so $p \leq n$. The case $p = n$ is trivial, so we shall assume $p < n$.

The finite field $\mathcal{F}$ gives prime p, so $p \bowtie X + c$ (Theorem 4). We have $k < p$, so $\gcd(p, k) = 1$. Assuming $\gcd(n, k) = 1$ and $n \bowtie X + c$, we have the ingredients for the introspective sets $\mathcal{N}$ and $\mathcal{P}$ (Sect. 4.1). Their finite counterparts, the modulo sets $\mathcal{M}$ and $\mathcal{Q}_h$ (Sect. 4.2), and reduced sets $\widehat{\mathcal{N}}$ and $\widehat{\mathcal{P}}$ (Sects. 4.3 and 4.4) can be set up accordingly.

Recall that the introspective relation is based on modulus $X^k - 1$. By the second useful fact in Sect. 4, in a finite field $\mathcal{F}$ it has a monic irreducible factor $h \neq X - 1$, i.e., $X^k - 1 \equiv \mathbf{0} \pmod{h}$. With prime k, we have $\mathrm{order}_h(X) = k$ (Theorem 14), giving the injective map from $\widehat{\mathcal{P}}$ to $\mathcal{Q}_h$ (Theorem 18), which is essential for the lower bound estimate of $\mathcal{Q}_h$.

In Sect. 4.6, we shall investigate the parameters k and ℓ. We shall show that $\ell < k$ (Theorem 27). By assumption, $k < p$, so $\ell < p$. Therefore $2^{\min(\ell, |\mathcal{M}|)} \leq |\mathcal{Q}_h|$ (Theorem 19, which invokes Theorem 18). We shall also show that $n^2\sqrt{|\mathcal{M}|} < 2^{\min(\ell, |\mathcal{M}|)}$ (Theorem 26). Hence $n^2\sqrt{|\mathcal{M}|} < |\mathcal{Q}_h|$. With $p < n$, these inequalities establish the injective map from $\widehat{\mathcal{N}}\ n\ p\sqrt{|\mathcal{M}|}$ to $\widehat{\mathcal{N}}$ (Theorem 22).

Now, given prime p and $p \mid n$, if n were not a perfect power of p, Theorem 24 applies, so that:

$$|\widehat{\mathcal{N}}\ p\ n\ \sqrt{|\mathcal{M}|}| = (\sqrt{|\mathcal{M}|} + 1)^2 = |\mathcal{M}| + (2\sqrt{|\mathcal{M}|}) + 1 > |\mathcal{M}|$$

This means the injective map from $\widehat{\mathcal{N}}\ p\ n\sqrt{|\mathcal{M}|}$ to $\mathcal{M}$, both finite sets, would violate the Pigeonhole Principle. Therefore, n must be a perfect power of $p = \chi$. $\square$

4.6 Parameters

The AKS Main Theorem contains a parameter k with the property: $\mathrm{order}_k(n) \geq (2\,(\log n + 1))^2$, from which a related parameter $\ell = 2\sqrt{k}\,(\log n + 1)$ is computed.

In the original AKS paper [2], parameter k is a prime (for a different set of conditions) while in the revised version [3] this prime requirement on k is dropped. Only the bound on k affects the conclusion "PRIMES is in P", a general k needs more advanced theory to establish. Our mechanisation effort is based on a prime k, following Dietzfelbinger [10]. We use a prime k to show $k = \mathrm{order}_h(X)$ in Theorem 14.

The existence of such a prime k can be established by generalizing the problem: given a number n, and a maximum m, find a prime modulus k such that $\mathrm{order}_k(n) \geq m$. This is applied in Theorem 8:

Theorem 25. *There is always a modulus k giving big enough order for n in $\mathbb{Z}_k$.*

$\vdash\ 1 < n \,\wedge\, 0 < m \,\Rightarrow\, \exists\,k.\ \text{prime } k \,\wedge\, \gcd(k, n) = 1 \,\wedge\, m \le \text{order}_k(n)$

Proof. First, we define a set of candidates:

$\vdash \text{candidates } n\ m = \{k \mid \text{prime } k \,\wedge\, k \nmid n \,\wedge\, \forall j.\ 0 < j \wedge j < m \Rightarrow k \nmid n^j - 1\}$

Pick a large prime $z > n^m$, then z cannot divide n or any of the factors $n^j - 1$ for $0 < j < m$, hence $z \in \text{candidates } n\ m$.

Thus candidates $n\ m \ne \emptyset$, and we can pick a candidate k, say the least value, from the set. Being an element, prime $k \wedge k \nmid n$. Since a prime is coprime to its non-multiples, $\gcd(k, n) = 1$. Thus n has nonzero order in $\mathbb{Z}_k$. Let $j = \text{order}_k(n)$, then $0 < j$ with $n^j \equiv 1 \,(\text{mod } k)$, or $k \mid n^j - 1$. If $j < m$, by the candidates definition $k \nmid n^j - 1$, a contradiction. Hence $\text{order}_k(n) = j \ge m$. $\qquad\square$

The parameters k and ℓ provide a crucial inequality involving $|\mathcal{M}|$, used in Theorem 11:

Theorem 26. *The AKS parameters meet the inequality condition.*

$\vdash\ \text{FiniteField } \mathcal{F} \wedge 1 < k \,\wedge\, 1 < n \,\wedge\, n \in \mathcal{N} \,\wedge\, (2\,(\log n + 1))^2 \le \text{order}_k(n) \,\wedge$
$\ \ell = 2\sqrt{k}\,(\log n + 1) \,\Rightarrow$
$\ \ n^{2\sqrt{|\mathcal{M}|}} < 2^{\min(\ell, |\mathcal{M}|)}$

Proof. Let $j = \text{order}_k(n)$, and $m = \log n + 1$, then $2^m > n$ for integer logarithm. By Theorem 12, $j \le |\mathcal{M}|$ and $|\mathcal{M}| < k$. By the given assumption, $(2\,m)^2 \le j$. Taking integer square roots, we have $\sqrt{|\mathcal{M}|} \ge \sqrt{j}$, $\sqrt{k} \ge \sqrt{|\mathcal{M}|}$ and $\sqrt{j} \ge 2\,m$. Note also $|\mathcal{M}| \ge \sqrt{|\mathcal{M}|}\,\sqrt{|\mathcal{M}|}$ by integer square root. Therefore:

- $\ell = 2\sqrt{k}\,m \ge m\,(2\sqrt{|\mathcal{M}|})$
- $|\mathcal{M}| \ge \sqrt{j}\,\sqrt{|\mathcal{M}|} \ge m\,(2\sqrt{|\mathcal{M}|})$

Thus $\min(\ell, |\mathcal{M}|) \ge m\,(2\sqrt{|\mathcal{M}|})$, and

$$2^{\min(\ell, |\mathcal{M}|)} \ge 2^{m\,(2\sqrt{|\mathcal{M}|})} = 2^{m2\sqrt{|\mathcal{M}|}} > n^{2\sqrt{|\mathcal{M}|}}. \qquad\square$$

Incidentally, the choice of k and ℓ ensures that $\ell \le k$, used in Theorems 10 and 11:

Theorem 27. *The AKS computed parameter does not exceed the modulus parameter.*

$\vdash\ 1 < n \,\wedge\, 1 < k \,\wedge\, \gcd(k, n) = 1 \,\wedge\, (2\,(\log n + 1))^2 \le \text{order}_k(n) \Rightarrow$
$\ \ 2\sqrt{k}\,(\log n + 1) \le k$

Proof. Since $\text{order}_k(n) \mid \varphi(k)$, and $\varphi(k) < k$ when $k > 1$, we have $\text{order}_k(n) < k$. Taking integer square-roots, with the given $\text{order}_k(n)$, deduce

$$k \ge \sqrt{k}\,\sqrt{k} \ge \sqrt{k}\,\sqrt{\text{order}_k(n)} \ge 2\sqrt{k}\,(\log n + 1). \qquad\square$$

5   Mechanisation and its Traps

The updated AKS proof [3] is contained within four pages. Mechanisation of such a proof is the process of unwinding the dense mathematics within those pages. It took us about a year to build up the basic libraries, another year to forge the advanced libraries, then about six months to adapt the libraries for the proof of AKS Main Theorem, during which time the missing pieces in the developed libraries were steadily being filled in.

There are always pitfalls during the mechanisation process. One example is the symbol X in various expositions of the AKS proof, *e.g.*, [4,7,8,10]. Usually X starts as an indeterminate or a degree one zero constant monomial, then switches to a root of unity, even to a primitive root of unity. While this is common practice, such changes mean that we needed to prove the switchings are valid.

The substitution by X is fundamental in the introspective relation (Sect. 2.2). These subtle changes in the role of X presented some difficulties in our initial effort to formalize the AKS proof. Indeed, we first used an inappropriate definition and got carried along until we found that shifting playgrounds (Sect. 3.3) is impossible with that definition.

Shifting of playgrounds in the AKS proof is pivotal. Most expositions just point this out without further elaboration.[4] After this shifting, where the playground is now $\mathbb{Z}_p$, the introspective relation is defined in $\mathbb{Z}_p[\mathsf{X}]$, side-stepping the issue. It was in the process of mechanisation that we realized a proper formulation should start by defining the introspective relation in a ring $\mathcal{R}$ (Sect. 2.2), and then prove that shifting is valid through ring homomorphisms from $\mathbb{Z}_n$ to $\mathbb{Z}_p$ (Sect. 3.3).

Lessons Learnt. Rather than attempting a direct transcription of the AKS proof, we came to understand the proof in the context of finite fields, identifying the key concepts involved in the proof, even comparing various expositions. By reformulations of polynomial theorems in number theory into their counterparts in rings and fields, a clear picture of the proof's logic emerged, resulting in this succinct presentation.

HOL4 and Abstract Algebra. This work demonstrates that HOL4's simple type theory, together with its proof machinery, are sufficient to allow the statement and proof of moderately complicated theorems in abstract algebra. Without dependent types (as in Coq) or locales (as in Isabelle), theorems are slightly more awkward to state, but our experience is that *ad hoc* overloading gets one a long way. Over-annotation of terms so that the parser chooses the "right" meaning of a symbol like + is only necessary occasionally. Exploiting overloading in this way requires a careful understanding of just what the parser is and is not capable of, and one is often on exactly that boundary. Nonetheless, the result gives terms that are not far removed from those that have been pretty-printed

---

[4] For example, [3] first stated the computational identity in $\mathbb{Z}_n$, then "this implies" the corresponding identity in $\mathbb{Z}_p$. Only [10] proved the shifting from $\mathbb{Z}_n$ to $\mathbb{Z}_p$ as a lemma.

in this paper. (Pretty-printing to LATEX adds niceties such as superscripts and juxtaposition for multiplication; these could not be handled by the parser.)

Nor should we forget that Campos *et al.* [5] proved half of the Main Theorem in ACL2, where the underlying logic is even simpler, and provides no static type-checking.

6   Related Work

Other Pen-and-Paper Proofs. The revised proof (2004) of the AKS team [3] takes this approach: use the injective map on $\mathcal{Q}_h$ to establish a lower bound for $|\mathcal{Q}_h|$; assuming that n is not a power of p, use the Pigeonhole Principle to show that a special nonzero polynomial has at least $|\mathcal{Q}_h|$ roots, thus giving an upper bound for $|\mathcal{Q}_h|$; manipulate inequalities to show that the chosen parameters will lead to the lower bound exceeding the upper bound, hence a contradiction.

Other expositions of the AKS Main Theorem [1, 11–13] take similar approaches, working mainly in $\mathbb{Z}_p[\mathsf{X}]$. Our method is equivalent, but is clean in that we: (*i*) emphasize the important role of shifting from $\mathbb{Z}_n$ to $\mathbb{Z}_p$ (Sect. 3.3); (*ii*) reformulate the AKS Main Theorem in the context of finite fields (Theorem 11); (*iii*) clarify that the choice of parameters gives injective maps between reduced sets and modulo sets (Theorems 18 and 22); (*iv*) bring in the assumption that n is not a power of prime p as late as possible; and (*v*) use the Pigeonhole Principle as a punch line to force n to be a power of prime p (Sect. 4.5).

Other Mechanisations. We believe that we are the first to mechanise both directions of the central theorem of AKS algorithm. As noted earlier, two other teams (Campos *et al.* [5] in ACL2, and de Moura and Tadeu [9] in Coq) have mechanised the fact that if the number being tested is prime, then the AKS algorithm will indeed report "yes".

We are also aware of preliminary work started by John Harrison, and carried out in HOL Light.[5]

7   Conclusion

It is well-known that the cardinality of a finite field must be a prime power, and it is elementary to check whether a number is power-free. In essence, the AKS team showed that primality testing can be reduced to finite field cardinality testing, and demonstrated that the latter can be done in polynomial time.

Through our mechanisation effort, especially in presenting the AKS proof as an introspective game (Sect. 4), we hope that this elementary proof of the AKS Main Theorem provides further appreciation of the AKS team's brilliant ideas.

Future Work. While the existence of parameter k in the AKS Main Theorem is assured, to show that it is bounded by a polynomial function of $\log n$ is harder. In future work, we intend to perform the necessary complexity analysis of the AKS algorithm to complete the mechanised proof that PRIMES is indeed in P.

---

[5] John was kind enough to share his approach with us *via* private communication.

References

1. Agrawal, M.: Primality tests based on Fermat's Little Theorem, December 2006. http://www.cse.iitk.ac.in/users/manindra/presentations/FLTBasedTests.pdf
2. Agrawal, M., Kayal, N., Saxena, N.: PRIMES is in P, August 2002. Original paper
3. Agrawal, M., Kayal, N., Saxena, N.: PRIMES is in P. Ann. Math. **160**(2), 781–793 (2004)
4. Bedodi, A.: Primality tests in polynomial time. Master's thesis, Universitá Degli Studi Roma TRE, February 2010
5. Campos, C., Modave, F., Roach, S.: Towards the verification of the AKS primality test in ACL2. In: Fifth International Conference on Intelligent Technologies, November 2004
6. Chan, H.-L., Norrish, M.: A string of pearls: proofs of Fermat's Little Theorem. J. Formalized Reason. **6**(1), 63–87 (2013)
7. Crandall, R., Pomerance, C.: Prime Numbers: A Computational Perspective. Springer, New York (2005)
8. Daleson, G.: Deterministic primality testing in polynomial time. Master's thesis, Portland State University, December 2006
9. de Moura, F.L. C., Tadeu, R.: The correctness of the AKS primality test in Coq. July 2008. http://www.cic.unb.br/~flavio/AKS.pdf
10. Dietzfelbinger, M.: Primality Testing in Polynomial Time: From Randomized Algorithms to 'PRIMES is in P'. Lecture Notes in Computer Science. Springer, Heidelberg (2004)
11. Domingues, R.: A polynomial time algorithm for prime recognition. Master's thesis, University of Pretoria, January 2006
12. Linowitz, B.: An exposition of the AKS polynomial time primality testing. Master's thesis, University of Pennsylvania, March 2006
13. Pomerance, C.: Primality testing, variations on a theme of Lucas (2008). http://cm.bell-labs.com/who/carlp/PS/primalitytalk5.ps

Machine-Checked Verification
of the Correctness and Amortized Complexity
# of an Efficient Union-Find Implementation

Arthur Charguéraud[1] and François Pottier[2]([⊠])

[1] Inria and LRI, Université Paris Sud, CNRS, Orsay, France
[2] Inria, Paris-Rocquencourt, France
francois.pottier@inria.fr

Abstract. Union-Find is a famous example of a simple data structure whose amortized asymptotic time complexity analysis is non-trivial. We present a Coq formalization of this analysis. Moreover, we implement Union-Find as an OCaml library and formally endow it with a modular specification that offers a full functional correctness guarantee as well as an amortized complexity bound. Reasoning in Coq about imperative OCaml code relies on the CFML tool, which is based on characteristic formulae and Separation Logic, and which we extend with time credits. Although it was known in principle that amortized analysis can be explained in terms of time credits and that time credits can be viewed as resources in Separation Logic, we believe our work is the first practical demonstration of this approach.

1   Introduction

The Union-Find data structure, also known as a disjoint set forest [12], is widely used in the areas of graph algorithms and symbolic computation. It maintains a collection of disjoint sets and keeps track in each set of a distinguished element, known as the representative of this set. It supports the following operations: *make* creates a new element, which forms a new singleton set; *find* maps an element to the representative of its set; *union* merges the sets associated with two elements (and returns the representative of the new set); *equiv* tests whether two elements belong to the same set. In OCaml syntax, this data structure offers the following signature, where the abstract type `elem` is the type of elements:

```
type elem
val make : unit -> elem
val find : elem -> elem
val union : elem -> elem -> elem
val equiv : elem -> elem -> bool
```

One could generalize the above signature by attaching a datum of type `'a` to every set, yielding a type `'a elem`. We have not done so for the moment.

Disjoint set forests were invented by Galler and Fischer [15]. In such a forest, an element either points to another element or points to no element (i.e., is a

C. Urban and X. Zhang (Eds.): ITP 2015, LNCS 9236, pp. 137–153, 2015.
DOI: 10.1007/978-3-319-22102-1_9

```
type rank = int                          let link x y =
                                           if x == y then x else
type elem = content ref                    match !x, !y with
                                           | Root rx, Root ry ->
and content = Link of elem | Root of rank    if rx < ry then begin
                                               x := Link y; y
let make () = ref (Root 0)                  end else if rx > ry then begin
                                               y := Link x; x
let rec find x =                            end else begin
  match !x with                              y := Link x; x := Root (rx+1); x
  | Root _ -> x                            end
  | Link y ->                            | _, _ -> assert false
      let z = find y in
      x := Link z;                      let union x y = link (find x) (find y)
      z                                 let equiv x y = (find x) == (find y)
```

Fig. 1. OCaml implementation of Union-Find

root). These pointers form a forest, where each tree represents a set, and where the root of the tree is the representative of the set. There is a unique path from every node in a tree to the root of this tree.

The *find* operation follows this unique path. For efficiency, it performs path compression: every node along the path is updated so as to point directly to the root of the tree. This idea is attributed by Aho *et al.* [1] to McIlroy and Morris.

The *union* operation updates the root of one tree so as to point to the root of the other tree. To decide which of the two roots should become a child of the other, we apply *linking-by-rank* [28,29]. A natural number, the *rank*, is associated with every root. The rank of a newly created node is zero. When performing a union, the root of lower rank becomes a child of the root of higher rank. In case of equality, the new root is chosen arbitrarily, and its rank is increased by one.

A complete OCaml implementation appears in Fig. 1.

Union-Find is among the simplest of the classic data structures, yet requires one of the most complicated complexity analyses. Tarjan [29] and Tarjan and van Leeuwen [28] established that the worst-case time required for performing m operations involving n elements is $O(m \cdot \alpha(n))$. The function α, an inverse of Ackermann's function, grows so slowly that $\alpha(n)$ does not exceed 5 in practice. The original analysis was significantly simplified over the years [21], ultimately resulting in a 2.5 page proof that appears in a set of course notes by Tarjan [27] and in the textbook *Introduction to Algorithms* [12].

In this paper, we present a machine-checked version of this mathematical result. In addition, we establish a machine-checked connection between this result and the code shown in Fig. 1. Our proofs are available online [8].

To assess the asymptotic time complexity of an OCaml program, we rely on the assumption that "it suffices to count the function calls". More precisely, if one ignores the cost of garbage collection and if one encodes loops as tail-recursive functions, then the number of function calls performed by the source program is an accurate measure of the number of machine instructions executed by the compiled program, up to a constant factor, which may depend (linearly) on the size of the program. Relying on this assumption is an old idea. For example, in the setting of a first-order functional programming language, Le Métayer [22] notes that "the asymptotic complexity and the number of recursive calls necessary for the evaluation of the program are of the same order-of-magnitude". In our higher-order setting, every function call is potentially a "recursive" call, so must be counted. Danielsson [13] does this. The cost of constructing and looking up the environment of a closure is not a problem: as noted by Blelloch and Greiner [5], it is a constant, which depends on the number of variables in the program.

We do not prove that the OCaml compiler respects our assumption. It is very likely that it does. If it did not, that would be a bug, which should and very likely could be fixed. On a related theme, the CerCo project [3] has built compilers that not only produce machine code, but also determine the actual cost (according to a concrete machine model) of every basic block in the source program. This allows carrying out concrete worst-case-execution-time analysis at the source level.

In order to formally verify the correctness and complexity of a program, we rely on an extension of Separation Logic with *time credits*. Separation Logic [26] offers a natural framework for reasoning about imperative programs that manipulate the heap. A time credit is a resource that represents a *right to perform one step of computation*. In our setting, every function call consumes one credit. A number of credits can be explicitly requested as part of the precondition of a function f, and can be used to justify the function calls performed by f and by its callees. A time credit can be stored in the heap for later retrieval and consumption: that is, amortization is permitted.

In short, the combination of Separation Logic and time credits is particularly attractive because it allows (1) reasoning about correctness and complexity at the same time, (2) dealing with dynamic memory allocation and mutation, and (3) carrying out amortized complexity analyses.

Time credits, under various forms, have been used previously in several type systems [13,17,25]. Atkey [4] has argued in favor of viewing credits as predicates in Separation Logic. However, Atkey's work did not go as far as using time credits in a general-purpose program verification framework.

We express the specification of every operation (*make, find, union,* etc.) in terms of an abstract predicate UF $N\,D\,R$, where N is an upper bound on the number of elements, D is the set of all elements, and R is a mapping of every element to its representative. The predicate UF captures: (1) the existence (and ownership) in the heap of a collection of reference cells; (2) the fact that the graph formed by these cells is a disjoint set forest; (3) the connection between this graph and the parameters N, D, and R, which the client uses in her reasoning; and (4) the ownership of a number of time credits that corresponds to the current "total potential", in Tarjan's terminology, of the data structure.

The precondition of an operation tells how many credits this operation requires. This is its amortized cost. Its actual cost may be lower or higher, as the operation is free to store credits in the heap, or retrieve credits from the heap, as long as it maintains the invariant encoded in the predicate UF. For instance, the precondition of *find* contains $\$(\alpha(N) + 2)$, which means "$\alpha(N)+2$ credits". We note that it might be preferable to state that *find* requires $O(\alpha(N))$ credits. However, we have not yet developed an infrastructure for formalizing the use of the big-O notation in our specifications.

To prove that the implementation of an operation satisfies its specification, we rely on the tool CFML [6,7], which is based on Separation Logic and on *characteristic formulae*. The characteristic formula of an OCaml term t is a higher-order logic formula $[\![t]\!]$ which describes the semantics of t. (This obviates the need for embedding the syntax of t in the logic.) More precisely, $[\![t]\!]$ denotes the set of all valid specifications of t, in the following sense: for any precondition H and postcondition Q, if the proposition $[\![t]\!] \, H \, Q$ can be proved, then the Separation Logic triple $\{H\} \, t \, \{Q\}$ holds. The characteristic formula $[\![t]\!]$ is built automatically from the term t by the CFML tool. It can then be used, in Coq, to formally establish that the term t satisfies a particular specification.

Our proofs are carried out in the Coq proof assistant. This allows us to perform in a unified framework a mathematical analysis of the Union-Find data structure and a step-by-step analysis of its OCaml implementation.

In addition to Coq, our trusted foundation includes the meta-theory and implementation of CFML. Indeed, the characteristic formulae produced by CFML are accepted as axioms in our Coq proofs. We trust these formulae because they are generated in a systematic way by a tool whose soundness has been proved on paper [6,7]. For convenience, we also rely on a number of standard logical axioms, including functional extensionality, predicate extensionality, the law of excluded middle, and Hilbert's ϵ operator. The latter allows, e.g., defining "the minimum element" of a subset of $\mathbb{N}$ before this subset has been proved nonempty (Sect. 4.4), or referring to "the parent" of a node before it has been established that this node has a parent (Sect. 4.5).

In summary, our contribution is to present:

- the first practical verification framework with support for heap allocation, mutation, and amortized complexity analysis;
- the first formalization of the potential-based analysis of Union-Find;
- the first integrated verification of correctness and time complexity for a Union-Find implementation.

The paper is organized as follows. We describe the addition of time credits to Separation Logic and to CFML (Sect. 2). We present a formal specification of Union-Find, which mentions time credits (Sect. 3). We present a mathematical analysis of the operations on disjoint set forests and of their complexity (Sect. 4). We define the predicate UF, which relates our mathematical view of forests with their concrete layout in memory, and we verify our implementation (Sect. 5). Finally, we discuss related work (Sect. 6) and future work (Sect. 7).

2   Time Credits, Separation, and Characteristic Formulae

2.1   Time Credits in Separation Logic

In Separation Logic, a heap predicate has type Heap $\rightarrow$ Prop and characterizes a portion of the heap. The fundamental heap predicates are defined as follows, where h is a heap, H is a heap predicate, and P is a Coq proposition.

$$
\begin{aligned}
[P] &\equiv \lambda h.\ h = \varnothing \wedge P \\
H_1 \star H_2 &\equiv \lambda h.\ \exists h_1 h_2.\ h_1 \perp h_2 \wedge h = h_1 \uplus h_2 \wedge H_1\, h_1 \wedge H_2\, h_2 \\
\exists x.\, H &\equiv \lambda h.\ \exists x.\, H\, h
\end{aligned}
$$

The pure heap predicate $[P]$ characterizes an empty heap and at the same time asserts that P holds. The empty heap predicate $[\,]$ is sugar for $[\mathsf{True}]$. The separating conjunction of two heap predicates takes the form $H_1 \star H_2$ and asserts that the heap can be partitioned in two disjoint parts, of which one satisfies H_1 and the other satisfies H_2. Its definition involves two auxiliary notions: $h_1 \perp h_2$ asserts that the heaps h_1 and h_2 have disjoint domains; $h_1 \uplus h_2$ denotes the union of two disjoint heaps. Existential quantification is also lifted to the level of heap predicates, taking the form $\exists x.\, H$.

Logical entailment between two heap predicates, written $H_1 \triangleright H_2$, is defined by $\forall h.\ H_1\, h \Rightarrow H_2\, h$. This relation is used in the construction of characteristic formulae (Sect. 2.2) and also appears in specifications (Sect. 3).

To assert the existence (and ownership) of a memory cell and to describe its content, Separation Logic introduces a heap predicate of the form $l \hookrightarrow v$, which asserts that the cell at location l contains the value v. Assuming the heap is a store, i.e., a map of locations to values, the predicate $l \hookrightarrow v$ is defined as $\lambda h.\ h = (l \mapsto v)$, where $(l \mapsto v)$ denotes the singleton map of l to v. More details on these definitions are given in the first author's description of CFML [6].

To accommodate time credits, we give a new interpretation of the type Heap. In traditional Separation Logic, a heap is a store, i.e., a map from location to values: $\mathsf{Heap} \equiv \mathsf{Store}$. We reinterpret a heap h as a pair (m, c) of a store and of a natural number: $\mathsf{Heap} \equiv \mathsf{Store} \times \mathbb{N}$. The second component represents a number of time credits that are available for consumption.

This new interpretation of Heap allows us to define the heap predicate $\$\,n$, which asserts the ownership of n time credits. Furthermore, the definitions of $\uplus$, $\perp$, $\varnothing$, and $l \hookrightarrow v$, are lifted as shown below.

$$
\begin{aligned}
\$\,n &\equiv \lambda(m, c).\ m = \varnothing \wedge c = n \\
l \hookrightarrow v &\equiv \lambda(m, c).\ m = (l \mapsto v) \wedge c = 0 \\
(m_1, c_1) \perp (m_2, c_2) &\equiv m_1 \perp m_2 \\
(m_1, c_1) \uplus (m_2, c_2) &\equiv (m_1 \uplus m_2,\ c_1 + c_2) \\
\varnothing_{:\mathsf{Heap}} &\equiv (\varnothing_{:\mathsf{Store}}, 0)
\end{aligned}
$$

In short, we view Heap as the product of the monoids Store and $(\mathbb{N}, +)$. The definitions of the fundamental heap predicates, namely $[P]$, $H_1 \star H_2$ and $\exists x.\, H$, are unchanged.

Two provable equalities are essential when reasoning about credits:

$$
\$(n + n') = \$\,n \star \$\,n' \qquad \text{and} \qquad \$\,0 = [\,].
$$

The left-hand equation, combined with the fact that the logic is affine, allows weakening $\$\,n$ to $\$\,n'$ when $n \geqslant n'$ holds. Technically, this exploits CFML's "garbage collection" rule [6], and can be largely automated using tactics.

2.2 Characteristic Formulae

Let t be an OCaml term. Its characteristic formula $[\![t]\!]$ is a higher-order predicate such that, for every precondition H and postcondition Q, if $[\![t]\!]\,H\,Q$ can be proved (in Coq), then the Separation Logic triple $\{H\}\,t\,\{Q\}$ holds. This implies that, starting in any state that satisfies H, the execution of t terminates and produces a value v such that the final state satisfies the heap predicate $Q\,v$. This informal sentence assumes that an OCaml value can be reflected as a Coq value; for the sake of simplicity, we omit the details of this translation. In the following, we also omit the use of a predicate transformer, called local [6], which allows applying the structural rules of Separation Logic. Up to these simplifications, characteristic formulae for a core subset of ML are constructed as follows:

$$[\![v]\!] \equiv \lambda HQ.\ H \triangleright Q\,v \tag{1}$$

$$[\![\text{let } x = t_1 \text{ in } t_2]\!] \equiv \lambda HQ.\ \exists Q'.\ [\![t_1]\!]\,H\,Q' \wedge \forall x.\ [\![t_2]\!]\,(Q'\,x)\,Q \tag{2}$$

$$[\![f\,v]\!] \equiv \lambda HQ.\ \mathsf{App}\,f\,v\,H\,Q \tag{3}$$

$$[\![\text{let } f = \lambda x.\,t_1 \text{ in } t_2]\!] \equiv \lambda HQ.\ \forall f.\,P \Rightarrow [\![t_2]\!]\,H\,Q \tag{4}$$

$$\text{where } P \equiv (\forall x H'Q'.\ [\![t_1]\!]\,H'\,Q' \Rightarrow \mathsf{App}\,f\,x\,H'\,Q')$$

In order to read Eqs. (3) and (4), one must know that an OCaml function is reflected in the logic as a value of abstract type func. Such a value is opaque: nothing is known a priori about it. The abstract predicate App is used to assert that a function satisfies a certain specification. Intuitively, $\mathsf{App}\,f\,v\,H\,Q$ stands for the triple $\{H\}\,f\,v\,\{Q\}$. When a function is defined, an App assumption is introduced Eq. (4); when a function is called, an App assumption is exploited Eq. (3). In short, Eq. (4) states that if the body t_1 of the function f satisfies a specification $\{H'\}\cdot\{Q'\}$, then one can assume that a call $f\,x$ satisfies the same specification. Equation (3) states that the only way of reasoning about a function call is to exploit such an assumption.

2.3 Combining Time Credits and Characteristic Formulae

To ensure that a time credit effectively represents "a right to perform one function call", we must enforce spending one credit at every function call. In principle, this can be achieved without any modification of the reasoning rules. All we need to do is transform the program before constructing its characteristic formula. We insert a call to an abstract function, pay, at the beginning of every function body (and loop body). This is analogous to Danielsson's "tick" [13]. We equip pay with a specification that causes one credit to be consumed when pay is invoked:

$$\mathsf{App}\,\mathsf{pay}\,\mathit{tt}\,(\$\,1)\,(\lambda \mathit{tt}.\,[\,])$$

Here, tt denotes the unit argument and unit result of pay. The precondition $\$\,1$ requests one time credit, while the postcondition $[\,]$ is empty. When reasoning about a call to pay, the user has no choice but to exploit the above specification and give away one time credit.

In practice, in order to reduce clutter, we simplify the characteristic formula for a sequence that begins with a call to pay. The simplified formula is as follows:

$$[\![\mathsf{pay}()\,;\,t]\!] \;\equiv\; \lambda HQ.\ \exists H'.\ H \triangleright \$1 \star H' \wedge [\![t]\!]\,H'\,Q$$

If desired, instead of performing a program transformation followed with the generation of a characteristic formula, one can alter Eq. (4) above to impose the consumption of one credit at the beginning of every function body:

$$[\![\mathsf{let}\ f = \lambda x.\,t_1 \ \mathsf{in}\ t_2]\!] \;\equiv\; \lambda HQ.\ \forall f.\ P \;\Rightarrow\; [\![t_2]\!]\,H\,Q$$
$$\text{where } P \equiv (\forall x H'H''Q'.\ H' \triangleright \$1 \star H'' \wedge [\![t_1]\!]\,H''\,Q' \;\Rightarrow\; \mathsf{App}\,f\,x\,H'\,Q')\,.$$

2.4 Meta-Theory

We revisit the informal soundness theorem for characteristic formulae [7] so as to account for time credits. The new theorem relies on a cost-annotated semantics of the programming language (a subset of OCaml). The judgment $t_{/m} \Downarrow^n v_{/m'}$ means that the term t, executed in the initial store m, terminates after n function calls and produces a value v and a final store m'. Our theorem is as follows.

Theorem 1 (Soundness of Characteristic Formulae with Time Credits).

$$\forall mc.\ \begin{cases} [\![t]\!]\,H\,Q \\ H\,(m,c) \end{cases} \;\Rightarrow\; \exists nvm'm''c'c''.\ \begin{cases} t_{/m} \Downarrow^n v_{/m' \uplus m''} & (1) \\ m' \perp m'' & (2) \\ Q\,v\,(m',c') & (3) \\ c = n + c' + c'' & (4) \end{cases}$$

Suppose we have proved $[\![t]\!]\,H\,Q$. Pick an initial heap (m,c) that satisfies the precondition H. Here, m is an initial store, while c is an initial number of time credits. (Time credits are never created out of thin air, so one must assume that they are given at the beginning.) Then, the theorem guarantees, the program t runs without error and terminates (1). The final heap can be decomposed into (m',c') and (m'',c'') (2), where (m',c') satisfies the postcondition $Q\,v$ (3) and (m'',c'') represents resources (memory and credits) that have been abandoned during reasoning by applying the "garbage collection" rule [7]. Our accounting of time is exact: the initial number of credits c is the sum of n, the number of function calls that have been performed by the program, and $c' + c''$, the number of credits that remain in the final heap (4). In other words, every credit either is spent to justify a function call, or is still there at the end. In particular, Eq. (4) implies $c \geqslant n$: the number of time credits that are initially supplied is an upper bound on the number of function calls performed by the program.

The proof of Theorem 1 follows the exact same structure as that of the original soundness theorem for characteristic formulae [7]. We have carried out the extended proof on paper [8]. As expected, only minor additions are required.

3   Specification of Union-Find

Our specification of the library is expressed in Coq. It relies on an abstract type
elem and an abstract representation predicate UF. Their definitions, which we
present later on (Sect. 5), are not publicly known. As far as a client is concerned,
elem is the type of elements, and UF $N\,D\,R$ is a heap predicate which asserts the
existence (and ownership) of a Union-Find data structure, whose current state
is summed up by the parameters N, D and R. The parameter D, whose type is
elem $\to$ Prop, is the domain of the data structure, that is, the set of all elements.
The parameter N is an upper bound on the cardinality of D. The parameter R,
whose type is elem $\to$ elem, maps every element to its representative.

Because R maps every element to its representative, we expect it to be an
idempotent function of the set D into itself. Furthermore, although this is in no
way essential, we decide that R should be the identity outside D. We advertise
this to the client via the first theorem in Fig. 2. Recall that $\rhd$ is entailment of
heap predicates. Thus, UF_properties states that, if one possesses UF $N\,D\,R$, then
certain logical properties of N, D and R hold.

The next theorem, UF_create, asserts that out of nothing one can create an
empty Union-Find data structure. UF_create is a "ghost" operation: it does not
exist in the OCaml code. Yet, it is essential: without it, the library would be
unusable, because UF appears in the pre-condition of every operation. When
one applies this theorem, one commits to an upper bound N on the number of
elements. N remains fixed thereafter.

The need for N is inherited from the proof that we follow [12,27]. Kaplan
et al. [20] and Alstrup et al. [2] have carried out more precise complexity analyses,
which lead to an amortized complexity bound of $\alpha(n)$, as opposed to $\alpha(N)$, where
n is the cardinality of D. In the future, we would like to formalize Alstrup et al.'s
argument, as it seems to require relatively minor adjustments to the proof that
we have followed. This would remove the need for fixing N in advance and would
thus make our specification easier to use for a client.

Next comes the specification of the OCaml function make. The theorem
make_spec refers to UnionFind_ml.make, which is defined for us by CFML and has
type func (recall Sect. 2.2). It states that UnionFind_ml.make satisfies a certain
specification, thereby describing the behavior of make. The condition card D < N
indicates that new elements can be created only as long as the number of elements remains under the limit N. Then comes an application of the predicate
App to the value UnionFind_ml.make, to the unit value tt, and to a pre- and
postcondition. The precondition is the conjunction of UF $N\,D\,R$, which describes
the pre-state, and of \$1, which indicates that make works in constant time. (We
view the OCaml function ref, which appears in the implementation of make, as
a primitive operation, so its use does not count as a function call.) In the postcondition, x denotes the element returned by make. The postcondition describes
the post-state via the heap predicate UF $N\,(D \cup \{x\})\,R$. It also asserts that x is
new, that is, distinct from all previous elements.

The next theorem provides a specification for find. The argument x must
be a member of D. In addition to UF $N\,D\,R$, the precondition requires $\alpha(N)+2$

(* UF : nat → (elem → Prop) → (elem → elem) → heap → Prop *)

Theorem UF_properties : ∀N D R, UF N D R ▷ UF N D R ⋆
  [(card D ⩽ N) ∧ ∀x, (R (R x) = R x) ∧ (x ∈ D → R x ∈ D) ∧ (x ∉ D → R x = x)].

Theorem UF_create : ∀N, [] ▷ UF N ∅ id.

Theorem make_spec : ∀N D R, card D < N →
  App UnionFind_ml.make tt
    (UF N D R ⋆ $1)
    (fun x ⇒ UF N (D ∪ {x}) R ⋆ [x ∉ D] ⋆ [R x = x]).

Theorem find_spec : ∀N D R x, x ∈ D →
  App UnionFind_ml.find x
    (UF N D R ⋆ $(alpha N + 2))
    (fun y ⇒ UF N D R ⋆ [R x = y]).

Theorem union_spec : ∀N D R x y, x ∈ D → y ∈ D →
  App UnionFind_ml.union x y
    (UF N D R ⋆ $(3*(alpha N)+6))
    (fun z ⇒ UF N D (fun w ⇒ If R w = R x ∨ R w = R y then z else R w)
      ⋆ [z = R x ∨ z = R y]).

Theorem equiv_spec : ∀N D R, x ∈ D → y ∈ D →
  App UnionFind_ml.equiv x y
    (UF N D R ⋆ $(2*(alpha N) + 5))
    (fun b ⇒ UF N D R ⋆ [b = true ↔ R x = R y]).

Fig. 2. Complete specification of Union-Find

credits. This reflects the amortized cost of **find**. The postcondition asserts that **find** returns an element y such that $R\,x = y$. In other words, **find** returns the representative of x. Furthermore, the postcondition asserts that UF $N\,D\,R$ still holds. Even though path compression may update internal pointers, the mapping of elements to representatives, which is all the client knows about, is preserved.

The precondition of **union** requires UF $N\,D\,R$ together with $3 \times \alpha(N) + 6$ time credits. The postcondition indicates that **union** returns an element z, which is either x or y, and updates the data structure to UF $N\,D\,R'$, where R' updates R by mapping to z every element that was equivalent to x or y. The construct If P then e1 else e2, where P is in Prop, is a non-constructive conditional. It is defined using the law of excluded middle and Hilbert's ϵ operator.

The postcondition of **equiv** indicates that **equiv** returns a Boolean result, which tells whether the elements x and y have a common representative.

The function **link** is internal, so its specification (given in Sect. 5) is not public.

4   Mathematical Analysis of Disjoint Set Forests

We carry out a mathematical analysis of disjoint set forests. This is a Coq formalization of textbook material [12, 27]. It is independent of the OCaml code (Fig. 1) and of the content of the previous sections (Sects. 2, 3). For brevity, we elide many details; we focus on the main definitions and highlight a few lemmas.

4.1   Disjoint Set Forests as Graphs

We model a disjoint set forest as a graph. The nodes of the graph inhabit a type V which is an implicit parameter throughout this section (Sect. 4). As in Sect. 3, the domain of the graph is represented by a set D of nodes. The edges of the graph are represented by a relation F between nodes. Thus, $F \, x \, y$ indicates that there is an edge from node x to node y.

The predicate path F is the reflexive, transitive closure of F. Thus, path $F \, x \, y$ indicates the existence of a path from x to y. A node x is a *root* iff it has no successor. A node x is *represented by* a node r iff there is a path from x to r and r is a root.

```
Definition is_root F x := ∀y, ¬ F x y.
Definition is_repr F x r := path F x r ∧ is_root F r.
```

Several properties express the fact that the graph represents a disjoint set forest. First, the relation F is *confined* to D: whenever there is an edge from x to y, the nodes x and y are members of D. Second, the relation F is *functional*: every node has at most one parent. Finally, the relation is_repr F is *defined*: every node x is represented by some node r. This ensures that the graph is acyclic. The predicate is_dsf is the conjunction of these three properties:

```
Definition is_dsf D F :=
  confined D F ∧ functional F ∧ defined (is_repr F).
```

4.2   Correctness of Path Compression

The first part of our mathematical analysis is concerned mostly with the functional correctness of linking and path compression. Here, we highlight a few results on compression. Compression assumes that there is an edge between x and y and a path from y to a root z. It replaces this edge with a direct edge from x to z. We write compress $F \, x \, z$ for the relation that describes the edges after this operation: it is defined as $F \setminus \{(x, \_)\} \cup \{(x, z)\}$.

We prove that, although compression destroys some paths in the graph (namely, those that used to go through x), it preserves the relationship between nodes and roots. More precisely, if v has representative r in F, then this still holds in the updated graph compress $F \, x \, z$.

```
Lemma compress_preserves_is_repr : ∀D F x y z v r,
  is_dsf D F → F x y → is_repr F y z →
  is_repr F v r → is_repr (compress F x z) v r.
```

It is then easy to check that compression preserves is_dsf.

4.3 Ranks

In order to perform linking-by-rank and to reason about it, we attach a rank to every node in the graph. To do so, we introduce a function K of type $V \rightarrow \mathbb{N}$. This function satisfies a number of interesting properties. First, because linking makes the node of lower rank a child of the node of higher rank, and because the rank of a node can increase only when this node is a root, ranks increase along graph edges. Furthermore, a rank never exceeds $\log|D|$. Indeed, if a root has rank p, then its tree has at least 2^p elements. We record these properties using a new predicate, called `is_rdsf`, which extends `is_dsf`. This predicate also records the fact that D is finite. Finally, we find it convenient to impose that the function K be uniformly zero outside of the domain D.

```
Definition is_rdsf D F K :=
    is_dsf D F ∧
    (∀ x y, F x y → K x < K y) ∧
    (∀ r, is_root F r → 2^(K r) ⩽ card (descendants F r)) ∧
    finite D ∧
    (∀ x, x ∉ D → K x = 0).
```

It may be worth noting that, even though at runtime only roots carry a rank (Fig. 1), in the mathematical analysis, the function K maps every node to a rank. The value of K at non-root nodes can be thought of as "ghost state".

4.4 Ackermann's Function and Its Inverse

For the amortized complexity analysis, we need to introduce Ackermann's function, written $A_k(x)$. According to Tarjan [27], it is defined as follows:

$$A_0(x) = x + 1 \qquad\qquad A_{k+1}(x) = A_k^{(x+1)}(x)$$

We write $f^{(i)}$ for the i-th iterate of the function f. In Coq, we write $\mathtt{iter}\, i\, f$. The above definition is transcribed very compactly:

```
Definition A k := iter k (fun f x ⇒ iter (x+1) f x) (fun x ⇒ x+1).
```

That is, A_k is the k-th iterate of $\lambda f.\lambda x.\, f^{(x+1)}(x)$, applied to A_0.

The inverse of Ackermann's function, written $\alpha(n)$, maps n to the smallest value of k such that $A_k(1) \geqslant n$. Below, `min_of le` denotes the minimum element of a nonempty subset of $\mathbb{N}$.

```
Definition alpha n := min_of le (fun k ⇒ A k 1 ⩾ n).
```

4.5 Potential

The definition of the potential function relies on a few auxiliary definitions. First, for every node x, Tarjan [27] writes $p(x)$ for the *parent* of x in the forest. If x is not a root, $p(x)$ is uniquely defined. We define $p(x)$ using Hilbert's ϵ operator:

```
Definition p F x := epsilon (fun y ⇒ F x y).
```

Thus, `p F x` is formally defined for every node `x`, but the characteristic property `F x (p F x)` can be exploited only if one can prove that `x` has a parent.

Then, Tarjan [27] introduces the *level* and the *index* of a non-root node x. These definitions involve the rank of x and the rank of its parent. The level of x, written $k(x)$, is the largest k for which $K(p(x)) \geqslant A_k(K(x))$ holds. It lies in the interval $[0, \alpha(N))$, if N is an upper bound on the number of nodes. The index of x, written $i(x)$, is the largest i for which $K(p(x)) \geqslant A_{k(x)}^{(i)}(K(x))$ holds. It lies in the interval $[1, K(x)]$. The formal definitions, shown below, rely on the function `max_of`, which is the dual of `min_of` (Sect. 4.4).

```
Definition k F K x :=
  max_of le (fun k ⇒ K (p F x) ⩾ A k (K x)).
Definition i F K x :=
  max_of le (fun i ⇒ K (p F x) ⩾ iter i (A (k F K x)) (K x)).
```

The potential of a node, written $\phi(x)$, depends on the parameter N, which is a pre-agreed upper bound on the number of nodes. (See the discussion of `UF_create` in Sect. 3.) Following Tarjan [27], if x is a root or has rank 0, then $\phi(x)$ is $\alpha(N) \cdot K(x)$. Otherwise, $\phi(x)$ is $(\alpha(N) - k(x)) \cdot K(x) - i(x)$.

```
Definition phi F K N x :=
  If (is_root F x) ∨(K x = 0)
    then (alpha N) ∗ (K x)
    else (alpha N − k F K x) ∗ (K x) − (i F K x).
```

The total potential Φ is obtained by summing ϕ over all nodes in the domain D.

```
Definition Phi D F K N := fold (monoid_ plus 0) phi D.
```

4.6 Rank Analysis

The definition of the representation predicate `UF`, which we present later on (Sect. 5), explicitly mentions Φ. It states that, between two operations, we have Φ time credits at hand. Thus, when we try to prove that every operation preserves `UF`, as claimed earlier (Sect. 3), we are naturally faced with the obligation to prove that the initial potential Φ, plus the number of credits brought by the caller, covers the new potential Φ', plus the number of credits consumed during the operation:

$$\Phi + \text{advertised cost of operation} \geqslant \Phi' + \text{actual cost of operation}$$

We check that this property holds for all operations. The two key operations are linking and path compression. In the latter case, we consider not just one step of compression (i.e., updating one graph edge, as in Sect. 4.2), but "iterated path compression", i.e., updating the parent of every node along a path, as performed by `find` in Fig. 1. To model iterated path compression, we introduce the predicate `ipc F x d F'`, which means that, if the initial graph is F and if one performs iterated path compression starting at node x, then one performs d individual compression steps and the final graph is F'.

$$\frac{\text{is\_root}\,F\,x}{\text{ipc}\,F\,x\,0\,F} \qquad \frac{F\,x\,y \quad \text{is\_repr}\,F\,y\,z \quad \text{ipc}\,F\,y\,d\,F' \quad F'' = \text{compress}\,F'\,x\,z}{\text{ipc}\,F\,x\,(d+1)\,F''}$$

On the one hand, the predicate ipc is a paraphrase, in mathematical language, of the recursive definition of find in Fig. 1, so it is easy to argue that "find implements ipc". This is done as part of the verification of find (Sect. 5). On the other hand, by following Tarjan's proof [27], we establish the following key lemma, which sums up the amortized complexity analysis of iterated path compression:

Lemma amortized_cost_of_iterated_path_compression : $\forall$D F K x N,
  is_rdsf D F K $\rightarrow$ x $\in$ D $\rightarrow$ card D $\leqslant$ N $\rightarrow$
  $\exists$d F', ipc F x d F' $\wedge$ (Phi D F K N + alpha N + 2 $\geqslant$ Phi D F' K N + d + 1).

This lemma guarantees that, in any initial state described by D, F, K and for any node x, (1) iterated path compression is possible, requiring a certain number of steps d and taking us to a certain final state F', and (2) more interestingly, the inequality "Φ + advertised cost $\geqslant \Phi'$ + actual cost" holds. Indeed, $\alpha(N) + 2$ is the advertised cost of find (Fig. 2), whereas $d + 1$ is its actual cost, because iterated path compression along a path of length d involves $d + 1$ calls to find.

5   Verification of the Code

To prove that the OCaml code in Fig. 1 satisfies the specification in Fig. 2, we first define the predicate UF, then establish each of the theorems in Fig. 2.

5.1   Definition of the Representation Predicate

In our OCaml code (Fig. 1), we have defined elem as content ref, and defined content as Link of elem | Root of rank. CFML automatically mirrors these type definitions in Coq. It defines elem as loc (an abstract type of memory locations, provided by CFML's library) and content as an inductive data type:

Definition elem := loc.
Inductive content := Link of elem | Root of rank.

To define the heap predicate UF $N\,D\,R$, we need two auxiliary predicates.

The auxiliary predicate Inv N D R F K includes the invariant is_rdsf D F K, which describes a ranked disjoint set forest (Sect. 4.3), adds the condition card D $\leqslant$ N (which appears in the complexity analysis, Sect. 4.6), and adds the requirement that the function R and the relation is_repr F agree, in the sense that $R\,x = r$ implies is_repr $F\,x\,r$. The last conjunct is needed because our specification exposes R, which describes only the mapping from nodes to representatives, whereas our mathematical analysis is in terms of F, which describes the whole graph.

```
Theorem link_spec : ∀D R x y, x ∈D → y ∈D → R x = x → R y = y →
  App UnionFind_ml.link x y
    (UF D R ⋆ $(alpha N+1))
    (fun z ⇒ UF D (fun w ⇒ If R w = R x ∨R w = R y then z else R w) ⋆ [z = x ∨z = y]).
Proof.
  xcf. introv Dx Dy Rx Ry. credits_split. xpay. xapps. xif.
  { xret. rewrite~ root_after_union_same. hsimpl~. }
  { unfold UF at 1. xextract as F K M HI HM. lets HMD: (Mem_dom HM). xapps~. xapps~. xmatch.
    { forwards* (HRx&Kx): Mem_root_inv x HM.
      forwards* (HRy&Ky): Mem_root_inv y HM. xif.
      { forwards* (F'&EF'&HI'&n&EQ): (>> Inv_link_lt D R F K x y). splits*. math.
        xchange (credits_eq EQ). chsimpl. xapp*. xret. unfold UF. chsimpl*. applys* Mem_link. }
      { xif.
        { forwards* (F'&EF'&HI'&n&EQ): (>> Inv_link_gt D R F K x y). splits*. math.
          xchange (credits_eq EQ). chsimpl. xapp*. xret. unfold UF. chsimpl*. applys* Mem_link. }
        { asserts: (rx = ry). math. asserts: (K x = K y). math.
          forwards* (F'&K'&EF'&EK'&HI'&n&EQ): (>> Inv_link_eq D R F K x y). splits*.
          xchange (credits_eq EQ). chsimpl. xapp~. xapp~. apply~ map_indom_update_already.
          xret. rewrite H2 in EK'. unfold UF. chsimpl*. applys* Mem_link_incr HM. } } }
    { xfail ((rm CO) (K x) (K y)). fequals; applys* Mem_root. }
Qed.
```

Fig. 3. Verification script for the link function

Definition Inv N D F K R :=
  (is_rdsf D F K) ∧ (card D ⩽ N) ∧ (rel_of_func R ⊂ is_repr F).

The auxiliary predicate Mem D F K M relates a model of a disjoint set forest, represented by D, F, K, and a model of the memory, represented by a finite map M of memory locations to values of type content. (This view of memory is imposed by CFML.) M maps a location x to either Link y for some location y or Root k for some natural number k. The predicate Mem D F K M explains how to interpret the contents of memory as a disjoint set forest. It asserts that M has domain D, that $M(x) =$ Link y implies the existence of an edge of x to y, and that $M(x) =$ Root k implies that x is a root and has rank k.

Definition Mem D F K M := (dom M = D) ∧ (∀ x, x ∈D →
  match M\(x) with Link y ⇒ F x y | Root k ⇒ is_root F x ∧ k = K x end).

At last, we are ready to define the heap predicate UF N D R:

Definition UF N D R := ∃F K M,
  Group (Ref Id) M ⋆ [Inv N D F K R] ⋆ [Mem D F K M] ⋆ $ (Phi D F K N).

The first conjunct asserts the existence in the heap of a group of reference cells, collectively described by the map M. (The predicates Group, Ref and Id are provided by CFML's library.) The second conjunct constrains the graph (D, F, K) to represent a valid disjoint set forest whose roots are described by R, while the third conjunct relates the graph (D, F, K) with the contents of memory (M). Recall that the brackets lift an ordinary proposition as a heap predicate (Sect. 2). The last conjunct asserts that we have Φ time credits at hand. The definition is existentially quantified over F, K, and M, which are not exposed to the client.

5.2 Verification Through Characteristic Formulae

Our workflow is as follows. Out of the file `UnionFind.ml`, which contains the OCaml source code, the CFML tool produces the Coq file `UnionFind_ml.v`, which contains characteristic formulae. Consider, for example, the OCaml function `link`. The tool produces two Coq axioms for it. The first axiom asserts the existence of a value `link` of type `func` (Sect. 2.2). The second axiom, `link_cf`, is a characteristic formula for `link`. It can be exploited to establish that `link` satisfies a particular specification. For instance, in Fig. 3, we state and prove a specification for `link`. The proof involves a mix of standard Coq tactics and tactics provided by CFML for manipulating characteristic formulae. It is quite short, because all we need to do at this point is to establish a correspondence between the operations performed by the imperative code and the mathematical analysis that we have already carried out.

The CFML library is about 5Kloc. The analysis of Union-Find (Sect. 4) is 3Kloc. The specification of Union-Find (Fig. 2) and the verification of the code (Sect. 5), together, take up only 0.4Kloc. Both CFML and our proofs about Union-Find rely on Charguéraud's TLC library, a complement to Coq's standard library. Everything is available online [8].

6 Related Work

Disjoint set forests as well as linking-by-size are due to Galler and Fischer [15]. Path compression is attributed by Aho *et al.* [1] to McIlroy and Morris. Hopcroft and Ullman [19] study linking-by-size and path compression and establish an amortized bound of $O(log^*N)$ per operation. Tarjan [29] studies linking-by-rank and path compression and establishes the first amortized bound in $O(\alpha(N))$. After several simplifications [21,28], this leads to the proof that we follow [12,27]. Kaplan *et al.* [20] and Alstrup *et al.* [2] establish a "local" bound: they bound the amortized cost of $find(x)$ by $O(\alpha(n))$, where n is the size of x's set.

We know of only one machine-checked proof of the functional correctness of Union-Find, due to Conchon and Filliâtre [11]. They reformulate the imperative algorithm in a purely functional style, where the store is explicit. They represent the store as a persistent array and obtain an efficient persistent Union-Find. We note that Union-Find is part of the VACID-0 suite of benchmark verification problems [23]. We did not find any solution to this particular benchmark problem online. We know of no machine-checked proof of the complexity of Union-Find.

The idea of using a machine to assist in the complexity analysis of a program goes back at least as far back as Wegbreit [30]. He extracts recurrence equations from the program and bounds their solution. More recent work along these lines includes Le Métayer's [22] and Danner *et al.*'s [14]. Although Wegbreit aims for complete automation, he notes that one could "allow the addition to the program of performance specifications by the programmer, which the system then checks for consistency". We follow this route.

There is a huge body of work on program verification using Separation Logic. We are particularly interested in embedding Separation Logic into an interactive proof assistant, such as Coq, where it is possible to express arbitrarily complex specifications and to perform arbitrarily complex proofs. Such an approach has been explored in several projects, such as Ynot [10], Bedrock [9], and CFML [6,7]. This approach allows verifying not only the implementation of a data structure but also its clients. In particular, when a data structure comes with an amortized complexity bound, we verify that it is used in a single-threaded manner.

Nipkow [24] carries out machine-checked amortized analyses of several data structures, including skew heaps, splay trees and splay heaps. As he seems to be mainly interested in the mathematical analysis of a data structure, as opposed to the verification of an actual implementation, he manually derives from the code a "timing function", which represents the actual time consumed by an operation.

The idea of extending a type system or program logic with time or space credits, viewed as affine resources, has been put forth by several authors [4,18,25]. The extension is very modest; in fact, if the source program is explicitly instrumented by inserting calls to pay, no extension at all is required. We believe that we are the first to follow this approach in practice to perform a modular verification of functional correctness and complexity for a nontrivial data structure.

A line of work by Hofmann et al. [16–18] aims to infer amortized time and space bounds. Because emphasis is on automation, these systems are limited in the bounds that they can infer (e.g., polynomial bounds) and/or in the programs that they can analyze (e.g., without side effects; without higher-order functions).

7   Future Work

We have demonstrated that the state of the art has advanced to a point where one can (and, arguably, one should) prove not only that a library is correct but also (and at the same time) that it meets a certain complexity bound.

There are many directions for future work. Concerning Union-Find, we would like to formalize Alstrup et al.'s proof [2] that the amortized cost can be expressed in terms of the current number n of nodes, as opposed to a fixed upper bound N. Concerning our verification methodology, we wish to use the big-O notation in our specifications, so as to make them more modular.

References

1. Aho, A.V., Hopcroft, J.E., Ullman, J.D.: The Design and Analysis of Computer Algorithms. Addison-Wesley, Reading (1974)
2. Alstrup, S., Thorup, M., Gørtz, I.L., Rauhe, T., Zwick, U.: Union-find with constant time deletions. ACM Trans. Algorithms 11(1), 6:1–6:28 (2014)
3. Amadio, rm, et al.: Certified complexity (CerCo). In: Lago, U.D., Peña, R. (eds.) FOPARA 2013. LNCS, vol. 8552, pp. 1–18. Springer, Heidelberg (2014)
4. Atkey, R.: Amortised resource analysis with separation logic. Logical Methods in Computer Science 7(2:17), 1–33 (2011)
5. Blelloch, G.E., Greiner, J.: Parallelism in sequential functional languages. In: Functional Programming Languages and Computer Architecture (FPCA) (1995)

6. Charguéraud, A.: Characteristic formulae for the verification of imperative programs in HOSC (2012, to appear)
7. Charguéraud, A.: Characteristic Formulae for Mechanized Program Verification. Ph.D. Thesis, Université Paris 7 (2010)
8. Charguéraud, A., Pottier, F.: Self-contained archive (2015). http://gallium.inria.fr/~fpottier/dev/uf/
9. Chlipala, A.: The Bedrock structured programming system: combining generative metaprogramming and Hoare logic in an extensible program verifier. In: International Conference on Functional Programming (ICFP) (2013)
10. Chlipala, A., Malecha, G., Morrisett, G., Shinnar, A., Wisnesky, R.: Effective interactive proofs for higher-order imperative programs. In: International Conference on Functional Programming (ICFP) (2009)
11. Conchon, S., Filliâtre, J.: A persistent union-find data structure. In: ACM Workshop on ML (2007)
12. Cormen, T.H., Leiserson, C.E., Rivest, R.L., Stein, C.: Introduction to Algorithms, 3rd edn. MIT Press, Cambridge (2009)
13. Danielsson, N.A.: Lightweight semiformal time complexity analysis for purely functional data structures. In: Principles of Programming Languages (POPL) (2008)
14. Danner, N., Paykin, J., Royer, J.S.: A static cost analysis for a higher-order language. In: Programming Languages Meets Program Verification (PLPV) (2013)
15. Galler, B.A., Fischer, M.J.: An improved equivalence algorithm. Commun. ACM **7**(5), 301–303 (1964)
16. Hoffmann, J., Aehlig, K., Hofmann, M.: Multivariate amortized resource analysis. ACM Trans. Program. Lang. Syst. **34**(3), 14:1–14:62 (2012)
17. Hoffmann, J., Hofmann, M.: Amortized resource analysis with polynomial potential. In: Gordon, A.D. (ed.) ESOP 2010. LNCS, vol. 6012, pp. 287–306. Springer, Heidelberg (2010)
18. Hofmann, M., Jost, S.: Static prediction of heap space usage for first-order functional programs. In: Principles of Programming Languages (POPL) (2003)
19. Hopcroft, J.E., Ullman, J.D.: Set merging algorithms. SIAM J. Comput. **2**(4), 294–303 (1973)
20. Kaplan, H., Shafrir, N., Tarjan, R.E.: Union-find with deletions. In: Symposium on Discrete Algorithms (SODA) (2002)
21. Kozen, D.C.: The Design and Analysis of Algorithms. Texts and Monographs in Computer Science. Springer, Heidelberg (1992)
22. Le Métayer, D.: ACE: an automatic complexity evaluator. ACM Trans. Program. Lang. Syst. **10**(2), 248–266 (1988)
23. Leino, K.R.M., Moskal, M.: VACID-0: Verification of ample correctness of invariants of data-structures, edition 0, manuscript KRML 209 (2010)
24. Nipkow, T.: Amortized complexity verified. In: Interactive Theorem Proving (2015)
25. Pilkiewicz, A., Pottier, F.: The essence of monotonic state. In: Types in Language Design and Implementation (TLDI) (2011)
26. Reynolds, J.C.: Separation logic: A logic for shared mutable data structures. In: Logic in Computer Science (LICS) (2002)
27. Tarjan, R.E.: Class notes: Disjoint set union (1999)
28. Tarjan, R.E., van Leeuwen, J.: Worst-case analysis of set union algorithms. J. ACM **31**(2), 245–281 (1984)
29. Tarjan, R.E.: Efficiency of a good but not linear set union algorithm. J. ACM **22**(2), 215–225 (1975)
30. Wegbreit, B.: Mechanical program analysis. Commun. ACM **18**(9), 528–539 (1975)

Formalizing Size-Optimal Sorting Networks: Extracting a Certified Proof Checker

Luís Cruz-Filipe[(⊠)] and Peter Schneider-Kamp

Department of Mathematics and Computer Science,
University of Southern Denmark, Campusvej 55, 5230 Odense M, Denmark
{lcf,petersk}@imada.sdu.dk

Abstract. Since the proof of the four color theorem in 1976, computer-generated proofs have become a reality in mathematics and computer science. During the last decade, we have seen formal proofs using verified proof assistants being used to verify the validity of such proofs.

In this paper, we describe a formalized theory of size-optimal sorting networks. From this formalization we extract a certified checker that successfully verifies computer-generated proofs of optimality on up to 8 inputs. The checker relies on an untrusted oracle to shortcut the search for witnesses on more than 1.6 million NP-complete subproblems.

1 Introduction

Sorting networks are hardware-oriented algorithms to sort a fixed number of inputs using a predetermined sequence of comparisons between them. They are built from a primitive operator, the *comparator*, which reads the values on two channels, and interchanges them if necessary to guarantee that the smallest one is always on a predetermined channel. Comparisons between independent pairs of values can be performed in parallel, and the two main optimization problems one wants to address are: how many comparators do we need to sort n inputs (the *optimal size* problem); and how many computation steps do we need to sort n inputs (the *optimal depth* problem). This paper focuses on the former problem.

Most results obtained on the optimal-size problem rely on exhaustive analysis of state spaces. For up to 4 inputs, a simple information-theoretical argument suffices, and for 5 inputs, the state space is small enough to be exhausted by manual inspection and symmetry arguments. However, all known optimality proofs for 7 inputs use computer programs to eliminate symmetries within the search space of almost 10^{20} comparator sequences of length 15 and show that no sorting network of this size exists [9]. It took 50 years before a similar approach [6] was able to settle that the known sorting network of size 25 on 9 inputs is optimal. Optimality results for 6, 8, and 10 are obtained by a theoretical argument [18].

The proof in [6] uses a generate-and-prune algorithm to show size optimality of sorting networks, and it not only established optimality of 25 comparisons for sorting 9 inputs, but it also directly confirmed all smaller cases. The pruning step includes an expensive test taking two sequences of comparators and deciding

© Springer International Publishing Switzerland 2015
C. Urban and X. Zhang (Eds.): ITP 2015, LNCS 9236, pp. 154–169, 2015.
DOI: 10.1007/978-3-319-22102-1_10

whether one of them can be ignored. This test, which often fails, is repeated an exponential number of times, each invocation requiring iteration over all permutations of n elements in the worst case. During execution, we took care to log the comparator sequences and permutations for which this test was *successful*.

In this paper we describe a Coq formalization of the theory of optimal-size sorting networks, from which we extract a certified checker implementing generate-and-prune with the goal of confirming the validity of these results. In order to obtain feasible runtimes, we bypass the expensive search process in the extracted certified checker by means of an oracle based on the logs of the original execution. The checker takes a skeptical approach towards the oracle: if the oracle provides wrong information at any step, the checker ignores it.

In the presentation, we discuss how we exploit the constructiveness of the theory to simplify the formalization and identify places where skepticism towards the oracle becomes relevant. Interestingly, the interactive process of formalization itself revealed minor gaps in the hand-written proofs of [6], underlining the importance of formal proof assistants in computer-aided mathematical proofs.

This paper is structured as follows. Section 2 summarizes the theory of sorting networks relevant for this formalization, and describes the generate-and-prune algorithm together with the information logged during its execution. Section 3 describes the formalization of the theory, with emphasis on the challenges encountered. Section 4 deals with the aspects of the formalization that have a direct impact on the extracted program, namely the specification of the oracle and the robustness needed to guarantee that unsound oracles do not compromise the final results. Section 5 addresses the implementation of the oracle and the execution of the extracted code. Finally, Sect. 6 presents some concluding remarks and directions in which this work can be extended.

1.1 Related Work

The proof of the four colour theorem from 1976 [1,2] was not the first computer-assisted proof, but it was the first to generate broad awareness of a new area of mathematics, sometimes dubbed "experimental" or "computational" mathematics. Since then, numerous theorems in mathematics and computer science have been proved by computer-assisted and computer-generated proofs. Besides obvious philosophical debates about what constitutes a mathematical proof, concerns about the validity of such proofs have been raised since. In particular, proofs based on exhausting the solution space have been met with skepticism.

During the last decade, we have seen an increasing use of verified proof assistants to create formally verified computer-generated proofs. This has been a success story, and it has resulted in a plethora of formalizations of mathematical proofs, a list too long to even start mentioning particular instances. *Pars pro toto*, consider the formal proof of the four colour theorem from 2005 [11].

Outside the world of formal proofs, computer-generated proofs are flourishing, too, and growing to tremendous sizes. The proof of Erdös' discrepancy conjecture for $C = 2$ from 2014 [13] has been touted as one of the largest mathematical proofs and produced approx. 13 GB of proof witnesses. Such large-scale proofs are extremely challenging for formal verification. Given the current state of theorem provers and computing equipment, it is unthinkable to use Claret et al.'s approach [5] of importing an *oracle* based on the proof witnesses into Coq, a process clearly prohibitive for such large-scale proofs as we consider.

The last years have seen the appearance of *untrusted oracles*, e.g. for a verified compiler [14] or for polyhedral analysis [10]. Here, the verified proof tool is relegated to a checker of the computations of the untrusted oracle, e.g., by using hand-written untrusted code to compute a result and verified (extracted) code to check it before continuing the computation.

The termination proof certification projects IsaFoR/CeTA [17], based on Isabelle/HOL, and A3PAT [7], based on Coq, go one step further, and use an *offline* untrusted oracle approach, where different termination analyzers provide proof witnesses, which are stored and later checked. However, a typical termination proof has 10-100 proof witnesses and totals a few KB to a few MB of data, and recent work [16] mentions problems were encountered when dealing with proofs using "several hundred megabytes" of oracle data. In contrast, the proof of size-optimality of sorting networks with 8 inputs requires dealing with 1.6 million proof witnesses, totalling more than 300 MB of oracle data.

2 Optimal-Size Sorting Networks

A *comparator network* C with n channels and size k is a sequence of *comparators* $C = (i_1, j_1); \ldots; (i_k, j_k)$, where each comparator (i_ℓ, j_ℓ) is a pair of channels $1 \le i_\ell < j_\ell \le n$. If C_1 and C_2 are comparator networks with n channels, then $C_1; C_2$ denotes the comparator network obtained by concatenating C_1 and C_2. An input $\boldsymbol{x} = x_1 \ldots x_n \in \{0, 1\}^n$ propagates through C as follows: $\boldsymbol{x}^0 = \boldsymbol{x}$, and for $0 < \ell \le k$, $\boldsymbol{x}^\ell$ is the permutation of $\boldsymbol{x}^{\ell-1}$ obtained by interchanging $\boldsymbol{x}_{i_\ell}^{\ell-1}$ and $\boldsymbol{x}_{j_\ell}^{\ell-1}$ whenever $\boldsymbol{x}_{i_\ell}^{\ell-1} > \boldsymbol{x}_{j_\ell}^{\ell-1}$. The output of the network for input $\boldsymbol{x}$ is $C(\boldsymbol{x}) = \boldsymbol{x}^k$, and $\mathsf{outputs}(C) = \{ C(\boldsymbol{x}) \mid \boldsymbol{x} \in \{0, 1\}^n \}$. The comparator network C is a *sorting network* if all elements of $\mathsf{outputs}(C)$ are sorted (in ascending order). The zero-one principle [12] implies that a sorting network also sorts sequences over any other totally ordered set, e.g. integers. Images (a) and (b) on the right depict sorting networks on 4 channels, each consisting of 6 comparators. The channels are indicated as horizontal lines (with channel 4 at the bottom), comparators are indicated as vertical lines connecting a pair of channels, and input values prop-

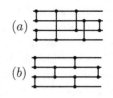

(a)

(b)

agate from left to right. The sequence of comparators associated with a picture representation is obtained by a left-to-right, top-down traversal. For example, the networks depicted above are: (a) $(1, 2); (3, 4); (1, 4); (1, 3); (2, 4); (2, 3)$ and (b) $(1, 2); (3, 4); (2, 3); (1, 2); (3, 4); (2, 3)$.

2.1 Optimal Size Sorting Networks

The optimal-size sorting network problem is about finding the smallest size, $S(n)$, of a sorting network on n channels. In 1964, Floyd and Knuth presented sorting networks of optimal size for $n \leq 8$ and proved their optimality [9]. Their proof required analyzing huge state spaces by means of a computer program, and the combinatorial explosion involved implied that there was no further progress on this problem in the next fifty years, until the proof that $S(9) = 25$ [6]. The best currently known upper and lower bounds for $S(n)$ are given in the following table.

n	1	2	3	4	5	6	7	8	9	10	11	12	13	14	15	16
Upper bound for $S(n)$	0	1	3	5	9	12	16	19	25	29	35	39	45	51	56	60
Lower bound for $S(n)$	0	1	3	5	9	12	16	19	25	29	33	37	41	45	49	53

Given an n channel comparator network $C = (i_1, j_1); \ldots; (i_k, j_k)$ and a permutation π on $\{1, \ldots, n\}$, $\pi(C)$ is the sequence $(\pi(i_1), \pi(j_1)); \ldots; (\pi(i_k), \pi(j_k))$. Formally, $\pi(C)$ is not a comparator network, but rather a *generalized comparator network*: a comparator network that may contain comparators (i, j) with $i > j$, which order their outputs in descending order instead of ascending. A generalized sorting network can always be transformed into a (standard) sorting network with the same size and depth by means of a procedure we will call *standardization*: if $C = C_1; (i, j); C_2$ is a generalized sorting network with C_1 standard and $i > j$, then standardizing C yields $C_1; (j, i); C_2'$, where C_2' is obtained by (i) interchanging i and j and (ii) standardizing the result. We write $C_1 \approx C_2$ (C_1 is equivalent to C_2) iff there is a permutation π such that C_1 is obtained by standardizing $\pi(C_2)$. The two networks (a) and (b) above are equivalent via the permutation $(1\,3)(2\,4)$.

Lemma 1. *If C is a generalized sorting network, then standardizing C yields a sorting network.*

The proof of this result, proposed as Exercise 5.3.4.16 in [12], is an induction proof that requires manipulating permutations and reasoning about the cardinality of $\mathsf{outputs}(C)$.

2.2 The Generate-and-Prune Approach

Conceptually, one could inspect all n-channel comparator networks of size k one-by-one to determine if any of them is a sorting network. However, even for small n such a naive approach is combinatorially infeasible. There are $n(n-1)/2$ comparators on n channels, and hence $(n(n-1)/2)^k$ networks with k comparators. For $n = 9$, aiming to prove that there does not exist a sorting network with 24 comparators would mean inspecting approximately 2.25×10^{37} comparator networks. Moreover, checking whether a comparator network is a sorting network is known to be a co-NP complete problem [15].

In [6] we propose an alternative approach, *generate-and-prune*, which is driven just as the naive approach, but takes advantage of the abundance of symmetries in comparator networks, formalized via the notion of *subsumption*.

Given two comparator networks on n channels C_a and C_b, we say that C_a subsumes C_b, and write $C_a \preceq C_b$, if there exists a permutation π such that $\pi(\mathsf{outputs}(C_a)) \subseteq \mathsf{outputs}(C_b)$. If we need to make π explicit, we will write $C_a \leq_\pi C_b$.

Lemma 2 ([6]). *Let C_a and C_b be comparator networks on n channels, both of the same size, and such that $C_a \preceq C_b$. Then, if there exists a sorting network $C_b; C$ of size k, there also exists a sorting network $C_a; C'$ of size k.*

Lemma 2 implies that, when adding a next comparator in the naive approach, we do not need to consider all possible positions to place it. In particular, we can omit networks that are subsumed by others.

The *generate-and-Prune* algorithm iteratively builds two sets R_k^n and N_k^n of n channel networks of size k. First, it initializes R_0^n to consist of a single element: the empty comparator network. Then, it repeatedly applies two types of steps, Generate and Prune, as follows.

1. **Generate:** Given R_k^n, construct N_{k+1}^n by adding one comparator to each element of R_k^n in all possible ways.
2. **Prune:** Given N_{k+1}^n, construct R_{k+1}^n such that every element of N_{k+1}^n is subsumed by an element of R_{k+1}^n.

The algorithm stops when a sorting network is found, which will make $|R_k^n| = 1$.

To implement Prune, we loop on N_k^n and check whether the current network is subsumed by any of the previous ones; if this is the case, we ignore it. Otherwise, we add it to R_k^n and remove any networks already in this set that are subsumed by it. This yields a double loop N_k^n where at each iteration we need to find out whether a subsumption exists – which, in the worst case, requires looping through all $n!$ permutations. For $n = 9$, the largest set N_k^n is N_{15}^9, with over 18 million elements, and there are potentially 300×10^{12} subsumptions to test.

These algorithms are straightforward to implement, test and debug. The implementation from [6], written in Prolog, can be applied to reconstruct all of the known values for $S(n)$ for $n \leq 6$ in under an hour of computation on a single core and, after several optimizations and parallelization as described in [6], was able to obtain the new value of $S(9)$.

Soundness of generate-and-prune follows from the observation that N_k^n (and R_k^n) are *complete* for the optimal size sorting network problem on n channels: if there exists an optimal size sorting network on n channels, then there exists one of the form $C; C'$ for some $C \in N_k^n$ (or $C \in R_k^n$), for every k.

2.3   Checking the Proof Using Proof Witnesses

Even though all of the mathematical claims underlying the design of the generate-and-prune algorithm were proved and the correctness of the Prolog implementation was carefully checked, it is reasonable to question the validity of the final result. In [6], use was made of the *de Bruijn criterion* [3]: every computer-generated proof should be verifiable by an independent small program (a "checker"). The code was therefore augmented to produce a log file of successful subsumptions during execution, and an independent Java verifier was able to re-check the result without needing to replicate the expensive search steps in just over 6 hours of computational time.[1]

However, we may again question the validity of the checker, and enter into an endless loop of validations. In this paper, we propose a different goal: to obtain a correct-by-design checker by extracting it from a formalized proof of the theory of sorting networks. The reason for aiming at extracting a checker, rather than the full generate-and-prune algorithm, is that by using the log as an (untrusted) oracle we again gain a speedup of several orders of magnitude, as we completely avoid all the search steps.

In total, for 9 inputs we have logged proof witnesses for approx. 70 million subsumptions, yielding a 27 GB log file. For the smaller case of 8 inputs, we logged 1.6 million subsumptions, yielding over 300 MB of data. Developing a formalization allowing the extraction of an efficient checker that uses an untrusted oracle of even this smaller magnitude is an exciting challenge that, to the best of our knowledge, has not been tackled before. We proceed in two stages. First, we formalize the theory of optimal-size sorting networks directly following [12], including the new results from [6,15]. Then we implement generate-and-prune with an oracle in Coq, prove its soundness, and extract a certified checker able to verify all results up to 8 inputs.

3   Formalizing Sorting Networks

Formalizing the theory of sorting networks presents some challenges relating to the very finite nature of the domain. All the relevant notions are parameterized on the number n of inputs, and thus the domain for most concepts is the finite set $\{0, 1, \ldots, n-1\}$.

Directly working with this set in Coq is very cumbersome due to the ensuing omnipresence of proof terms – every number required as argument has to be accompanied by a proof that it is in the adequate range. Furthermore, these proof terms are completely trivial, since the order relations on natural numbers are all decidable. Therefore, we chose to define all relevant concepts in terms of natural numbers, and define additional properties specifying that particular instances fall in the appropriate range. For example, a comparator is simply defined as a pair of natural numbers:

```
Definition comparator : Set := (prod nat nat).
```

and we define with predicates stating that a particular comparator is a (standard) comparator on n channels:

```
Definition comp_channels (n:nat) (c:comparator) :=
  let (i,j) := c in (i<n) /\ (j<n) /\ (i<>j).
```

```
Definition comp_std (n:nat) (c:comparator) :=
  let (i,j) := c in (i<n) /\ (j<n) /\ (i<j).
```

---

[1] The logs and the Java verifier are available at http://imada.sdu.dk/~petersk/sn/.

Likewise, the type CN of comparator networks is defined as `list comparator`, and there are predicates stating that a comparator network spans n channels, and that it is standard (which simply state that all comparators in the list have the corresponding comparator property).

Comparator networks act on binary sequences, and we define these as a dependent type, similar to `Vector`.

```
Inductive bin_seq : nat -> Set :=
  | empty : bin_seq 0
  | zero : forall n:nat, bin_seq n -> bin_seq (S n)
  | one : forall n:nat, bin_seq n -> bin_seq (S n).
```

We then define two operations `get` and `set`, such that (`get n s`) returns the element (0 or 1) in position n of s, and (`set n s k`) sets position n of s to 0, if k is 0, and 1 otherwise. Setting an index larger than the length of a sequence leaves the sequence unchanged, while attempting to get the value in an index out of range returns 2. The contexts where these functions are used ensure that these situations do not occur, so these options are immaterial for the formalization.

A sequence is sorted if its first element is 0 and the remaining sequence is sorted, or if it consists entirely of 1s.

```
Fixpoint all_ones (n:nat) (x:bin_seq n) : Prop := match x with
  | empty => True
  | zero _ _ => False
  | one _ y => all_ones _ y
  end.
```

```
Fixpoint sorted (n:nat) (x:bin_seq n) : Prop := match x with
  | empty _ => True
  | zero _ y => sorted _ y
  | one _ y => all_ones _ y
  end.
```

Sequences propagate through comparator networks as expected; a sorting network is a comparator network that sorts all inputs.

```
Fixpoint apply (c:comparator) n (s:bin_seq n) :=
  let (i,j):=c in let x:=(get s i) in let y:=(get s j) in
    match (le_lt_dec x y) with
    | left _ => s
    | right _ => set (set s j x) i y
    end.
```

```
Fixpoint full_apply (S:CN) n (s:bin_seq n) := match S with
  | nil => s
  | cons c S' => full_apply S' _ (apply c s)
  end.
```

```
Definition sorting_network (n:nat) (S:CN) :=
  (channels n S) /\ forall s:bin_seq n, sorted (full_apply S s).
```

We also define an alternative characterization of sorting networks in terms of their sets of outputs and prove its equivalence to this one, but for space constraints we will not present it here.

A posteriori, this characterization is actually quite close to the corresponding mathematical definition: the comparator network $\{(0,2); (2,5)\}$ is a network on 6 channels, but it is also a network on 9 or 11 channels, or indeed on n channels for any $n \geq 6$. So this implementation option does not only make the formalization easier, but it is also a good model of the way we think about these objects.

3.1   Proof Methodology

As discussed above, we completely separate between objects and proof terms. This option is dictated by the constructive nature of the theory of sorting networks. The key results underlying the pruning step of the generate-and-prune algorithm are all of the form "If there is a sorting network N, then there is another sorting network N' such that...", and the published proofs of these results [6,12,15] *all* proceed by explicitly describing how to construct N' from N. We formalize these results as operators in order to simplify their reuse: instead of proving statements of the form $\forall N.\varphi(N) \rightarrow \exists N'.\psi(N')$ we first define some transformation T and prove that $\forall N.\varphi(N) \rightarrow \psi(T(N))$ from which we can straightforwardly prove the original statement.

Keeping the proof terms separated from the definitions and formalizing existential proofs as operators has several advantages:

- it is much easier to define these operators and then prove that they satisfy the required properties than including proof terms in their definition;
- the hypotheses underlying the properties themselves become much more explicit – e.g. renumbering channels yields a network on the same number of channels and of the same size, and these two results are independent;
- later, we can use the operators directly together with the relevant lemmas, rather than repeatedly applying inversion on an existential statement;
- additional properties of the operators that are needed later are easy to add as new lemmas, instead of requiring changes to the original theorem;
- we automatically get proof irrelevance, as proof terms are universally quantified in lemmas.

As an example, recall the definition of standardization: given a comparator network C, pick the first comparator (i,j) for which $i > j$, replace it with (j,i) and exchange i with j in all subsequent comparators, then iterate the process until a fixpoint is reached. Standardization is the key ingredient in Lemma 2, and it is formalized using well-founded recursion as follows.

```
Function std (S:CN) {measure length S} : CN := match S with
  | nil => nil
  | cons c S' => let (x,y) := c in
        match (le_lt_dec x y) with
        | left _ => (x[<]y :: std S')
        | right Hxy => (y[<]x :: std (permute x y) S'))
end end.
```

We then prove that standardizing a comparator network on n channels yields a standard comparator network on n channels. This is not completely trivial, because of the permutation in the recursive step: permuting channel labels x and y preserves the total number channels n only if $x < n$ and $y < n$. While these hypotheses trivially hold (x and y are channels in an n-channel network), requiring the corresponding proof terms in the definition of comparator would make the definition of std unreadable. In later results, we can use the std operator when needed. There are in total seven different properties that are needed in different places in the formalization, and all but two are proven independently. If we did not have std as an operator, they would have to be formalized together as one gigantic result stating that for every network C there exists another network C' satisfying the seven necessary properties.

3.2 Permutations

The soundness of the checker essentially depends on the proof of Lemma 2. This lemma is proved directly: if $C_a \leq_\pi C_b$ and $C_b; C$ is a sorting network, then $C_a; \text{std}(\pi(C))$ is a sorting network. The key ingredient in this proof is comparing the effect of C and $\pi(C)$ on the same input string, which requires extensive manipulation of permutations.

Permutations are therefore an essential part of the formalization. Representing them in Coq is a challenging problem, and there are several common alternatives. The standard library includes an inductive type stating that two lists are permutations of each other; but manipulating it is cumbersome. Furthermore, we are not interested in developing a theory of permutations, but rather in proving results about comparator networks whose channels are renamed by means of a permutation of the numbers $0, \ldots, n - 1$, which we will hereafter refer to as "a permutation of $[n]$". Therefore we want a definition that makes it easy and efficient to apply permutations to objects.

For this reason, we chose to represent permutations as finite functions. A permutation P is a list of pairs of natural numbers, with the intended meaning that $(i, j) \in P$ corresponds to P mapping i into j. We assume that P does not change i if there is no pair $(i, j) \in P$ – this makes it much simpler to represent transpositions, which are the only permutations we need to represent explicitly in the formalization. In order for P to be a valid permutation of $[n]$, several conditions have to hold:

1. all pairs $(i, j) \in P$ must satisfy $i < n$ and $j < n$;
2. no number may occur twice either as the first or as the second element of distinct pairs in P; and
3. the sets of numbers occuring as first or second elements of the pairs in P must coincide.

As before, we separate the datatype of permutations from the property of being a permutation. Here, NoDup is the Coq standard library predicate stating that a list does not have duplicate elements, and all_lt(n,l) is an inductive predicate stating that all elements of l are smaller than n.

```
Definition permut := list (nat*nat).

Definition dom (P:permut) := map (fst (A:=nat) (B:=nat)) P.
Definition cod (P:permut) := map (snd (A:=nat) (B:=nat)) P.

Definition permutation n (P:permut) :=
  NoDup (dom P) /\ all_lt n (dom P) /\
  forall i, In i (dom P) <-> In i (cod P).
```

As a sanity-check, we prove the relationship with the permutations in the Coq standard library.

```
Lemma permutation_Permutation : Permutation (dom P) (cod P).
```

All properties of permutations are added to the **core** hint database, so that Coq can automatically prove most properties of permutations required during the formalization.

We provide mechanisms to define permutations in four different ways, three of which correspond to the usage of permutations in proofs, and another one which will be necessary for interacting with the oracle. The former are as follows.

1. The identity permutation is simply the empty list, and it is easily shown to be a permutation of $[n]$ for any n.
2. Given a permutation P, we construct its inverse (**inverse_perm P**) by reversing all pairs in P. If P is a permutation of $[n]$, then so is (**inverse_perm P**).
3. The transposition (**transposition i j**) is the permutation that switches i and j, leaving all other values unchanged. This transposition is defined as the list $\{(i,j),(j,i)\}$, and in order for it to be a permutation it is necessary that $i \neq j$ (otherwise the list $\{i,j\}$ contains duplicate elements). This condition therefore shows up in some results about transpositions; it is never a problem, though, as all transpositions arise from the **standardization** function, where i and j are obtained from a comparator (i,j).

We also need to get permutations from the oracle, and here we use a different representation for efficiency reasons. The log files record permutations as their output on the set $[n]$, so for example the transposition, when $n = 4$, exchanging 0 and 2 would be represented as $\{2,1,0,3\}$. Since we want to be skeptic about the oracle, we do not assume anything about the lists we are given; rather, we show that the property of a list of natural numbers corresponding to a permutation on $[n]$ is decidable. We then define a function **make_perm** to translate lists of natural numbers into (syntactic) permutations, and show that the resulting object satisfies **permutation** if the original list corresponds to a permutation.

```
Variable n:nat.
Variable l:list nat.

Definition pre_perm := NoDup l /\ all_lt n l /\ length l = n.

Lemma pre_perm_dec : {pre_perm} + {~pre_perm}.

Lemma pre_perm_lemma : pre_perm -> permutation n (make_perm l).
```

Thus, our checker will be able to get a list of natural numbers from the oracle, test whether it corresponds to a permutation, and in the affirmative case use this information.

One might question whether we could not have represented permutations uniformly throughout. The reason for not doing so is that we have two distinct objectives in mind. While formalizing results, we are working with an unknown number n of channels, and it is much simpler to represent permutations by only explicitly mentioning the values that are changed, as this allows for uniform representations of transpositions and the identity permutation. Also, computing the inverse of a permutation is very simple with the finite function representation, but not from the compact list representation given by the oracle. When running the extracted checker, however, we are concerned with efficiency. The oracle will provide information on millions of subsumptions, so it is of the utmost importance to minimize its size.

4   Formalizing Generate-and-Prune

Soundness of the generate-and-prune algorithm relies on the notion of a complete set of filters. When formalizing this concept, we needed to make two changes: the element C of R being extended must be a standard comparator network with no redundant comparators[2]; and (as a consequence) the size of the sorting network extending C is *at most* k, since there is an upper bound on how many non-redundant comparisons we can make on n inputs.

```
Definition size_complete (R:list CN) (n:nat) := forall k:nat,
  (exists C:CN, sorting_network n C /\ length C = k) ->
  exists C' C'':CN, In C' R /\ standard n (C'++C'')
       /\ (forall C1 c C2, (C'++C'') = (C1++c::C2) -> ~redundant n C1 c)
       /\ sorting_network n (C'++C'') /\ length (C'++C'') <= k.
```

These changes were discovered during the formalization of the original soundness proof [6], which implicitly used the fact that the elements of the complete sets of filters constructed were not redundant.

We prove that the set $\{\emptyset\}$ is complete, and that if there is a complete set of filters R whose elements all have size k, then all sorting networks on n channels have size at least k. This key property does not hold for the previous informal definition of size completeness.

```
Lemma empty_complete : forall n, size_complete (nil::nil) n.

Lemma complete_size : forall R n k, size_complete R n ->
                      (forall C, In C R -> length C = k) ->
                      forall S, sorting_network n S -> length S >= k.
```

4.1   The Generation Step

The formalization of the generation step proceeds in two phases. First, we define the simple function adding a comparator at the end of a comparator network

---

[2] Comparator (i, j) in comparator network $C; (i, j); C'$ is redundant if $x_i < x_j$ for all $x \in \mathsf{outputs}(C)$ – in other words, (i, j) never changes its inputs.

in all possible ways, and **Generate** simply maps it into a set. The function
all_st_comps produces a list of all standard comparators on n channels.

```
Definition add_to_all (cc:list comparator) (C:CN) :=
  map (fun c => (C ++ (c :: nil))) cc.

Fixpoint Generate (R:list CN) (n:nat) := match R with
  | nil => nil
  | cons C R' => (add_to_all (all_st_comps n) C) ++ Generate R' n
  end.
```

Then, we use the fact that redundancy of the last comparator is decidable
to define an optimized version that removes redundant networks.

```
Definition last_red (n:nat) (C:CN) : Prop :=
  exists C' c, redundant n C' c /\ C = (C' ++ c :: nil).

Lemma last_red_dec : forall n C, {last_red n C} + {~last_red n C}.

Fixpoint filter_nred (n:nat) (R:list CN) := match R with
  | nil => nil
  | (C :: R') => match last_red_dec n C with
                 | left _ => filter_nred n R'
                 | right _ => C :: filter_nred n R'
  end end.

Definition OGenerate (R:list CN) (n:nat) := filter_nred n (Generate R n).
```

Both **Generate** and its optimized version map size complete sets into size
complete sets, as long as the input set does not already contain a sorting network
(in which case **OGenerate** would return an empty set).

```
Theorem OGenerate_complete : forall R n, size_complete R n ->
                        (forall C, In C R -> ~sorting_network n C) ->
                                   size_complete (OGenerate R n) n.
```

The extracted code for these two functions coincides with their Coq definition, since they use no proof terms, and matches the pseudo-code in [6].

4.2   The Pruning Step

For the pruning step, we need to work with the untrusted oracle. We define an
oracle to be a list of subsumption triples $\langle C, C', \pi \rangle$, with intended meaning that
$C \leq_\pi C'$. Using the oracle, we then define the pruning step as follows.

```
Function Prune (O:Oracle) (R:list CN) (n:nat) {measure length R}
              : list CN := match O with
  | nil => R
  | cons (C,C',pi) O' => match (CN_eq_dec C C') with
      | left _ => R
      | right _ => match (In_dec CN_eq_dec C R) with
          | right _ => R
```

```
        | left _ => match (pre_perm_dec n pi) with
          | right _ => R
          | left A => match (subsumption_dec n C C' pi' Hpi) with
            | right _ => R
            | left _ => Prune O' (remove CN_eq_dec C' R) n
end end end end end.
```

The successive tests in **Prune** verify that: $C \neq C'$; $C \in R$; π is a permutation; and $C \leq_\pi C'$. If any of these fail, this subsumption is skipped, else C' is removed from R. For legibility, we wrote pi' for the actual permutation generated by pi and Hpi for the proof term stating that this is indeed a permutation. The Java verifier from [6] did not validate that the permutations in the log files were correct permutations.

The key result states that this is a mapping from complete sets of filters into complete sets of filters, *regardless of the correctness of the oracle*, as long as the input set contains only standard comparator networks with no redundant comparators, and all are of the same size.

```
Theorem Prune_complete : forall O R n, size_complete R n ->
  (forall C, In C R -> standard n C) ->
  (forall C C' c C'', In C R -> C = C'++c::C'' -> ~redundant n C' c) ->
  (forall C C', In C R -> In C' R -> length C = length C') ->
  size_complete (Prune O R n) n.
```

This implementation is simpler than the pseudo-code in [6], as the oracle allows us to bypass all search steps – both for permutations and for possible subsumptions.

4.3   Coupling Everything Together

We now want to define the iterative generate-and-prune algorithm and prove its correctness. Here we deviate somewhat from the original presentation. Our algorithm will receive as inputs two natural numbers (the number of channels n and the number of iterations m) and return one of three possible answers: (**yes n k**), meaning that a sorting network of size k was found and that no sorting network of size $(k-1)$ exists; (**no n m R H1 H2 H3**), meaning that R is a set of standard (H3) comparator networks of size m (H2), with no duplicates (H1); or **maybe**, meaning that an error occurred. The proof terms in **no** are necessary for the correctness proof, but they are all removed in the extracted checker. They make the code quite complex to read, so we present a simplified version where they are omitted.

```
Inductive Answer : Set :=
  | yes : nat -> nat -> Answer
  | no : forall n k:nat, forall R:list CN, NoDup R ->
                    (forall C, In C R -> length C = k) ->
                    (forall C, In C R -> standard n C) -> Answer
  | maybe : Answer.

Fixpoint Generate_and_Prune (n m:nat) (O:list Oracle) := match m with
  | 0 => match n with
```

```
        | 0 => yes 0 0
        | 1 => yes 1 0
        | _ => no n 0 (nil :: nil) _ _ _
        end
| S k => match O with
    | nil => maybe
    | X :: O' => let GP := (Generate_and_Prune n k O') in
        match GP with
        | maybe => maybe
        | yes p q => yes p q
        | no p q R _ _ _ => let GP' := Prune X (OGenerate R p) p in
                            match (exists_SN_dec p GP' _) with
                            | left _ => yes p (S q)
                            | right _ => no p (S q) GP' _ _ _
end end end end.
```

In the case of a positive answer, the network constructed in the original proof is guaranteed to be a sorting network; therefore we do not need to return it. Note the elimination over exists_SN_dec, which states that we can decide whether a set contains a sorting network.

```
Lemma exists_SN_dec : forall n l, (forall C, In C l -> channels n C) ->
                {exists C, In C l /\ sorting_network n C} +
                {forall C, In C l -> ~sorting_network n C}.
```

The correctness of the answer is shown in the two main theorems: if the answer is (yes n k), then the smallest sorting network on n channels has size $k \leq m$; and if the answer is (no n m), then there is no sorting network on n channels with size m or smaller. These results universally quantify over O, thus holding regardless of whether the oracle gives right or wrong information.

```
Theorem GP_yes : forall n m O k, Generate_and_Prune n m O = yes n k ->
                (forall C, sorting_network n C -> length C >= k) /\
                exists C, sorting_network n C /\ length C = k.
```

```
Theorem GP_no : forall n m O R HR0 HR1 HR2,
                Generate_and_Prune n m O = no n m R HR0 HR1 HR2 ->
                forall C, sorting_network n C -> length C > m.
```

The full Coq formalization consists of 102 definitions and 405 lemmas, with a total size of 206 KB, and the extracted program is around 650 lines of Haskell code. The formalization and its generated documentation are available from http://imada.sdu.dk/~petersk/sn/.

5   Running the Extracted Program

We extracted the certified checker to Haskell using Coq's extraction mechanism. The result is a file Checker.hs containing among others a Haskell function generate_and_Prune :: Nat -> Nat -> (List Oracle) -> Answer. In order to run this extracted certified checker, we wrote an interface that calls generate_and_Prune function with the number of channels, the maximum size of the networks, and the list of the oracle information, and then prints the answer. This interface includes conversion functions from Haskell integers to the extracted naturals and a function implementing the oracle, as well as a definition of `Checker.Answer` as an instance of the type class `Show` for printing the result.

It is important to stress that we do *not* need to worry about soundness of almost any function defined in the interface, as the oracle is untrusted. For example, a wrong conversion from natural numbers to their Peano representation will not impact the correctness of the execution (although it will definitely impact the execution *time*, as all subsumptions will become invalid). We only need to worry about the function printing the result, but this is straightforward to verify.

The extracted checker was able to validate the proofs of optimal size up to and including $n = 8$ in around one day – roughly the same time it took to produce the original proof, albeit without search. This required processing more than 300 MB of proof witnesses for the roughly 1.6 million subsumptions. To the best of our knowledge, it constitutes the first formal proof of the results in [9].

Experiments suggested that, using this extracted checker, the verification of the proof that $S(9) = 25$ would take around 20 years. Subsequent work in optimizing its underlying algorithm [8], without significantly changing the formalization herein described, was able to reduce this time to just under one week.

6   Conclusions

We have presented a formalization of the theory of size-optimal sorting networks, extracted a verified checker for size-optimality proofs, and used it to show that informal results obtained in previous work are correct.

Our main contribution is a *formalization* of the theory of size-optimal sorting networks, including an intuitive and reusable formalization of comparator networks and a new representation of permutations more suitable for computation.

Another immediate contribution is a *certified checker* that directly confirmed all the values of $S(n)$ quoted in [9], and that was subsequently able also to verify that $S(9) = 25$ [8]. We plan to apply the same technique – first formalize, then optimize – to other computer-generated proofs where formal verification has been prohibitively expensive so far. We also plan to formalize the results from [18] in order to obtain a formal proof of $S(10) = 29$.

Acknowledgements. We would like to thank Femke van Raamsdonk, whose initial skepticism about our informal proof inspired this work, and Michael Codish for his support and his enthusiasm about sorting networks. The authors were supported by the Danish Council for Independent Research, Natural Sciences. Computational resources were generously provided by the Danish Center for Scientific Computing.

References

1. Appel, K., Haken, W.: Every planar map is four colorable. Part I: discharging. Ill. J. Math. **21**, 429–490 (1977)
2. Appel, K., Haken, W., Koch, J.: Every planar map is four colorable. Part II: reducibility. Ill. J. Math. **21**, 491–567 (1977)

3. Barendregt, H., Wiedijk, F.: The challenge of computer mathematics. Trans. A Roy. Soc. **363**(1835), 2351–2375 (2005)
4. Blazy, S., Paulin-Mohring, C., Pichardie, D. (eds.): ITP 2013. LNCS, vol. 7998. Springer, Heidelberg (2013)
5. Claret, G., González-Huesca, L.C., Régis-Gianas, Y., Ziliani, B.: Lightweight proof by reflection using a posteriori simulation of effectful computation. In Blazy et al. [4], pp. 67–83
6. Codish, M., Cruz-Filipe, L., Frank, M., Schneider-Kamp, P.: Twenty-five comparators is optimal when sorting nine inputs (and twenty-nine for ten). In: ICTAI 2014, pp. 186–193. IEEE (2014)
7. Contejean, E., Courtieu, P., Forest, J., Pons, O., Urbain, X.: Automated certified proofs with CiME3. In: Schmidt-Schauß, M., (ed.) RTA 2011. LIPIcs, vol. 10, pp. 21–30. Schloss Dagstuhl (2011)
8. Cruz-Filipe, L., Schneider-Kamp, P.: Optimizing a certified proof checker for a large-scale computer-generated proof. In: Kerber, M., Carette, J., Kaliszyk, C., Rabe, F., Sorge, V. (eds.) CICM 2015. LNCS, vol. 9150, pp. 55–70. Springer, Heidelberg (2015)
9. Floyd, R.W., Knuth, D.E.: The Bose-Nelson sorting problem. In: Srivastava, J.N. (ed.) A Survey of Combinatorial Theory, pp. 163–172. North-Holland, Amsterdam (1973)
10. Fouilhe, A., Monniaux, D., Périn, M.: Efficient generation of correctness certificates for the abstract domain of polyhedra. In: Logozzo, F., Fähndrich, M. (eds.) Static Analysis. LNCS, vol. 7935, pp. 345–365. Springer, Heidelberg (2013)
11. Gonthier, G.: Formal proof - the four-color theorem. Not. AMS **55**(11), 1382–1393 (2008)
12. Knuth, D.E.: The Art of Computer Programming. Sorting and Searching, vol. 3. Addison-Wesley, Reading (1973)
13. Konev, B., Lisitsa, A.: A SAT attack on the Erdős discrepancy conjecture. In: Sinz, C., Egly, U. (eds.) SAT 2014. LNCS, vol. 8561, pp. 219–226. Springer, Heidelberg (2014)
14. Leroy, X.: Formal verification of a realistic compiler. Commun. ACM **52**(7), 107–115 (2009)
15. Parberry, I.: A computer-assisted optimal depth lower bound for nine-input sorting networks. Math. Syst. Theor. **24**(2), 101–116 (1991)
16. Sternagel, C., Thiemann, R.: The certification problem format. In: Benzmüller, C., Paleo, B.W. (eds.) UITP 2014. EPTCS, vol. 167, pp. 61–72 (2014)
17. Thiemann, R.: Formalizing bounded increase. In: Blazy et al. [4], pp. 245–260
18. van Voorhis, D.C.: Toward a lower bound for sorting networks. In: Miller, R.E., Thatcher, J.W. (eds.) Complexity of Computer Computations. The IBM Research Symposia Series, pp. 119–129. Plenum Press, New York (1972)

Proof-Producing Reflection for HOL
With an Application to Model Polymorphism

Benja Fallenstein[1] and Ramana Kumar[2]([✉])

[1] Machine Intelligence Research Institute, Berkeley, USA
[2] Computer Laboratory, University of Cambridge, Cambridge, UK
ramana.kumar@cl.cam.ac.uk

Abstract. We present a reflection principle of the form "If $\ulcorner \varphi \urcorner$ is provable, then φ" implemented in the HOL4 theorem prover, assuming the existence of a large cardinal. We use the large-cardinal assumption to construct a model of HOL within HOL, and show how to ensure φ has the same meaning both inside and outside of this model. Soundness of HOL implies that if $\ulcorner \varphi \urcorner$ is provable, then it is true in this model, and hence φ holds. We additionally show how this reflection principle can be extended, assuming an infinite hierarchy of large cardinals, to implement *model polymorphism*, a technique designed for verifying systems with self-replacement functionality.

1 Introduction

Reflection principles of the form[1] "if $\ulcorner \varphi \urcorner$ is provable, then φ" have long been of interest in logic and set theory (see, e.g., Franzén [5] and Jech [13]). In this paper, we show how to implement a reflection principle for HOL in the HOL4 theorem prover [21], using a novel approach for establishing a correspondence between the logic and an internal model. Such a reflection principle does not come for free, since by Gödel's second incompleteness theorem, HOL cannot prove itself consistent [6]. We pay with an assumption about the existence of a "large enough" HOL type. But we endeavour to ensure that this assumption has the same content as the assumption, commonly studied in set theory [13], of the existence of a strongly inaccessible cardinal.

Reflection is about trying to fit a logic inside itself, so one has to keep two instances of the logic in mind separately. We use the term *inner HOL* to refer to the object logic (the HOL that is formalised) and *outer HOL* for the meta logic (the HOL in which inner HOL is formalised). We build upon Harrison's formalisation of HOL in itself [11] that was extended in our previous work [15,16] to support defined constants. The first reflection principle we implement uses a model of inner HOL provided by a large-cardinal assumption, together with the previously-proved soundness theorem, to show that provability of a proposition in inner HOL implies its truth in outer HOL.

---

[1] $\ulcorner \varphi \urcorner$ refers to φ as a syntactic object, represented e.g. by its Gödel number or by an abstract syntax tree.

© Springer International Publishing Switzerland 2015
C. Urban and X. Zhang (Eds.): ITP 2015, LNCS 9236, pp. 170–186, 2015.
DOI: 10.1007/978-3-319-22102-1_11

This kind of reflection principle is asymmetric: the large-cardinal assumption used in outer HOL to justify reflection cannot be used again in inner HOL to justify further reflection. One setting in which it is useful to lessen this asymmetry is in constructing and verifying a system that can replace itself (including the replacement mechanism) by a new version while nevertheless always satisfying some safety property. We have extended the reflection principle above to an implementation of *model polymorphism* [3], which enables the use of reflective reasoning multiple times to verify such self-modifying systems.

Our specific contributions in this paper are as follows:

- A simple approach to defining, in outer HOL, a partial embedding of outer HOL in inner HOL (Sect. 3). We re-use a single polymorphic constant, to_inner, to embody the embedding at (almost) all outer HOL types.
- An algorithm for producing theorems that relate the semantics of inner HOL terms to their outer HOL counterparts via the embedding (Sect. 4). The method applies to terms that include user-defined constants and types, and therefore supports construction of a semantic interpretation of constants based on replaying outer definitions in inner HOL (Sect. 5). Combined with the soundness theorem for inner HOL, this gives us a reflection principle for HOL (Sect. 7).
- An extension of the reflection principle above to support model polymorphism [3] (Sect. 8).
- A reduction of the unavoidable assumption on the proof (in previous work) of soundness and consistency of HOL within HOL to a traditionally-stated large-cardinal assumption (Sect. 6).

All the work we present has been implemented in the HOL4 theorem prover, and is available online at https://github.com/CakeML/hol-reflection.

Synopsis. Our reflection principle has two legs: (1) a soundness theorem for HOL (from previous work) asserting that a provable formula is true in all models, and (2) a formal correspondence between objects in a particular, reflective model (constructed assuming a large cardinal) and their counterparts outside. The reflection principle works as follows: if a formula $\ulcorner \varphi \urcorner$ is provable, the soundness theorem asserts that $\ulcorner \varphi \urcorner$ is true in the reflective model of HOL, and via the correspondence we may conclude that φ is true outside the model. The bulk of our work in this paper is in building the reflective model and establishing the correspondence, particularly in supporting user-defined constants.

2   Background: Inner HOL

The starting point for studying reflection principles is a formalisation of a logic inside itself. We use higher-order logic (HOL), the logic of the HOL Light theorem prover [12]. Our formalisation of HOL within itself is described carefully in previous work [15, 16], but we summarise it briefly in this section.

HOL has the syntax of the simply-typed lambda calculus[2] and a well-known semantics in Zermelo set theory. The judgements of the logic are sequents, (thy,hs) |- c, where c is a formula (a term of Boolean type), hs is a set of formulas, and thy specifies the current set of axiom formulas and the current signature of constants and type operators. The semantics of each type is a non-empty set; the Boolean type is specified as a distinguished two-element set (Boolset, with elements True and False) and function types are specified as function spaces (a set of pairs forming a functional relation). The semantics of each term is an element of the semantics of its type. The semantics of each sequent is true, written $(thy,hs) \models c$, if the semantics of c is true whenever the semantics of all hs are true, in any model (an interpretation of constants satisfying the axioms) of thy.

Since the semantic objects are sets, we need a model of set theory to define the semantics of inner HOL. Our semantics does not pin down a particular model of set theory. Instead, it is polymorphic: we use an outer HOL type variable, usually μ, for the universe of sets, and encode the axioms of set theory as a pair of assumptions, is_set_theory mem and $\exists inf.$ is_infinite mem inf, about a membership relation, $(mem : \mu \rightarrow \mu \rightarrow bool)$, on sets. All our semantic functions take the membership relation mem (and with it the universe μ) as a parameter, and most theorems about the semantics assume the set-theory axioms hold. However, for brevity we usually hide[3] these parameters and assumptions when they are obvious in context. We write $x \lessdot y$ for infix application of the mem relation (i.e., mem x y).

The important semantic function in this paper is the semantics of terms, that is, the function that assigns an element of μ to each term of inner HOL. We illustrate how it works with an example, considering the semantics of the inner HOL version of the term $\lambda x.$ foo y, where the constant (foo : $\alpha \rightarrow \beta$) is instantiated at the types bool and ind list. The four parameters governing the semantics are: the membership relation mem (hidden), the signature (s) of constants and type operators in the theory, the interpretation (i) of constants and type operators, and the valuation (v) of free term and type variables. We denote the valuation of type variables by v_{ty} whereas for term variables we use v_{tm}, and similar subscripts apply to the other semantic parameters. Our use of interpretations and valuations is intended to be conventional and unsurprising; we show the example here mainly to make our notation clear.[4]

```
termsem s_tm i v
        (Abs (Var "x" Bool)
            (Comb (Const "foo" (Fun Bool (Tyapp "list" [Tyapp "ind" []])))
            (Var "y" Bool))) =
        Abstract Boolset sil (λ sx. sfoo ' sy)
```

---

[2] Including, as in Gordon [9], polymorphism (type variables and polymorphic constants) and defined constants and type operators.

[3] This is akin to working within an Isabelle [23] locale which fixes μ and mem and assumes is_set_theory mem. (We assume infinity only when necessary).

[4] The abstract syntax here is the inner HOL rendition of our example term. Bool and Fun a b are abbreviations for Tyapp "bool" [] and Tyapp "fun" [a; b].

The semantics of the given lambda abstraction is a set-theoretic function (created with Abstract) – in particular the set of two pairs, True $\mapsto$ *sfoo* ' *sy* and False $\mapsto$ *sfoo* ' *sy*. The following equations show how we obtain the semantics of the applied type operator, the instantiated constant, and the free variable:

$$sil = i_{\mathsf{ty}} \text{ "list" } [i_{\mathsf{ty}} \text{ "ind" } []]$$
$$sfoo = i_{\mathsf{tm}} \text{ "foo" } [\mathsf{Boolset};\ sil]$$
$$sy = v_{\mathsf{tm}} (\text{"y"}, \mathsf{Bool})$$

Inner HOL supports defined constants and type operators via the *current theory*, which is attached to each sequent. The inference system uses a concrete implementation of each theory as a list of *updates*. Such a list is called a *context* and every context can be viewed abstractly as a theory (written: thyof *ctxt*). Contexts are made from five kinds of update: new type operator, new constant, new axiom, new defined type operator, and new defined constant. The updates for defined type operators and constants have preconditions that require sequents proved in the previous context. If an update *upd* satisfies all the preconditions to be a valid update of the previous context *ctxt*, we say *upd* updates *ctxt*.

Both the within-theory inference rules and the theory-extending rules for making updates (except new axiom) have been proved sound with respect to the semantics. For the inference rules, the soundness theorem states[5] that every provable sequent is true:

$$\vdash (thy,hs) \mathrel{|\!-} c \Rightarrow (thy,hs) \models c$$

For the extension rules, the soundness theorem states that there exists an extended interpretation of the new theory that continues to be a model of the theory. Both theorems require the is_set_theory *mem* assumption.

3   An Inner Copy of Outer HOL

Given an outer HOL term, it is straightforward to write a function in ML—our function is called term_to_deep—that walks the structure of the term and builds the corresponding inner HOL term. For example, term_to_deep

turns   Suc x into   Comb (Const "Suc" (Fun Num Num)) (Var "x" Num).

This syntactic connection is straightforward. But what is the relationship between the semantics of an outer HOL term and its inner counterpart? Let $\ulcorner tm \urcorner$ stand for term_to_deep(tm). Ideally, we would like a connection between Suc x and termsem s_{tm} i v $\ulcorner$ Suc x $\urcorner$. Such a connection would mean that the structure of outer HOL terms and types is replicated (indeed reflected) within

---

[5] To understand the three turnstiles: $\vdash$ denotes provability in outer HOL (and applies to the whole formula), infix $\mathrel{|\!-}$ denotes provability in inner HOL, and infix $\models$ states that a sequent is valid according to the semantics of inner HOL.

the type denoted by μ. We cannot expect to reflect everything: in particular, we cannot reflect μ within itself, nor anything depending on μ. But we can cover all other outer HOL types and terms generically.

Our solution is to define a polymorphic constant, to_inner, which sends an outer HOL term to the element of μ to which it is supposed to correspond. We show how to prove theorems of the form termsem s_{tm} i v $\ulcorner tm \urcorner$ = to_inner ty tm. The type of to_inner ty is $\alpha \rightarrow \mu$.[6] (We explain the ty argument shortly.)

We do not want to have to specify exactly how to_inner creates its (partial) copy of outer HOL. What is important is that the copy is faithful, which means to_inner is injective. We formalise this injectivity property precisely by saying that to_inner at type α should be a bijection between $\mathcal{U}(:\alpha)$ (everything of type α) and the set[7] of elements of some set x in μ. Formally, we define a well-formedness condition:

$$\text{wf\_to\_inner } f \iff \exists x. \text{ BIJ } f\, \mathcal{U}(:\alpha) \ \{\, a \mid a \ll x \,\}$$

Then we define to_inner as an arbitrary well-formed injection, using Hilbert choice[8] as follows:

$$\text{to\_inner } ty = \text{tag } ty \circ \varepsilon f. \text{ wf\_to\_inner } f$$

The tag ty part of the definition wraps the set produced by the well-formed injection with a tag for the given inner HOL type, ty, thereby avoiding the possibility of inadvertently sending outer HOL terms with different types to the same element in μ. The need for this tagging is explained in Sect. 5.

Since Hilbert choice only picks a well-formed injection *if one exists*, the usefulness of to_inner at any particular outer HOL type depends on our assuming (or being able to prove) wf_to_inner (to_inner ty). The automation we describe in Sect. 4 produces theorems that assume these well-formedness conditions, and the automation in Sect. 5 proves almost all of them.

Of course, since the well-formedness condition on to_inner says it is not just an injection but a bijection, we can also go in the other direction, from inner HOL terms to their outer counterparts. Given a well-formed injection ina, we denote the set of terms of type μ that are in its range—that is, the inner representation of the domain of ina—by range ina. The faithfulness of the representation is summarised in the following two theorems exhibiting invertibility.

$$\vdash \text{wf\_to\_inner } ina \Rightarrow \forall x. \ ina^{-1} \, (ina\ x) = x$$
$$\vdash \text{wf\_to\_inner } ina \Rightarrow \forall x. \ x \ll \text{range } ina \Rightarrow ina \, (ina^{-1}\ x) = x$$

---

[6] To be pedantic, to_inner also depends on the pervasive *mem* relation of the set theory.

[7] We follow the convention of treating predicates in outer HOL as sets. To be clear $\mathcal{U}(:\alpha)$ is a term of type $\alpha \rightarrow$ *bool*, and $\{\, a \mid a \ll x \,\}$ is a term of type $\mu \rightarrow$ *bool*. These sets-as-predicates are distinct from the Zermelo sets, i.e., the terms of type μ.

[8] Also known as indefinite choice, the Hilbert choice principle in HOL provides a constant (ε), usually written as a binder, together with the axiom $(\exists x. \ P\ x) \Rightarrow P\ (\varepsilon x. \ P\ x)$ which holds for any predicate P.

Usually, but not always, *ina* will be to_inner *ty* for some *ty*. We use to_inner to reflect outer HOL at all types except for those that depend on μ and except for the two primitive types of HOL: Booleans and function types. The primitive types must be treated differently because the semantics of HOL requires them to be interpreted by a distinguished two-element set and by set-theoretic function spaces, respectively. We provide specialised constants that map these types to their intended interpretations instead of an arbitrary set:

bool_to_inner $b =$ if b then True else False
fun_to_inner *ina inb* $f =$
    Abstract (range *ina*) (range *inb*) $(\lambda x.\ inb\ (f\ (ina^{-1}\ x)))$

We have now seen how we intend to reflect outer HOL terms into inner HOL, using to_inner *ty* injections. We next describe an algorithm that produces a certificate theorem for (almost) any HOL term, asserting that the semantics of the inner version of the term matches the reflection of the outer version. We develop this algorithm in two stages: first (next section), we ignore the interpretation of constants and the valuation of variables in the semantics and simply make assumptions about them; then (Sect. 5), we build an interpretation and valuation that satisfies these assumptions.

4   Proof-Producing Reflection

Let us begin with an example of the kind of theorem produced by the first stage of our automation. Given as input the term Suc x, we produce the following theorem that looks very long but whose components we will explain one by one. We call each such theorem a *certificate theorem* for the input term.

$\vdash$ good_context *mem s i* $\wedge$ wf_to_inner (to_inner Num) $\wedge$
    lookup s_{tm} "Suc" $=$ Some (Fun Num Num) $\wedge$ lookup s_{ty} "num" $=$ Some 0 $\wedge$
    i_{ty} "num" $[\,] =$ range (to_inner Num) $\wedge$
    i_{tm} "Suc" $[\,] =$ fun_to_inner (to_inner Num) (to_inner Num) Suc $\wedge$
    v_{tm} ("x",Num) $=$ to_inner Num x $\Rightarrow$
    termsem s_{tm} i v $\ulcorner$Suc $x\urcorner =$ to_inner Num (Suc x)

First, look at the conclusion[9] (the last line): we have produced an equality between the semantics of the inner version of our input term and its reflection created with to_inner. Under what assumptions? The assumptions come in five categories:

1. The assumption stated using good_context, which represents the pervasive is_set_theory *mem* assumption as well as some basic well-formedness conditions on the signature and interpretation.

---

[9] Num is an abbreviation for Tyapp "num" $[\,]$.

2. wf_to_inner assumptions on all *base types* appearing in the input term (in this case, just Num). A base type is any type variable or any application of a non-primitive type operator. (For the primitives, Booleans and functions, we prove wf_to_inner once and for all.)

3. Signature (s) assumptions stating that all non-primitive constants and type operators in the input term (in this case Suc and *num*) have the same type/arity in inner HOL as in outer HOL. (The only primitive constant, equality, is assumed to have the correct type as part of good_context.)

4. Interpretation (i) assumptions stating that the inner versions of all the base types and constants of the input term are mapped to their reflections.

5. Valuation (v) assumptions stating that the inner versions of all type and term variables in the input term are mapped to their reflections.

The algorithm for producing such a theorem works by recursively traversing the structure of the input term. We build a theorem like the one above, equating the semantics of an inner term to the reflection of its outer counterpart, for each subterm in the input term starting at the leaves and progressing bottom-up. When we encounter non-primitive type operators and constants we add signature and interpretation assumptions as required, and similarly add valuation assumptions for free variables. The substitution of type variables used to instantiate a polymorphic constant is easily reflected from outside to inside.

This algorithm works because the semantics itself is recursive. For example, shown below is a theorem about the semantics that our algorithm uses when it encounters a combination (function application) term. Notice that the theorem makes two assumptions of the same form as its conclusion – these correspond to recursive calls in the algorithm.

$\vdash$ termsem s_{tm} i v *ftm* $=$ fun_to_inner *ina inb f* $\wedge$ termsem s_{tm} i v *xtm* $=$ *ina x* $\wedge$
wf_to_inner *ina* $\wedge$ wf_to_inner *inb* $\Rightarrow$
termsem s_{tm} i v (Comb *ftm xtm*) $=$ *inb* (f x)

The analogous theorem for lambda abstractions requires us to prove connections between inner HOL types and their reflections (via range). Therefore, we have a similar recursive algorithm for types as the one described above for terms.

So far, our certificate theorems leave the semantic parameters (s, i, and v) as free variables but make various assumptions about them. Our aim in the next section is to show how we can instantiate these parameters in such a way that most of the assumptions become provable. In other words, we show how to build a reflective model that satisfies the assumptions in each of the categories above.

5   Building a Reflective Interpretation and Valuation

Signature Assumptions. In outer HOL, types like *num* and constants like Suc are all user-defined. Logically speaking, there is a context (typically not represented explicitly in the theorem prover) that lists the sequence of updates made to produce the current environment of defined constants. An appropriate signature (s)

for a certificate theorem is one that corresponds to some context for the input term, which will naturally satisfy all the signature assumptions. We require the user of our automation to build an explicit representation of the desired context (although we provide tools to help with this), and with that information proceed to construct an interpretation and valuation to satisfy the assumptions of the certificate theorem.

Algorithm for Building the Model. The idea is to reflect the updates from the outer context into the inner inference system, creating an inner context, and then, update by update, build an interpretation of the inner context. The interpretation we build must be reflective, which means it must satisfy the certificate theorem's assumptions, namely: all the defined constants and types appearing in the input term are mapped to the reflections of their outer versions. We build the reflective interpretation by: (1) starting with the interpretation asserted to exist by the soundness theorem, and (2) making a finite number of modifications for each update in the context. We work recursively up the context applying the relevant modifications for the last-added update at each stage.

Interpretation Assumptions. Each update introduces some number of type operators, constants, and axioms[10] to the theory. Each update thereby induces some *constraints* on an interpretation for the interpretation to be reflective. For example, the update that defines ⌜Suc⌝ has an associated constraint that this constant is interpreted as to_inner (Fun Num Num) Suc.

For polymorphic constants and type operators with arguments, the constraints are more involved. In general, a constraint is induced by each instance of the update and constrains the corresponding instance of the introduced constants. Importantly, we only consider instances of the update that correspond to the finitely many instances of the defined constants in the input term. We require to_inner to produce distinct reflections of distinct types, because otherwise we cannot reliably distinguish different instances of an update.

To show that an interpretation continues to be a model after being constrained, we require that at each constrained instance the axioms of the update are satisfied. We can prove this for the constraints we build (which map things to their reflections) because the axioms of the update being replayed are true in outer HOL. For each constrained instance of an axiom, we use the algorithm from the previous section to generate a certificate theorem with assumptions that are all easy to discharge because the constraint is reflective.

wf_to_inner *Assumptions.* To prove the wf_to_inner assumptions (for defined type operators) on our certificate theorems, we use the fact that in outer HOL every defined type is represented by a predicate on a previously defined (or primitive) type. The same is true in inner HOL as we build up our interpretation, so the wf_to_inner assumption on a defined type reduces to the wf_to_inner assumption

---

[10] When a new constant or type operator is defined, theorems produced by the definition are considered axioms of the resulting theory. The other source of axioms is new-axiom updates, which we do not support since they are not sound in general.

on its representing type. These assumptions propagate back recursively to the base case. The only wf_to_inner assumptions left after this process are for type variables in the input term, and for the type of individuals (*ind*) – this last assumption may be introduced if not present.

Base Case of the Algorithm. To finish describing how we build a constrained model that satisfies the assumptions of the certificate theorem, only the base case of the recursive algorithm remains. What model do we start with before replaying all the updates in the context? In our previous work [15,16] on the soundness and consistency of HOL, we showed that there is a model for the base context of HOL (assuming the set-theory axioms, including infinity). We use this model in the base case of the algorithm for building a reflective interpretation, modulo some subtleties concerning the Hilbert choice operator.

The base context, hol_ctxt, includes the primitive types and constants (Booleans, functions, equality) and the three axioms of HOL (extensionality, choice, and infinity). To state the axioms, the base context also includes some defined constants (conjunction, implication, etc.), and, importantly, the Hilbert choice constant, and the type of individuals. Although most constraints are dealt with before the base case, constraints on Hilbert choice remain. Therefore, we extend the proof that hol_ctxt has a model to show that in fact there is a model satisfying any finite number of constraints on the intepretation of Hilbert choice. We also modify the model to use range (to_inner Ind) for the type of individuals.

Valuation Assumptions. So far we have described the algorithm for building an interpretation (*i*) to satisfy the interpretation assumptions on certificate theorems. To build a valuation (*v*) satisfying the valuation assumptions, we follow a similar approach based on constraining an arbitrary valuation so that it maps the free variables that occur in the input term to their reflections. We have not covered all the gory details of these algorithms, but hope to have shared the important insights. The ML code for our automation is around 1800 lines long, supported by around 1000 lines of proof script about constrained interpretations.

6   Set Theory from a Large-Cardinal Assumption

Up until now, we have been working under the assumption is_set_theory *mem*, as used in previous work [15,16], which consists of the set-theoretic axioms of extensionality, specification, pairing, union, and power sets. To produce models of hol_ctxt and all its extensions, we also require the set-theoretic axiom of infinity (which is implied by wf_to_inner (to_inner Ind)). One of our contributions is to clarify that these assumptions are implied by a more traditionally-stated large-cardinal assumption.

Our large-cardinal assumption is strictly stronger than what is necessary for a model of HOL. We leave proving an equivalence between is_set_theory *mem* plus infinity and a large-cardinal assumption for future work, and here just prove implication from a strongly inaccessible cardinal. The point is to build confidence in our specification of set theory, by showing it is implied by a traditional specification (of a strong inaccessible).

We make use of Norrish and Huffman's formalisation [20] of cardinals and ordinals in HOL4. We write $s \prec t$ to mean the predicate s has smaller cardinality than the predicate t, i.e., there is an injection from s to t but not vice versa. Remember that sets and predicates are usually identified in HOL; we use the usual notation for sets like $x \in s$ (x is an element of s) and $f" s$ (the image of s under f) when s is a predicate.

The reduction theorem we have proved is that if some (outer HOL) type, say μ, is a *strongly inaccessible cardinal*, then there is a membership relation *mem* on that type that satisfies is_set_theory and is_infinite. Furthermore, according to this *mem*, for every predicate s on the type μ which is strictly smaller than $\mathcal{U}(:\mu)$, there is a Zermelo set x whose elements are exactly the extension of s. In this section, we are explicit with all uses of the type variable μ and the variable *mem*. The formal statement of our reduction theorem is:

$$\vdash \mathsf{strongly\_inaccessible}\ \mathcal{U}(:\mu) \Rightarrow$$
$$\exists\, mem.$$
$$\mathsf{is\_set\_theory}\ mem\ \wedge$$
$$(\forall\, s.\ s \prec \mathcal{U}(:\mu) \Rightarrow \exists x.\ s = \{\ a \mid a < x\ \})\ \wedge$$
$$\exists\, inf.\ \mathsf{is\_infinite}\ mem\ inf$$

It is well known that the existence of a strongly inaccessible cardinal gives rise to a model of set theory [13, Lemma 12.13], and the proof of our reduction theorem contains no surprises. (The main lemma is that a strongly inaccessible cardinal is in bijection with its smaller subsets.) Our focus here is on the definition of strongly_inaccessible, aiming to accurately capture traditional usage.

A cardinal is called "strongly inaccessible" if it is uncountable, a regular cardinal, and a strong limit cardinal [13]. Assuming the axiom of choice (which we do as we are working in HOL), a cardinal X is regular iff it cannot be expressed as the union of a set smaller than X all of whose elements are also smaller than X [13, Lemma 3.10]. Formally:

$$\mathsf{regular\_cardinal}\ X \iff$$
$$\forall x\, f.$$
$$x \subseteq X \wedge x \prec X \wedge (\forall a.\ a \in x \Rightarrow f\, a \subseteq X \wedge f\, a \prec X) \Rightarrow$$
$$\bigcup (f" x) \prec X$$

A cardinal is a strong limit if it is larger than the power set of any smaller cardinal. This is straightforward to formalise:

$$\mathsf{strong\_limit\_cardinal}\ X \iff \forall x.\ x \subseteq X \wedge x \prec X \Rightarrow \mathcal{P}(x) \prec X$$

A cardinal is countable if it can be injected into the natural numbers (this is already defined in HOL4's standard library). With these three ideas formalised, we define strongly_inaccessible as follows:

$$\mathsf{strongly\_inaccessible}\ X \iff$$
$$\mathsf{regular\_cardinal}\ X \wedge \mathsf{strong\_limit\_cardinal}\ X \wedge \neg\mathsf{countable}\ X$$

7   Proving Reflection Principles

We now have enough machinery in place to exhibit a reflection principle for HOL. In the examples that follow, we use a two-place predicate Safe t v to construct our input propositions, to match the example in Sect. 8. However, Safe can be considered as a placeholder for any two-place predicate. As a concrete, simple example, set Safe t v $\iff$ $v \neq$ Suc t.

First, let us see how the reflection principle works on the input proposition $\forall t.$ Safe t 0. Combining the automation from Sects. 4 and 5, we can construct a certificate theorem with almost all of the assumptions proved:

$\vdash$ is_set_theory mem $\wedge$ wf_to_inner (to_inner Ind) $\Rightarrow$
    termsem (sigof inner_ctxt)$_{\text{tm}}$ constrained_model constrained_valuation
        $\ulcorner\forall t.$ Safe t $0\urcorner =$
    bool_to_inner ($\forall t.$ Safe t 0)

Here constrained_model and constrained_valuation are built as described in Sect. 5: they are a reflective model for the context (inner_ctxt) that defines $\ulcorner$Safe$\urcorner$, and a reflective valuation[11] for the input proposition.

Next, consider the inner HOL sequent corresponding to provability of our input proposition:

$$(\text{thyof inner\_ctxt}, []) \mid - \ulcorner\forall t.\ \text{Safe } t\ 0\urcorner$$

The soundness theorem for inner HOL states that if this sequent is provable, then the semantics of its conclusion is true. Our certificate theorem above already tells us that the semantics of the conclusion is equal to the reflection of the input proposition. By definition of bool_to_inner, if the reflection of a proposition is true then that proposition holds, hence we obtain:

$\vdash$ is_set_theory mem $\wedge$ wf_to_inner (to_inner Ind) $\Rightarrow$
    (thyof inner_ctxt, []) $\mid$- $\ulcorner\forall t.$ Safe t $0\urcorner \Rightarrow \forall t.$ Safe t 0

There would be additional wf_to_inner assumptions for each additional type variable in the input term.

The final step is to replace the is_set_theory mem assumption with the large-cardinal assumption that provides us with a model of set theory. In this model, we can replace the wf_to_inner assumption on ind by a lower bound on the cardinality of μ. Similarly, any wf_to_inner assumption on a type variable, say α, could be replaced by an assumption of the form $\mathcal{U}(:\alpha) \prec \mathcal{U}(:\mu)$. Such assumptions can be satisfied by large enough μ, except in the case $\alpha = \mu$; hence, it is important that μ not occur in the input proposition. The resulting theorem no longer contains any occurrences of mem, not even hidden ones.

$\vdash$ strongly_inaccessible $\mathcal{U}(:\mu) \wedge \mathcal{U}(:ind) \prec \mathcal{U}(:\mu) \Rightarrow$
    (thyof inner_ctxt, []) $\mid$- $\ulcorner\forall t.$ Safe t $0\urcorner \Rightarrow \forall t.$ Safe t 0

---

[11] Any valuation would do for this input since it has no free type or term variables.

Thus we have shown that provability of our input proposition implies its truth, assuming the existence of a large cardinal that is larger than the type of individuals in outer HOL. We can prove a theorem like this for any input proposition (that does not mention μ), including less obvious and even false propositions.

The reflection principle above can be generalised to a *uniform reflection principle* [4], which, for an input proposition with a free natural-number variable, requires an inner-HOL proof about only the relevant value of the variable. We write $\ulcorner \overline{v} \urcorner$ for the inner HOL numeral corresponding to the value of the outer HOL variable v; for example, if $v = 1$ in outer HOL, then $\ulcorner$ Safe $t\ \overline{v} \urcorner$ denotes the term $\ulcorner$ Safe t (Suc 0) $\urcorner$. The uniform reflection principle for the predicate $\forall t.$ Safe $t\ v$ is:

$$\vdash \mathsf{strongly\_inaccessible}\ \mathcal{U}(:\mu) \wedge \mathcal{U}(:ind) \prec \mathcal{U}(:\mu) \Rightarrow$$
$$\forall v.\ (\mathsf{thyof}\ \mathsf{inner\_ctxt}, [])\ \vdash\ \ulcorner \forall t.\ \mathsf{Safe}\ t\ \overline{v} \urcorner \Rightarrow \forall t.\ \mathsf{Safe}\ t\ v$$

Since v is quantified in outer HOL, this theorem encapsulates infinitely many reflection theorems of the previous kind. To implement uniform reflection, we define an outer-HOL function quote : *num* $\rightarrow$ *term* which can be spliced into the result of term_to_deep to provide terms of the form $\ulcorner \overline{v} \urcorner$. It is straightforward to show that the semantics of $\ulcorner \overline{v} \urcorner$ in a reflective valuation is equal to the reflection of v, and uniform reflection follows.

8   An Implementation of Model Polymorphism

One application of reflection principles is in designing and verifying systems that include mechanisms for self-replacement. For concreteness, consider an operating system intended to satisfy a certain safety property (e.g., a certain file is never overwritten), but also with a mechanism for replacing itself by an arbitrary updated version. For this system to be safe, the replacement mechanism must be restricted to prevent replacement by an unsafe update. A simple restriction would be to require a proof that the updated version is safe until replacement is invoked, together with a syntactic check that the updated version's replacement mechanism (including the proof checker) is unchanged (or that the replacement mechanism is removed altogether). To verify this system, we need to know that the system's proof checker is sound (only admits valid proofs) and we need a reflection principle for its logic (the conclusions of valid proofs are true). But we only need to establish these properties once.

Things get more interesting if we want to allow updates that might change the replacement mechanism. A more general replacement mechanism simply requires a proof that the updated version (including any new replacement mechanisms) is safe. To verify this system, we need a reflection principle to conclude that any updated version is safe assuming only that it was proved safe before the update was made. Furthermore, if we want to leave open the possibility that later updated versions retain a general replacement mechanism, we need a reflection principle that can be iterated.

The reflection principles in the previous section do not iterate. Using the assumption that a strongly inaccessible cardinal exists and is larger than the type of individuals they let us prove results of the form "If $\ulcorner \varphi \urcorner$ is provable, then φ" for every φ not containing type variables. However, the proof of $\ulcorner \varphi \urcorner$ may only make use of the ordinary axioms of HOL: in particular, it cannot in turn assume that there is a strongly inaccessible cardinal. It is tempting to choose φ to be the formula strongly_inaccessible $\mathcal{U}(:\mu) \Longrightarrow \psi$, for some formula ψ. But this contains the type variable μ, leading our automation to produce the unsatisfiable assumption $\mathcal{U}(:\mu) \prec \mathcal{U}(:\mu)$. We could instead try to use a different type variable, as in strongly_inaccessible $\mathcal{U}(:\nu) \Longrightarrow \psi$; this would lead to a theorem showing ψ under the assumptions that strongly_inaccessible $\mathcal{U}(:\mu)$, strongly_inaccessible $\mathcal{U}(:\nu)$, $\mathcal{U}(:\nu) \prec \mathcal{U}(:\mu)$, and the provability of $\ulcorner$strongly_inaccessible $\mathcal{U}(:\nu) \Longrightarrow \psi \urcorner$. However, while the proof in inner HOL may now assume that there is *one* inaccessible (ν), the theorem in outer HOL now assumes that there are *two* strongly inaccessible cardinals (ν and μ), one of which is larger than the other.

Indeed, there must always be a stronger assumption in outer HOL than in inner HOL, since by Gödel's second incompleteness theorem, no consistent proof system as strong as Peano Arithmetic can prove the reflection principle for itself: Choosing $\varphi \equiv \mathsf{F}$, the identically false proposition, the assertion "If $\ulcorner \mathsf{F} \urcorner$ is provable, then F" is equivalent to "$\ulcorner \mathsf{F} \urcorner$ is not provable", which asserts consistency of the proof system. But by the second incompleteness theorem, a sufficiently strong proof system which can show its own consistency is inconsistent.

Nevertheless, the construction outlined above can be repeated any finite number of times, showing that if there are $(n + 1)$ nested strongly inaccessible cardinals, then a proposition φ holds if it is provable under the assumption that there are n inaccessibles. Formally, we can define a term LCA of type $num \rightarrow (\mu \rightarrow bool) \rightarrow bool$ such that LCA 0 $\mathcal{U}(:\mu)$ indicates that *ind* fits inside μ, and such that LCA (Suc t) $\mathcal{U}(:\mu)$ indicates that μ is a strongly inaccessible cardinal and there is a strictly smaller subset Q of $\mathcal{U}(:\mu)$ which satisfies LCA t Q:

LCA 0 $P \iff \mathcal{U}(:ind) \preccurlyeq P$
LCA (Suc n) $P \iff$ strongly_inaccessible $P \wedge \exists Q.\ Q \subseteq P \wedge Q \prec P \wedge$ LCA n Q

Then, we can show a reflection principle of the form "If $\ulcorner$LCA $\bar{t}$ $\mathcal{U}(:\mu) \Rightarrow \varphi \urcorner$ is provable, then LCA (Suc t) $\mathcal{U}(:\mu) \Rightarrow \varphi$". This approach systematises the idea of allowing stronger systems to reflect the reasoning of weaker systems. As described below, we can strengthen this principle so that it requires proofs not about particular numerals $\ulcorner \bar{t} \urcorner$ but about a universally quantified variable t.

To see the relevance to problems like verifying the self-replacing operating system, consider interpreting Safe t v as "candidate system v, and any updated versions it permits, behave safely for t updates", hence we want to show $\forall t.$ Safe t v_0 for the initial system v_0. Suppose we have proved Safe 0 v_0, that is, we have verified the initial system except for its replacement mechanism, and suppose the initial replacement mechanism requires a proof of the proposition $\ulcorner \forall t.$ LCA t $\mathcal{U}(:\mu) \Rightarrow$ Safe t $\bar{v} \urcorner$ before installing update v. We need, for all t, to show Safe (Suc t) v_0, but this is equivalent to showing Safe t v for all v that satisfy the replacement mechanism's proof requirement. The following reflection principle enables us to do just that, provided we assume LCA (Suc t) $\mathcal{U}(:\mu)$.

$$\vdash \forall v.$$
$$(\text{thyof inner\_ctxt}, [\,]) \mid - \ulcorner \forall t. \text{ LCA } t \, \mathcal{U}(:\mu) \Rightarrow \text{Safe } t \, \overline{v} \urcorner \Rightarrow$$
$$\forall t. \text{ LCA (Suc } t) \, \mathcal{U}(:\mu) \Rightarrow \text{Safe } t \, v$$

Furthermore, the same reflection principle can be used within the proof required by the replacement mechanism, because the increase in the value of t passed to LCA is cancelled out by the decrease in the argument to Safe when considering a new version.

To prove the theorem above (for any predicate Safe t v not containing type variables), our automation shows that for every natural number t, LCA (Suc t) $\mathcal{U}(:\mu)$ implies the existence of a model of inner HOL in which $\ulcorner$LCA t $\mathcal{U}(:\mu)\urcorner$ is true (with the inner variable t being interpreted to equal its outer value). Thus, if the implication in the assumption of the theorem is provable, then $\ulcorner$Safe t $\overline{v}\urcorner$ is true in this model as well, and by our link between inner and outer HOL, it follows that Safe t v. [12]

At the top level, we are still left with the assumption of LCA (Suc t) $\mathcal{U}(:\mu)$ for every t. A model of HOL in which this assumption is true can easily be constructed, for example, in ZFC extended with the much stronger assumption that there is a *Mahlo cardinal* [13, Chapter 8], a strongly inaccessible cardinal κ such that there are exactly κ strongly inaccessibles $\prec \kappa$. The assumption that a Mahlo cardinal exists is not uncommon in set-theoretical work, and has been used in studies of dependent type theory to justify inductive-recursive definitions [2].

9   Related Work

Reflection principles of the form "If $\ulcorner\varphi\urcorner$ is provable, then φ", as well as the uniform reflection principle "If $\ulcorner\varphi(\overline{n})\urcorner$ is provable, then $\varphi(n)$", have been studied by Turing [22] and Feferman [4]. These authors consider sequences of theories obtained by starting with a theory T_0 (Peano Arithmetic, say), and repeatedly constructing theories $T_{\alpha+1}$ by adding to T_α the reflection principle for all theories T_β, $\beta \le \alpha$. One can extend these sequences transfinitely by letting $T_\lambda := \bigcup_{\alpha<\lambda} T_\alpha$ at limit ordinals λ. Turing and Feferman showed that, in a rather technical sense, sequences constructed in this way prove every true sentence of arithmetic (see Franzén [5] for an introduction, including an explanation why this statement is not as strong as it may appear). Set theorists have been interested in reflection principles, provable in ZF, which show that a sentence φ is true if it is provable from a fixed finite subset of ZF [13].

In the interactive theorem proving community, interest in reflection principles has mainly come from the perspective of *computational* reflection [10], which

---

[12] Our proof constructs a potentially slightly different model of inner HOL for each value of t; this is, roughly, the origin of the term *model polymorphism*.

justifies the use of an efficient decision procedure by proving that if the decision procedure declares a sentence to be true, the sentence is in fact provable. For example, Allen et al. [1] extend Nuprl with a *reflection rule*, an inference rule to infer φ from "$\ulcorner\varphi\urcorner$ is provable". To avoid inconsistency, they stratify their rule in a manner similar to the transfinite progressions of Turing and Feferman: each invocation of the reflection rule is annotated with a level, ℓ, and an invocation at level ℓ requires a proof that $\ulcorner\varphi\urcorner$ is provable using the reflection rule only at levels $< \ell$. Perhaps the most systematic use of reflection to justify more complex inference rules from simpler ones is in Milawa [18], which allows its entire proof checker to be replaced by a new version when given a proof that all sentences which are provable according to the new proof checker were also provable according to the old one.

Harrison [10], reviewing a large number of arguments and proposals for computational reflection, finds no evidence that it ever makes an otherwise infeasible proof technique feasible, though he concedes that in some cases, the speed-up can be significant. More recently, Coq's `ssreflect` library [8] for computational reflection has been instrumental in the formal verification of the four-color theorem [7].

The algorithm in Sect. 4 bears a strong resemblance to the proof-producing translation (or code generation) algorithm presented by Myreen and Owens [19]. The input term in their algorithm is also an outer HOL term, but the target, instead of inner HOL, is the functional programming language CakeML. The semantics of CakeML is very different from the semantics of inner HOL, but the overall approach of producing certificate theorems bottom-up works similarly.

10   Conclusion

We have described automation for proving reflection principles of the form "If $\ulcorner\varphi\urcorner$ is provable, then φ" from assumptions about the existence of large cardinals. Based on this work, we have discussed an implementation of *model polymorphism* [3], which allows reflective reasoning to be used in the verification of self-replacing systems.

In this paper, we have focused on the automation for proving reflection principles. In future work, we plan to apply this to an implementation of a self-modifying, self-verifying system. It would also be interesting to apply these techniques to create an extensible version of HOL, similar to Milawa [18]; in particular, where Milawa requires a proof that a new version of the proof checker is *conservative* (i.e., only accepts proofs of propositions that were also provable according to the old proof checker), our semantic approach would allow us to instead require a proof that the new version is *sound* (i.e., whenever it accepts a proof of a proposition, that proposition is semantically true). It would be interesting to explore how these syntactic and semantic extension principles compare in practice.

Acknowledgements. We thank Magnus Myreen for feedback on a draft of this paper. We also thank the anonymous reviewers for their helpful criticism.

References

1. Allen, S.F., Constable, R.L., Howe, D.J., Aitken, W.E.: The semantics of reflected proof. In: Proceedings of the LICS, pp. 95–105, IEEE Computer Society (1990)
2. Dybjer, P., Setzer, A.: A finite axiomatization of inductive-recursive definitions. In: Girard, J.-Y. (ed.) TLCA 1999. LNCS, vol. 1581, pp. 129–146. Springer, Heidelberg (1999)
3. Fallenstein, B., Soares, N.: Vingean reflection. Technical report, Machine Intelligence Research Institute, Berkeley, CA (2015)
4. Feferman, S.: Transfinite recursive progressions of axiomatic theories. J. Symb. Log. **27**(3), 259–316 (1962)
5. Franzén, T.: Transfinite progressions: a second look at completeness. B. Symb. Log. **10**(3), 367–389 (2004). http://www.math.ucla.edu/~asl/bsl/1003/1003-003.ps
6. Gödel, K.: Über formal unentscheidbare Sätze der Principia Mathematica und verwandter Systeme I. Monatshefte fr Mathematik und Physik **38**(1), 173–198 (1931)
7. Gonthier, G.: The four colour theorem: engineering of a formal proof. In: Kapur, D. (ed.) ASCM 2007. LNCS (LNAI), vol. 5081, pp. 333–333. Springer, Heidelberg (2008)
8. Gonthier, G., Mahboubi, A.: An introduction to small scale reflection in Coq. J. Form. Reasoning **3**(2), 95–152 (2010)
9. Gordon, M.: From LCF to HOL: a short history. In: Plotkin, G.D., Stirling, C., Tofte, M. (eds.) Proof, Language, and Interaction, Essays in Honour of Robin Milner, pp. 169–186. The MIT Press, Cambridge (2000)
10. Harrison, J.: Metatheory and reflection in theorem proving: A survey and critique. Technical report CRC-053, SRI, Cambridge, UK (1995). http://www.cl.cam.ac.uk/~jrh13/papers/reflect.dvi.gz
11. Harrison, J.: Towards self-verification of HOL Light. In: Furbach, U., Shankar, N. (eds.) IJCAR 2006. LNCS (LNAI), vol. 4130, pp. 177–191. Springer, Heidelberg (2006)
12. Harrison, J.: HOL Light: an overview. In: Berghofer, S., Nipkow, T., Urban, C., Wenzel, M. (eds.) TPHOLs 2009. LNCS, vol. 5674, pp. 60–66. Springer, Heidelberg (2009)
13. Jech, T.: Set Theory. The Third Millenium Edition, Revised and Expanded. Springer Monographs in Mathematics. Springer, Heidelberg (2003)
14. Klein, G., Gamboa, R. (eds.): Interactive Theorem Proving. Springer, Heidelberg (2014)
15. Kumar, R., Arthan, R., Myreen, M.O., Owens, S.: HOL with definitions: semantics, soundness, and a verified implementation. In: Klein, G., Gamboa, R. (eds.) ITP 2014. LNCS, vol. 8558, pp. 308–324. Springer, Heidelberg (2014)
16. Kumar, R., Arthan, R., Myreen, M.O., Owens, S.: Self-formalisation of higher-order logic. J. Autom. Reasoning (2015), submitted. Preprint at https://cakeml.org
17. Mohamed, O.A., Muñoz, C.A., Tahar, S. (eds.): Theorem Proving in Higher Order Logics. Springer, Heidelberg (2008)
18. Myreen, M.O., Davis, J.: The reflective Milawa theorem prover is sound - (down to the machine code that runs it). In: Klein and Gamboa [14], pp. 421–436
19. Myreen, M.O., Owens, S.: Proof-producing translation of higher-order logic into pure and stateful ML. J. Funct. Program. **24**(2–3), 284–315 (2014)

20. Norrish, M., Huffman, B.: Ordinals in HOL: transfinite arithmetic up to (and beyond) ω_1. In: Blazy, S., Paulin-Mohring, C., Pichardie, D. (eds.) ITP 2013. LNCS, vol. 7998, pp. 133–146. Springer, Heidelberg (2013)
21. Slind, K., Norrish, M.: A brief overview of HOL4. In: Mohamed et al. [17], pp. 28–32
22. Turing, A.M.: Systems of logic based on ordinals. Proc. LMS **2**(1), 161–228 (1939)
23. Wenzel, M., Paulson, L.C., Nipkow, T.: The Isabelle framework. In: Mohamed et al. [17], pp. 33–38

Improved Tool Support
for Machine-Code Decompilation in HOL4

Anthony Fox[✉]

University of Cambridge, Cambridge, UK
anthony.fox@cl.cam.ac.uk

Abstract. The HOL4 interactive theorem prover provides a sound logical environment for reasoning about machine-code programs. The rigour
of HOL's LCF-style kernel naturally guarantees very high levels of assurance, but it does present challenges when it comes implementing efficient
proof tools. This paper presents improvements that have been made
to our methodology for soundly *decompiling* machine-code programs to
functions expressed in HOL logic. These advancements have been facili-
tated by the development of a domain specific language, called L3, for the
specification of Instruction Set Architectures (ISAs). As a result of these
improvements, decompilation is faster (on average by one to two orders
of magnitude), the instruction set specifications are easier to write, and
the proof tools are easier to maintain.

Traditional formal software verification has primarily focussed on developing and
using formalisations of high-level programming languages, with formal reasoning
occurring at the level of the programmer. However, in some high-assurance applications, such language formalisations could be too abstract or unrealistic, and the
trustworthiness of compilers may come into play. These issues can be addressed
by using a verified compiler, see [8,9]. However, an alternative approach is to
work directly with machine-code, which could be generated by any compiler for a
particular platform. This has the advantage that one does not have to formalise
the semantics of high-level source languages, and formal reasoning relates more
directly to the code that is actually being run. The success of this approach
hinges upon the ability to accurately formalise a processor's instruction set and
on the ability to overcome the challenges of working with low-level code, which
is less structured and replete with platform specific details. To this end, Magnus
Myreen has developed an approach for soundly *decompiling* machine-code using
the HOL4 interactive theorem prover, see [12].

Commercial instruction set architectures are large and complex, with reference manuals running to thousands of pages in length.[1] We use the L3 domain
specific language to formally specify ISAs, see [4]. Details of our current ISA
formalisations can be found in Sects. 2 and 8. Each architecture has its own idio-
syncrasies, which must be accommodated when writing proof automation for a

---

[1] For example, 2736 pages for ARMv7-A, 5242 pages for ARM-v8 (which contains a
full description of the legacy AArch32 mode) and 3020 pages for x86.

© Springer International Publishing Switzerland 2015
C. Urban and X. Zhang (Eds.): ITP 2015, LNCS 9236, pp. 187–202, 2015.
DOI: 10.1007/978-3-319-22102-1_12

theorem prover. In this paper our main working instruction set is ARMv7-A but we also support other architectures.[2] In particular, we have recently added support for the new 64-bit ARMv8 architecture.

Papers [12,13] of Myreen et al. present proof tools for decompiling machine-code programs into logic using the HOL4 interactive theorem prover. The second paper (from 2012) presented a new version of this proof-producing decompiler, which provided significant improvements in the speed of decompilation. Benchmark figures showed that the overall run time had become dominated by model evaluation, i.e. the time taken to generate Hoare triples that capture the semantics of each individual machine-code instruction. These Hoare triples take the form of *spec* theorems, which provide an interface between ISA models and the decompiler, see Sect. 1. This paper presents substantial improvements in our treatment of ISA specification and model evaluation. The techniques presented here supersede those of [5]. For comparison purposes, updated benchmark figures are provided in Sect. 7 for the examples listed in [13]. The HOL4 tools for generating *spec* theorems have been enhanced to be roughly one hundred times faster than previous versions (those presented in [5]). This speedup comes from utilising dynamic databases of generic *spec* theorems, which capture the semantics of multiple instruction instances (for multiple operating modes). These theorems are derived by partially evaluating ISA models, see Sects. 3 and 4.

The improvements described in this paper have made it more tractable to work with larger machine-code programs. The largest decompilation undertaken to date has been for the seL4 microkernel (see [15]), where the code size is roughly 12,000 ARM instructions. Our models have also been used by other organisations and research groups, including at the KTH Royal Institute of Technology, see [2].

The approach described here has matured to the point where specifying a new ISA, and linking the generated HOL model to the decompiler, is mostly routine. Experience indicates that the effort of supporting a new ISA is split almost evenly between model development (in L3) and implementing tool support (in HOL4). For simple RISC architectures, such as MIPS, preliminary support (without model validation) has been provided within a few weeks. Although our tools have been implemented in HOL4, the overall approach is applicable to other LCF-style provers, such as HOL Light, Isabelle/HOL and ProofPower.

1   Decompilation of Machine-Code to HOL Logic

In [12] Myreen describes methods for soundly decompiling machine-code into HOL functions. The decompiler outputs a collection of definitions, as well as a *certificate* theorem, which proves that the definitions correctly capture the semantics of the supplied machine-code. For example, the ARM code

```
e0010291    (*  loop: mul   r1, r1, r2    *)
e2500001    (*        subs  r0, r0, #1     *)
1afffffc    (*        bne   loop          *)
```

---

[2] L3 specifications are available at www.cl.cam.ac.uk/-acjf3/l3/isa-models.tar.bz2 and the HOL4 developments can be viewed at the Github repository www.github.com/HOL-Theorem-Prover/HOL under the directory www.examples/l3-machine-code.

is decompiled into the following HOL function:[3]

```
power (r0,r1,r2) =
    let r1 = r1 * r2 in
    let r0' = r0 - 1w
    in
        if r0 = 1w then (r0',r1,r2) else power (r0',r1,r2)
```

Certificate theorems are instances of a total correctness Hoare triple assertion SPEC *model p code q*, see [12]. The certificate theorem for our example is

```
⊢ SPEC ARM_MODEL
    (~aS * arm_OK m * arm_PC p * arm_REG (R_mode m 0w) r0 *
     arm_REG (R_mode m 1w) r1 * arm_REG (R_mode m 2w) r2 *
     cond (power_pre (r0,r1,r2)))
    {(p,0xE0010291w); (p + 4w,0xE2500001w); (p + 8w,0x1AFFFFFCw)}
    (let (r0,r1,r2) = power (r0,r1,r2)
     in
         ~aS * arm_OK m * arm_PC (p + 12w) * arm_REG (R_mode m 0w) r0 *
         arm_REG (R_mode m 1w) r1 * arm_REG (R_mode m 2w) r2)
```

Here ARM_MODEL is a 5-tuple that incorporates an ARM semantics relation. A set (the *code* pool) associates memory addresses with machine-code values. The pre- and post-conditions (*p* and *q*) are split into assertions, combined using Myreen's machine-code logic *separating conjunction operator* (*), see [12]. For example,

```
arm_OK m * arm_PC (p + 12w) * arm_REG (R_mode m 0w) r0
```

asserts that the program-counter has value p + 12w and the zeroth general-purpose register (R_mode m 0w) has value r0. The processor mode is constrained to be valid with the assertion arm_OK m, e.g. m = 16w (user mode) is valid.

The decompiler works by deriving and composing SPEC theorems for single machine-code instructions. The semantics of each instruction is determined by symbolically evaluating the next-state function of an ISA. This is implemented in HOL with *step* and *spec* tools, which are described in Sects. 3 and 4. This paper focusses on improvements made to the ISA models and associated model evaluation tools. A list of these improvements is contained in Table 1.

The implementation of ISA specific *step* and *spec* tools – which effectively link ISA models to the decompiler – is aided through the use of special purpose HOL libraries. The primitive assertion predicates (such as arm_REG) and associated lemmas are now automatically generated by a tool which examines the state type of the ISA. This automation makes it much easier to accommodate changes to specifications and to support new ISAs. However, the process of developing ISA specific tools is not fully automatic. An understanding of each ISA is required, and there are places where cognisant choices are made, i.e. which parts of the ISA model are pertinent. Naïve symbolic evaluation of sizeable ISA formalisations typically leads to the generation of unwieldy terms (perhaps taking minutes to compute in HOL). Also, it is frequently necessary to manually prove simplification rewrites, as steadfastly following ISA reference manuals can result in unwanted expressions arising within assertions and decompiler output.

---

[3] This computes r1 := r1 * pow(r2, r0). Note that 1w denotes a machine word (bit-vector) with value one.

Table 1. Comparison of the old and new approaches.

	Old	New
Specification and formal model	Native HOL using a state-transformer monad with exceptions	Specified in an imperative style using the L3 domain specific language. The exported HOL models employ let-expressions and have an 'exception status' state component. [See Sect. 2]
Step tool	Requires a full machine-code value (opcode). Evaluates the instruction set model directly. The ARM *step* tool employs a call-by-value based conversion	Mainly based on machine-code bit-patterns (named instructions). Uses a database of pre-computed 'instruction semantics' theorems, with the *step* theorem being derived using HOL's MATCH_MP rule. [See Sect. 3]
Spec tool	State assertions and associated theorems are all defined and proved manually. The theorems are based on concrete opcode values. The derivation tactic is hard-coded for each ISA	Assertions are automatically defined and various theorems are automatically derived. Supports *generic* Hoare triples, based on instruction bit-patterns. The derivation tactic is customisable, so as to support multiple ISAs. [See Sect. 4]
Improved decompiler support [13]	The Hoare *triple* theorems are derived direct from *step* theorems. This derivation is hard-coded for each ISA	The *triple* theorem derivation uses *spec* theorems. This method is generic (readily customisable), so it is easy to support new ISAs. [See Sect. 5]
Assembly code support	Written mostly in Standard ML, using a parser generator or parser combinators	Mostly written in L3, with a small amount of Standard ML. [See Sect. 6]
ISAs	ARMv4 through to ARMv7, PowerPC, x86-32, x86-64	ARMv4 through to ARMv7, ARMv6-M, ARMv8, x86-64, MIPS, CHERI. [See Sect. 8]

2   ISA Specification Using L3

The domain specific language L3 [4] provides an effective means to specify and maintain a diverse selection of instruction set models. The language has been designed to be simple and intuitive, with specifications being compact and easy to comprehend. To date, L3 models have been exported to HOL4, Standard ML, Isabelle/HOL and TSL [10].

The primary component of a formal ISA specification is a *next-state* function, which defines the architecture's operational semantics at the *programmer's model* level. One machine-code instruction is run for every application of the next-state function. A HOL4 formalisation of the ARMv7 architecture was presented in [5]. This model was based on defining a state-transformer monad (with exceptions) directly in HOL. Our new L3 source specifications now look much more like the pseudo-code found in architecture reference manuals. The exported HOL4 version is treated as a reference formal specification (trusted model), whereas exported SML code is used to implement emulators and assemblers. There is no formal or verified connection between each of these models. We primarily carry out model validation with respect to the trusted HOL model, typically using *step* tools to determine the behaviour of machine-code instructions (see Sect. 3). Additional validation and development work may be performed using the much faster ML-based emulators. In particular, our MIPS model is now capable of booting FreeBSD, see Sect. 8. This emulation work has also demonstrated that L3 models are well suited to rapidly prototyping new architectural designs.

Two styles of HOL specification can be exported from L3: one uses a generic state-transformer monad; and the other uses let-expressions, which directly manipulate (pass through) the state of the architecture. The monadic style is suited to security oriented proofs, such as [2]. This paper is primarily concerned with the let-expression style, since this is the version that is used by the decompiler. The following two sections describe this new style of specification.

Model Exceptions. Model exceptions[4] are normally used to handle various instances of *unpredictable* (under-specified) behaviour.[5] With the use of L3, our treatment of model exceptions has changed so as to make it easier to write efficient automated proof tools.

In our old monadic approach, the type of each (impure or state transforming) operation is roughly:

$$[args \rightarrow]state \rightarrow (value \times state) + exception$$

where $+$ denotes the disjoint union type operator and square parenthesis are used to indicate an optional type component. Such operations either return a value together with an updated state, or they return an exception value. A monadic *bind* operation (>>=) is used to compose sequential computations. Note that performing symbolic evaluation in the context of >>= frequently leads to the generation of unwieldy terms, since the bind definition includes a case split over the *value-state* (exception free) and *exception* cases.

The following operation type is used in L3 generated models:

$$[args \rightarrow]state' \rightarrow value \times state'$$

---

[4] This should not be confused with ISA exceptions (e.g. software interrupts), which are typically modelled explicitly, following the ISA reference manual semantics.

[5] The ARM manual describes the *unpredictable* as meaning "the behaviour cannot be relied upon". For example, the shift instruction ASR r1,r2,pc is unpredictable.

where $state' = state[\times\ exception - status]$. The range now consists of a value paired with a state, which may incorporate an exception status component. The *value* and *state* parts of the result should be regarded as meaningless when an exception occurs, which is flagged using the exception status component of $state'$. Although this representation is less principled – since junk values and states are returned when an exception has occurred – it does have the advantage that we can express operational semantics using a standard state-transformer monad or using ordinary let-expressions. This makes the generated models easier to work with.

Example Specification. Consider the following pedagogical L3 specification:

```
exception UNPREDICTABLE {- declare a new exception type -}

declare { A :: bits(8), B :: bits(8) } {- global variables -}

bits(8) example (c :: bits(8)) =
{
   A <- B;                           {- assignment -}
   when A < 4 do #UNPREDICTABLE;  {- raise an exception -}
   return (A * c + 1)
}
```

This L3 specification declares a new model exception UNPREDICTABLE, two global state components A and B, and a unary function example. The generated HOL script for this specification defines the following function:

```
example c =
(λstate.
   (let s = state with A := state.B in
    let s = if s.A < 4w then SND (raise'exception UNPREDICTABLE s)
            else s
    in
       (s.A * c + 1w,s)))
```

Note that the state is explicitly modified using let-expressions. A record type is used, with state components A and B being sub-fields of the global state. The helper function raise'exception is used to tag the state when an exception occurs (it ensures earlier exceptions are not overridden).

3   Model Evaluation: *Step* Theorems

A *step* tool uses forward proof to derive a theorem that characterises the next-state behaviour of a particular instruction. In [5], *step* theorems are of the form:

$$\vdash \forall s.\ P(s) \Rightarrow (\text{next}(s) = \text{SOME}\ (f_0(f_1(\ldots f_n(s)))))$$

where s is a state and predicate P consists of various conjuncts, usually including assertions of the from $mem(pc + i) = b_i$ for $i \in 0, \ldots, 3$. Each byte b_i forms part of the machine-code opcode of the instruction being run, which is located in memory mem at the program-counter address pc. Other clauses determine the operating mode, e.g. big- or little-endian byte ordering and so forth. Additional clauses are also required to avoid unpredictable or undesired behaviour; for example, we assert that the program-counter address is word aligned (divisible by four). The functions f_i denote state updates for particular components; for example, one function might write a value to a register. The next-state function returns an *option* type, which is a standard HOL datatype.[6]

Generic *Step* Theorems. With the move to L3 specifications, we now use *step* theorems of the form:

$$P_0, \ldots, P_n \vdash \text{next}(s) = \text{SOME } (s \text{ with } \ldots).$$

The hypotheses P_i correspond with clauses of the old *spec* theorem predicate P. The 'with' syntax denotes updates to a record, e.g. the next-state might be

$$s \text{ with } pc := s.pc + 4.$$

Here the program-counter is updated so as to point to the next instruction in memory. A key improvement is that we now generate *step* theorems that are not restricted to concrete (ground) machine-code opcodes. Instead, each byte of the fetched instruction can be represented by a list of Booleans, which form part of the instruction's opcode pattern. For example, the four bytes for the 'Multiply Accumulate' instruction MLA (with little-endian ordering) are

```
i=0    v2w [T; F; F; T; x₁₃; x₁₄; x₁₅; x₁₆]
i=1    v2w [x₅; x₆; x₇; x₈; x₉; x₁₀; x₁₁; x₁₂]
i=2    v2w [F; F; T; F; x₁; x₂; x₃; x₄]
i=3    v2w [T; T; T; F; F; F; F; F]
```

Here T represents true, F is false and the HOL function v2w : bool list $\rightarrow$ word8 constructs a bit-vector from a list. Variables are used to encode register and immediate argument values, as well as various instruction configuration options. A particular instance of this ARM instruction is MLA r1, r2, r3, r4, which has opcode 0xE0214392. This instance corresponds with the substitution:

$$x_1, x_2, x_3, x_4 \mapsto \text{F, F, F, T,} \quad (\text{r1}) \qquad x_5, x_6, x_7, x_8 \mapsto \text{F, T, F, F,} \quad (\text{r4})$$
$$x_9, x_{10}, x_{11}, x_{12} \mapsto \text{F, F, T, T,} \quad (\text{r3}) \qquad x_{13}, x_{14}, x_{15}, x_{16} \mapsto \text{F, F, T, F} \quad (\text{r2}) \ .$$

This substitution effectively specialises the instruction pattern for our instruction instance (choice of registers), i.e. r2 * r3 + r4 is written to register r1.

By developing tools that generate generic *step* theorems (which represent partial evaluations of a next-state function with respect to opcode bit-patterns), we are then able to derive generic *spec* theorems, i.e. Hoare triples for a class of instructions. This generalisation provides a means to greatly speed-up the model evaluation phase of decompilation. Rather than generate Hoare triple theorems

---

[6] A state option value is either NONE (when there is an exception) or otherwise it is SOME s for some state s. We are only interested in exception free cases here.

from scratch for every machine-code opcode (which is expensive), we can instead dynamically build up a database of generic triples that can be quickly specialised when needed.

Derivation. Generic *step* theorems are derived using HOL's MATCH_MP rule.[7] For ARM, we first derive antecedent theorems (roughly) of the form:

$$F_0, \ldots, F_f \vdash \texttt{Fetch}(s) = (v, s_0) \tag{1}$$

$$D_0, \ldots, D_d \vdash \texttt{DecodeARM}(v, s_0) = (ast, s_1) \tag{2}$$

$$\vdash \forall s.\ \texttt{Run}(ast)(s) = defn(x)(s) \tag{3}$$

$$I_0, \ldots, I_i \vdash defn(x)(s_1) = s \text{ with } \ldots \tag{4}$$

which together imply the following *step* theorem

$$F_0, \ldots, F_f, D_0, \ldots, D_d, I_0, \ldots, I_i \vdash \texttt{NextARM}(s) = \text{SOME } (s \text{ with } \ldots). \tag{5}$$

In the above, v represents a machine-code bit-pattern; ast is an instruction datatype value; x is an instruction's arguments (e.g. register indices and immediate values); and *defn* represents an instruction semantics function. The various hypotheses relate to different stages of execution, e.g. the F hypotheses assert that the bytes of the opcode v are located in main memory, starting at the program-counter address.

The functions Fetch, DecodeARM, Run and NextARM all come from the current L3 specification of ARM. Equations (1) to (4) deconstruct the next-state function NextARM, and the implication above is proved in HOL with a one-off theorem. Similar theorems are proved for each ISA that we support. It is hard to fully automate the process of decomposing next-state specifications, so as to construct and verify suitable MATCH_MP theorems. These theorems are very architecture specific, for example, the MIPS theorems have to take the branch-delay slot into consideration and the x86 model has to accommodate variable width instruction opcodes and an instruction cache.

In deriving a *step* theorem for a particular instruction class (bit-pattern), various parts of Eqs. (1) to (4) are specialised, prior to applying the MATCH_MP rule. For example, for a particular instruction instance the semantics function *defn* might be dfn'StoreByte or dfn'MultiplyLong and Eq. (4) will give the next-state semantics for that type of instruction. Equations (1) to (3) can be readily derived on the fly for any particular opcode bit-pattern. Equations (2) and (4) are precomputed for a fixed set of *supported* instruction bit-patterns. These theorems are stored in databases that are based on Michael Norrish's LVTermNet structure, which implements *local variable term nets*.[8] Using this method, the resulting *step* tool is extremely efficient. The four antecedent theorems can be

---

[7] This is the Modus Ponens inference rule with automatic matching. For example, given a theorem $A_0 \vdash t_1 \wedge \cdots \wedge t_n \Rightarrow t_0$ and a list of theorems $A_1 \vdash t_1, \ldots, A_n \vdash t_n$, we can use this rule to derive $A_0, \ldots, A_n \vdash t_0$.

[8] These are a form of *discrimination net*. A similar structure Net (credited to Larry Paulson) has been used for many years in HOL's term-rewriting conversions.

deduced very quickly (since database lookup does not require any *additional* logical inference) and an application of MATCH_MP rule is fast as well.

The practicability of this new approach hinges upon the ability to precompute all of the required instruction semantics theorems (instances of Eq. (4)) in an "acceptable" amount of time. The ARM model is complex and, at the time of writing, there are 318 of these theorems. They are produced by expanding definitions to a canonical form that consists of primitive state (record field) updates. This symbolic evaluation involves: considering instruction sub-cases; eliminating let-expressions, avoiding unpredictable cases (by adding new conditions to the set of hypotheses); and applying simplifications. A custom tool has been developed to aid this process. Where appropriate, simplification rules are manually identified and verified, e.g. they may involve reasoning about machine arithmetic and bit-vector manipulations. Here, HOL4's bit-blasting procedure for deciding bit-vector problems is of great use, see [3]. The HOL library arm_stepLib implements the ARM *step* tool; it consists of 4014 lines of hand-written code and takes just under two minutes to build. This library uses theorems from arm_stepTheory, which is built using 1498 lines of hand-written proof script.

4 Model Evaluation: Machine-Code Hoare Triples

A *spec* tool derives *spec* theorems (Hoare triples) for machine-code instructions. The performance of these tools has been greatly enhanced through the use of generic *spec* theorems, which can be instantiated to obtain triples for concrete opcodes (where instruction arguments become fixed). This new approach is illustrated in Fig. 1. A feature of this algorithm is that it incorporates a form of memoization. The performance of the tool improves with use, since the costly "no" branch in Fig. 1 only occurs when new instruction types are encountered.

Multiple generic *spec* theorems (up to sixteen for ARM) are derived for each generic *step* theorem. These theorems cover various instruction forms, e.g. MOV Rm, Rn (with Rm ≠ Rn) and MOV Rm, Rm are distinct forms. The pre- and post-conditions are determined by syntactically examining the supplied *step* theorem. The *spec* theorem derivation is based on using a carefully crafted tactic; see www.HOL/examples/l3-machine-code/common/stateScript.sml for an overview of the approach and for proofs of key lemmas.

The 'reject vacuous' stage in Fig. 1 is used to select the appropriate specialised *spec* theorems. In practice, we also apply post-processing stages, which support different state assertions, e.g. viewing registers and/or memory as maps.

To illustrate the improvements in performance, consider the 'Store Register Dual' instruction STRD r0, r1, [r2, r3]!, which has opcode 0xE1A200F3. The timings for this instruction (on a 3.4 GHz Core i7, 32 GB) are as follows:

Old *step* tool:	0.35 s
Old *spec* tool:	2.54 s
New *step* tool ("STRD (+reg,pre,wb)" instruction class):	0.0017 s
New *spec* tool (first call):	0.91 s
New *spec* tool (subsequent calls within class):	0.0027 s

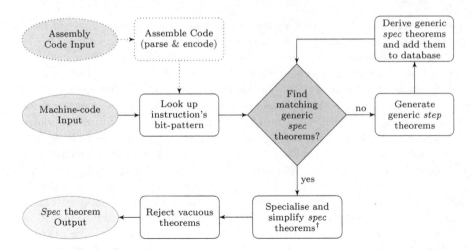

Fig. 1. Generation of *spec* theorems. † Note that theorem specialisation is fast and the simplification phase has been fine-tuned for performance.

5   Supporting the Improved Decompiler

The improved decompiler [13] uses a new Hoare triple predicate, wherein the pre- and post-condition assertions are more hardwired, i.e. fixed for a particular (manually determined) processor configuration and state 'view'. The uniformity of these *triple* theorems helps in speeding up the decompilation process. A tool has been developed to derive *triple* theorems from *spec* theorems. There is a small overhead (typically a few thousandths of a second) for this conversion. As such, it is relatively easy to support both versions of the decompiler.

6   Assembly Code Support

When working with an instruction set model, it is helpful to provide support for some standard assembly code representation of instructions (which humans can more readily comprehend). In particular, it is useful to be able to input assembly code programs (or single instructions) and then output machine-code opcodes. We achieve this by implementing light-weight assemblers, which consist of two parts: a *parser*, which maps assembly code syntax (strings) into an instruction datatype (AST); and an *encoder*, which maps instruction datatype values into machine-code opcodes. We also define *pretty-printers*, which map instruction datatype values back into assembly code syntax (strings). An example of these mappings is shown below for an ARM load-multiple instruction:

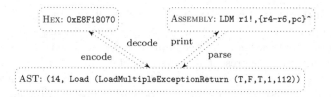

Round-trip validation is used to test the consistency of the decode, encode, parse and print specifications; this is illustrated below.

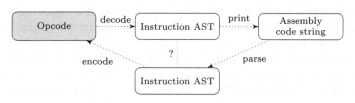

A successful round-trip occurs when the opcode produced by the encoder is the same as an initial opcode. Note that the original opcode may not be 'canonical' (e.g. ARM immediate values do not have unique encodings) — in such cases the round-trip will fail and a check is made on whether the abstract syntax for the two instructions (from decoding and parsing) are equivalent. This approach has been highly effective in terms of detecting inconsistencies and bugs in instruction representation logic. The parsing and encoding logic has also been checked laboriously with test vectors, e.g. to ensure that syntax and bounds errors are handled correctly.

We have found little need for assembly code parsers and pretty-printers to be formalised within a theorem prover, since we normally work directly with machine-code opcodes when considering the semantics of low-level programs. As such, these components are implemented at the meta-level using Standard ML. There are some use cases for formalised instruction encoders, as these can be used in compiler backends; for example, in the CakeML project [8].

Previously, encoders, parsers and printers have been written in Standard ML (HOL's meta-language), see [5]. In particular, parsing has been implemented using parser generators and later with parsing combinators. We now specify encoding, parsing and printing using L3, which helps in keeping these components consistent and synchronised with the core model. L3 is well-suited to specifying instruction encoders, since the language provides excellent support for working with bit-vectors. The complete specification is exported to ML and this is then used to write simple assemblers, using a relatively small piece of hand coded ML. The parser and printer specifications are not exported to HOL.

7   Performance

In presenting a faster decompilation algorithm, Myreen et al. provided some benchmarks figures, see Table 2. Following the changes presented in this paper, the latest performance figures are shown in Table 3. Column one shows that the performance of the decompiler itself, which excludes the model evaluation phase, has improved since 2012.[9] This has been achieved by implementing simple coding improvements within the decompiler and associated library code. The underlying algorithms, as presented by Myreen et el., have not been modified. Of most interest is the model evaluation figures (column group two). Three timings are presented: the first column contains updated figures for the old model and tools; the second corresponds with generating *spec* theorems from a 'cold' state (where no instructions have been encountered before) using the new L3 model and tools; and the third is from a 'warm' state, where all of the instructions have been encountered before.[10] It is clear that significant performance improvements have been achieved. The old tools (see [5]) have again improved due to tweaks within library code. For the new tools, the performance is at worst just a bit slower than before (i.e. in the 'sum of array' example), however this corresponds with an outlying case where just four instruction are processed from a cold state. From a 'warm' state, model evaluation is now faster than the main decompilation phase itself. As such, this new technology has successfully overcome a bottleneck in scalability. The faster and more general *step* tools also help in areas such as model validation and compiler verification.

Table 2. Performance figures found in [13]. Examples: (1) sum of array; (2) copying garbage collector; (3) 1024-bit multiword addition; and (4) 256-bit Skein hash function.

| Example | Code size (instructions) | Decompilation time (and inferences) | | Model evaluation time |
		Original version	2012 version	(and inferences)
1	4	2.5 s (73039 i)	0.3 s (4019 i)	7.8 s (1.5 Mi)
2	89	50 s (1526281 i)	6.0 s (53301 i)	173 s (40 Mi)
3	224	70 s (1029685 i)	1.2 s (10802 i)	37 s (8.9 Mi)
4	1352	5366 s (21432926 i)	56 s (1,842,642 i)	500 s (105 Mi)

Table 3. Latest performance figures for the same ARM examples.

| Decompilation time (and inferences) | | Model evaluation time (and inferences) | | |
Original version	2012 version	Old	Cold	Warm
1.47 s (78,688 i)	0.12 s (16481 i)	3.2 s (0.74 Mi)	9.8 s (3.9 Mi)	0.02 s (15,521 i)
32.5 s (1,598,271 i)	2.0 s (349740 i)	51.3 s (16.2 Mi)	19.8 s (13.5 Mi)	0.35 s (273725 i)
50.0 s (1,085,104 i)	0.3 s (45949 i)	20.3 s (3.3 Mi)	9.9 s (9 Mi)	0.03 s (8435 i)
11786 s (19,921,648 i)	8.0 s (4756617 i)	350 s (44.6 Mi)	23.7 s (28 Mi)	4.0 s (2.8 Mi)

---

[9] The Skein example under the 'original' decompiler is an anomaly here.

[10] Caching on hexadecimal opcode values has not been enabled, so these run times correspond with specialising generic *spec* theorems.

Table 4. L3 instruction set models. The 'Lines of L3' figure is split into *core model* and *additional logic* (for instruction encoding and assembly code support).

ISA	Operating Modes	System Levels	Instruction Width	General-Purpose Registers	Flags	Endianness	Coverage	Lines of L3
ARMv4 to ARMv7	ARM, Thumb	User, System, (Hypervisor), Abort, Undefined, (Monitor), IRQ, FIQ	32-bit, 16-bit	32-bit 16† (banked)	N, Z, C, V, Q, GE	Big, Little	Partial VFP No Adv. SIMD No CP	9238+7687
ARMv6-M	Thumb	Main, Process	16-bit	32-bit 16† (SP banked)	N, Z, C, V	Big, Little	No CP	1996+2095
ARMv8	AArch64 only	EL0, EL1, EL2, EL3	32-bit	64-bit 32 (reg. 31 is 0)	N, Z, C, V	Big, Little	No VFP No Adv. SIMD Partial System	2434+4097
x86-64	64-bit only	-	Variable (bytes)	64-bit, 16	CF, PF, AF, ZF, SF, OF	Little	40 basic instructions	1357+1579
MIPS (RS4000)	-	User, Supervisor, Kernel	32-bit	64-bit 32 (reg. 0 is 0)	-	Big, Little	Partial CP Partial System	2080+700
CHERI	-	User, Supervisor, Kernel	32-bit	64-bit 32 (reg. 0 is 0)	-	Big	Partial CP	5299

8   Instructions Sets

A summary of our ISA formalisations in L3 is shown in Table 4. The following sections give a brief overview of these architectures.

ARMv4 through to ARMv7. Although nominally a RISC architecture, ARM is challenging from a specification and verification perspective. Areas of complexity are: the number of system levels and the use of banked general-purpose registers; the program-counter is a general-purpose register,[11] which leads to *unpredictable* behaviour (see Sect. 2); and the LDM and STM block data transfer instructions are remarkably elaborate. As part of previous work, described in [5], a fairly large set of single instruction test vectors were developed for the purposes of ARMv7 validation. These tests have helped identify a handful of bugs in the new L3 specification, which were all trivial to fix. The new model is now considered to be as trustworthy as the previous version, which was specified directly in HOL. We have no plans to formally verify a correspondence between the two models.

A notable change to the new specification is with regard to the specification of *unpredictable* and *undefined*[12] instruction instances. This logic has moved from instruction semantics functions to decoders. As such, the decoders can be used to check the validity of instruction encodings, which is useful when writing an assembler, see Sect. 6. The instruction semantics functions have also become easier to work with.

---

[11] When an instruction explicitly reads the PC this *normally* gives the value of that instruction's address plus eight (in ARM mode) or plus four (in Thumb mode). This is because early ARM processors employed a 3-stage pipeline.

[12] An instruction opcode is *undefined* when it is not supported by an architecture version or configuration. For example, CLZ opcodes on ARMv4 will raise an undefined exception trap.

ARMv6-M. This architecture is implemented by the Cortex-M0 micro-controller and our model includes processor timing information (a cycle counter). Extensive model validation has been carried out by Brian Campbell [1].

ARMv8. This is a completely new 64-bit architecture. Although compatibility with ARMv7 is provided with an AArch32 operating mode, our L3 formalisation omits this functionality and just supports the new AArch64 mode. The instruction set is completely new; as is the underlying programmer's model. Our formalisation is currently awaiting validation. Due to some rationalisations in the ARMv8 architecture (in AArch64 mode), this ISA is actually cleaner and easier to work with when compared to ARMv7. There are still some complexities, e.g. the various encoding schemes for immediate values are somewhat elaborate.

x86-64. Being an older CISC architecture, the x86 family is renowned for its size and complexity. However, we only provide a comparatively simple model of x86-64 in L3, which covers just 64-bit operating mode for a core set of about forty instructions (providing adequate coverage for case studies). This was ported from a native HOL4 specifications. Some limited model validation with respect to hardware has been carried out with the assistance of Magnus Myreen.

MIPS64 and CHERI. MIPS64 is a relatively clean RISC architecture. The CHERI research architecture extends MIPS with capabilities for implementing security management, see [17]. One source of complexity for MIPS is the presence of a branch-delay slot, which affects the semantics of jump instructions. The CHERI model permits multi-core operation and also adds support for: interrupts; UART I/O; a translation lookaside buffer (TLB); and more coprocessor instructions. This advanced, high-fidelity model is primarily used for emulation, validation and rapid prototyping. The model is mature enough that it can boot an unmodified development version of FreeBSD (which has a 7.2 MB image size) in about ten minutes on a modern machine. Booting the OS in multi-core mode is supported as well. Alexandre Joannou, Matthew Taylor and Mike Roe have worked on the extended MIPS and CHERI models, and this development can be found on Github at www.github.com/acjf3/l3mips.

9   Related Work

Other recent work on reasoning about machine-code programs has mainly been undertaken using the ACL2 and Coq theorem provers. Most of this work has focussed on the x86 architecture. Related work is carried out in the field of binary-analysis (using flow-based approaches), where instruction set models are developed and used in less formal settings, i.e. where machine-code programs are not formally verified against specifications.

Warren Hunt's group have developed a high quality model of x86-64 using ACL2, see [6]. Shilpi et al. report that their ACL2 model covers 121 user-level instructions (much more than our L3 specification) and they note that there is work in progress (headed by Moore) on an ACL2 based automated tool, called Code Walker, that is comparable with Myreen's decompiler. By design, the ACL2 programming language naturally supports fast model evaluation. Their x86 model can be tested in an *execution mode* and proofs can be constructed in a *logical mode*. With the former evaluation mode, they achieve a performance of nearly a million instructions per second, with a two-level page table enabled. By contrast, HOL4 is an LCF-style theorem prover that is not designed for high performance model evaluation. As such, when carrying out emulation work we generate Standard ML versions of our models. When reasoning in the HOL logic, the techniques presented in this paper provide sufficient performance for formal verification work. We consider the main advantages of our approach to be that our ISA models are compact and accessible (through the use of a domain specific language); and also that our infrastructure for supporting automated decompilation to logic (for multiple ISAs) is now relatively mature.

As part of the Rocksalt project (for a software-based fault isolation checker), Morrisett et al. have developed an x86 model in Coq, see [11]. Other x86 models have been developed in Coq as part of research into devising logical frameworks for reasoning about low-level code, see [7,14]. In addition, simple assembly-level Coq models of x86, PowerPC and ARM have also been used within the CompCert verified compiler, see [9].

Within the area of binary-analysis, the work of Reps et al. is of note, see [10]. They use a domain specific language TSL to specify ISAs, including PowerPC and x86. Recently, they have customised L3, so as to generate TSL code from our ARMv7 model. These TSL specifications are used to generate a range of binary-analysis tools. Related work includes the GDSL toolkit of Simon et al., see [16]. They have a used a domain specific language to specify (fast) decoders, as well as semantics translators, e.g. for x86 and Atmel AVR. At present it is unclear how easy it would be to adapt their work for the purposes of formal verification using an interactive theorem prover.

10 Summary

This paper describes the current status of our models, tools and methodology for the formal verification of machine-code programs. Our approach is based on using three programming environments: L3 for developing ISA specifications; Standard ML (compiled using Poly/ML or MLton) for efficient emulation; and HOL4 for formal reasoning. L3 has eased the task of developing instruction set specifications, and it has also helped ensure that generated HOL models are of a known and manageable form. Improved techniques for model evaluation are presented and proof tools have been implemented. The performance of machine-code decompilation has been greatly enhanced, see Sect. 7. These gains have been achieved by maintaining various databases of pre-proved theorems, see Sects. 3 and 4. In particular, the *spec* tool now maintains a database of generic *spec* theorems. These methods are applicable to other LCF-style theorem provers.

Thanks to Mike Gordon, Magnus Myreen and the reviewers for providing helpful comments on drafts of this paper.

References

1. Campbell, B., Stark, I.: Randomised testing of a microprocessor model using SMT-solver state generation. In: Lang, F., Flammini, F. (eds.) FMICS 2014. LNCS, vol. 8718, pp. 185–199. Springer, Heidelberg (2014)
2. Dam, M., Guanciale, R., Nemati, H.: Machine code verification of a tiny ARM hypervisor. In: Sadeghi, A., Armknecht, F., Seifert, J. (eds.) TrustED 2013, pp. 3–12. ACM, New York (2013)
3. Fox, A.C.J.: LCF-Style bit-blasting in HOL4. In: van Eekelen, M., Geuvers, H., Schmaltz, J., Wiedijk, F. (eds.) ITP 2011. LNCS, vol. 6898, pp. 357–362. Springer, Heidelberg (2011)
4. Fox, A.: Directions in ISA specification. In: Beringer, L., Felty, A. (eds.) ITP 2012. LNCS, vol. 7406, pp. 338–344. Springer, Heidelberg (2012)
5. Fox, A., Myreen, M.O.: A trustworthy monadic formalization of the ARMv7 instruction set architecture. In: Kaufmann, M., Paulson, L.C. (eds.) ITP 2010. LNCS, vol. 6172, pp. 243–258. Springer, Heidelberg (2010)
6. Goel, S., Hunt, Jr., W.A., Kaufmann, M., Ghosh, S.: Simulation and formal verification of x86 machine-code programs that make system calls. In: FMCAD 2014, pp. 91–98. IEEE (2014)
7. Jensen, J.B., Benton, N., Kennedy, A.: High-level separation logic for low-level code. In: Giacobazzi, R., Cousot, R. (eds.) POPL 2013, pp. 301–314. ACM, New York (2013)
8. Kumar, R., Myreen, M.O., Norrish, M., Owens, S.: CakeML: a verified implementation of ML. In: Jagannathan, S., Sewell, P. (eds.) POPL 2014, pp. 179–192. ACM, New York (2014)
9. Leroy, X.: Formal verification of a realistic compiler. Commun. ACM $52(7)$, 107–115 (2009)
10. Lim, J., Reps, T.W.: TSL: a system for generating abstract interpreters and its application to machine-code analysis. ACM Trans. Program. Lang. Syst. $35(1)$, 4 (2013)
11. Morrisett, G., Tan, G., Tassarotti, J., Tristan, J., Gan, E.: Rocksalt: better, faster, stronger SFI for the x86. In: Vitek, J., Lin, H., Tip, F. (eds.) PLDI 2012, pp. 395–404. ACM, New York (2012)
12. Myreen, M.O., Gordon, M.J.C.: Hoare logic for realistically modelled machine code. In: Grumberg, O., Huth, M. (eds.) TACAS 2007. LNCS, vol. 4424, pp. 568–582. Springer, Heidelberg (2007)
13. Myreen, M.O., Gordon, M.J.C., Slind, K.: Decompilation into logic - improved. In: Cabodi, G., Singh, S. (eds.) FMCAD, pp. 78–81. IEEE (2012)
14. Ni, Z., Yu, D., Shao, Z.: Using XCAP to certify realistic systems code: machine context management. In: Schneider, K., Brandt, J. (eds.) TPHOLs 2007. LNCS, vol. 4732, pp. 189–206. Springer, Heidelberg (2007)
15. Sewell, T.A.L., Myreen, M.O., Klein, G.: Translation validation for a verified OS kernel. In: Boehm, H.J., Flanagan, C. (eds.) PLDI, pp. 471–482. ACM, New York (2013)
16. Simon, A., Kranz, J.: The GDSL toolkit: Generating frontends for the analysis of machine code. In: Jagannathan, S., Sewell, P. (eds.) PPREW 2014, p. 7. ACM, New York (2014)
17. Woodruff, J., Watson, R.N.M., Chisnall, D., Moore, S.W., Anderson, J., Davis, B., Laurie, B., Neumann, P.G., Norton, R., Roe, M.: The CHERI capability model: revisiting RISC in an age of risk. In: ISCA 2014, pp. 457–468. IEEE Computer Society (2014)

A Formalized Hierarchy of Probabilistic System Types
Proof Pearl

Johannes Hölzl[1][✉], Andreas Lochbihler[2][✉], and Dmitriy Traytel[1][✉]

[1] Fakultät Für Informatik, Technische Universität München, Munich, Germany
{hoelzl,traytel}@in.tum.de
[2] Department of Computer Science, Institute of Information Security, ETH Zurich, Zurich, Switzerland
andreas.lochbihler@inf.ethz.ch

Abstract. Numerous models of probabilistic systems are studied in the literature. Coalgebra has been used to classify them into system types and compare their expressiveness. In this work, we formalize the resulting hierarchy of probabilistic system types in Isabelle/HOL by modeling the semantics of the different systems as codatatypes. This approach yields simple and concise proofs, as bisimilarity coincides with equality for codatatypes. On the way, we develop libraries of bounded sets and discrete probability distributions and integrate them with the facility for (co)datatype definitions.

1 Introduction

The framework of coalgebra provides a unified view on various ways of modeling (probabilistic) systems [2,20,21,24]. A system is represented as a function of type $\sigma \Rightarrow \sigma\,\mathsf{F}$ that describes the possible evolutions of a state of type σ. Here, the functor F (written postfix) determines the type of the system. For example, a non-deterministic labeled transition system corresponds to a function $\sigma \Rightarrow (\alpha \Rightarrow \sigma\,\mathsf{set})$, which returns the set of the possible successor states for each label of type α. Similarly, a Markov chain can be characterized by a function from a state to the probability distribution over the successor states. More complicated types combine non-deterministic and probabilistic aspects in different ways.

Bartels et al. [2] and Sokolova [21] compare the expressiveness of system types found in the literature and arrange them in a hierarchy. They define a type of systems to be at least as expressive as another if every system of the latter can be mapped to a system of the former such that the mapping preserves and reflects bisimilarity, where two systems are bisimilar iff they cannot be distinguished by finite observations [17].

In this paper, we formalize the probabilistic system types (Sect. 5) and their hierarchy (Sect. 6) in Isabelle/HOL. The salient feature is that we model the system types as codatatypes (Sect. 2) rather than functions as done in the original

© Springer International Publishing Switzerland 2015
C. Urban and X. Zhang (Eds.): ITP 2015, LNCS 9236, pp. 203–220, 2015.
DOI: 10.1007/978-3-319-22102-1_13

proofs [21]. On codatatypes, bisimilarity coincides with equality, which allows for convenient equational reasoning. This makes the proofs simple and concise, i.e., highly automated and without lots of technical clutter. Our formalization is publicly available [14].

To construct the codatatypes, we introduce new types to express non-deterministic and probabilistic choice, namely bounded (non-empty) powerset (Sect. 3) and discrete probability distributions (Sect. 4). We integrate them with Isabelle's new package for (co)datatypes [4, 6, 22]. Thus, we can define the codatatypes directly, which demonstrates the versatility of the new package. Moreover, future formalizations [18, 25] benefit, too, as recursion in (co)datatypes may now occur under discrete distributions and bounded sets.

Our work is more than just an exercise in formalization. We extend the original hierarchy with additional standard system types and discover new interconnections by considering systems extended with an additional label (Sect. 6.3). Moreover, the formalization has revealed a flaw in the original hierarchy proof. We show that Vardi systems (also known as concurrent Markov chains [23]) do not satisfy the assumptions required in [21] and therefore must be (partially) dismissed from the hierarchy (Sect. 6.4).

2   Preliminaries: Codatatypes via Bounded Natural Functors

The flexibility of Isabelle's (co)datatype package originates from a semantic criterion that defines where (co)recursion may appear on the right-hand side of a (co)datatype declaration (in contrast to syntactic criteria employed by most if not all other proof assistants including past versions of Isabelle).

The core of the semantic criterion relies on the notion of a *bounded natural functor* (BNF) [4, 22]. Here, we shortly introduce BNFs targeted at our application. A (unary) BNF is a type expression α F with a type parameter α equipped with a polymorphic map function $\mathsf{map}_F :: (\alpha \Rightarrow \beta) \Rightarrow \alpha\ F \Rightarrow \beta\ F$, a polymorphic set function $\mathsf{set}_F :: \alpha\ F \Rightarrow \alpha\ \mathsf{set}$, and an infinite cardinal bound bd_F on the number of elements returned by set_F. Additionally, these constants must satisfy certain properties (e.g., map_F is *functorial*, i.e., preserves identities and composition, and set_F is a *natural transformation*, i.e., commutes with map_F). For example, the type of (finite) lists α list forms a BNF with the standard map function map and the function set returning the set of the list's elements.

The semantic criterion allows (co)recursion to occur nested under BNFs. For example, the (co)datatypes α tree and α ltree of finitely branching trees nest the (co)recursive occurrences of α tree and α ltree under the BNF list:

> **datatype** α tree = Node α (α tree list)
> **codatatype** α ltree = Node α (α ltree list)

While only trees of finite depth inhabit the datatype α tree, the codatatype α ltree also hosts trees of infinite depth. For example, the full binary

tree z containing 0 everywhere is defined by primitive corecursion [4]:
primcorec z :: nat ltree **where** z = Node 0 [z, z].

Users can register custom types as BNFs by supplying the required constants and discharging the proof obligations for the BNF properties. Newly registered BNFs can then participate in further (co)datatype declarations. For example, after registering Isabelle's type of finite sets α fset as a BNF, we can define unordered finitely branching trees of potentially infinite depth:
codatatype α lftree = Node α (α lftree fset).

In general, BNFs can have arbitrary arity and may depend on additional *dead* type variables that are ignored by the map function. For example, the sum and product types are binary BNFs, while the function type $\alpha \Rightarrow \beta$ is a unary BNF with the dead variable α (BNFs thereby disallow recursion through negative positions [10]). Compositions of BNF are again BNFs. We say that a BNF α F *induces* a codatatype C_F

$$\textbf{codatatype } C_F = Ctr_F\ (C_F\ F)$$

with a single bijective *constructor* $Ctr_F :: C_F\ F \Rightarrow C_F$, its inverse *destructor* $Dtr_F :: C_F \Rightarrow C_F\ F$ and the associated *coiterator* $unfold_F$ defined by primitive corecursion:

primcorec $unfold_F :: (\alpha \Rightarrow \alpha\ F) \Rightarrow \alpha \Rightarrow C_F$ **where**
  $Dtr_F\ (unfold_F\ s\ a) = map_F\ (unfold_F\ s)\ (s\ a)$

Finally, induced codatatypes are equipped with a *coinduction* rule for proving equality by exhibiting a *bisimulation relation witness R*:

$$\frac{R\ x\ y \quad \forall x\ y.\ R\ x\ y \longrightarrow rel_F\ R\ (Dtr_F\ x)\ (Dtr_F\ y)}{(x :: C_F) = (y :: C_F)}$$

where the *relator* $rel_F :: (\alpha \Rightarrow \beta \Rightarrow bool) \Rightarrow \alpha\ F \Rightarrow \beta\ F \Rightarrow bool$ lifts binary relations over elements to binary relations over the functor F. It is defined for each BNF canonically in terms of map_F and set_F (where π_1 and π_2 denote the standard product projections):

$$rel_F\ R\ x\ y = \exists z.\ set_F\ z \subseteq \{(x, y) \mid R\ x\ y\} \land map_F\ \pi_1\ z = x \land map_F\ \pi_2\ z = y \quad (1)$$

3 Bounded Powerset

In this and the next section we define three new types and register them as BNFs. We start with the simpler two: bounded sets and non-empty bounded sets, with which we will model non-determinism on a state space. Our new type and its BNF structure generalize the existing BNFs for finite sets α fset and countable sets α cset in Isabelle/HOL's library. Note that Isabelle's standard type of unbounded sets α set is not a BNF, due to the absence of a cardinal bound on the number of elements contained in a set.

As for bounded sets, we cannot directly express the dependence of a type on a cardinal bound constant within the simply typed logic of Isabelle. A standard trick [11] is to let the type depend on a type κ (and thereby on κ's cardinality) instead. We obtain the following type definitions for the type α set^κ of strictly κ-bounded sets:

typedef α $\mathsf{set}^\kappa = \{A :: \alpha \text{ set} \mid |A| <_\mathsf{o} |\mathsf{UNIV} :: \kappa \text{ set}| +_\mathsf{c} \aleph_0\}$

The operators $|-|$, $<_\mathsf{o}$, $+_\mathsf{c}$ and the constant $\aleph_0 = |\mathsf{UNIV} :: \text{nat set}|$ are part of Isabelle's library of cardinals [5]—their exact definition is irrelevant; they encode the intuition that α set^κ contains all sets of strictly smaller cardinality than κ if κ is an infinite type (in which case $|\mathsf{UNIV} :: \kappa \text{ set}| +_\mathsf{c} \aleph_0 =_\mathsf{o} |\mathsf{UNIV} :: \kappa \text{ set}|$) and all finite sets otherwise (since $|\mathsf{UNIV} :: \kappa \text{ set}| +_\mathsf{c} \aleph_0 =_\mathsf{o} \aleph_0$ for finite κ). In other words: If we instantiate κ with a finite or countable type, then α set^κ is isomorphic to α fset, and if we instantiate κ with the cardinal successor of $\aleph_0$ [5], then α set^κ is isomorphic to α cset.

It is easy to define the map and set function for α set^κ using the Lifting tool [15]:

lift-definition $\mathsf{map}_\mathsf{set} :: (\alpha \Rightarrow \beta) \Rightarrow \alpha \ \mathsf{set}^\kappa \Rightarrow \beta \ \mathsf{set}^\kappa$ **is** image
lift-definition $\mathsf{set}_\mathsf{set} :: \alpha \ \mathsf{set}^\kappa \Rightarrow \alpha \ \text{set}$ **is** id

The map function only acts on the element type α, which implies that κ will be a dead type variable of the following BNF structure. The bound for the set size in the above **typedef** command serves as bound for the BNF, too.

bnf α set^κ map: $\mathsf{map}_\mathsf{set}$ set: $\mathsf{set}_\mathsf{set}$ bd: $|\mathsf{UNIV} :: \kappa \text{ set}| +_\mathsf{c} \aleph_0$

To finish the registration of α set^κ as a BNF, the **bnf** command requires the user to discharge the following proof obligations. (The proofs of these properties are straightforward generalizations of the ones for α fset.)

$$\mathsf{map}_\mathsf{set} \ \mathsf{id} = \mathsf{id} \qquad\qquad\qquad \mathsf{map}_\mathsf{set} \ (f \circ g) = \mathsf{map}_\mathsf{set} \ f \circ \mathsf{map}_\mathsf{set} \ g$$

$$|\mathsf{set}_\mathsf{set} \ X| \leq_\mathsf{o} |\mathsf{UNIV} :: \kappa \text{ set}| +_\mathsf{c} \aleph_0 \qquad \mathsf{set}_\mathsf{set} \circ \mathsf{map}_\mathsf{set} \ f = \text{image} \ f \circ \mathsf{set}_\mathsf{set}$$

$$(\forall x \in \mathsf{set}_\mathsf{set} \ X. \ f \ x = g \ x) \longrightarrow \mathsf{map}_\mathsf{set} \ f \ X = \mathsf{map}_\mathsf{set} \ g \ X$$

$$\mathsf{rel}_\mathsf{set} \ R \ \infty \ \mathsf{rel}_\mathsf{set} \ S \sqsubseteq \mathsf{rel}_\mathsf{set} \ (R \infty S)$$

The first five being easy to discharge, the last proof obligation requires some explanation: $\sqsubseteq$ denotes implication lifted to binary predicates and ∞ denotes the relational composition of binary predicates. With this definition the last proof obligation is equivalent to what in categorical jargon is called *weak pullback preservation*. We can show that bounded sets preserve weak pullbacks iff the bound on the number of elements is infinite or ≤ 2. In our case, the bound is infinite due to the addition of $\aleph_0$, therefore α set^κ is a BNF. This corrects an earlier claim that α set^κ is a BNF for all κ [22].

Similarly to α set^κ, we define (and prove being a BNF) the type α set_1^κ of nonempty strictly κ-bounded sets which will be used to model Markov decision processes.

typedef α $\mathsf{set}_1^\kappa = \{A :: \alpha \text{ set} \mid A \neq \varnothing \wedge |A| <_\mathsf{o} |\mathsf{UNIV} :: \kappa \text{ set}| +_\mathsf{c} \aleph_0\}$

4 Probability Mass Functions

We introduce a type of *probability mass functions* (*pmf*) on a type α, representing distributions of discrete random variables on α. There are two views on a pmf: (1) as a non-negative real-valued function which sums up to 1, and (2) as a discrete probability measure which has a countable set S which has probability 1. Both views are available in our formalization. In this paper, however, we only present the measure view, as we lift all presented definitions from the existing formalization of measure theory [13].

A measure $M :: \alpha$ measure consists of a σ-algebra of measurable sets sets M and a measure function $\mu\ M$ that is non-negative and countably-additive on sets M. A probability distribution M assigns 1 to the whole space ($\mu\ M$ UNIV $= 1$). It is discrete iff every set is measurable (sets $M =$ UNIV) and there exists a countable set S with $\mu\ M\ S = 1$.

> **typedef** α pmf $= \{M :: \alpha$ measure $|$
> $\quad \mu\ M$ UNIV $= 1 \wedge$ sets $M =$ UNIV $\wedge (\exists S.$ countable $S \wedge \mu\ M\ S\ = 1)\}$

The command **typedef** generates a representation function measure$_{\text{pmf}} :: \alpha$ pmf$\Rightarrow \alpha$ measure. By declaring it as a coercion function, we can omit it in most cases. In particular, we write $\mu\ p\ A$ for μ (measure$_{\text{pmf}}\ p$) A. So, the probability mass of a value x is the measure of its singleton set $\{x\}$. We lift the support set from the measure definition:

> **lift-definition** set$_{\text{pmf}} :: \alpha$ pmf $\Rightarrow \alpha$ set is $\lambda M.\ \{x \mid \mu\ M\ \{x\} \neq 0\}$

Next, we lift the monadic operators bind$_{\text{pmf}}$ and return$_{\text{pmf}}$ from the Giry monad on measure spaces [8] to pmfs. The map function map$_{\text{pmf}}$ is then defined in a standard way as a combination of these monadic operators.

> **lift-definition** bind$_{\text{pmf}} :: \alpha$ pmf $\Rightarrow (\alpha \Rightarrow \beta$ pmf$) \Rightarrow \beta$ pmf is bind
> **lift-definition** return$_{\text{pmf}} :: \alpha \Rightarrow \alpha$ pmf is return (count-space UNIV)
> **definition** map$_{\text{pmf}} :: (\alpha \Rightarrow \beta) \Rightarrow \alpha$ pmf $\Rightarrow \beta$ pmf **where**
> $\quad$ map$_{\text{pmf}}\ f\ M =$ bind$_{\text{pmf}}\ M\ (\lambda x.$ return$_{\text{pmf}}\ (f\ x))$

When working with general measure spaces, all functions must be shown to be measurable. In our restricted discrete setting all function are trivially measurable, hence characteristic theorems about bind$_{\text{pmf}}$ and return$_{\text{pmf}}$ carry no measurability assumptions:

> bind$_{\text{pmf}}$ (bind$_{\text{pmf}}\ M\ f$) $g =$ bind$_{\text{pmf}}\ M\ (\lambda x.$ bind$_{\text{pmf}}\ (f\ x)\ g)$
> bind$_{\text{pmf}}$ (return$_{\text{pmf}}\ x$) $f = f\ x$
> bind$_{\text{pmf}}\ M$ return$_{\text{pmf}} = M$

The behavior of bind$_{\text{pmf}}$ and return$_{\text{pmf}}$ under set$_{\text{pmf}}$ is as expected:

> set$_{\text{pmf}}$ (bind$_{\text{pmf}}\ M\ f$) $= \bigcup_{x \in \text{set}_{\text{pmf}}\ M}$ set$_{\text{pmf}}\ (f\ x)$
> set$_{\text{pmf}}$ (return$_{\text{pmf}}\ x$) $= \{x\}$
> $(\forall x \in$ set$_{\text{pmf}}\ M.\ f\ x = g\ x) \longrightarrow$ bind$_{\text{pmf}}\ M\ f =$ bind$_{\text{pmf}}\ M\ g$

Another standard construction in probability theory is the conditional probability $\Pr(X \in A \mid X \in B) = \Pr(X \in A \wedge X \in B) / \Pr(X \in B)$, i.e. the probability that the random variable X has a result in A under the assumption that X is in B. This requires that X being in B has positive probability. In Isabelle's measure theory, the function uniform-measure expresses a conditional probability. It returns a probability space when the measure of the set B is positive. Clearly, lifting uniform-measure to pmfs works only if we restrict B to such sets. Therefore, we fix a pmf p and a set B with the assumption $\mathsf{set}_{\mathsf{pmf}}\, p \cap B \neq \varnothing$, which is equivalent to $\mu\, p\, B \neq 0$.

lift-definition $\mathsf{cond}_{\mathsf{pmf}} :: \alpha$ pmf **is** uniform-measure ($\mathsf{measure}_{\mathsf{pmf}}\, p$) B

Whenever $\mathsf{set}_{\mathsf{pmf}}\, p \cap B \neq \varnothing$ holds we will from now on write $\mathsf{cond}_{\mathsf{pmf}}\, p\, B$. We then have $\mu\, (\mathsf{cond}_{\mathsf{pmf}}\, p\, B)\, A = \mu\, p\, (A \cap B)\, /\, \mu\, p\, B$ and hence $\mathsf{set}_{\mathsf{pmf}}(\mathsf{cond}_{\mathsf{pmf}}\, p\, B) = \mathsf{set}_{\mathsf{pmf}}\, p \cap B$.

Probability Mass Functions as a BNF. We now prove that α pmf is a BNF such that the codatatypes for the probabilistic systems can recurse through α pmf. To that end, we define the relator $\mathsf{rel}_{\mathsf{pmf}}$ on pmfs and prove that $\mathsf{set}_{\mathsf{pmf}}$, $\mathsf{map}_{\mathsf{pmf}}$, and $\mathsf{rel}_{\mathsf{pmf}}$ satisfy the BNF properties. The definition of $\mathsf{rel}_{\mathsf{pmf}}\, R\, p\, q$ is canonical as in (1). The existentially quantified z corresponds to a matrix of non-negative reals with a row and a column for each element in the support of p and q, respectively, such that (i) summing over a row i or a column j yields the mass of p or q concentrated in i or j, and (ii) positive entries are only at cells (i, j) for which $R\, i\, j$ holds. We call such a matrix an R-lifting matrix for p and q.

With the lemmas about $\mathsf{bind}_{\mathsf{pmf}}$, $\mathsf{return}_{\mathsf{pmf}}$, $\mathsf{set}_{\mathsf{pmf}}$ and the definition of $\mathsf{map}_{\mathsf{pmf}}$ we immediately derive the functorial BNF properties for pmfs with the cardinal bound $\aleph_0$. Only distributivity with composition has interesting proof, i.e., $\mathsf{rel}_{\mathsf{pmf}}\, R \infty \mathsf{rel}_{\mathsf{pmf}}\, S \sqsubseteq \mathsf{rel}_{\mathsf{pmf}}\, (R \infty S)$. That is, given an R-lifting matrix z_1 for pmfs p and q and an S-lifting matrix z_2 for q and r, we have to construct an $(R \infty S)$-lifting matrix z for p and r. In the course of this work, we have formalized a series of three different constructions for z, each of which made the previous proof simpler and more concise. The steps are recorded in the changesets (mentioned below) of the Isabelle repository at http://isabelle.in.tum.de/repos/isabelle. This process illustrates how pmfs provide abstraction and lead to shorter proofs.

Initially, we followed Sokolova's construction [21]. She defines the matrix z as the sum of matrices z^j over $j \in \mathsf{set}_{\mathsf{pmf}}\, q$ where each z^j is a T^j-lifting matrix for the jth column of z_1 and the jth row of z_2 (we ignore that the rows and columns do not sum to 1) where $T^j\, i\, k = R\, i\, j \wedge S\, j\, k$. An iterative algorithm constructs the matrix z^j by walking on a path from the upper left corner to the lower right and setting each entry to the maximum such that neither the row sum nor the column sum is exceeded. If the row sum is matched after setting the entry, the path continues down, if the column sum is matched, it goes to the left, if both are matched, it goes diagonally to the right and down. Her proof that z is an $(R \infty S)$-lifting matrix for p and r requires five pages on paper [21, Lemmas 3.5.5, 3.5.6]. Our HOL formalization of a recursive version of the algorithm and the proof of the distributivity property is arduous and takes 577 lines (4999a616336c). By switching from real-valued functions to pmfs and using $\mathsf{map}_{\mathsf{pmf}}$ in the construction of z from the z^j, we were able to shorten the proof to 406 lines (43e07797269b). Still, most of the proof script dealt with showing the equality of different summations.

Next, we realized that taking a path through the matrix and setting the entries to maximum values was needlessly convoluted. Instead, we fill the ith row of z^j by distributing z_1's value at (i, j) over the columns according to the jth row of z_2. This eliminates all the inductions and several bijections between the support sets and natural numbers, which were needed for the recursion. This is the proof by Jonsson et al. [16] formalized in 101 lines (922d31f5c3f5, 922d31f5c3f5). Zanella [26] has previously formalized this proof for CertiCrypt using Audebaud's and Mohring's library [1]. His proof script needs 337 lines of Coq.

Finally, we noted that the distribution over the columns and the summation over the z^j yields a conditional probability. So, we now define z simply as

$$z = \mathsf{bind}_{\mathsf{pmf}}\, z_1\ (\lambda(i,\, j).\, \mathsf{bind}_{\mathsf{pmf}}\ (\mathsf{cond}_{\mathsf{pmf}}\, z_2\ \{(j',\, k) \mid j' = j\})\ (\lambda(\_,\, k).\, \mathsf{return}_{\mathsf{pmf}}(i,\, k)))$$

Thus, only one conjunct is shown with summations, namely of $\mathsf{map}_{\mathsf{pmf}}\, \pi_1\, z = p$. The others are discharged by reasoning with the laws about $\mathsf{set}_{\mathsf{pmf}}$, $\mathsf{bind}_{\mathsf{pmf}}$, $\mathsf{cond}_{\mathsf{pmf}}$, and $\mathsf{return}_{\mathsf{pmf}}$. The following law is particularly useful. It generalizes the *law of total probability*, which states $\Pr(A) = \sum_n \Pr(A \mid B_n) \cdot \Pr(B_n)$ for a countably indexed partition B. Note that $\mathsf{bind}_{\mathsf{pmf}}$ expresses the sum and R relates the events of a and b.

$$\frac{\mathsf{rel}_{\mathsf{set}}\, R\ (\mathsf{set}_{\mathsf{pmf}}\, a)\ (\mathsf{set}_{\mathsf{pmf}}\, b)\qquad \forall x \in \mathsf{set}_{\mathsf{pmf}}\, a.\ \forall y \in \mathsf{set}_{\mathsf{pmf}}\, b.\ R\, x\, y \longrightarrow \mu\, a\, \{x \mid R\, x\, y\} = \mu\, b\, \{y \mid R\, x\, y\}}{\mathsf{bind}_{\mathsf{pmf}}\, b\ (\lambda y.\, \mathsf{cond}_{\mathsf{pmf}}\, a\, \{x \mid R\, x\, y\}) = a} \tag{2}$$

Here, the set relator $\mathsf{rel}_{\mathsf{set}}\, R\, A\, B$ denotes $(\forall a \in A.\, \exists b \in B.\, R\, a\, b) \wedge (\forall b \in B.\, \exists a \in A.\, R\, a\, b)$. (Using the same notation for bounded and unbounded sets, this characterization also holds for the relator of bounded sets.) We use this law to show $\mathsf{map}_{\mathsf{pmf}}\, \pi_2\, z = r$. Observe that $\mathsf{map}_{\mathsf{pmf}}\, \pi_2\, z = \mathsf{map}_{\mathsf{pmf}}\, \pi_2\ (\mathsf{bind}_{\mathsf{pmf}}\, q\ (\lambda y.\, \mathsf{cond}_{\mathsf{pmf}}\, z_2\, \{(y',z) \mid y' = y\}))$. Applying the law to the right-hand side yields $\mathsf{map}_{\mathsf{pmf}}\, \pi_2\, z_2$, which equals r by assumption.

In the end, our proof is just 46 lines, which includes 18 lines for the proof of Eq. 2 (224741ede5ae). This confirms in our eyes that our pmf library is well designed. Also, we argue that the proof has gained in clarity from the reduction in size. We eliminated most of the technical transformations of sums and express them more abstractly.

5   Probabilistic Systems

Probabilistic Systems as Probabilistic Automata. First, we review the approach of modeling probabilistic systems as probabilistic automata. These automata fall into different classes depending on whether they make use of probabilistic and non-deterministic choice, where labels are placed, and whether transitions generate output for the environment or receive input from it.

Labeled Markov chains are a very simple class of probabilistic automata. Here each state has a label and specifies a probability distribution over the successor

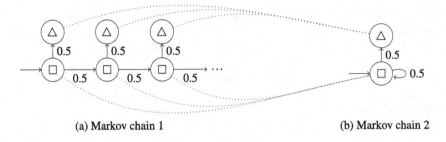

(a) Markov chain 1   (b) Markov chain 2

Fig. 1. Two labeled Markov chains (the dotted lines represent a bisimulation relation)

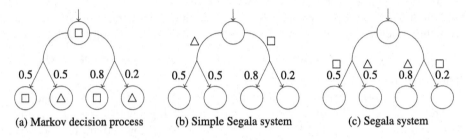

(a) Markov decision process   (b) Simple Segala system   (c) Segala system

Fig. 2. Three probabilistic automata with non-deterministic and probabilistic choice

states. Figure 1 shows two labeled Markov chains. In our figures, □ or △ denote labels and numbers between 0 and 1 denote probabilities. The Markov chain on the right stores in the state only whether the system has reached the label △. In contrast, the one on the left additionally records in its states how many steps have been taken to reach △.

When modeling systems with probabilistic automata, we usually care only about the labels, not the states. In that respect, both Markov chains produce the same observations with the same probabilities. Thus, it is sensible to consider the two chains as being equivalent. Bisimulation captures this equivalence by identifying states which cannot be distinguished by observing the labels in any behavior originating from these states. For labeled Markov chains, a bisimulation relation is an equivalence relation R on the states such that whenever s and t are related by R, then their labels are the same and for all equivalence classes C of R, the probabilities of going to C from s and t are the same. In Fig. 1, the dotted lines show a bisimulation relation between the two chains. We say that labeled Markov chains on the same state space are bisimilar if the initial states are related in some bisimulation relation.

Combining non-deterministic and probabilistic choice, we get more complicated models. Figure 2 shows three examples. In a Markov decision process, each state has a label, but it may choose non-deterministically a distribution of the successor states. Graphically, the transition edges split after having taken the non-deterministic choice. In a simple Segala system, the label is attached to the non-deterministic transitions rather than the states. So, the transition generates the label instead of the state. And in a Segala system, the label is attached to the

probabilistic choice rather than the non-deterministic one. For these more complicated systems, the definition of bisimulation is analogous to Markov chains, but
more involved.

Coalgebraic View on Probabilistic Systems. Next, we switch perspective and outline the coalgebraic approach to modeling probabilistic systems [2, 21, 24]. We rec-
ollect the basic coalgebraic vocabulary (an in-depth introduction can be found else-
where [20]) and show how these notions are reflected in Isabelle/HOL.

Given a functor F, an F-*coalgebra* is defined as a pair (A, s) with the carrier set
A and the structural mapping $s : A \to F A$. In our typed environment of HOL, we
restrict our attention to bounded natural functors and require the carrier set of a
coalgebra to be the universe of a certain type σ. Therefore, for us a coalgebra is just a
function $s :: \sigma \Rightarrow \sigma$ F for a BNF σ F. Intuitively, a coalgebra $s :: \sigma \Rightarrow \sigma$ F describes
a transition system whose states are in σ and each state $x :: \sigma$ evolves into $s\ x :: \sigma$ F.
For example, if σ F $= \sigma$ pmf, then $s\ x :: \sigma$ pmf is a discrete probability distribution
over the next states and s taken as a whole denotes an unlabeled Markov chain.

Bisimilarity can be defined uniformly on coalgebras [12]: states x and y of
two systems s_1 and s_2 are F-*bisimilar* (written $x \ {}^{s_1}\!\sim_F^{s_2} y$) iff there exists a relation $R :: \alpha \Rightarrow \beta \Rightarrow$ bool (called bisimulation) that relates x and y and for all
related pairs of states x and y their evolutions $s_1\ x$ and $s_2\ y$ are related by
$\mathsf{rel}_F R :: \alpha$ F $\Rightarrow \beta$ F $\Rightarrow$ bool. Formally:

$$x \ {}^{s_1}\!\sim_F^{s_2} y = \exists R.\ R\ x\ y \wedge (\forall x\, y.\ R\ x\ y \longrightarrow \mathsf{rel}_F R\ (s_1\ x)\ (s_2\ y))$$

It turns out that this generic notion coincides with the known concrete bisimilarity notions for all systems F that we consider [2, 21]. We should note that
for σ F $= \sigma$ pmf all states of all systems are bisimilar: $\forall s_1\ s_2\ x\ y.\ x \ {}^{s_1}\!\sim_{\mathsf{pmf}}^{s_2} y$—the
bisimulation relation witness is $R = \lambda x\ y.$ True. This fact corresponds to the intu-
ition that bisimilarity can only distinguish states through observations along their
evolutions, while an unlabeled Markov chain does not produce anything observable. For labeled Markov chains and other systems bisimilarity is a more interesting
concept.

The last important notion is that of a *final* F-*coalgebra*: an F-coalgebra for which
there exists a unique morphism from any other coalgebra. In our context, the final
coalgebra for a BNF F is the destructor $\mathsf{Dtr}_F :: C_F \Rightarrow C_F$ F of the codatatype C_F
induced by F (states of the final coalgebra are of type C_F) and the finality is witnessed by the coiterator unfold_F mapping a coalgebra $s :: \sigma \Rightarrow \sigma$ F to the (unique)
function $\mathsf{unfold}_F s :: \sigma \Rightarrow C_F$ satisfying the characteristic equation of a coalgebra
morphism: $\mathsf{Dtr}_F \circ \mathsf{unfold}_F s = \mathsf{map}_F (\mathsf{unfold}_F s) \circ s$. The similarity of the coinduction rule for C_F to the definition of bisimilarity is not a coincidence: codatatypes
are quotients modulo bisimilarity.

Modeled Systems. Table 1 lists the systems we consider and their BNFs. (The sta-
ndard type **datatype** α option $=$ None | Some α is another BNF.) This list con-
tains all the systems from the original probabilistic hierarchy [21], except for
Vardi systems, which must be treated separately (Sect. 6.4). Moreover, popular
systems—labeled Markov chains and Markov decision processes—are new

Table 1. List of formalized probabilistic systems

Name	BNF	Induced codatatype
Markov chain	σ pmf	MC
Labeled Markov chain	$\alpha \times \sigma$ pmf	α LMC
Labeled Markov decision process	$\alpha \times \sigma$ pmf set$_1^\kappa$	α LMDP$^\kappa$
Deterministic automaton	$\alpha \Rightarrow \sigma$ option	α DLTS
Non-deterministic automaton[a]	$(\alpha \times \sigma)$ set$^\kappa$	α LTS$^\kappa$
Reactive system	$\alpha \Rightarrow \sigma$ pmf option	α React
Generative system	$(\alpha \times \sigma)$ pmf option	α Gen
Stratified system	σ pmf $+ (\alpha \times \sigma)$ option	α Str
Alternating system	σ pmf $+ (\alpha \times \sigma)$ set$^\kappa$	α Alt$^\kappa$
Simple Segala system	$(\alpha \times \sigma$ pmf$)$ set$^\kappa$	α SSeg$^\kappa$
Segala system	$(\alpha \times \sigma)$ pmf set$^\kappa$	α Seg$^\kappa$
Bundle system	$(\alpha \times \sigma)$ set$^\kappa$ pmf	α Bun$^\kappa$
Pnueli-Zuck system	$(\alpha \times \sigma)$ set$^{\kappa_1}$ pmf set$^{\kappa_2}$	α PZ$^{\kappa_1, \kappa_2}$
Most general system	$(\alpha \times \sigma + \sigma)$ set$^{\kappa_1}$ pmf set$^{\kappa_2}$	α MG$^{\kappa_1, \kappa_2}$

[a] The type $(\alpha \times \sigma)$ set$^\kappa$ is isomorphic to the more standard $\alpha \Rightarrow \sigma$ set$^\kappa$ for α set $\leq \kappa$

additions. The third column assigns a name to the induced codatatype (e.g., for labeled Markov decision processes, we write **codatatype** α LMDP$^\kappa$ = Ctr$_{\mathsf{LMDP}}$ $(\alpha \times \alpha$ LMDP$^\kappa$ pmf set$_1^\kappa)$ in Isabelle).

6   The Formalized Hierarchy

How can one compare the expressiveness of the different probabilistic system types? A natural criterion [21] is to exhibit a mapping between the types of systems that preserves and reflects bisimilarity as a witness for an increase in expressiveness along the mapping. Figure 3 shows our formalized hierarchy where arrows represent such mappings. New systems, not analyzed by Sokolova [21], are highlighted with a gray background. Some arrows are annotated with necessary conditions on the bounds of the involved bounded set types. We refer to our formalization [14] for the definitions of all mappings.

Below, we first sketch our proof of the preservation and reflection of bisimilarity abstractly for any mapping. Then we present our formal Isabelle proof for one particular pair of system types and compare our formalized hierarchy with the original [21].

6.1   The Abstract Proof

Formally, for two types of systems given as BNFs F and G, we consider G to be at least as expressive as F, if there is a mapping G_of_F :: σ F $\Rightarrow \sigma$ G that preserves and reflects bisimilarity, i.e., satisfies

$$x \, {}^{s_1}\!\sim_\mathsf{F}^{s_2} y \longleftrightarrow x \, {}^{(\mathsf{G\_of\_F} \circ s_1)}\!\sim_\mathsf{G}^{(\mathsf{G\_of\_F} \circ s_2)} y \tag{3}$$

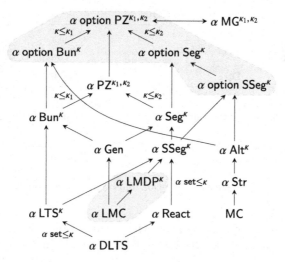

Fig. 3. Probabilistic hierarchy

for all F-coalgebras s_1, s_2 :: $\sigma \Rightarrow \sigma$ F and states x, y :: σ. Note that by composing the F-coalgebras with G_of_F we obtain G-coalgebras: G_of_F $\circ$ s_1, G_of_F $\circ$ s_2 :: $\sigma \Rightarrow \sigma$ G.

For any mapping G_of_F :: σ F $\Rightarrow \sigma$ G, we prove Eq. 3 in four steps starting from the right-hand side:

$$x \overset{(\text{G\_of\_F} \circ s_1)}{\underset{\text{G}}{\sim}} \overset{(\text{G\_of\_F} \circ s_2)}{} y \overset{1}{\longleftrightarrow} \text{unfold}_\text{G}\,(\text{G\_of\_F} \circ s_1)\,x = \text{unfold}_\text{G}\,(\text{G\_of\_F} \circ s_2)\,y$$
$$\overset{2}{\longleftrightarrow} \overline{\text{G\_of\_F}}\,(\text{unfold}_\text{F}\,s_1\,x) = \overline{\text{G\_of\_F}}\,(\text{unfold}_\text{F}\,s_2\,y)$$
$$\overset{3}{\longleftrightarrow} \text{unfold}_\text{F}\,s_1\,x = \text{unfold}_\text{F}\,s_2\,y$$
$$\overset{1}{\longleftrightarrow} x \overset{s_1}{\underset{\text{F}}{\sim}} \overset{s_2}{} y$$

where $\overline{\text{G\_of\_F}}$ :: $C_\text{F} \Rightarrow C_\text{G}$ abbreviates unfold$_\text{G}$ (G_of_F $\circ$ Dtr$_\text{F}$). The first and the last step (labeled with a 1) are both instances of the general fact that for any BNF F, bisimilarity is equivalent to equality on the induced codatatype C_F. Formally, $x \overset{s_1}{\underset{\text{F}}{\sim}} \overset{s_2}{} y \longleftrightarrow$ unfold$_\text{F}$ s_1 x = unfold$_\text{F}$ s_2 y.

In step 2 we perform equational reasoning. The diagram in Fig. 4 illustrates the situation. Essentially it shows three commutative diagrams for the characteristic property of the coiterators unfold$_\text{F}$ and unfold$_\text{G}$: one for the F-coalgebra s in the lower left rectangle; one for the G-coalgebra G_of_F $\circ$ s using the outermost arrows; and one for the G-coalgebra G_of_F $\circ$ Dtr$_\text{F}$ in the upper rectangle.

To make the overall diagram commute, the mapping G_of_F has to be a natural transformation, i.e., commute with the map functions for F and G (lower right rectangle). Once this is ensured we can deduce unfold$_\text{G}$ (G_of_F $\circ$ s) = $\overline{\text{G\_of\_F}}\circ$ unfold$_\text{F}$ s (leftmost "triangle") and use this equation as a rewrite rule.

Step 3 holds universally iff $\overline{\text{G\_of\_F}}$ is injective (note that unfold$_\text{F}$ is surjective, since, e.g., unfold$_\text{F}$ Dtr$_\text{F}$ = id). In principle, injectivity of $\overline{\text{G\_of\_F}}$ can be further

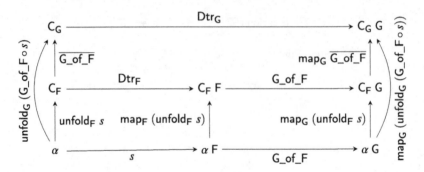

Fig. 4. Bisimilarity preservation and reflection via codatatypes

reduced to injectivity of G_of_F, which yields the nice abstract characterization from the original hierarchy [2,21]: if G_of_F is an injective natural transformation then it preserves and reflects bisimilarity. Instead of formalizing the reduction of injectivity of $\overline{\text{G\_of\_F}}$ to the injectivity of G_of_F (which must be done for all concrete instances of G_of_F), we found it easier to prove the injectivity of $\overline{\text{G\_of\_F}}$ directly by coinduction. Likewise, instead of chasing the above commutative diagram, we also prove directly the equation $\text{unfold}_G\ (\text{G\_of\_F} \circ s) = \overline{\text{G\_of\_F}} \circ \text{unfold}_F\ s$ by coinduction.

6.2 A Concrete Example

We consider a particular instantiation for the BNFs F and G: simple Segala systems $(\text{F} = (\alpha \times \sigma\ \text{pmf})\ \text{set}^\kappa)$ and Segala systems $(\text{G} = (\alpha \times \sigma)\ \text{pmf}\ \text{set}^\kappa)$ and define the mapping G_of_F $sseg = \text{map}_{\text{set}}\ (\lambda(a,\ p).\ \text{map}_{\text{pmf}}\ (\lambda s.\ (a,\ s))\ p)\ sseg$.

Next, we formally prove the properties of $\overline{\text{G\_of\_F}}$ outlined in the previous section by straightforward coinductions. We start with the detailed manual proof of the commutation property (leftmost "triangle" in Fig. 4).

The proof shown in Fig. 5 gives a flavor of the proof obligations that arise with coinduction. The *coinduction* method instantiates the free variable R from the coinduction rule for $\alpha\ \text{Seg}^\kappa$ with the canonical bisimulation witness $\lambda seg\ seg'.\ \exists x.\ seg = \text{unfold}_G\ (\text{G\_of\_F} \circ s)\ x \wedge seg' = (\overline{\text{G\_of\_F}} \circ \text{unfold}_F\ s)\ x$ and performs some minimal postprocessing [4]. We are left to prove that $\text{Dtr}_G\ (\text{unfold}_G\ (\text{G\_of\_F} \circ s)\ x)$ and $\text{Dtr}_G\ (\text{unfold}_G\ (\text{G\_of\_F} \circ s)\ x)$ are related by the bisimulation witness lifted to the $\alpha\ \text{Seg}^\kappa$-inducing BNF via its relator. This subgoal is easy to discharge by unfolding and resolution. All the used theorems with a dot in their name are generated by the **primrec** and **bnf** commands. The theorem *unfold_F . simps* is the characteristic property of the coiterator; theorems *rel-map* (two theorems) and *rel-refl* follow from the BNF properties and are given below for $\alpha\ \text{pmf}$:

$\text{rel}_{\text{pmf}}\ R\ (\text{map}_{\text{pmf}}\ f\ p)\ q = \text{rel}_{\text{pmf}}\ (\lambda x\ y.\ R\ (f\ x)\ y)\ p\ q$
$\text{rel}_{\text{pmf}}\ R\ p\ (\text{map}_{\text{pmf}}\ g\ q) = \text{rel}_{\text{pmf}}\ (\lambda x\ y.\ R\ x\ (g\ y))\ p\ q$ $\quad\quad (\forall x.\ R\ x\ x) \longrightarrow \text{rel}_{\text{pmf}}\ R\ p\ p$

The proof can be automated by registering the appropriate rules as simplification and introduction rules. Furthermore, it can be seen as a proof template: we have

lemma unfold$_G$ (G_of_F $\circ$ s) = $\overline{G\_of\_F}$ $\circ$ unfold$_F$ s
proof (*rule ext*)
  fix x
  show unfold$_G$ (G_of_F $\circ$ s) x = ($\overline{G\_of\_F}$ $\circ$ unfold$_F$ s) x
  proof (*coinduction arbitrary: x*)
    fix x
    show rel$_{set}$ (rel$_{pmf}$ (rel$_{prod}$ (=) (λseg seg'. $\exists x.$ seg = unfold$_G$ (G_of_F$\circ$s) x $\wedge$
                                         seg' = $\overline{(G\_of\_F}$ $\circ$ unfold$_F$ s) x)))
      (Dtr$_G$ (unfold$_G$ (G_of_F$\circ$s) x)) (Dtr$_G$ (($\overline{G\_of\_F}$ $\circ$ unfold$_F$ s) x))
    unfolding unfold$_G$.simps unfold$_F$.simps bset.rel-map pmf.rel-map
      o-def split-beta map-prod-def rel-prod-apply id-apply fst-conv snd-conv
    by (*rule bset.rel-refl pmf.rel-refl conjI exI refl*)+
  qed
qed

Fig. 5. Isar proof of the commutation property from simple Segala to Segala systems

to perform the same reasoning for all concrete mappings that we consider and the only part that is changing is the relator. Fortunately, the Eisbach proof method language [19] helps us to avoid repeating the proof by creating a dedicated proof method, where we replace the manual **unfolding** and *rule* steps by *fastforce*. The proof then collapses to a one-liner.

method-definition *commute-prover* =
  rule ext,
  match conclusion in u_1 s_1 x = $(f \circ u_2$ $s_2)$ x for f u_1 u_2 s_1 s_2 x $\Rightarrow$
    (*coinduction arbitrary: x, fastforce*)
lemma unfold$_G$ (G_of_F $\circ$ s) = $\overline{G\_of\_F}$ $\circ$ unfold$_F$ s **by** *commute-prover*

We treat the injectivity of $\overline{G\_of\_F}$ and the fact that bisimilarity coincides with equality on codatatypes for F and G in a similar fashion. As before, we omit some essential simplification and introduction rules given as arguments to *fastforce* that make the following degree of automation possible.

method-definition *inj-prover* =
  rule injI,
  match conclusion in x = y for x y $\Rightarrow$ (*coinduction arbitrary: x y, fastforce*)
lemma inj $\overline{G\_of\_F}$ **by** *inj-prover*

method-definition $\sim$-*alt-prover* =
  intro iffI, elim exE conjE,
  match conclusion in u_1 s_1 x = u_2 s_2 y for u_1 u_2 s_1 s_2 x y $\Rightarrow$
    (*coinduction arbitrary: x y, fastforce*), *fastforce*
lemma x $^{s_1}\sim_F^{s_2}$ y $\longleftrightarrow$ unfold$_F$ s_1 x = unfold$_F$ s_2 y **by** $\sim$-*alt-prover*
lemma x $^{s_1}\sim_G^{s_2}$ y $\longleftrightarrow$ unfold$_G$ s_1 x = unfold$_G$ s_2 y **by** $\sim$-*alt-prover*

Overall, for each of the 14 considered probabilistic system types we prove the alternative bisimilarity characterization **by** $\sim$-*alt-prover* and for each of the 22 mappings (there are 25 arrows in Fig. 3, but e.g., the mapping from α option SSeg^κ to α option Seg^κ is the same as the one from α SSeg^κ to α Seg^κ) we prove two statements by a one-liner with one of our dedicated methods: *commute-prover* and *inj-prover*. Finally, we state the 25 bisimilarity preservation and reflection properties (Eq. 3) and prove all of them by equational reasoning (i.e., one line of unfolding). The whole hierarchy is formalized in 450 lines (including the codatatype declarations).

6.3   Comparison to the Original Hierarchy

Our formalized hierarchy differs structurally from the original hierarchy [21] in three aspects. First, ours omits the Vardi systems (also known as concurrent Markov chains) for reasons we outline in a separate section (Sect. 6.4). Conversely, we have added two popular types of systems, namely labeled Markov chains and Markov decision processes. Furthermore, we observe that the Most General systems α $\mathsf{MG}^{\kappa_1,\kappa_2}$, specifically introduced in the original hierarchy in order to have a top element, are isomorphic to Pnuelli-Zuck systems extended with a single additional label (which we model by using α option instead of just α, i.e., α option $\mathsf{PZ}^{\kappa_1,\kappa_2}$). In other words, no new structurally different probabilistic system is needed to get a top element if one allows additional labels. Following up on this idea, we investigated adding new labels to various other systems in the hierarchy. As a result, Alternating systems α Alt^κ are placed below label-extended simple Segala systems α option SSeg^κ and Bundle systems α option Bun^κ, instead of just below the top element α $\mathsf{MG}^{\kappa_1,\kappa_2}$ as in the original hierarchy.

Our usage of codatatypes (final coalgebras) caters for highly automatable proofs. However, the resulting conciseness comes at a price: final coalgebras need to exist. Concretely, this means that all our systems must be BNFs, in particular bounded and weak pullback preserving. In contrast, the original hierarchy does not require a boundedness assumption (basically allowing to use α set instead of α set^κ) and requires for each mapping only the system functor of the mapping's domain to preserve weak pullbacks. While we acknowledge the latter as a limitation of our approach, we point out that the boundedness assumption is not a restriction in the setting of the hierarchy, since the mappings are polymorphic in the type κ used as the bound. That is, for any concrete system with unbounded non-determinism (α set) expressible in HOL we can find an isomorphic bounded one, and the mapping will show how to transform this bounded system into one that is higher in the hierarchy.[1] In contrast, the bounds being part of the types is in some sense more precise—for example, we see that there are two ways of embedding α Alt^κ in α $\mathsf{MG}^{\kappa_1,\kappa_2}$ transitively via α option SSeg^κ or α option Bun^κ and the cardinal bounds give a hint which route was taken.

---

[1] Clearly, this discussion is somewhat esoteric, since in practice one barely is interested to look beyond countable sets. Still, we are interested in keeping the results as general as possible.

6.4   Vardi Systems

Vardi systems, also known as concurrent Markov chains [23], blend non-deterministic and probabilistic transitions in a rather symmetric fashion. They are similar to coalgebras of the binary BNF $(\alpha, \sigma)\, \mathsf{Var}_0^\kappa = (\alpha \times \sigma)\, \mathsf{pmf} + (\alpha \times \sigma)\, \mathsf{set}^\kappa$; however there is a twist: Vardi systems identify the singleton bounded set $\{(a, s)\}$ with the singleton discrete distribution $\mathsf{return}_{\mathsf{pmf}}\,(a, s)$. Formally, we define the equivalence relation $\bowtie$ inductively by the following three rules, where Inl and Inr are the sum type embeddings.

$$v \bowtie v \qquad \mathsf{Inl}\,(\mathsf{return}_{\mathsf{pmf}}\,(a, s)) \bowtie \mathsf{Inr}\,\{(a, s)\} \qquad \mathsf{Inr}\,\{(a, s)\} \bowtie \mathsf{Inl}\,(\mathsf{return}_{\mathsf{pmf}}\,(a, s))$$

The type $(\alpha, \sigma)\, \mathsf{Var}^\kappa$ is then defined as a quotient of $(\alpha, \sigma)\, \mathsf{Var}_0^\kappa$ by $\bowtie$. Lifting the functorial structure of $(\alpha, \sigma)\, \mathsf{Var}_0^\kappa$ to the quotient $(\alpha, \sigma)\, \mathsf{Var}^\kappa$ is straightforward and we omit the definitions. However, it turns out that the resulting quotient is not a BNF: its canonical relator $\mathsf{rel}_{\mathsf{Var}}$ does not distribute over relation composition. We only try to convey the intuition behind this fact—a formal proof can be found in our formalization. Figure 6 shows on the left two Vardi automata that use only non-deterministic transitions and are related by $\mathsf{rel}_{\mathsf{set}}\, R$ (lifted to the sum type $(\alpha, \sigma)\, \mathsf{Var}_0^\kappa$ and further to the quotient type $(\alpha, \sigma)\, \mathsf{Var}^\kappa$) where $R\, x\, y \longleftrightarrow y = \triangle$. Similarly, the two automata on the right are related by $\mathsf{rel}_{\mathsf{pmf}}\, S$ where $S\, x\, y \longleftrightarrow x = \triangle$. The two middle automata are related by $\bowtie$, i.e., they are equal on the quotient type $(\alpha, \sigma)\, \mathsf{Var}^\kappa$. Distributivity of the relator requires the two outermost automata to be related by $\mathsf{rel}_{\mathsf{Var}}$, but this is not the case.

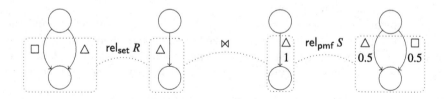

Fig. 6. Intransitivity of Vardi systems

Since $(\alpha, \sigma)\, \mathsf{Var}^\kappa$ is not a BNF, our proof approach is not applicable. Not only that, the above counterexample, found in the course of formalization, is easily transferable into the general coalgebraic setting, allowing us to prove that the functor used in the original hierarchy [21] does not preserve weak pullbacks, and as a consequence bisimilarity of Vardi systems is not an equivalence relation. The weak pullback preservation, however, is a necessary criterion for the original proof method for two mappings from Vardi to Segala and Bundle systems. Those outgoing "arrows" must be purged: there is no such bisimilarity preserving and reflecting mapping.

In contrast to our approach, the original proof still covers the two incoming "arrows" from non-deterministic automata and generative systems to Vardi systems. We have formalized those bisimilarity preserving and reflecting mappings separately, without going through codatypes. The proofs are significantly longer (overall 145 lines for just two mappings, contrasting 450 lines for 25 mappings in our hierarchy) and less suited for automation, because they require several non-trivial quantifier instantiations. In summary, equational reasoning on codatatypes proved superior whenever applicable.

7   Further Related Work

In the other sections of the paper, we have already referenced existing work we build on, in particular Sokolova's [2,21] and the Isabelle packages and tools [4–6,13,15,19].

Our formalization of pmfs is similar to the work in Coq presented by Audebaud and Paulin-Mohring [1]. They introduce a monadic structure on subprobability measures. They use integrals as representations of measures, while in our case we directly lift measures from Isabelle's measure theory. As their measures are subprobabilities they also provide a fixed-point operator which is not available in general for α pmf. Their formalization is also directed towards program verification; they do not provide a functorial structure (i.e. $\mathsf{map}_{\mathsf{pmf}}$ and $\mathsf{set}_{\mathsf{pmf}}$ in our case) on their type of measures.

Apart from process algebra [16], the relator $\mathsf{rel}_{\mathsf{pmf}}$ is used in probabilistic relational Hoare logic, too [3]. In this context, Zanella [26] proved distributivity with composition in Coq; see Sect. 4 for a comparison. Deng [7] collects further results on the relator and its applications. Beyond (strong) bisimilarity, weak bisimilarity compares systems modulo certain irrelevant *invisible* observations. Sokolova [21] recasts weak bisimilarity as bisimilarity of translated systems, which in turn can be hierarchized as presented here.

Mechanizations of category theory abound (see [9] for an overview), and the hierarchy result could probably be formalized with some of them. Yet, we do not formalize the general theory, but its application to concrete instances. Thus, our system types are proper HOL types and can be used directly for modeling concrete systems.

8   Conclusion

We have presented a formalization of the hierarchy of probabilistic system types in Isabelle/HOL. The hierarchy stems from the coalgebraic framework, which presents the various systems in a uniform fashion and caters for simple and concise proofs. We model probabilistic systems as codatatypes, which enables convenient equational reasoning and makes the proofs even more concise. This modeling requires nested corecursion through bounded sets and discrete probability distributions—a perfect match for demonstrating the flexibility of Isabelle's new codatatype facility. Finally, we have learned that weak pullback preservation is an important but subtle property, by uncovering two mistakes in informal proofs.

Acknowledgment. We thank Tobias Nipkow for supporting this collaboration and Ana Sokolova for confirming our findings regarding Vardi systems. Jasmin Blanchette, Ondřej

Kunčar, and anonymous reviewers helped to improve the presentation through numerous comments and offered stylistic advice. Hölzl is supported by the DFG project Verification of Probabilistic Models in Interactive Theorem Provers (grant Ni 491/15-1). Traytel is supported by the DFG program Program and Model Analysis (doctorate program 1480). The authors are listed alphabetically.

References

1. Audebaud, P., Paulin-Mohring, C.: Proofs of randomized algorithms in Coq. Sci. Comput. Program. **74**(8), 568–589 (2009)
2. Bartels, F., Sokolova, A., de Vink, E.P.: A hierarchy of probabilistic system types. Theor. Comput. Sci. **327**(1–2), 3–22 (2004)
3. Barthe, G., Fournet, C., Grégoire, B., Strub, P.Y., Swamy, N., Zanella Béguelin, S.: Probabilistic relational verification for cryptographic implementations. In: Jagannathan, S., Sewell, P. (eds.) POPL 2014, pp. 193–205. ACM, New York (2014)
4. Blanchette, J.C., Hölzl, J., Lochbihler, A., Panny, L., Popescu, A., Traytel, D.: Truly modular (co)datatypes for Isabelle/HOL. In: Klein, G., Gamboa, R. (eds.) ITP 2014. LNCS, vol. 8558, pp. 93–110. Springer, Heidelberg (2014)
5. Blanchette, J.C., Popescu, A., Traytel, D.: Cardinals in Isabelle/HOL. In: Klein, G., Gamboa, R. (eds.) ITP 2014. LNCS, vol. 8558, pp. 111–127. Springer, Heidelberg (2014)
6. Blanchette, J.C., Popescu, A., Traytel, D.: Witnessing (Co)datatypes. In: Vitek, J. (ed.) ESOP 2015. LNCS, vol. 9032, pp. 359–382. Springer, Heidelberg (2015)
7. Deng, Y.: Semantics of Probabilistic Processes. Springer, Heidelberg (2014)
8. Eberl, M., Hölzl, J., Nipkow, T.: A verified compiler for probability density functions. In: Vitek, J. (ed.) ESOP 2015. LNCS, vol. 9032, pp. 80–104. Springer, Heidelberg (2015)
9. Gross, J., Chlipala, A., Spivak, D.I.: Experience implementing a performant category-theory library in Coq. In: Klein, G., Gamboa, R. (eds.) ITP 2014. LNCS, vol. 8558, pp. 275–291. Springer, Heidelberg (2014)
10. Gunter, E.L.: Why we can't have SML-style `datatype` declarations in HOL. In: TPHOLs 1992. IFIP Transactions, vol. A-20, pp. 561–568. North-Holland/Elsevier (1993)
11. Harrison, J.V.: A HOL theory of Euclidean space. In: Hurd, J., Melham, T. (eds.) TPHOLs 2005. LNCS, vol. 3603, pp. 114–129. Springer, Heidelberg (2005)
12. Hermida, C., Jacobs, B.: Structural induction and coinduction in a fibrational setting. Inf. Comput. **145**(2), 107–152 (1998)
13. Hölzl, J.: Construction and Stochastic Applications of Measure Spaces in Higher-Order Logic. Ph.D. thesis, Institut für Informatik, Technische Universität München (2013)
14. Hölzl, J., Lochbihler, A., Traytel, D.: A zoo of probabilistic systems. In: Klein, G., Nipkow, T., Paulson, L. (eds.) Archive of Formal Proofs (2015). http://afp.sf.net/entries/Probabilistic_System_Zoo.shtml
15. Huffman, B., Kunčar, O.: Lifting and Transfer: a modular design for quotients in Isabelle/HOL. In: Gonthier, G., Norrish, M. (eds.) CPP 2013. LNCS, vol. 8307, pp. 131–146. Springer, Heidelberg (2013)
16. Jonsson, B., Larsen, K.G., Yi, W.: Probabilistic extensions of process algebras. In: Bergstra, J.A., Ponse, A., Smolka, S.A. (eds.) Handbook of Process Algebras Chap. 11, pp. 685–710. Elsevier, Amsterdam (2001)

17. Larsen, K.G., Skou, A.: Bisimulation through probabilistic testing. Inf. Comp. **94**(1), 1–28 (1991)
18. Lochbihler, A.: Measure definition on streams, 24 February 2015. Archived at https://lists.cam.ac.uk/pipermail/cl-isabelle-users/2015-February/msg00112.html
19. Matichuk, D., Wenzel, M., Murray, T.: An Isabelle proof method language. In: Klein, G., Gamboa, R. (eds.) ITP 2014. LNCS, vol. 8558, pp. 390–405. Springer, Heidelberg (2014)
20. Rutten, J.J.M.M.: Universal coalgebra: a theory of systems. Theor. Comput. Sci. **249**, 3–80 (2000)
21. Sokolova, A.: Coalgebraic Analysis of Probabilistic Systems. Ph.D. thesis, Technische Universiteit Eindhoven (2005)
22. Traytel, D., Popescu, A., Blanchette, J.C.: Foundational, compositional (co)datatypes for higher-order logic–Category theory applied to theorem proving. In: LICS 2012, pp. 596–605. IEEE (2012)
23. Vardi, M.Y.: Automatic verification of probabilistic concurrent finite-state programs. In: FOCS 1985, pp. 327–338. IEEE (1985)
24. de Vink, E.P., Rutten, J.J.: Bisimulation for probabilistic transition systems: a coalgebraic approach. Theor. Comput. Sci. **221**(1–2), 271–293 (1999)
25. Weber, T.: Introducing a BNF for sets of bounded cardinality, 14 March 2015. Archived at https://lists.cam.ac.uk/pipermail/cl-isabelle-users/2015-March/msg00116.html
26. Zanella Béguelin, S.: Formal Certification of Game-Based Cryptographic Proofs. Ph.D. thesis, École Nationale Supérieure des Mines de Paris (2010)

A Verified Enclosure for the Lorenz Attractor (Rough Diamond)

Fabian Immler[✉]

Institut für Informatik, Technische Universität München, München, Germany
immler@in.tum.de

Abstract. A rigorous numerical algorithm, formally verified with Isabelle/HOL, is used to compute an accurate enclosure for the Lorenz attractor.

Accurately enclosing the attractor is highly relevant: a similar non verified computation is part of Tucker's proof that the Lorenz attractor is chaotic in a rigorous mathematical sense. This proof settled a conjecture that Fields medalist Stephen Smale has put on his list of eighteen important mathematical problems for the twenty-first century.

Keywords: Isabelle/HOL · Ordinary differential equation · Lorenz attractor · Rigorous numerics

1 Introduction

The Lorenz system of ordinary differential equation (ODEs) has become famous as a classical example of chaotic dynamics since its introduction as a model for atmospheric flows by Edward Lorenz in 1963. Numerical experiments suggested chaotic behavior; in dynamical systems parlance, the existence of a *strange attractor*. However, the existence of a strange attractor for the Lorenz equations could not be proved until 1999 – shortly after Fields medalist Stephen Smale put it on his list of eighteen unsolved mathematical problems for the 21st century.

The proof was accomplished by Warwick Tucker [8], and an interesting aspect is that his proof relies on the output of a numerical program. His program computes enclosures for the attractor and analytical properties of solutions on the attractor.

These programs were written in C++ and not formally verified, Tucker even discovered (and fixed) some bugs in it [7]. Formally verifying the numerical results needed for the proof is therefore a worthwhile goal.

The contribution of this work is the computation of an accurate enclosure for the Lorenz attractor with a formally verified ODE solver. The development is available in the Archive of Formal Proofs [5].

F.Immler—Supported by the DFG RTG 1480 (PUMA).

C. Urban and X. Zhang (Eds.): ITP 2015, LNCS 9236, pp. 221–226, 2015.
DOI: 10.1007/978-3-319-22102-1_14

2   The Lorenz Attractor

We consider (like Tucker) the classical parameter values, for which the Lorenz equations in Jordan normal form are approximately given by the right hand side $f(x, y, z) = (11.8x - 0.29(x+y)z, -22.8y + 0.29(x+y)z, -2.67z + (x+y)(2.2x - 1.3y))$. It can be shown that the properties of interest are robust under small perturbations of the parameters.

A solution $\varphi : \mathbb{R} \to \mathbb{R}^3$ is any function with derivative $\dot{\varphi}(t) = f(\varphi(t))$. With an *initial condition* $\varphi(0) = x_0$, the solution is unique. We denote with *flow* the solution $\varphi(x_0, t)$ depending on initial condition x_0 at time t.

Numerical simulations suggest the existence of a set $\mathcal{A}$, the *Lorenz attractor* (enclosures of which is depicted in Fig. 1) with the following 3 properties, which make it a *strange* attractor.

Property 1. Solutions (of the Lorenz system) tend towards $\mathcal{A}$.

The dynamics on $\mathcal{A}$ can be described as follows: solutions starting from $\Sigma = \{(x, y, z) \mid z = 27 \wedge x \in [-5.5; 5.5]\}$ flow downwards (towards lower values of z) and enter either the left or right branch of the attractor in order to circle around the "holes" around $(\pm 6, \cdot, 27)$ and return back to Σ. Depending on where they return, they either stay in the same branch or switch to the other side in the next revolution. Small initial sets approaching the fixed point $(0, 0, 0)$ exhibit strong expansion in the x-direction, which causes Property 2.

Property 2. Solutions on $\mathcal{A}$ exhibit sensitive dependence on initial conditions.

The expansion is strong enough for Property 3.

Property 3. Small initial sets eventually spread over the whole attractor $\mathcal{A}$.

A standard approach in the analysis of dynamical systems is to provide sufficient conditions for properties 1,2,3 by studying a so-called *Poincaré map* R, which simplifies reasoning about the three-dimensional flow to reasoning about discrete iterations of the two-dimensional map R. To define R for Tucker's proof, consider the *return plane* Σ as defined before. For any point $x \in \Sigma$, $\tau(x)$ is the first time when x flows through Σ from above. The Poincaré map R is then defined as $R(x) := \varphi(x, \tau(x))$. The significance is that the equivalents of properties 1,2,3 for the iterations of the map R and its attractor $\mathcal{A} \cap \Sigma$ carry over to the flow φ and $\mathcal{A}$. Tucker proved those with a combination of rigorous numerics and analytical reasoning locally around the origin $(0, 0, 0)$.

Rigorous Numerics. Property 1 can be shown by exhibiting a forward invariant (i.e. $R(N) \subseteq N$) subset N of the return plane that contains the attractor: $\mathcal{A} \cap \Sigma \subseteq N \subseteq \Sigma$.

Tucker proves this by computing enclosures for R with *rigorous numerics*: a function $step(X)$ computes some set (by safely including e.g. round-off errors into the result) that is reachable via the flow.

$$\forall x \in X. \, \exists h > 0. \, \varphi(x, h) \in step(X)$$

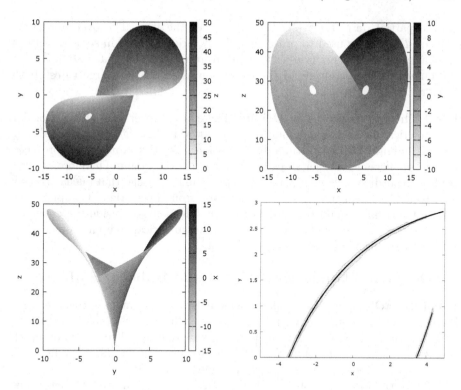

Fig. 1. *Top left, top right, bottom left*: projections of enclosures for Lorenz attractor. *Bottom right*: forward invariant subset N in black of the return plane Σ with Tucker's enclosure in gray; one half is omitted due to the symmetry $(x, y, z) \in N \longleftrightarrow (-x, -y, z) \in N$

Overapproximations to the Poincaré map R can then be obtained by iterating *step*(X) until the return plane Σ is reached:

poincare$(X) :=$
    let $Y = step(X)$ **in if** $Y \cap \Sigma \neq \emptyset$ **then** *step-to-sigma*(X) **else** *poincare*(Y)

Here *step-to-sigma* is like *step* but needs to satisfy *step-to-sigma*$(X) \subseteq \Sigma$. Now if $X \subseteq \Sigma$, then $R(X) \subseteq$ *poincare*(X). Tucker represents a candidate for the forward invariant subset as a union of small rectangles $N = \bigcup_{i < k} N_i$, for which computations confirm that *poincare*$(N_i) \subseteq N$ for all i and therefore $R(N) \subseteq N$.

Tucker implements *step* as a function that propagates axis-parallel rectangles by overapproximating the flow with the Euler method. Tucker implements *step-to-sigma* by choosing an appropriate interval for the step size and projecting the result onto the plane.

To quantify the dependence on initial conditions for Property 2, it is a standard approach to study the derivative DR of R. This can be done by adding to *step*, which computes overapproximations to the flow φ, overapproximations

for the partial derivatives $\frac{\partial\varphi}{\partial x}$, $\frac{\partial\varphi}{\partial y}$, $\frac{\partial\varphi}{\partial z}$, $\frac{\partial\varphi}{\partial t}$ of the flow. The overapproximations on DR can be used to prove that the N_i are expanded in some directions
and contracted in others (via a "forward invariant cone field" for DR). Tucker
can quantify the magnitude of the expansion as sufficiently large to establish
Property 3.

Local Theory Around the Origin. There is, however, one obstruction for
the numerical methods: some solutions tend towards the origin $(0, 0, 0)$, and they
do so in infinite time. Any time discretization algorithm would therefore need
to take smaller and smaller steps but never reach the origin. Therefore, Tucker
derived a coordinate change (in about 25 pages in his article) that makes the flow
approximately linear in a cube with width 0.1 around the origin. Computations
can be interrupted upon reaching that cube. The solution inside the cube can
be propagated via an explicit formula, and the numerical computations can be
continued afterwards.

3   ODEs and Numerical Solutions in Isabelle/HOL

In Isabelle/HOL [6], ordinary differential equations are formalized together with
basic theorems for local existence/uniqueness (the Picard-Lindelöf theorem),
global unique solutions and basic properties of the flow, like continuity with
respect to initial conditions. Differentiability with respect to initial conditions is
still missing and would be needed for reasoning about DR.

We use the formalization in Isabelle/HOL of rigorous numerical algorithms
for ODEs [4]. The method we employ for *step* is slightly different from Tucker's
approach: instead of the Euler method, we use the method of Heun, a two-
stage Runge-Kutta method (where the error in one step is cubic in the step
size) with adaptive step size control. Instead of rectangles we use zonotopes
(our algorithm is based on affine arithmetic [1] instead of interval arithmetic).
Numerical computations are carried out with software floating point numbers
$m \cdot 2^e$ for (unbounded) integers $m, e \in \mathbb{Z}$. Explicit round-off operations restrict
the size of m during the computations.

As detailed in the earlier paper [4], it turns out that reducing the reachable sets to two dimensions from time to time is crucial for maintaining precise enclosures and acceptable performance. For these reductions as well as for
step-to-sigma, it is necessary to compute intersections of reachable zonotopes
with intermediate planes or the return plane Σ, for which we use our formalization [3] of Girard/Le Guernic's geometric algorithm [2]. Tucker's algorithm
achieves that implicitly because it propagates rectangles exclusively from plane
to plane.

Tucker's and our implementation have in common that reachable sets are
split when their size exceeds some given threshold.

The following theorem for partial correctness, proved in Isabelle/HOL assures
that if *poincare* returns a result, this result is a safe overapproximation for the
flow of the Lorenz equations:

Theorem 4. $poincare(X) \subseteq \Sigma \wedge \forall x \in X \, \exists t > 0. \, \varphi(x, t) \in poincare(X)$

4 Computing a Verified Enclosure for the Lorenz Attractor

We only tackle the numerical computations needed for Property 1: we verify a forward invariant set N for R (and in the process an enclosure for the Lorenz attractor $\mathcal{A}$, forward invariant under φ).

The set N used for our computations is plotted in the bottom right of Fig. 1 in black. The enclosure that was verified by Tucker is depicted with gray rectangles and one can see that our computations are at least as accurate. In our case, N is a collection of 14816 squares N_i with width 2^{-8}.

4.1 Parallelization on Supercomputer

The overall computation is embarassingly parallel, since the computations for the 14816 initial rectangles N_i are independent. We extracted code for our verified algorithm poincare to Standard ML (SML) and compiled it with MLTon. With this setup, integers $\mathbb{Z}$ from the formalization are mapped to the arbitrary-precision integers from *The GNU Multiple Precision Arithmetic Library* (GMP).

Then we distributed the program for the different input data on 1024 cores of the computer cluster SuperMUC. With a wall-clock time limit of 7 h, this amounts to a total computation time of around 7000 h. Tucker's original computations (more than fifteen years ago) have been distributed on 20 computers for about 100 h.

During reachability analysis, the program outputs a trace containing information like enclosures during propagation, which allowed us to plot enclosures of the Lorenz attractor (see Fig. 1) generated from the formally verified program. Since every rectangle N_i returns within N, N is verified as forward invariant under R.

Note that a small part of the attractor is missing: we interrupt computations close to the origin (as is necessary in Tucker's proof as well), but we do not continue with a symbolic propagation from there. This only affects 16 reachable sets (which do not expand much anymore after leaving the cube around the origin) so the impact on overall computation time is negligible.

4.2 Parallelization with Isabelle/ML

If one wants to avoid running independent instances compiled outside of Isabelle, it is also possible to compile and evaluate the code from within Isabelle. Then Isabelle trusts the outcome of the computations after reconstructing Isabelle/HOL terms from the result of the SML program. To exploit the parallelization, Isabelle/HOL's library provides special combinators, for which the generated code uses parallel combinators of Isabelle/ML (i.e. Par_List.map). Using these combinators, we tested running the initial rectangles N_i for $0 \leq i \leq 32$ on a 8 core machine, which gave a speedup of factor 6.1 compared to serial execution. Evaluating in Isabelle gives us the following theorem (trusting the code generation oracle).

Theorem 5. $\forall x \in \bigcup_{i \leq 32} N_i.\ \exists t > 0.\ \varphi(x, t) \in N$

5   Conclusion

We took a first step towards a formal verification of the numerical part of Tucker's proof. However we do not track the derivative DR, because we have no formalization of differentiability of the flow (and therefore R). It should increase the computational efforts only by a constant factor (since in every step, one propagates in addition to the flow on a reachable set just the derivative of the flow on a cone field). Furthermore, we ignore the symbolic propagation at the origin but this does not impact the overall computational effort too much.

Nevertheless we managed to obtain formally verified results on an important and computationally intensive part of the proof, which we hope to be able to extend with reasonable effort towards a propagation of DR.

Acknowledgments. I would like to thank Florian Haftmann for providing the theories for parallelization with Isabelle/HOL and Makarius Wenzel for the underlying infrastructure for parallel combinators in Isabelle/ML.

References

1. de Figueiredo, L., Stolfi, J.: Affine arithmetic: concepts and applications. Numer. Algorithms **37**(1–4), 147–158 (2004)
2. Girard, A., Le Guernic, C.: Zonotope/hyperplane intersection for hybrid systems reachability analysis. In: Egerstedt, M., Mishra, B. (eds.) HSCC 2008. LNCS, vol. 4981, pp. 215–228. Springer, Heidelberg (2008)
3. Immler, F.: A verified algorithm for geometric zonotope/hyperplane intersection. In: Proceedings of the 2015 Conference on Certified Programs and Proofs, CPP 2015, pp. 129–136. ACM, New York (2015)
4. Immler, F.: Verified reachability analysis of continuous systems. In: Baier, C., Tinelli, C. (eds.) TACAS 2015. LNCS, vol. 9035, pp. 37–51. Springer, Heidelberg (2015). http://dx.doi.org/10.1007/978-3-662-46681-0_3
5. Immler, F., Hölzl, J.: Ordinary differential equations. Archive of Formal Proofs, August 2015, Formal proof development. http://afp.sf.net/devel-entries/Ordinary_Differential_Equations.shtml
6. Nipkow, T., Paulson, L.C., Wenzel, M. (eds.): Isabelle/HOL. LNCS, vol. 2283. Springer, Heidelberg (2002)
7. Tucker, W.: My thesis: the Lorenz attractor exists. http://www2.math.uu.se/~warwick/main/pre_thesis.html
8. Tucker, W.: A rigorous ODE solver and Smale's 14th problem. Found. Comput. Math. **2**(1), 53–117 (2002)

Learning to Parse on Aligned Corpora
(Rough Diamond)

Cezary Kaliszyk[1], Josef Urban[2]($^{(\boxtimes)}$), and Jiří Vyskočil[3]

[1] University of Innsbruck, Innsbruck, Austria
cezary.kaliszyk@uibk.ac.at
[2] Radboud University Nijmegen, Nijmegen, The Netherlands
Josef.Urban@gmail.com
[3] Czech Technical University, Prague, Czech Republic
vyskoj1@fel.cvut.cz

Abstract. One of the first big hurdles that mathematicians encounter when considering writing formal proofs is the necessity to get acquainted with the formal terminology and the parsing mechanisms used in the large ITP libraries. This includes the large number of formal symbols, the grammar of the formal languages and the advanced mechanisms instrumenting the proof assistants to correctly understand the formal expressions in the presence of ubiquitous overloading.

In this work we start to address this problem by developing approximate probabilistic parsing techniques that autonomously train disambiguation on large corpora. Unlike in standard natural language processing, we can filter the resulting parse trees by strong ITP and AR semantic methods such as typechecking and automated theorem proving, and even let the probabilistic methods self-improve based on such semantic feedback. We describe the general motivation and our first experiments, and build an online system for parsing ambiguous formulas over the Flyspeck library.

1 Introduction

Is it possible to automatically parse informal mathematical texts into formal ones and formally verify them? Four out of five ITP (interactive theorem proving) practitioners say no.[1] Even Andrzej Trybulec – an accomplished linguist by one of his professions and the father of human-like formal mathematical notation, linguistic typing mechanisms and proof style – used to quote the past work (e.g., by Zinn [10]) as discouraging from such efforts. We however believe that it is a good time to try, and in particular to try to automatically *learn* how to formalize ("semanticize") informal texts, based on the knowledge available in existing large formal corpora. There are several reasons [6].

C. Kaliszyk—Supported by the Austrian Science Fund (FWF): P26201.

J. Urban—Supported by NWO grant nr. 612.001.208.

[1] Approximate results of an opinion poll run by the second author since 2000.

C. Urban and X. Zhang (Eds.): ITP 2015, LNCS 9236, pp. 227–233, 2015.
DOI: 10.1007/978-3-319-22102-1_15

First, statistical machine learning (data-driven algorithm design) has been responsible for several recent AI breakthroughs, including machine translation systems like Google Translate that automatically train on large aligned bilingual corpora. Similar successes are in query answering (IBM Watson) and autonomous car driving, which are arguably much more semantic domains than just "simple" natural language alignment. It seems today that as soon as there are sufficiently large datasets, data-driven algorithms can automatically learn complicated sets of rules – thus perhaps also the nontrivial mapping of informal to formal – that would be otherwise hard to program and maintain manually.

Second, recent formalization projects have produced large corpora that can – perhaps after additional annotation – be used for such experiments with machine learning of formalization. Further growth of such corpora is only a matter of time, and assisted formalization might help "bootstrap" this process, making it faster and faster due to the positive feedback loop from more data becoming available.

Third, statistical machine learning methods have recently really turned out to be useful in deductive AI domains such as automated reasoning in large theories [1] (ARLT). This shows that in practice, its inherent undecidability does not make mathematics into some special field where statistical techniques cannot apply. Quite the opposite: formal mathematical corpora seem to largely obey similar statistical laws as other texts produced by humans, and statistical and information-retrieval algorithms such as TF-IDF, naive Bayes, k-nearest neighbor, and kernel methods, are indispensable parts of the ARLT methods [4, 7].

Finally, we believe that strong ARLT methods are a new useful weapon in auto-formalization, that can complement the statistical translation methods. This could result in hybrid understanding/thinking AI methods that self-improve on large annotated corpora by cycling between (i) statistical prediction of the text disambiguation based on learning from existing annotations and knowledge, and (ii) improving such knowledge by confirming or rejecting the predictions by the semantic ARLT methods. This point is quite unique to the domain of (informal/formal) mathematics, and a good independent reason for this AI research.

2   Contributions

Below we briefly present the first significant effort in statistical learning of parsing ambiguous formulas over a very large formal mathematical corpus – the Flyspeck project. The main result of this effort is a large-scale evaluation of the methods (Sect. 6), and the first version of an online system[2] (Sect. 5) that allows HOL Light and Flyspeck users to write ambiguous bracket-free formulas using many common ambiguous symbols, skipping disambiguation mechanisms such as casting functors. Such formulas are probabilistically parsed, using an efficiently implemented parsing system (Sect. 4) trained on the correct parse trees of all (about 22000) toplevel Flyspeck theorems (Sect. 3). The trained parsing system produces a required number of most probable parse trees, which are then further filtered by parsing and type checking in HOL Light, presenting the most

---

[2] http://colo12-c703.uibk.ac.at/hh/parse.html.

probable filtered parses in a disambiguated HOL Light notation. Simultaneously, these typechecked formulas are given to the HOL(y)Hammer system which then further marks those that can be automatically proved using the whole Flyspeck library and thus are much more likely to have the intended meaning.

In some sense we thus implement the first version of "jumping" between probabilistic and semantic parsing used by informal mathematicians, as fittingly described by Dijkstra [2]:

> The bulk of traditional mathematics is highly informal: formulae are not manipulated in their own right, they are all the time viewed as denoting something, as standing for something else. The bulk of traditional mathematics is characterized by a constant jumping back and forth between the formulae and their interpretation and the latter has to carry the burden of justifying the manipulations. The manipulations of the formulae are not justified by an appeal to explicitly stated rules but by the appeal to the interpretation in which the manipulations are "obviously" OK. By and large, the mathematicians form a much more informal lot than they are aware of.

3   Making Ambiguous Data

While our ultimate goal is to parse the informal LATEX formulas that have been aligned by Hales with the formal Flyspeck formulas [3,8], our initial research approach is to explore parsing of increasingly ambiguous versions of the formal HOL Light and Flyspeck theorems. Making the formal notation more ambiguous turns out to be relatively easy, allowing us to experiment with different kinds of ambiguities and their amount. We did the initial development and evaluations on a subset of 550 Flyspeck trigonometric theorems.[3] This subset is interesting and suitable, because it contains complex and real versions of trigonometric functions (e.g. csin instead of sin) and frequently uses casting functions such as e.g. Cx which casts a real number to a complex number.

It could be however argued that this subset is a toy domain which does not differ much from manually prepared examples, and where manual tweaking of the algorithms is easy and not particularly useful to interested Flyspeck users. That is why we have tried to scale the parsing methods to the whole Flyspeck, making a large number of ambiguous symbols and sentences, hopefully in a way that makes writing such sentences an interesting experiment for some users. The transformations (*informalizing*) consist of:

- Using 72 overloaded instances defined in HOL Light/Flyspeck, like ("+", "vector_add"). The result sentence would use + instead of vector_add.
- Getting the (currently 108) infix operators from HOL Light, and printing them as infix in the informalized sentences. Since + is declared as infix, vector_add u v, would thus result in u + v.

---

[3] Exactly, the theorems containing substrings sin, cos and tan.

- Getting all "prefixed" symbols from the list of 1000 most frequent symbols by searching for: `real_`, `int_`, `vector_`, `nadd_`, `treal_`, `hreal_`, `matrix_`, `complex_` and making them ambiguous by forgetting the prefix.
- Similar overloading of various other symbols that disambiguate overloading, for example the "c"-versions of functions such as `ccos cexp clog csin`, similarly for `vsum`, `rpow`, `nsum`, `list_sum`, etc. In the end the above steps yield a list of about 70 overloaded symbols corresponding to some 200 nonambiguous symbols used very frequently throughout HOL Light and Flyspeck.
- Deleting all brackets, type annotations, and the 10 most frequent casting functors such as `Cx` and `real_of_num` (which alone is used 17152 times).

4   Probabilistic Parsing and its Extensions

Our task is to assign to each informalized sentence (a list of often ambiguous symbols) its most probable HOL parse tree, with all terms annotated by types. For example, the correct parse tree for `REAL_NEGNEG:  ! A0 -- -- A0 = A0` is:

```
(Comb (Const "!" (Tyapp "fun" (Tyapp "fun" (Tyapp "real") (Tyapp "bool")) (Tyapp "bool")))
(Abs "A0" (Tyapp "real") (Comb (Comb (Const "=" (Tyapp "fun" (Tyapp "real") (Tyapp "fun"
(Tyapp "real") (Tyapp "bool")))) (Comb (Const "real_neg" ... (Var "A0" (Tyapp "real")))))))
```

For this, after initial tries with the Stanford Statistical Parser,[4] we wrote our custom OCaml implementation of the CYK chart parsing algorithm [9] for probabilistic context-free grammars (PCFG), and a custom tree transformation tool that enables us to create ambiguous sentences and annotated training input ("grammar") trees for the parser from the HOL parse trees. These grammar trees treat each (possibly complicated) type as the resulting nonterminal assigned to parsing each term, and additionally each ambiguous symbol (terminal) such as ``--'' is wrapped in its disambiguating nonterminal, such as `$#real_neg`. This is analogous to annotating with word-sense disambiguations for linguistic PCFG tools, however our "semantic concepts" are not word senses but HOL types and unambiguous symbols. The tool also applies infix notations and replaces casting terminals with their corresponding nonterminals. For example the complex casting terminal `Cx` disappears, since we do not want the user to write it explicitly, and is replaced in the grammar tree by the corresponding "semantic" nonterminal `$#Cx` applied to the corresponding ambiguous subterm.

We produce two versions of the parser, a standard one and a HOL-specialized one that prunes the parse space by additional fast lightweight semantic restrictions, such as compatibility constraints on types of free variables in parsed subtrees. Both versions have a training and testing phase. In the training phase all the grammar trees (we use all 22000 trees for Flyspeck formulas by default, but this can be further limited) are used to generate grammar rules about the terminals and nonterminals and their probabilities, generating a binarized PCFG. In the testing (evaluation) phase the PCFG is used to parse a given ambiguous sentence with a required number of best parses. Efficient indexing is used to

---

[4] http://nlp.stanford.edu/software/lex-parser.shtml.

prune the search space, and the parse limit is used to prune improbable parsing subtrees, making it reasonably fast (on average 4 s for a Flyspeck theorem) to get the 20 most probable parses. The resulting grammar trees are again transformed back into a HOL parse tree, to which HOL parsing and typechecking is applied as an additional filter. Since all these three parts (CYK, transformations, and HOL Light routines) are written in OCaml, their tight integration is possible, offering further future options such as full HOL-based pruning of untypable subtrees during the CYK parsing, etc. The so far implemented HOL-based extensions of CYK, include the variable typing constraints, special treatment of lambda abstractions, and allowing all unknown symbols to have small nonzero probability of being a variable.

5   Online Parsing System

Since we are very interested in seeing the probabilistic parsing in action, we deploy the whole parsing toolchain as an online service[5] that further uses the HOL(y)Hammer AI/ATP system [5] for even stronger semantic filtering. The service allows HOL Light and Flyspeck users to write ambiguous formulas using many common ambiguous symbols and omitting brackets and casting functors. For example, the top two parses out of allowed 16 for ``sin 0 * x = cos pi / 2'' are

```
sin (&0) * A0 = cos (pi / &2) where A0:real
sin (&0) * A0 = cos pi / &2 where A0:real
```

where only the first one can be automatically proved by HOL(y)Hammer. The user can add brackets to limit the parses, and then for example ``sin (0 * x) = cos pi / 2'' produces 16 parses of which 11 get type-checked by HOL Light as follows, with all but three being proved by HOL(y)Hammer:

```
sin (&0 * A0) = cos (pi / &2) where A0:real
sin (&0 * A0) = cos pi / &2 where A0:real
sin (&0 * &A0) = cos (pi / &2) where A0:num
sin (&0 * &A0) = cos pi / &2 where A0:num
sin (&(0 * A0)) = cos (pi / &2) where A0:num
sin (&(0 * A0)) = cos pi / &2 where A0:num
csin (Cx (&0 * A0)) = ccos (Cx (pi / &2)) where A0:real
csin (Cx (&0) * A0) = ccos (Cx (pi / &2)) where A0:real^2
Cx (sin (&0 * A0)) = ccos (Cx (pi / &2)) where A0:real
csin (Cx (&0 * A0)) = Cx (cos (pi / &2)) where A0:real
csin (Cx (&0) * A0) = Cx (cos (pi / &2)) where A0:real^2
```

The HOL-specialized probabilistic parsing and HOL typechecking phases are fast (given that the input sentence is not too long), because we limit the number of required parses to 16, and preselect only the 1024 closest grammar trees for the grammar training by running a k-nearest neighbor (k-NN) filter using n-gram (unigram, bigram and trigram) representations of all Flyspeck theorems in their ambiguous form. Thus the four first phases – k-NN filtering, grammar induction, probabilistic parsing, and HOL typechecking – typically take several seconds,

---

[5] http://colo12-c703.uibk.ac.at/hh/parse.html.

giving real-time feedback to the user. The AI/ATP phase is slow, because for maximal semantic performance it typically runs parallelized (14-CPU) AI/ATP methods selecting relevant premises from the whole Flyspeck, and this needs to be done for a dozen of the most probable and typechecked parse trees. This is however a "mere hardware" issue: If we had a 200-core server rather than the current one, the 16 best parses could be attacked by HOL(y)Hammer in parallel, and the AI/ATP phase would feel much more real-time too. Some screenshots of the service in action are available on our web page.[6]

6   Evaluation on Flyspeck

Once the methods were reasonably scaled up to the whole Flyspeck, we have done a large-scale training/testing evaluation (100-fold cross-validation) on the whole corpus of 22000 theorems. It proceeds as follows:

1. We create the ambiguous sentences and the disambiguated grammar trees from all 22k Flyspeck theorems as described in Sect. 3. These sets are permuted randomly and split into 100 equally sized chunks of about 220 trees or sentences. The trees serve for training and the sentences for evaluation.
2. For each testing chunk C_i ($i \in 1..100$) of 220 sentences we take the union of the 99 chunks of grammar trees (altogether about 21800 trees) that correspond to the remaining sentences and build the probabilistic grammar on them - this is fast, taking several seconds. This way we avoid training on the parse trees of the testing sentences.
3. Then we try to get the best 20 parse trees for all the 220 sentences in C using that grammar. This takes on average 4 s for each sentence, i.e. the whole parsing takes about 90000 CPU seconds = 25 CPU hours.
4. The parse trees are again transformed into HOL syntax trees, typechecked in HOL, and a single AI/ATP method is run on each typechecked tree for 30 s (using Vampire and 128 most relevant Flyspeck premises). This is weaker than using the full HOL(y)Hammer online system, but we cannot afford the 14-fold AI/ATP parallelization due to the number of parse trees.
5. 698549 of the parse trees typecheck (221145 do not), resulting in 302329 distinct (modulo alpha) HOL formulas. These are subjected to ATP, i.e., we run for ca 9000000 CPU seconds = 2500 CPU hours. This is done on a large server with 100-fold parallelization, taking about one day of real time.

We can automatically prove about 70957 (23.5 %) of the 302329 typechecked formulas.[7] However, first analysis shows that many of them are provable only because they are parsed incorrectly, for example when the antecedent of an implication becomes trivially false. In this first experiment we do not recognize such cases, however it should not be too difficult to remove such cases with another ATP round that checks for the unsatisfiability of antecedents. Such

---

[6] http://colo12-c703.uibk.ac.at/hh/parseimg.html.

[7] The exact list is at http://mizar.cs.ualberta.ca/~mptp/i2f/00proved2.

additional semantic checks could also eventually become a part of the (more tightly integrated) semantic-parsing toolchain.

In 39.4 % of the 22000 cases, the HOL formula resulting from one of the sentence's 20 parse trees is alpha-equal to the correct (training) original HOL formula, and its average rank there is 9.34. This is quite encouraging statistics, given that this runs efficiently over whole Flyspeck with quite a high number of introduced ambiguities, and many more sophisticated probabilistic parsing tricks (such as full-scale lexicalization) have not been used yet.

Interestingly, 0.2 % of the 22000 cases produce a parse tree that is the same as an existing training tree, but of a different theorem. This means – as can already be seen from the online system parses above – that thanks to the probabilistic behavior the system also (quite necessarily) functions as a conjecture maker. Given a seed of symbols, the system tries to figure out the most probable ways how to give meaning to them, a bit like the Dijkstra's "informal mathematical lot" does. Quite likely one of the many interesting future directions is to evolve one version of the system in such a way that the conjectures are as interesting as possible, using our probabilistic setting to avoid today's brute-force methods.

References

1. Blanchette, J. C., Kaliszyk, C., Paulson, L.C., Urban, J.: Hammering towards QED. Accepted to J. Formalized Reasoning (2015). Preprint at http://www4.in. tum.de/~blanchet/h4qed.pdf
2. Dijkstra, E.W.: The fruits of misunderstanding. Elektronische Rechenanlagen **25**(6), 10–13 (1983)
3. Hales, T.: Dense Sphere Packings: A Blueprint for Formal Proofs. London Mathematical Society Lecture Note Series, vol. 400. Cambridge University Press, Cambridge (2012)
4. Kaliszyk, C., Urban, J.: Stronger automation for Flyspeck by feature weighting and strategy evolution. In: Blanchette, J.C., Urban, J. (eds.) PxTP 2013. EPiC Series, vol. 14, pp. 87–95. EasyChair (2013)
5. Kaliszyk, C., Urban, J.: HOL(y)Hammer: online ATP service for HOL Light. Math. Comput. Sci. **9**(1), 5–22 (2015)
6. Kaliszyk, C., Urban, J., Vyskočil, J., Geuvers, H.: Developing corpus-based translation methods between informal and formal mathematics: project description. In: Watt, S.M., Davenport, J.H., Sexton, A.P., Sojka, P., Urban, J. (eds.) CICM 2014. LNCS, vol. 8543, pp. 435–439. Springer, Heidelberg (2014)
7. Kühlwein, D., van Laarhoven, T., Tsivtsivadze, E., Urban, J., Heskes, T.: Overview and evaluation of premise selection techniques for large theory mathematics. In: Gramlich, B., Miller, D., Sattler, U. (eds.) IJCAR 2012. LNCS, vol. 7364, pp. 378–392. Springer, Heidelberg (2012)
8. Tankink, C., Kaliszyk, C., Urban, J., Geuvers, H.: Formal mathematics on display: a wiki for flyspeck. In: Carette, J., Aspinall, D., Lange, C., Sojka, P., Windsteiger, W. (eds.) CICM 2013. LNCS, vol. 7961, pp. 152–167. Springer, Heidelberg (2013)
9. Younger, D.H.: Recognition and parsing of context-free languages in time n^3. Inf. Control **10**(2), 189–208 (1967)
10. Zinn, C.: Understanding informal mathematical discourse. Ph.D. thesis, University of Erlangen-Nuremberg (2004)

A Consistent Foundation for Isabelle/HOL

Ondřej Kunčar[1]($\boxtimes$) and Andrei Popescu[2]

[1] Fakultät für Informatik, Technische Universität München, Munich, Germany
kuncar@in.tum.de
[2] Department of Computer Science, School of Science and Technology,
Middlesex University, London, UK
a.popescu@mdx.ac.uk

Abstract. The interactive theorem prover Isabelle/HOL is based on the well understood Higher-Order Logic (HOL), which is widely believed to be consistent (and provably consistent in set theory by a standard semantic argument). However, Isabelle/HOL brings its own personal touch to HOL: *overloaded constant definitions*, used to achieve *Haskell-like type classes* in the user space. These features are a delight for the users, but unfortunately are not easy to get right as an extension of HOL—they have a history of inconsistent behavior. It has been an open question under which criteria overloaded constant definitions and type definitions can be combined together while still guaranteeing consistency. This paper presents a solution to this problem: non-overlapping definitions and termination of the definition-dependency relation (tracked not only through constants but also through types) ensures relative consistency of Isabelle/HOL.

1 Introduction

Polymorphic HOL, more precisely, Classic Higher-Order Logic with Infinity, Hilbert Choice and Rank-1 Polymorphism, endowed with a mechanism for constant and type definitions, was proposed in the nineties as a logic for interactive theorem provers by Mike Gordon, who also implemented the seminal HOL theorem prover [12]. This system has produced many successors and emulators known under the umbrella term "HOL-based provers" (e.g., [2,3,5,14,27]), launching a very successful paradigm in interactive theorem proving.

A main strength of HOL-based provers is a sweet spot in expressiveness versus complexity: HOL on the one hand is sufficient for most mainstream mathematics and computer science applications, and on the other is a well-understood logic. In particular, the consistency of HOL has a standard semantic argument, comprehensible to any science graduate: one interprets its types as sets, in particular the function types as sets of functions, and the terms as elements of these sets, in a natural way; the rules of the logic are easily seen to hold in this model. The definitional mechanism has two flavors:

- New constants c are introduced by equations $c \equiv t$, where t is a closed term not containing c.

© Springer International Publishing Switzerland 2015
C. Urban and X. Zhang (Eds.): ITP 2015, LNCS 9236, pp. 234–252, 2015.
DOI: 10.1007/978-3-319-22102-1_16

– New types τ are introduced by "typedef" equations $\tau \equiv t$, where $t : \sigma \Rightarrow$ bool is a predicate on an existing type σ (not containing τ anywhere in the types of its subterms)—intuitively, the type τ is carved out as the t-defined subset of σ.

Again, this mechanism is manifestly consistent by an immediate semantic argument [30]; alternatively, its consistency can be established by regarding definitions as mere abbreviations (which here are non-cyclic by construction).

Polymorphic HOL with Ad Hoc Overloading. Isabelle/HOL [26,27] adds its personal touch to the aforementioned sweet spot: it extends polymorphic HOL with a mechanism for (ad hoc) overloading. As an example, consider the following Nominal-style [33] definitions, where prm is the type of finite-support bijections on an infinite type atom, and where we write apply pi a for the application of a bijection pi to an atom a:

Example 1. consts perm : prm $\Rightarrow \alpha \Rightarrow \alpha$
defs perm_atom: perm pi (a : atom) $\equiv$ apply pi a
defs perm_nat: perm pi (n : nat) $\equiv$ n
defs perm_list: perm pi (xs : α list) $\equiv$ map (perm pi) xs

Above, the constant perm is declared using the keyword "consts"—its intended behavior is the application of a permutation to all atoms contained in an element of a type α. Then, using the keyword "defs", several overloaded definitions of perm are performed for different instances of α. For atoms, perm applies the permutation; for numbers (which don't have atoms), perm is the identity function; for α list, the instance of perm is defined in terms of the instance for the component α. All these definitions are non-overlapping and their type-based recursion is terminating, hence Isabelle is fine with them.

Inconsistency. Of course, one may not be able to specify all the relevant instances immediately after declaring a constant like perm—at a later point, a user may define their own atom-container type, such as[1]

```
datatype myTree = Atom atom | LNode atom list | FNode nat => atom
```

and instantiate perm to this type. (In fact, the Nominal package automates instantiations for user-requested datatypes, including terms with bindings.) To support such delayed instantiations, which are also crucial for the implementation of type classes [13,35], Isabelle/HOL allows intermixing definitions of instances of an overloaded constant with definitions of other constants and types. Unfortunately, the improper management of the intermixture leads to inconsistency: Isabelle/HOL accepts the following definitions[2].

Example 2. consts c : α
typedef T = {True, c} by blast
defs c_bool_def: c:bool $\equiv \neg (\forall(x:T) \ y. \ x = y)$

---

[1] In Isabelle/HOL, as in any HOL-based prover, the "datatype" command is not primitive, but is compiled into "typedef."

[2] This example works in Isabelle2014; our correction patch [1] based on the results of this paper and in its predecessor [19] is being evaluated at the Isabelle headquarters.

which immediately yield a proof of False:

lemma L: (∀(x:T) y. x = y) ↔ c
using Rep_T Rep_T_inject Abs_T_inject by (cases c:bool) force+

theorem False
using L unfolding c_bool_def by auto

The inconsistency argument takes advantage of the circularity $T \rightsquigarrow c_{bool} \rightsquigarrow T$ in the dependency relation introduced by the definitions: one first defines T to contain only one element just in case c_{bool} is True, and then defines c to be True just in case T contains more than one element.

Our Contribution. In this paper, we provide the following, in the context of polymorphic HOL extended with ad hoc overloading (Sect. 3):

- A definitional dependency relation that factors in both constant and type definitions in a sensible fashion (Sect. 4.1).
- A proof of consistency of any set of constant and type definitions whose dependency relation satisfies reasonable conditions, which accept Example 1 and reject Example 2 (Sect. 4).
- A new semantics for polymorphic HOL (Sect. 4.4) that guides both our definition of the dependency relation and our proof of consistency.

More details on our constructions and proofs can be found in the technical report [20]. We hope that our work settles the consistency problem for Isabelle/HOL's extension of HOL, while showing that the mechanisms of this logic admit a natural and well-understandable semantics. We start with a discussion of related work, including previous attempts to establish consistency (Sect. 2). Later we also show how this work fits together with previous work by the first author (Sect. 5).

2   Related Work

Type Classes and Overloading. Type classes were introduced in Haskell by Wadler and Blott [34]—they allow programmers to write functions that operate generically on types endowed with operations. For example, assuming a type α which is a semigroup (i.e., comes with a binary associative operation +), one can write a program that computes the sum of all the elements in an α-list. Then the program can be run on any concrete type T that replaces α provided T has this binary operation +. Prover-powered type classes were introduced by Nipkow and Snelting [28] in Isabelle/HOL and by Sozeau and Oury [32] in Coq—they additionally feature verifiability of the type-class conditions upon instantiation: a type T is accepted as a member of the semigroup class only if associativity can be proved for its + operation.

Whereas Coq implements type classes directly by virtue of its powerful type system, Isabelle/HOL relies on arbitrary ad hoc overloading: to introduce the semigroup class, the system declares a "global" constant $+ : \alpha \Rightarrow \alpha \Rightarrow \alpha$ and defines the associativity predicate; then different instance types T are registered after defining the corresponding overloaded operation $+ : T \Rightarrow T \Rightarrow T$ and verifying the condition. Our current paper focuses not on the Isabelle/HOL type classes, but on the consistency of the mechanism of ad hoc overloading which makes them possible.

Previous Consistency Attempts. The settling of this consistency problem has been previously attempted by Wenzel [35] and Obua [29]. In 1997, Wenzel defined a notion of safe theory extension and showed that overloading conforms to this notion. But he did not consider type definitions and worked with a simplified version of the system where all overloadings for a constant c are provided at once. Years later, when Obua took over the problem, he found that the overloadings were almost completely unchecked—the following trivial inconsistency was accepted by Isabelle2005:

Example 3. consts c : $\alpha \Rightarrow$ bool
defs c (x : α list $\times$ α) $\equiv$ c (snd x # fst x)
defs c (x : α list) $\equiv \neg$ c (tl x, hd x)

lemma c [x] $= \neg$ c([], x) $= \neg$ c[x]

Obua noticed that the rewrite system produced by the definitions has to terminate to avoid inconsistency, and implemented a private extension based on a termination checker. He did consider intermixing overloaded constant definitions and type definitions but his syntactic proof sketch misses out inconsistency through type definitions.

Triggered by Obua's observations, Wenzel implemented a simpler and more structural solution based on work of Haftmann, Obua and Urban: fewer overloadings are accepted in order to make the consistency/termination check decidable (which Obua's original check is not). Wenzel's solution has been part of the kernel since Isabelle2007 without any important changes—parts of this solution (which still misses out dependencies through types) are described by Haftmann and Wenzel [13].

In 2014, we discovered that the dependencies through types are not covered (Example 2), as well as an unrelated issue in the termination checker that led to an inconsistency even without exploiting types. Kunčar [19] amended the latter issue by presenting a modified version of the termination checker and proving its correctness. The proof is general enough to cover termination of the definition dependency relation through types as well. Our current paper complements this result by showing that termination leads to consistency.

Inconsistency Club. Inconsistency problems arise quite frequently with provers that step outside the safety of a simple and well-understood logic kernel. The various proofs of False in the early PVS system [31] are folklore. Coq's [8] current stable version[3] is inconsistent in the presence of Propositional Extensionality;

---

[3] Namely, Coq 8.4pl5; the inconsistency is fixed in Coq 8.5 beta.

this problem stood undetected by the Coq users and developers for 17 years; interestingly, just like the Isabelle/HOL problem under scrutiny, it is due to an error in the termination checker [11]. Agda [10] suffers from similar problems [23]. The recent Dafny prover [21] proposes an innovative combination of recursion and corecursion whose initial version turned out to be inconsistent [9].

Of course, such "dangerous" experiments are often motivated by better support for the users' formal developments. The Isabelle/HOL type class experiment was practically successful: substantial developments such as the Nominal [17, 33] and HOLCF [24] packages and Isabelle's mathematical analysis library [16] rely heavily on type classes. One of Isabelle's power users writes [22]: "Thanks to type classes and refinement during code generation, our light-weight framework is flexible, extensible, and easy to use."

Consistency Club. Members of this select club try to avoid inconsistencies by impressive efforts of proving soundness of logics and provers by means of interactive theorem provers themselves. Harisson's pioneering work [15] uses HOL Light to give semantic proofs of soundness of the HOL logic without definitional mechanisms, in two flavors: either after removing the infinity axiom from the object HOL logic, or after adding a "universe" axiom to HOL Light; a proof that the OCaml implementation of the core of HOL Light correctly implements this logic is also included. Kumar et al. [18] formalize in HOL4 the semantics and the soundness proof of HOL, with its definitional principles included; from this formalization, they extract a verified implementation of a HOL theorem prover in CakeML, an ML-like language featuring a verified compiler. None of the above verified systems factor in ad-hoc overloading, the starting point of our work.

Outside the HOL-based prover family, there are formalizations of Milawa [25], Nuprl [4] and fragments of Coq [6,7].

3   Polymorphic HOL with Ad Hoc Overloading

Next we present syntactic aspects of our logic of interest (syntax, deduction and definitions) and formulate its *consistency problem*.

3.1   Syntax

In what follows, by "countable" we mean "either finite or countably infinite." All throughout this section, we fix the following:

- A countably infinite set TVar, of *type variables*, ranged over by α, β
- A countably infinite set Var, of *(term) variables*, ranged over by x, y, z
- A countable set K of symbols, ranged over by k, called *type constructors*, containing three special symbols: "bool", "ind" and "$\Rightarrow$" (aimed at representing the type of booleans, an infinite type and the function type constructor, respectively)

We also fix a function arOf : $K \rightarrow \mathbb{N}$ associating an arity to each type constructor, such that arOf(bool) = arOf(ind) = 0 and arOf($\Rightarrow$) = 2. We define the set Type, ranged over by σ, τ, of *types*, inductively as follows:

- $\mathsf{TVar} \subseteq \mathsf{Type}$
- $(\sigma_1, \ldots, \sigma_n)k \in \mathsf{Type}$ if $\sigma_1, \ldots, \sigma_n \in \mathsf{Type}$ and $k \in K$ such that $\mathsf{arOf}(k) = n$

Thus, we use postfix notation for the application of an n-ary type constructor k to the types $\sigma_1, \ldots, \sigma_n$. If $n = 1$, instead of $(\sigma)k$ we write $\sigma\, k$ (e.g., σ list).

A *typed variable* is a pair of a variable x and a type σ, written x_σ. Given $T \subseteq \mathsf{Type}$, we write Var_T for the set of typed variables x_σ with $\sigma \in T$. Finally, we fix the following:

- A countable set Const, ranged over by c, of symbols called *constants*, containing five special symbols: "$\rightarrow$", "$=$", "some" "zero", "suc" (aimed at representing logical implication, equality, Hilbert choice of "some" element from a type, zero and successor, respectively).
- A function $\mathsf{tpOf} : \mathsf{Const} \rightarrow \mathsf{Type}$ associating a type to every constant, such that:

$$
\begin{aligned}
\mathsf{tpOf}(\rightarrow) &= \mathsf{bool} \Rightarrow \mathsf{bool} \Rightarrow \mathsf{bool} & \qquad \mathsf{tpOf}(\mathsf{zero}) &= \mathsf{ind} \\
\mathsf{tpOf}(=) &= \alpha \Rightarrow \alpha \Rightarrow \mathsf{bool} & \qquad \mathsf{tpOf}(\mathsf{suc}) &= \mathsf{ind} \Rightarrow \mathsf{ind} \\
\mathsf{tpOf}(\mathsf{some}) &= (\alpha \Rightarrow \mathsf{bool}) \Rightarrow \alpha
\end{aligned}
$$

We define the type variables of a type, $\mathsf{TV} : \mathsf{Type} \rightarrow \mathcal{P}(\mathsf{TVar})$, as expected. A type σ is called *ground* if $\mathsf{TV}(\sigma) = \emptyset$. We let GType be the set of ground types.

A *type substitution* is a function $\rho : \mathsf{TVar} \rightarrow \mathsf{Type}$; we let TSubst denote the set of type substitutions. Each $\rho \in \mathsf{TSubst}$ extends to a homonymous function $\rho : \mathsf{Type} \rightarrow \mathsf{Type}$ by defining $\rho((\sigma_1, \ldots, \sigma_n)k) = (\rho(\sigma_1), \ldots, \rho(\sigma_n))k$. We let $\mathsf{GTSubst}$ be the set of all ground type substitutions $\theta : \mathsf{TVar} \rightarrow \mathsf{GType}$, which again extend to homonymous functions $\theta : \mathsf{Type} \rightarrow \mathsf{GType}$.

We say that σ is an *instance* of τ, written $\sigma \leq \tau$, if there exists $\rho \in \mathsf{TSubst}$ such that $\rho(\tau) = \sigma$. Two types σ and τ are called *orthogonal*, written $\sigma \# \tau$, if they have no common instance.

Given $c \in \mathsf{Const}$ such that $\sigma \leq \mathsf{tpOf}(c)$, we call the pair (c, σ), written c_σ, an *instance of* c. A *constant instance* is therefore any such pair c_σ. We let CInst be the set of all constant instances, and GCInst the set of constant instances whose type is ground. We extend the notions of being an instance ($\leq$) and being orthogonal ($\#$) from types to constant instances, as follows:

$$
c_\tau \leq d_\sigma \text{ iff } c = d \text{ and } \tau \leq \sigma \qquad\qquad c_\tau \# d_\sigma \text{ iff } c \neq d \text{ or } \tau \# \sigma
$$

We also define tpOf for constant instances by $\mathsf{tpOf}(c_\sigma) = \sigma$.

The tuple $(K, \mathsf{arOf}, C, \mathsf{tpOf})$, which will be fixed in what follows, is called a *signature*. This signature's *pre-terms*, ranged over by s, t, are defined by the grammar:

$$
t = x_\sigma \mid c_\sigma \mid t_1\, t_2 \mid \lambda x_\sigma.t
$$

Thus, a pre-term is either a typed variable, or a constant instance, or an application, or an abstraction. As usual, we identify pre-terms modulo alpha-equivalence.

Typing of pre-terms is defined in the expected way (by assigning the most general type possible); a *term* is a well-typed pre-term, and Term denotes the set of terms. Given $t \in$ Term, we write tpOf(t) for its (uniquely determined, most general) type and FV(t) for the set of its free (term) variables. We call t *closed* if FV$(t) = \emptyset$.

We let TV(t) denote the set of type variables occurring in t. A term t is called *ground* if TV$(t) = \emptyset$. Thus, closedness refers to the absence of free (term) variables in a term, whereas groundness refers to the absence of type variables in a type or a term. Note that, for a term, being ground is a stronger condition than having a ground type: $(\lambda x_\alpha. c_{\text{bool}}) x_\alpha$ has the ground type bool, but is not ground.

We can apply a type substitution ρ to a term t, written $\rho(t)$, by applying ρ to all the type variables occurring in t; and similarly for ground type substitutions θ; note that $\theta(t)$ is always a ground term.

A *formula* is a term of type bool. We let Fmla, ranged over by φ, denote the set of formulas. The formula connectives and quantifiers are defined in the standard way, starting from the implication and equality primitives.

When writing concrete terms or formulas, we often omit indicating the type in occurrences of bound variables—e.g., we may write $\lambda x_\alpha. x$ instead of $\lambda x_\alpha. x_\alpha$.

3.2   Built-Ins and Non-Built-Ins

The distinction between built-in and non-built-in types and constants will be important for us, since we will employ a slightly non-standard semantics only for the latter.

A *built-in type* is any type of the form bool, ind, or $\sigma_1 \Rightarrow \sigma_2$ for $\sigma_1, \sigma_2 \in$ Type. We let Type$^\bullet$ denote the set of types that are *not* built-in. Note that a non-built-in type can have a built-in type as a subtype, and vice versa; e.g., if list is a type constructor, then bool list and $(\alpha \Rightarrow \beta)$ list are non-built-in types, whereas $\alpha \Rightarrow \beta$ list is a built-in type. We let GType$^\bullet =$ GType $\cap$ Type$^\bullet$ denote the set of ground non-built-in types.

Given a type σ, we define types$^\bullet(\sigma)$, the *set of non-built-in types* of σ, as follows:

types$^\bullet$(bool) $=$ types$^\bullet$(ind) $= \emptyset$
types$^\bullet((\sigma_1, \ldots, \sigma_n) k) = \{(\sigma_1, \ldots, \sigma_n) k\}$, if k is different from $\Rightarrow$
types$^\bullet(\sigma_1 \Rightarrow \sigma_2) =$ types$^\bullet(\sigma_1) \cup$ types$^\bullet(\sigma_2)$

Thus, types$^\bullet(\sigma)$ is the smallest set of non-built-in types that can produce σ by repeated application of the built-in type constructors. E.g., if the type constructors prm (0-ary) and list (unary) are in the signature and if σ is (bool $\Rightarrow$ α list) $\Rightarrow$ prm $\Rightarrow$ (bool $\Rightarrow$ ind) list, then types$^\bullet(\sigma)$ has three elements: α list, prm and (bool $\Rightarrow$ ind) list.

A built-in constant is a constant of the form $\rightarrow$, $=$, some, zero or suc. We let CInst$^\bullet$ be the set of constant instances that are *not* instances of built-in constants, and GCInst$^\bullet \subseteq$ CInst$^\bullet$ be its subset of ground constants.

As a general notation rule: the prefix "G" indicates ground items, whereas the superscript • indicates non-built-in items, where an item can be either a type or a constant instance. In our semantics (Sect. 4.4), we will stick to the standard interpretation of built-in items, whereas for non-built-in items we will allow an interpretation looser than customary. The standardness of the bool, ind and function-type interpretation will allow us to always automatically extend the interpretation of a set of non-built-in types to the interpretation of its built-in closure.

Given a term t, we let $\mathsf{consts}^\bullet(t) \subseteq \mathsf{CInst}^\bullet$ be the set of all non-built-in constant instances occurring in t and $\mathsf{types}^\bullet(t) \subseteq \mathsf{Type}^\bullet$ be the set of all non-built-in types that compose the types of non-built-in constants and (free or bound) variables occurring in t. Note that the $\mathsf{types}^\bullet$ operator is overloaded for types and terms.

$$\mathsf{consts}^\bullet(x_\sigma) = \emptyset \qquad\qquad \mathsf{types}^\bullet(x_\sigma) = \mathsf{types}^\bullet(\sigma)$$

$$\mathsf{consts}^\bullet(c_\sigma) = \begin{cases} \{c_\sigma\} & \text{if } c_\sigma \in \mathsf{CInst}^\bullet \\ \emptyset & \text{otherwise} \end{cases} \qquad \mathsf{types}^\bullet(c_\sigma) = \mathsf{types}^\bullet(\sigma)$$

$$\mathsf{consts}^\bullet(t_1\,t_2) = \mathsf{consts}^\bullet(t_1) \cup \mathsf{consts}^\bullet(t_2) \qquad \mathsf{types}^\bullet(t_1\,t_2) = \mathsf{types}^\bullet(t_1) \cup \mathsf{types}^\bullet(t_2)$$

$$\mathsf{consts}^\bullet(\lambda x_\sigma.\,t) = \mathsf{consts}^\bullet(t) \qquad\qquad \mathsf{types}^\bullet(\lambda x_\sigma.\,t) = \mathsf{types}^\bullet(\sigma) \cup \mathsf{types}^\bullet(t)$$

Note that the $\mathsf{consts}^\bullet$ and $\mathsf{types}^\bullet$ operators commute with ground type substitutions (and similarly with type substitutions, of course):

Lemma 4. (1) $\mathsf{consts}^\bullet(\theta(t)) = \{c_{\theta(\sigma)} \mid c_\sigma \in \mathsf{consts}^\bullet(t)\}$
(2) $\mathsf{types}^\bullet(\theta(t)) = \{\theta(\sigma) \mid \sigma \in \mathsf{types}^\bullet(t)\}$.

3.3 Deduction

If D is a finite set of closed formulas, a.k.a. a *theory*, and φ is a closed formula, we write $D \vdash \varphi$ for the deducibility of φ from D according to the standard deduction rules of polymorphic HOL [12,27].[4] A theory D is called *consistent* if there exists φ such that $D \not\vdash \varphi$ (or equivalently if $D \not\vdash \mathsf{False}$, where the formula False is defined in a standard way from the built-in constants).

3.4 Definitional Theories

We are interested in the consistency of theories arising from constant-instance and type definitions, which we call definitional theories.

Given $c_\sigma \in \mathsf{CInst}^\bullet$ and a closed term $t \in \mathsf{Term}_\sigma$, we let $c_\sigma \equiv t$ denote the formula $c_\sigma = t$. We call $c_\sigma \equiv t$ a *constant-instance definition* provided $\mathsf{TV}(t) \subseteq \mathsf{TV}(c_\sigma)$ (i.e., $\mathsf{TV}(t) \subseteq \mathsf{TV}(\sigma)$).

Given the types $\tau \in \mathsf{Type}^\bullet$ and $\sigma \in \mathsf{Type}$ and the closed term t whose type is $\sigma \Rightarrow \mathsf{bool}$, we let $\tau \equiv t$ denote the formula

---

[4] The deduction in polymorphic HOL takes place using open formulas in contexts. In addition, Isabelle/HOL distinguishes between theory contexts and proof contexts. We ignore these aspects in our presentation here, since they do not affect our consistency argument.

$$(\exists x_\sigma.\, t\, x) \rightarrow$$
$$\exists rep_{\tau\Rightarrow\sigma}.\,\exists abs_{\sigma\Rightarrow\tau}.$$
$$(\forall x_\tau.\, t\,(rep\, x)) \wedge (\forall x_\tau.\, abs\,(rep\, x) = x) \wedge (\forall y_\sigma.\, t\, y \rightarrow rep\,(abs\, y) = y).$$

We call $\tau \equiv t$ a *type definition*, provided $\mathsf{TV}(t) \subseteq \mathsf{TV}(\tau)$ (which also implies $\mathsf{TV}(\sigma) \subseteq \mathsf{TV}(\tau)$).

Note that we defined $\tau \equiv t$ *not* to mean:

(∗): *The type τ is isomorphic, via abs and rep, to the subset of σ given by t*

as customary in most HOL-based systems, but rather to mean:

If t gives a nonempty subset of σ, then (∗) *holds*

Moreover, note that we do *not* require τ to have the form $(\alpha_1, \ldots, \alpha_n)k$, as is currently required in Isabelle/HOL and the other HOL provers, but, more generally, allow any $\tau \in \mathsf{Type}^\bullet$. (To ensure consistency, we will also require that τ has no common instance with the left-hand side of any other type definition.) This enables an interesting feature: ad hoc overloading for type definitions. For example, given a unary type constructor tree, we can have totally different definitions for nat tree, bool tree and α list tree.

In general, a *definition* will have the form $u \equiv t$, where u is either a constant instance or a type and t is a term (subject to the specific constraints of constant-instance and type definitions). u and t are said to be the left-hand and right-hand sides of the definition. A *definitional theory* is a finite set of definitions.

3.5 The Consistency Problem

An Isabelle/HOL development proceeds by:

1. declaring constants and types,
2. defining constant instances and types,
3. stating and proving theorems using the deduction rules of polymorphic HOL.

Consequently, at any point in the development, one has:

1. a signature $(K, \mathsf{arOf} : K \rightarrow \mathbb{N}, \mathsf{Const}, \mathsf{tpOf} : \mathsf{Const} \rightarrow \mathsf{Type})$,
2. a definitional theory D,
3. other proved theorems.

In our abstract formulation of Isabelle/HOL's logic, we do not represent explicitly point 3, namely the stored theorems that are not produced as a result of definitions, i.e., are not in D. The reason is that, in Isabelle/HOL, the theorems in D are not influenced by the others. Note that this is not the case of the other HOL provers, due to the type definitions: there, $\tau \equiv t$, with $\mathsf{tpOf}(t) = \sigma \Rightarrow \mathsf{bool}$, is introduced in the unconditional form (∗), and only after the user has proved that t gives a nonempty subset (i.e., that $\exists x_\sigma.\, t\, x$ holds). Of course, Isabelle/HOL's behavior converges with standard HOL behavior since the user is also required to prove nonemptiness, after which (*) is inferred by the system—however, this last inference step is normal deduction, having nothing to do with the definition itself. This very useful trick, due to Wenzel, cleanly separates definitions from proofs. In summary, we only need to guarantee the consistency of D:

> **The Consistency Problem:** Find a sufficient criterion for a definitional theory D to be consistent (while allowing flexible overloading, as discussed in the introduction).

4   Our Solution to the Consistency Problem

Assume for a moment we have a proper dependency relation between defined items, where the defined items can be types or constant instances. Obviously, the closure of this relation under type substitutions needs to terminate, otherwise inconsistency arises immediately, as shown in Example 3. Moreover, it is clear that the left-hand sides of the definitions need to be orthogonal: defining $c_{\alpha \times \text{ind} \Rightarrow \text{bool}}$ to be $\lambda x_{\alpha \times \text{ind}}.\text{False}$ and $c_{\text{ind} \times \alpha \Rightarrow \text{bool}}$ to be $\lambda x_{\text{ind} \times \alpha}.\text{True}$ yields $\lambda x_{\text{ind} \times \text{ind}}.\text{False} = c_{\text{ind} \times \text{ind} \Rightarrow \text{bool}} = \lambda x_{\text{ind} \times \text{ind}}.\text{True}$ and hence $\text{False} = \text{True}$.

It turns out that these necessary criteria are also *sufficient* for consistency. This was also believed by Wenzel and Obua; what they were missing was a proper dependency relation and a transparent argument for its consistency, which is what we provide next.

4.1   Definitional Dependency Relation

Given any binary relation R on $\text{Type}^\bullet \cup \text{CInst}^\bullet$, we write R^+ for its transitive closure, R^* for its reflexive-transitive closure and $R^\downarrow$ for its (type-)substitutive closure, defined as follows: $p\, R^\downarrow q$ iff there exist p', q' and a type substitution ρ such that $p = \rho(p')$, $q = \rho(q')$ and $p'\, R\, q'$. We say that a relation R is *terminating* if there exists no sequence $(p_i)_{i \in \mathbb{N}}$ such that $p_i\, R\, p_{i+1}$ for all i.

Let us fix a definitional theory D. We say D is *orthogonal* if for all distinct definitions $u \equiv t$ and $u' \equiv t'$ in D, we have one of the following cases:

- either one of $\{u, u'\}$ is a type and the other is constant instance,
- or both u and u' are types and are orthogonal $(u \# u')$,
- or both u and u' are constant instances and are orthogonal $(u \# u')$.

We define the binary relation $\rightsquigarrow$ on $\text{Type}^\bullet \cup \text{CInst}^\bullet$ by setting $u \rightsquigarrow v$ iff one of the following holds:

1. there exists a (constant-instance or type) definition in D of the form $u \equiv t$ such that $v \in \text{consts}^\bullet(t) \cup \text{types}^\bullet(t)$,
2. there exists $c \in \text{Const}^\bullet$ such that $u = c_{\text{tpOf}(c)}$ and $v \in \text{types}^\bullet(\text{tpOf}(c))$.

We call $\rightsquigarrow$ the *dependency relation* (associated to D).

Thus, when defining an item u by means of t (as in $u \equiv t$), we naturally record that u depends on the constants and types appearing in t (clause 1); moreover, any constant c should depend on its type (clause 2). But notice the bullets! We only record dependencies on the non-built-in items, since intuitively the built-in items have a pre-determined semantics which cannot be redefined or overloaded, and hence by themselves cannot introduce inconsistencies. Moreover, we do not dig for dependencies under any non-built-in type constructor—this can be seen from the definition of the types$^{\bullet}$ operator on types which yields a singleton whenever it meets a non-built-in type constructor; the rationale for this is that a non-built-in type constructor has an "opaque" semantics which does not expose the components (as does the function type constructor). These intuitions will be made precise by our semantics in Sect. 4.4.

Consider the following example, where the definition of $\alpha\ k$ is omitted:

Example 5. consts c : α d : α
typedef α k = ...
defs c : ind k $\Rightarrow$ bool $\equiv$ (d : bool k k $\Rightarrow$ ind k $\Rightarrow$ bool) (d : bool k k)

We record that $c_{\mathsf{ind}\ k \Rightarrow \mathsf{bool}}$ depends on the non-built-in constants $d_{\mathsf{bool}\ k\ k \Rightarrow \mathsf{ind}\ k \Rightarrow \mathsf{bool}}$ and $d_{\mathsf{bool}\ k\ k}$, and on the non-built-in types bool $k\ k$ and ind k. We do *not* record any dependency on the built-in types bool $k\ k \Rightarrow$ ind $k \Rightarrow$ bool, ind $k \Rightarrow$ bool or bool. Also, we do *not* record any dependency on bool k, which can only be reached by digging under k in bool $k\ k$.

4.2   The Consistency Theorem

We can now state our main result. We call a definitional theory D *well-formed* if it is orthogonal and the substitutive closure of its dependency relation, $\leadsto^{\downarrow}$, is terminating.

Note that a well-formed definitional theory is allowed to contain definitions of two different (but orthogonal) instances of the same constant—this ad-hoc overloading facility is a distinguishing feature of Isabelle/HOL among the HOL provers.

Theorem 6. If D is well-formed, then D is consistent.

Previous attempts to prove consistency employed syntactic methods [29,35]. Instead, we will give a semantic proof:

1. We define a new semantics of Polymorphic HOL, suitable for overloading and for which standard HOL deduction is sound (Sect. 4.4).
2. We prove that D has a model according to our semantics (Sect. 4.5).

Then 1 and 2 immediately imply consistency.

4.3   Inadequacy of the Standard Semantics of Polymorphic HOL

But why define a new semantics? Recall that our goal is to make sense of definitions as in Example 1. In the standard (Pitts) semantics [30], one chooses a "universe" collection of sets $\mathcal{U}$ closed under suitable set operations (function space, an infinite set, etc.) and interprets:

1. the built-in type constructors and constants as their standard counterparts in $\mathcal{U}$:
   - [bool] and [ind] are some chosen two-element set and infinite set in $\mathcal{U}$
   - $[\Rightarrow]$: $\mathcal{U} \to \mathcal{U} \to \mathcal{U}$ takes two sets $A_1, A_2 \in \mathcal{U}$ to the set of functions $A_1 \to A_2$
   - [True] and [False] are the two distinct elements of [bool], etc.
2. the non-built-in type constructors similarly:
   - a defined type prm or type constructor list as an element [prm] $\in \mathcal{U}$ or operator [list] : $\mathcal{U} \to \mathcal{U}$, produced according to their "typedef"
   - a polymorphic constant such as perm : prm $\to \alpha \to \alpha$ as a family [perm] $\in$ $\prod_{A \in \mathcal{U}} [\text{prm}] \to A \to A$.

In standard polymorphic HOL, perm would be either completely unspecified, or completely defined in terms of previously existing constants—this has a faithful semantic counterpart in $\mathcal{U}$. But now how to represent the overloaded definitions of perm from Example 1? In $\mathcal{U}$, they would become:

$[\text{perm}]_{[\text{atom}]} \; pi \; a = [\text{apply}] \; pi \; a$
$[\text{perm}]_{[\text{nat}]} \; pi \; n = n$
$[\text{perm}]_{[\text{list}](A)} \; pi \; xs = [\text{map}]_A \; ([\text{perm}]_A \; pi) \; xs$

There are two problems with these semantic definitions. First, given $B \in \mathcal{U}$, the value of $[\text{perm}]_B$ varies depending on whether B has the form [atom], or [nat], or [list](A) for some $A \in \mathcal{U}$; hence the interpretations of the type constructors need to be non-overlapping—this is not guaranteed by the assumptions about $\mathcal{U}$, so we would need to perform some low-level set-theoretic tricks to achieve the desired property. Second, even though the definitions are syntactically terminating, their semantic counterparts may not be: unless we again delve into low-level tricks in set theory (based on the axiom of foundation), it is not guaranteed that decomposing a set A_0 as [list](A_1), then A_1 as [list](A_2), and so on (as prescribed by the third equation for [perm]) is a terminating process.

Even worse, termination is in general a global property, possibly involving both constants and type constructors, as shown in the following example where c and k are mutually defined (so that a copy of $e_{\text{bool } k^n}$ is in bool k^{n+1} iff n is even):

Example 7. consts c : $\alpha \Rightarrow$ bool   d : α e : α
typedef α k = {d:α} $\cup$ {e : α . c (d : α)}
c (x : α k) $\equiv \neg$ c (d : α)
c (x : bool) $\equiv$ True

The above would require a set-theoretic setting where such fixpoint equations have solutions; this is in principle possible, provided we tag the semantic equations with enough syntactic annotations to guide the fixpoint construction. However, such a construction seems excessive given the original intuitive justification: the definitions are "OK" because they do not overlap and they terminate. On the other hand, a purely syntactic (proof-theoretic) argument also seems difficult due to the mixture of constant definitions and (conditional) type definitions.

Therefore, we decide to go for a natural syntactic-semantic blend, which avoids stunt performance in set theory: we do *not* semantically interpret the polymorphic types, but only the ground types, thinking of the former as "macros" for families of the latter. For example, $\alpha \Rightarrow \alpha$ list represents the family $(\tau \Rightarrow \tau \text{ list})_{\tau \in \mathsf{GType}}$. Consequently, we think of the meaning of $\alpha \Rightarrow \alpha$ list not as $\prod_{A \in \mathcal{U}} A \to [\text{list}](A)$, but rather as $\prod_{\tau \in \mathsf{GType}}[\tau] \to [\tau \text{ list}]$. Moreover, a polymorphic formula φ of type, say, $(\alpha \Rightarrow \alpha \text{ list}) \Rightarrow \mathsf{bool}$, will be considered true just in case all its ground instances of types $(\tau \Rightarrow \tau \text{ list}) \Rightarrow \mathsf{bool}$ are true.

Another (small) departure from standard HOL semantics is motivated by our goal to construct a model for a well-formed definitional theory. Whereas in standard semantics one first interprets all type constructors and constants and only afterwards extends the interpretation to terms, here we need to interpret some of the terms eagerly, *before* some of the types and constants. Namely, given a definition $u \equiv t$, we interpret t before we interpret u (according to t). This requires a straightforward refinement of the notion of semantic interpretation: to interpret a term, we only need the interpretations for a sufficient fragment of the signature containing all the items appearing in t.

4.4   Ground, Fragment-Localized Semantics for Polymorphic HOL

Recall that we have a fixed signature $(K, \mathsf{arOf}, \mathsf{Const}, \mathsf{tpOf})$, that $\mathsf{GType}^\bullet$ is the set of ground non-built-in types and $\mathsf{GCInst}^\bullet$ the set of ground non-built-in constant instances.

Given $T \subseteq \mathsf{Type}$, we define $\mathsf{Cl}(T) \subseteq \mathsf{Type}$, the *built-in closure* of T, inductively:

- $T \cup \{\mathsf{bool}, \mathsf{ind}\} \subseteq \mathsf{Cl}(T)$
- $\sigma_1 \Rightarrow \sigma_2 \in \mathsf{Cl}(T)$ if $\sigma_1 \in \mathsf{Cl}(T)$ and $\sigma_2 \in \mathsf{Cl}(T)$.

I.e., $\mathsf{Cl}(T)$ is the smallest set of types built from T by repeatedly applying built-in type constructors.

A *(signature) fragment* is a pair (T, C) with $T \subseteq \mathsf{GType}^\bullet$ and $C \subseteq \mathsf{GCInst}^\bullet$ such that $\sigma \in \mathsf{Cl}(T)$ for all $c_\sigma \in C$.

Let $F = (T, C)$ be a fragment. We write:

- Type^F, for the set of types generated by this fragment, namely $\mathsf{Cl}(T)$
- Term^F, for the set of terms that fall within this fragment, namely $\{t \in \mathsf{Term} \mid \mathsf{types}^\bullet(t) \subseteq T \wedge \mathsf{consts}^\bullet(t) \subseteq C\}$
- Fmla^F, for $\mathsf{Fmla} \cap \mathsf{Term}^F$.

Lemma 8. The following hold:

(1) $\mathsf{Type}^F \subseteq \mathsf{GType}$

(2) $\mathsf{Term}^F \subseteq \mathsf{GTerm}$

(3) If $t \in \mathsf{Term}^F$, then $\mathsf{tpOf}(t) \in \mathsf{Type}^F$

(4) If $t \in \mathsf{Term}^F$, then $\mathsf{FV}(t) \subseteq \mathsf{Term}^F$

(5) If $t \in \mathsf{Term}^F$, then each subterm of t is also in Term^F

(6) If $t_1, t_2 \in \mathsf{Term}^F$ and $x_\sigma \in \mathsf{Var}_{\mathsf{Type}^F}$, then $t_1[t_2/x_\sigma] \in \mathsf{Term}^F$

The above straightforward lemma shows that fragments F include only ground items (points (1) and (2)) and are "autonomous" entities: the type of a term from F is also in F (3), and similarly for the free (term) variables (4), subterms (5) and substituted terms (6). This autonomy allows us to define semantic interpretations for fragments.

For the rest of this paper, we fix the following:

- a singleton set $\{*\}$
- a two-element set $\{\mathsf{true}, \mathsf{false}\}$
- a global choice function, choice, that assigns to each nonempty set A an element $\mathsf{choice}(a) \in A$.

Let $F = (T, C)$ be a fragment. An F-*interpretation* is a pair $\mathcal{I} = (([\tau])_{\tau \in T}, ([c_\tau])_{c_\tau \in C})$ such that:

1. $([\tau])_{\tau \in T}$ is a family such that $[\tau]$ is a non-empty set for all $\tau \in T$. We extend this to a family $([\tau])_{\tau \in \mathsf{Cl}(T)}$ by interpreting the built-in type constructors as expected:

$[\mathsf{bool}] = \{\mathsf{true}, \mathsf{false}\}$

$[\mathsf{ind}] = \mathbb{N}$ (the set of natural numbers)[5]

$[\sigma \Rightarrow \tau] = [\sigma] \to [\tau]$ (the set of functions from $[\sigma]$ to $[\tau]$)

2. $([c_\tau])_{c_\tau \in C}$ is a family such that $[c_\tau] \in [\tau]$ for all $c_\tau \in C$.

(Note that, in condition 2 above, $[\tau]$ refers to the extension described at point 1.)

Let GBI^F be the set of ground built-in constant instances c_τ with $\tau \in \mathsf{Type}^F$. We extend the family $([c_\tau])_{c_\tau \in C}$ to a family $([c_\tau])_{c_\tau \in C \cup \mathsf{GBI}^F}$, by interpreting the built-in constants as expected:

- $[\rightarrow_{\mathsf{bool} \Rightarrow \mathsf{bool} \Rightarrow \mathsf{bool}}]$ as the logical implication on $\{\mathsf{true}, \mathsf{false}\}$
- $[=_{\tau \Rightarrow \tau \Rightarrow \mathsf{bool}}]$ as the equality predicate in $[\tau] \to [\tau] \to \{\mathsf{true}, \mathsf{false}\}$
- $[\mathsf{zero}_{\mathsf{ind}}]$ as 0 and $[\mathsf{suc}_{\mathsf{ind} \Rightarrow \mathsf{ind}}]$ as the successor function for $\mathbb{N}$
- $[\mathsf{some}_{(\tau \Rightarrow \mathsf{bool}) \Rightarrow \tau}]$ as the following function, where, for each $f : [\tau] \to \{\mathsf{true}, \mathsf{false}\}$, we let $A_f = \{a \in [\tau] \mid f(a) = \mathsf{true}\}$:

$$[\mathsf{some}_{(\tau \Rightarrow \mathsf{bool}) \Rightarrow \tau}](f) = \begin{cases} \mathsf{choice}(A_f) & \text{if } A_f \text{ is non-empty} \\ \mathsf{choice}([\tau]) & \text{otherwise} \end{cases}$$

---

[5] Any infinite (not necessarily countable) set would do here; we only choose $\mathbb{N}$ for simplicity.

In summary, an interpretation $\mathcal{I}$ is a pair of families $(([\tau])_{\tau \in T}, ([c_\tau])_{c_\tau \in C})$, which in fact gives rise to an extended pair of families $(([\tau])_{\tau \in \mathsf{Cl}(T)}, ([c_\tau])_{c_\tau \in C \cup \mathsf{GBI}^F})$.

Now we are ready to interpret the terms in Term^F according to $\mathcal{I}$. A valuation $\xi : \mathsf{Var}_{\mathsf{Type}^F} \to \mathcal{S}et$ is called $\mathcal{I}$-*compatible* if $\xi(x_\sigma) \in [\sigma]^{\mathcal{I}}$ for each $x_\sigma \in \mathsf{Var}_{\mathsf{GType}}$. We write $\mathsf{Comp}^{\mathcal{I}}$ for the set of compatible valuations. For each $t \in \mathsf{Term}^F$, we define a function $[t] : \mathsf{Comp}^{\mathcal{I}} \to [\mathsf{tpOf}(t)]$ recursively over terms as expected:

$$[x_\sigma](\xi) = \xi(x_\sigma) \qquad\qquad [\lambda x_\sigma . t](\xi) \text{ is the function sending each}$$
$$[c_\sigma](\xi) = [c_\sigma] \qquad\qquad\qquad a \in [\sigma] \text{ to } [t](\xi(x_\sigma \leftarrow a)), \text{ where}$$
$$[t_1 t_2](\xi) = [t_1](\xi) \, ([t_2](\xi)) \qquad\quad \xi(x_\sigma \leftarrow a) \text{ is } \xi \text{ updated with } a \text{ at } x_\sigma$$

(Note that this recursive definition is correct thanks to Lemma 8.(5).)

If t is a closed term, then $[t]$ does not truly depend on ξ, and hence we can assume $[t] \in [\mathsf{tpOf}(t)]$. In what follows, we only care about the interpretation of closed terms.

The above concepts are parameterized by a fragment F and an F-interpretation $\mathcal{I}$. If $\mathcal{I}$ or F are not clear from the context, we may write, e.g., $[t]^{\mathcal{I}}$ or $[t]^{F,\mathcal{I}}$. If $\varphi \in \mathsf{Fmla}^F$, we say that $\mathcal{I}$ is a *model* of φ, written $\mathcal{I} \models \varphi$, if $[\varphi]^{\mathcal{I}} = \mathsf{true}$.

Note that the pairs $(F, \mathcal{I})$ are naturally ordered: Given fragments $F_1 = (T_1, C_1)$ and $F_2 = (T_2, C_2)$, F_1-interpretation $\mathcal{I}_1$ and F_2-interpretation $\mathcal{I}_2$, we define $(F_1, \mathcal{I}_1) \leq (F_2, \mathcal{I}_2)$ to mean $T_1 \subseteq T_2$, $C_1 \subseteq C_2$ and $[u]^{\mathcal{I}_1} = [u]^{\mathcal{I}_2}$ for all $u \in T_1 \cup C_1$.

Lemma 9. If $(F_1, \mathcal{I}_1) \leq (F_2, \mathcal{I}_2)$, then the following hold:

(1) $\mathsf{Type}^{F_1} \subseteq \mathsf{Type}^{F_2}$ (3) $[\tau]^{F_1, \mathcal{I}_1} = [\tau]^{F_2, \mathcal{I}_2}$ for all $\tau \in \mathsf{Type}^{F_1}$

(2) $\mathsf{Term}^{F_1} \subseteq \mathsf{Term}^{F_2}$ (4) $[t]^{F_1, \mathcal{I}_1} = [t]^{F_2, \mathcal{I}_2}$ for all $t \in \mathsf{Term}^{F_1}$

The *total fragment* $\top = (\mathsf{GType}^\bullet, \mathsf{GCInst}^\bullet)$ is the top element in this order. Note that $\mathsf{Type}^\top = \mathsf{GType}$ and $\mathsf{Term}^\top = \mathsf{GTerm}$.

So far, $\mathcal{I} \models \varphi$, the notion of $\mathcal{I}$ being a model of φ, was only defined for formulas φ that belong to Term^F, in particular, that are ground formulas. As promised, we extend this to polymorphic formulas by quantifying universally over all ground type substitutions. We only care about such an extension for the total fragment: Given a polymorphic formula φ and a $\top$-interpretation $\mathcal{I}$, we say $\mathcal{I}$ is a *model* of φ, written $\mathcal{I} \models \varphi$, if $\mathcal{I} \models \theta(\varphi)$ for all ground type substitutions θ. This extends to sets E of (polymorphic) formulas: $\mathcal{I} \models E$ is defined as $\mathcal{I} \models \varphi$ for all $\varphi \in E$.

Theorem 10 (Soundness). Let E be a set of formulas that has a total-fragment model, i.e., there exists a $\top$-interpretation $\mathcal{I}$ such that $\mathcal{I} \models E$. Then E is consistent.

Proof. It is routine to verify that the deduction rules for polymorphic HOL are sound w.r.t. our ground semantics. □

4.5   The Model Construction

The only missing piece from the proof of consistency is the following:

Theorem 11. Assume D is a well-formed definitional theory. Then it has a total-fragment model, i.e., there exists a $\top$-interpretation $\mathcal{I}$ such that $\mathcal{I} \models D$.

Proof. For each $u \in \mathsf{GType}^\bullet \cup \mathsf{GCInst}^\bullet$, we define $[u]$ by well-founded recursion on $\leadsto^{\downarrow +}$, the transitive closure of $\leadsto^\downarrow$; indeed, the latter is a terminating (well-founded) relation by the well-formedness of D, hence the former is also terminating.

We assume $[v]$ has been defined for all $v \in \mathsf{GType}^\bullet \cup \mathsf{GCInst}^\bullet$ such that $u \leadsto^{\downarrow +} v$. In order to define $[u]$, we first need some terminology: We say that a definition $w \equiv s$ matches u if there exists a type substitution θ with $u = \theta(w)$. We distinguish the cases:

1. There exists no definition in D that matches u. Here we have two subcases:
   - $u \in \mathsf{GType}^\bullet$. Then we define $[u] = \{*\}$.
   - $u \in \mathsf{GCInst}^\bullet$. Say u has the form c_σ. Then $u \leadsto^\downarrow \sigma$, and hence $[\sigma]$ is defined; we define $[u] = \mathsf{choice}([\sigma])$.
2. There exists a definition $w \equiv s$ in D that matches u. Then let θ be such that $u = \theta(w)$, and let $t = \theta(s)$. Let $V_u = \{v \mid u \leadsto^{\downarrow +} v\}$, $T_u = V_u \cap \mathsf{Type}$ and $C_u = V_u \cap \mathsf{CInst}$. It follows from the definition of $\leadsto$ that $F_u = (T_u, C_u)$ is a fragment; moreover, from the definition of $\leadsto$ and Lemma 4, we obtain that $\mathsf{types}^\bullet(t) \subseteq T_u$ and $\mathsf{consts}^\bullet(t) \subseteq C_u$, which implies $t \in \mathsf{Term}^{F_u}$; hence we can speak of the value $[t]^{F_u, \mathcal{I}_u}$ obtained from the F_u-interpretation $\mathcal{I}_u = (([v])_{v \in T_u}, ([v])_{v \in C_u})$. We have two subcases:
   - $u \in \mathsf{GCInst}^\bullet$. Then we define $[u] = [t]^{F_u, \mathcal{I}_u}$.
   - $u \in \mathsf{GType}^\bullet$. Then the type of t has the form $\sigma \Rightarrow \mathsf{bool}$; and since $\sigma \in \mathsf{types}^\bullet(t) \subseteq \mathsf{Type}^{F_u}$, it follows that $[\sigma]^{F_u, \mathcal{I}_u}$ is also defined. We have two subsubcases:
     - $[\exists x_\sigma.\, t\, x] = \mathsf{false}$. Then we define $[u] = \{*\}$.
     - $[\exists x_\sigma.\, t\, x] = \mathsf{true}$. Then we define $[u] = \{a \in [\sigma]^{F_u, \mathcal{I}_u} \mid [t](a) = \mathsf{true}\}$.

Having defined the $\top$-interpretation $\mathcal{I} = (([u])_{u \in \mathsf{GType}^\bullet}, ([u])_{u \in \mathsf{GCInst}^\bullet})$, it remains to show that $\mathcal{I} \models D$. To this end, let $w \equiv s$ be in D and let θ' be a ground type substitution. We need to show $\mathcal{I} \models \theta'(w \equiv s)$, i.e., $\mathcal{I} \models \theta'(w) \equiv \theta'(s)$.

Let $u = \theta'(w)$; then u matches $w \equiv s$, and by orthogonality this is the only definition in D that it matches. So the definition of $[u]$ proceeds with case 2 above, using $w \equiv s$—let θ be the ground type substitution considered there. Since $\theta'(w) = \theta(w)$, it follows that θ' and θ coincide on the type variables of w, and hence on the type variables of s (because, in any definition, the type variables of the right-hand side are included in those of the left-hand side); hence $\theta'(s) = \theta(s)$.

Now the desired fact follows from the definition of $\mathcal{I}$, by a case analysis matching the subcases of the above case 2. (Note that the definition operates with $[t]^{F_u, \mathcal{I}_u}$, whereas we need to prove the fact for $[t]^{\top, \mathcal{I}}$; however, since $(F_u, \mathcal{I}_u) \le (\top, \mathcal{I})$, by Lemma 9 the two values coincide; and similarly for $[\sigma]^{F_u, \mathcal{I}_u}$ vs. $[\sigma]^{\top, \mathcal{I}}$.) $\qquad\square$

5 Deciding Well-Formedness

We proved that every well-formed theory is consistent. From the implementation perspective, we can ask ourselves how difficult it is to check that the given theory is well-formed. We can check that D is definitional and orthogonal by simple polynomial algorithms. On the other hand, Obua [29] showed that a dependency relation generated by overloaded definitions can encode the Post correspondence problem and therefore termination of such a relation is not even a semi-decidable problem.

Kunčar [19] takes the following approach: Let us impose a syntactic restriction, called *compositionality*, on accepted overloaded definitions which makes the termination of the dependency relation decidable while still permitting all use cases of overloading in Isabelle. Namely, let $\rightsquigarrow\!\!\!\twoheadrightarrow$ be the substitutive and transitive closure of the dependency relation $\rightsquigarrow$ (which is in fact equal to $\rightsquigarrow^{\downarrow+}$). Then D is called *composable* if for all u, u' that are left-hand sides of some definitions from D and for all v such that $u \rightsquigarrow\!\!\!\twoheadrightarrow v$, it holds that either $u' \leq v$, or $v \leq u'$, or $u' \mathbin{\#} v$. Under composability, termination of $\rightsquigarrow\!\!\!\twoheadrightarrow$ is equivalent to acyclicity of $\rightsquigarrow$, which is a decidable condition.[6]

Theorem 12. The property of D of being composable and well-formed is decidable.

Proof. The above-mentioned paper [19] presents a quadratic algorithm (in the size of $\rightsquigarrow$), CHECK, that checks that D is definitional, orthogonal and composable, and that $\rightsquigarrow\!\!\!\twoheadrightarrow$ terminates.[7] Notice that $\rightsquigarrow\!\!\!\twoheadrightarrow\ =\rightsquigarrow^{\downarrow+}$ terminates iff $\rightsquigarrow^{\downarrow}$ terminates. Thus, CHECK decides whether D is composable and well-formed. □

For efficiency reasons, we optimize the size of the relation that the quadratic algorithm works with. Let $\rightsquigarrow_1$ be the relation defined like $\rightsquigarrow$, but only retaining clause 1 in its definition. Since $\rightsquigarrow^{\downarrow}$ is terminating iff $\rightsquigarrow_1^{\downarrow}$ is terminating, it suffices to check termination of the latter.

---

[6] Composability reduces the search space when we are looking for the cycle—it tells us that there exist three cases on how to extend a path (to possibly close a cycle): in two cases we can still (easily) extend the path ($v \leq u'$ or $u' \leq v$) and in one case we cannot ($v \mathbin{\#} u'$). The fourth case (v and u' have a non-trivial common instance; formally $u' \not\leq v$ and $v \not\leq u'$ and there exists w such that $w \leq u'$, $w \leq v$), which complicates the extension of the path, is ruled out by composability. More about composability can be found in the original paper.

[7] The correctness proof is relatively general and works for any $\rightsquigarrow : \mathcal{U}_\Sigma \rightarrow \mathcal{U}_\Sigma \rightarrow \mathsf{bool}$ on a set $\mathcal{U}_\Sigma$ endowed with a certain structure—namely, three functions $= : \mathcal{U}_\Sigma \rightarrow \mathcal{U}_\Sigma \rightarrow \mathsf{bool}$, $\mathsf{App} : (\mathsf{Type} \rightarrow \mathsf{Type}) \rightarrow \mathcal{U}_\Sigma \rightarrow \mathcal{U}_\Sigma$ and $\mathsf{size} : \mathcal{U}_\Sigma \rightarrow \mathbb{N}$, indicating how to compare for equality, type-substitute and measure the elements of $\mathcal{U}_\Sigma$. In this paper, we set $\Sigma = (K, \mathsf{arOf}, C, \mathsf{tpOf})$ and $\mathcal{U}_\Sigma = \mathsf{Type}^\bullet \cup \mathsf{CInst}^\bullet$. The definition of $=, \mathsf{App}$ and size is then straightforward: two elements of $\mathsf{Type}^\bullet \cup \mathsf{CInst}^\bullet$ are equal iff they are both constant instances and they are equal or they are both types and they are equal; $\mathsf{App}\,\rho\,\tau = \rho(\tau)$ and $\mathsf{App}\,\rho\,c_\tau = c_{\rho(\tau)}$; finally, $\mathsf{size}(\tau)$ counts the number of type constructors in τ and $\mathsf{size}(c_\tau) = \mathsf{size}(\tau)$.

6   Conclusion

We have provided a solution to the consistency problem for Isabelle/HOL's logic, namely, polymorphic HOL with ad hoc overloading. Consistency is a crucial, but rather weak property—a suitable notion of conservativeness (perhaps in the style of Wenzel [35], but covering type definitions as well) is left as future work. Independently of Isabelle/HOL, our results show that Gordon-style type definitions and ad hoc overloading can be soundly combined and naturally interpreted semantically.

Acknowledgments. We thank Tobias Nipkow, Larry Paulson and Makarius Wenzel for inspiring discussions and the anonymous referees for many useful comments. This paper was partially supported by the DFG project Security Type Systems and Deduction (grant Ni 491/13-3) as part of the program Reliably Secure Software Systems (RS^3, priority program 1496).

References

1. http://www21.in.tum.de/~kuncar/documents/patch.html
2. The HOL4 Theorem Prover. http://hol.sourceforge.net/
3. Adams, M.: Introducing HOL Zero. In: Fukuda, K., Hoeven, J., Joswig, M., Takayama, N. (eds.) ICMS 2010. LNCS, vol. 6327, pp. 142–143. Springer, Heidelberg (2010)
4. Anand, A., Rahli, V.: Towards a formally verified proof assistant. In: Klein, G., Gamboa, R. (eds.) ITP 2014. LNCS, vol. 8558, pp. 27–44. Springer, Heidelberg (2014)
5. Arthan, R.D.: Some mathematical case studies in ProofPower-HOL. In: TPHOLs 2004 (2004)
6. Barras, B.: Coq en Coq. Technical report 3026, INRIA (1996)
7. Barras, B.: Sets in Coq, Coq in Sets. J. Formalized Reasoning **3**(1), 29–48 (2010)
8. Bertot, Y., Casteran, P.: Interactive Theorem Proving and Program Development: Coq'Art: The Calculus of Inductive Constructions. Springer, Heidelberg (2004)
9. Blanchette, J.C., Popescu, A., Traytel, D.: Foundational extensible corecursion. In: ICFP 2015. ACM (2015)
10. Bove, A., Dybjer, P., Norell, U.: A brief overview of Agda – a functional language with dependent types. In: Berghofer, S., Nipkow, T., Urban, C., Wenzel, M. (eds.) TPHOLs 2009. LNCS, vol. 5674, pp. 73–78. Springer, Heidelberg (2009)
11. Dénès, M.: [Coq-Club] Propositional extensionality is inconsistent in Coq, archived at https://sympa.inria.fr/sympa/arc/coq-club/2013-12/msg00119.html
12. Gordon, M.J.C., Melham, T.F. (eds.): Introduction to HOL: A Theorem Proving Environment for Higher Order Logic. Cambridge University Press, New York (1993)
13. Haftmann, F., Wenzel, M.: Constructive type classes in Isabelle. In: Altenkirch, T., McBride, C. (eds.) TYPES 2006. LNCS, vol. 4502, pp. 160–174. Springer, Heidelberg (2007)
14. Harrison, J.: HOL Light: a tutorial introduction. In: Srivas, M., Camilleri, A. (eds.) FMCAD 1996. LNCS, vol. 1166, pp. 265–269. Springer, Heidelberg (1996)
15. Harrison, J.: Towards self-verification of HOL Light. In: Furbach, U., Shankar, N. (eds.) IJCAR 2006. LNCS (LNAI), vol. 4130, pp. 177–191. Springer, Heidelberg (2006)

16. Hölzl, J., Immler, F., Huffman, B.: Type classes and filters for mathematical analysis in Isabelle/HOL. In: Blazy, S., Paulin-Mohring, C., Pichardie, D. (eds.) ITP 2013. LNCS, vol. 7998, pp. 279–294. Springer, Heidelberg (2013)

17. Huffman, B., Urban, C.: Proof pearl: a new foundation for Nominal Isabelle. In: Kaufmann, M., Paulson, L.C. (eds.) ITP 2010. LNCS, vol. 6172, pp. 35–50. Springer, Heidelberg (2010)

18. Kumar, R., Arthan, R., Myreen, M.O., Owens, S.: HOL with definitions: semantics, soundness, and a verified implementation. In: Klein, G., Gamboa, R. (eds.) ITP 2014. LNCS, vol. 8558, pp. 308–324. Springer, Heidelberg (2014)

19. Kunčar, O.: Correctness of Isabelle's cyclicity checker: implementability of overloading in proof assistants. In: CPP 2015. ACM (2015)

20. Kunčar, O., Popescu, A.: A consistent foundation for Isabelle/HOL. Technical report (2015). www.eis.mdx.ac.uk/staffpages/andreipopescu/pdf/IsabelleHOL.pdf

21. Leino, K.R.M., Moskal, M.: Co-induction simply–automatic co-inductive proofs in a program verifier. In: FM 2014 (2014)

22. Lochbihler, A.: Light-Weight Containers for Isabelle: Efficient, Extensible, Nestable. In: Blazy, S., Paulin-Mohring, C., Pichardie, D. (eds.) ITP 2013. LNCS, vol. 7998, pp. 116–132. Springer, Heidelberg (2013)

23. McBride, C., et al.: [HoTT] Newbie questions about homotopy theory and advantage of UF/Coq, archived at http://article.gmane.org/gmane.comp.lang.agda/6106

24. Müller, O., Nipkow, T., von Oheimb, D., Slotosch, O.: HOLCF = HOL + LCF. J. Funct. Program. 9, 191–223 (1999)

25. Myreen, M.O., Davis, J.: The reflective Milawa theorem prover is sound. In: Klein, G., Gamboa, R. (eds.) ITP 2014. LNCS, vol. 8558, pp. 421–436. Springer, Heidelberg (2014)

26. Nipkow, T., Klein, G.: Concrete Semantics - With Isabelle/HOL. Springer, New York (2014)

27. Nipkow, T., Paulson, L.C., Wenzel, M.: Isabelle/HOL. LNCS, vol. 2283. Springer, Heidelberg (2002)

28. Kang, J., Adibi, S.: Type classes and overloading resolution via order-sorted unification. In: Doss, R., Piramuthu, S., ZHOU, W. (eds.) Functional Programming Languages and Computer Architecture. LNCS, vol. 523, pp. 1–14. Springer, Heidelberg (1991)

29. Obua, S.: Checking conservativity of overloaded definitions in higher-order logic. In: Pfenning, F. (ed.) RTA 2006. LNCS, vol. 4098, pp. 212–226. Springer, Heidelberg (2006)

30. Pitts, A.: Introduction to HOL: a theorem proving environment for higher order logic. Chapter The HOL Logic, pp. 191–232. In: Gordon and Melham [12] (1993)

31. Shankar, N., Owre, S., Rushby, J.M.: PVS Tutorial. Computer Science Laboratory, SRI International (1993)

32. Sozeau, M., Oury, N.: First-class type classes. In: Mohamed, O.A., Muñoz, C., Tahar, S. (eds.) TPHOLs 2008. LNCS, vol. 5170, pp. 278–293. Springer, Heidelberg (2008)

33. Urban, C.: Nominal techniques in Isabelle/HOL. J. Autom. Reason. 40(4), 327–356 (2008)

34. Wadler, P., Blott, S.: How to make ad-hoc polymorphism less ad-hoc. In: POPL (1989)

35. Wenzel, M.: Type classes and overloading in higher-order logic. In: Gunter, E.L., Felty, A.P. (eds.) TPHOLs 1997. LNCS, vol. 1275, pp. 307–322. Springer, Heidelberg (1997)

Refinement to Imperative/HOL

Peter Lammich[(⊠)]

Technische Universität München, Munich, Germany
lammich@in.tum.de

Abstract. Many algorithms can be implemented most efficiently with imperative data structures that support destructive update. In this paper we present an approach to automatically generate verified imperative implementations from abstract specifications in Isabelle/HOL. It is based on the Isabelle Refinement Framework, for which a lot of abstract algorithms are already formalized.

Based on Imperative/HOL, which allows to generate verified imperative code, we develop a separation logic framework with automation to make it conveniently usable. On top of this, we develop an imperative collection framework, which provides standard implementations for sets and maps like hash tables and array lists. Finally, we define a refinement calculus to refine abstract (functional) algorithms to imperative ones.

Moreover, we have implemented a tool to automate the refinement process, replacing abstract data types by efficient imperative implementations from our collection framework. As a case study, we apply our tool to automatically generate verified imperative implementations of nested depth-first search and Dijkstra's shortest paths algorithm, which are considerably faster than the corresponding functional implementations. The nested DFS implementation is almost as fast as a C++ implementation of the same algorithm.

1 Introduction

Using the Isabelle Refinement Framework (IRF) [13,17], we have verified several graph and automata algorithms (e.g. [14,23]), including a fully verified LTL model checker [6]. The IRF features a stepwise refinement approach, where an abstract algorithm is refined, in possibly many steps, to a concrete implementation. This approach separates the correctness proof of the abstract algorithmic ideas from the correctness proof of their implementation. This reduces the proof complexity, and makes larger developments manageable in the first place.

The IRF only allows refinement to purely *functional* code, while the most efficient implementations of (model checking) algorithms typically require imperative features like destructive update of arrays.

The goal of this paper is to verify *imperative* algorithms using a stepwise refinement approach, and automate the canonical task of replacing abstract by concrete data structures. We build on Imperative/HOL [2], which introduces a heap monad in Isabelle/HOL and supports code generation for several target platforms (currently OCaml, SML, Haskell, and Scala). However, the automation provided by

© Springer International Publishing Switzerland 2015
C. Urban and X. Zhang (Eds.): ITP 2015, LNCS 9236, pp. 253–269, 2015.
DOI: 10.1007/978-3-319-22102-1_17

Imperative/HOL is limited, and it has rarely been used for verification projects so far. Thus, we have developed a separation logic framework that comes with a verification condition generator and some automation that greatly simplifies reasoning about programs in Imperative/HOL.

Based on the separation logic framework, we formalize imperative data structures and integrate them into an imperative collection framework, which, similar to the Isabelle Collection Framework [15], defines interfaces for abstract data types and instantiates them with concrete implementations.

Next, we define a notion of data refinement between an IRF program and an Imperative/HOL program, which supports mixing of imperative and functional data structures, and provide proof rules for the standard combinators (return, bind, recursion, while, foreach, etc.). We implement a tool which automatically synthesizes an imperative program from an abstract functional one, selecting appropriate data structures from both our imperative collection framework and the (functional) Isabelle Collection Framework.

Finally, we present some case studies. We can use already existing abstract formalizations for the IRF unchanged. In particular we use our tool to automatically synthesize an imperative implementation of a nested DFS algorithm from an existing abstract formalization, which is considerably faster than the original purely functional implementation, and almost as fast as a C++ implementation of the same algorithm. Our tool and the case studies are available at http:// www21.in.tum.de/~lammich/refine_imp_hol.

The remainder of this paper is structured as follows: In Sect. 2 we present our separation logic framework for Imperative/HOL. In Sect. 3, we describe the imperative collection framework. The refinement from the IRF to Imperative/HOL and its automation is described Sect. 4. Section 5 contains the case studies, and, finally, Sect. 6, contains the conclusions, related work, and an outlook to future research.

2   A Separation Logic for Imperative/HOL

Imperative/HOL provides a heap monad formalized in Isabelle/HOL, as well as a code generator extension to generate imperative code in several target languages (currently OCaml, SML, Haskell, and Scala). However, Imperative/HOL itself only comes with minimalistic support for reasoning about programs. In this section, we report on our development of a separation logic framework for Imperative/HOL. A preliminary version of this, which did not include frame inference nor other automation, was formalized by Meis [19]. A more recent version is available in the Archive of Formal Proofs [16].

2.1   Basics

We formalize separation logic [25] along the lines of Calcagno et al. [4].

We define a type *pheap* for a *partial heap*, which describes the content of a heap at a specific set of addresses. An *assertion* is a predicate on partial heaps that satisfies a well-formedness condition[1]. We define the type $assn \subset pheap \Rightarrow bool$ of assertions, and write $h \models P$ if the partial heap h satisfies the assertion P.

We define the basic assertions $true$, $false$, emp for the empty heap, $p \mapsto_r v$ for a heap storing value v at address p, and $p \mapsto_a l$ for a heap storing an array[2] with values $l :: \alpha\ list$ at address p. Moreover we define the $pure$ assertion $\uparrow \Phi$, which holds if the heap is empty and the predicate Φ holds.

On assertions, we define the standard Boolean connectives, and show that they form a Boolean algebra. We also define universal and existential quantification. Moreover, we define the $separation\ conjunction$ $P * Q$, which holds if the heap can be split into two disjoint parts, such that P holds on the first, and Q on the second part. Finally, we define entailment as $P \Longrightarrow_A Q$ iff $\forall h.\ h \models P \Longrightarrow h \models Q$.

We prove standard properties on assertions and use them to set up Isabelle's proof tools to work seamlessly with assertions. For example, the simplifier pulls existential quantifiers to the front, groups together pure assertions, and checks pointer assertions ($\mapsto_r$ and $\mapsto_a$) for consistency.

Example 1. The simplifier rewrites the assertion $P* \uparrow \Phi * (\exists p.\ p \mapsto_r v* \uparrow \Psi)$ to $\exists p.\ P * p \mapsto_r v* \uparrow (\Phi \wedge \Psi)$, and the assertion $P* \uparrow \Phi * (\exists p.\ p \mapsto_r v* \uparrow \Psi * p \mapsto_r w)$ is rewritten to False (as p would have to point to two separate locations).

2.2 Hoare Triples

Having defined assertions, we are ready to define a separation logic on programs. Imperative/HOL provides a shallow embedding of heap-manipulating programs into Isabelle/HOL. A program is encoded in a heap-exception monad, i.e. it has type $\alpha\ Heap = heap \Rightarrow (\alpha \times heap)\ option$. Intuitively, a program takes a heap and either produces a result of type α and a new heap, or fails.

We define the *Hoare triple* $\langle P \rangle\ c\ \langle Q \rangle$ to hold, iff for all heaps that satisfy P, program c returns a result x such that the new heap satisfies $Q\ x$.[3] When reasoning about garbage collected languages, one has to frequently specify that an operation may allocate some heap space for internal use. For this purpose, we define $\langle P \rangle\ c\ \langle Q \rangle_t$ as a shortcut for $\langle P \rangle\ c\ \langle \lambda x.\ Q\ x * true \rangle$.

For Hoare triples, we prove rules for the basic heap commands (allocation, load/store from/to pointer and array, get array length), rules for the standard combinators (return, bind, recursion, if, case, etc.), as well as a consequence and a frame rule. Note that the frame rule, $\langle P \rangle\ c\ \langle Q \rangle \Longrightarrow \langle P * F \rangle\ c\ \langle \lambda x.\ Q\ x * F \rangle$, is crucial for modular reasoning in separation logic. Intuitively, it states that a program does not depend on the content of the heap that it does not access.

---

[1] For technical reasons, we formalize a partial heap as a full heap with an address range. Assertions must not depend on heap content outside this address range.

[2] The distinction between values and arrays is dictated by Imperative/HOL.

[3] Again, for technical reasons, we additionally check that the program does not modify addresses outside the heap's address range, and that it does not deallocate memory.

2.3 Automation

Application of these rules can be automated. We implement a verification condition generator that transforms a Hoare triple to verification conditions, which are plain HOL propositions and do not contain separation logic. The basic strategy for verification condition generation is simple: Compute a strong enough postcondition of the precondition and the program, and show that it entails the postcondition of the Hoare triple. In Isabelle/HOL, this is implemented by using a schematic variable as postcondition (i.e. a unification variable that can be instantiated on rule application). However, there are two cases that cannot be fully automated: frame inference and recursion.

Frame Inference. When applying a rule for a command c, say $\langle P \rangle c \langle Q \rangle$, the current goal has the form $\langle P' \rangle c \langle ?R \rangle$, where P' is the precondition describing the current heap, and $?R$ is the unification variable that shall take the postcondition. In order to apply the rule for c, we have to find a part of the heap that satisfies P. In other words, we have to find a *frame* F such that $P' \Longrightarrow_A P * F$. Then, we use the frame rule to prove $\langle P * F' \rangle c \langle \lambda x.\ Qx * F' \rangle$, and with the consequence rule we instantiate $?R$ to $\lambda x.\ Q\ x * F$. A detailed discussion about automating frame inference can be found in [28]. We implement a quite simple but effective method: After some initial simplifications to handle quantifiers and pure predicates, we split P and P' into $P = P_1 * \ldots * P_n$ and $P' = P'_1 * \ldots * P'_n$. Then, for every P_i, we find the first P'_j that can be unified with P_i. If we succeed to match up all P_is, without using a P'_j twice, we have found a valid frame, otherwise the heuristic fails and frame inference has to be performed manually.

Recursion. Recursion over the heap monad is modeled as least fixed point over an adequate CCPO [10]. Proving a Hoare triple for a recursive function requires to perform induction over a well-founded ordering that is compatible with the recursion scheme. Coming up with an induction ordering and finding a generalization that is adequate for the induction proof to go through is undecidable in general. We have not attempted to automate this, although there exist some heuristics [3].

2.4 All-in-one Method

Finally, we combine the verification condition generator, frame inference and Isabelle/HOL's *auto* tactic into a single proof tactic *sep_auto*, which is able to solve many subgoals involving separation logic completely automatically. Moreover, if it cannot solve a goal it returns the proof state at which it got stuck. This is a valuable tool for proof exploration, as this stuck state usually hints to missing lemmas. The *sep_auto* method allows for very straightforward and convenient proofs. For example, the original Imperative/HOL formalization [2] contains an example of in-place list reversal. The correctness proof requires about 100 lines of quite involved proof text. Using *sep_auto*, the proof reduces to 6 lines of straightforward proof text [16].

3   Imperative Collection Framework

We use our separation logic framework to implement an imperative collection framework along the lines of [15]: For each abstract data type (e.g. set, map) we define a locale that fixes a *refinement assertion* that relates abstract values with concrete values (which may be on the heap). This locale is polymorphic in the concrete data type $'s$, and is later instantiated for every implementation. For each operation, we define a locale that includes the locale of the abstract data type, fixes a parameter for the operation, and assumes a Hoare triple stating the correctness of the operation.

Example 2. The abstract set data type is specified by the following locale:

locale *imp_set* = **fixes** *is_set* :: $'a$ *set* $\Rightarrow 's \Rightarrow$ *assn*
  assumes *precise: precise is_set*

Here, the predicate *precise* describes that the abstract value is uniquely determined by the concrete value.

The insert operation on sets is specified as follows:

locale *imp_set_ins* = *imp_set* + **fixes** *ins* :: $'a \Rightarrow 's \Rightarrow 's$ *Heap*
  assumes *ins_rule:* $\langle is\_set\ s\ p \rangle\ ins\ a\ p\ \langle \lambda r.\ is\_set\ (\{a\} \cup s)\ r \rangle_t$

Note that this specifies a destructive update of the set, as the postcondition does not contain the original set p any more.

Example 3. Finite sets of natural numbers can be implemented by bitvectors. We define a corresponding refinement assertion, and instantiate the set locale:

definition *is_bv* :: *nat set* $\Rightarrow$ *bool array* $\Rightarrow$ *assn* [...]
interpretation *bv: imp_set is_bv* [...]

Then, we define the insert operation, and instantiate the locale *imp_set_ins*:

definition *bv_ins* :: *nat* $\Rightarrow$ *bool array* $\Rightarrow$ *bool array Heap* [...]
interpretation *bv: imp_set_ins is_bv bv_ins* [...]

Using the approach sketched above, we have defined more standard data structures for sets and maps, including hash sets, hash maps and array maps.

4   Refinement to Imperative/HOL

In the last section, we have described how to formalize imperative collection data structures. In order to use these data structures in efficient algorithms, we develop a refinement technique that allows us to refine a formalization of the algorithm over abstract data types to one that uses efficient data structures.

With the Isabelle Refinement Framework [17], we have already developed a formalism to describe and prove correct algorithms on an abstract level and then refine them to use efficient *purely functional* data structures. With the Autoref tool [13], we have even automated this process. In this section, we extend these techniques to imperative data structures.

4.1 Isabelle Refinement Framework

We briefly review the Isabelle Refinement Framework. For a more detailed description, we refer to [12,17]. Programs are described via a nondeterminism monad over the type $'a\,nres$, which is defined as follows:

datatype $'a\ nres =$ **res** $('a\ set)$ | **fail**
fun $\le$ $::\, 'a\ nres \Rightarrow\, 'a\ nres \Rightarrow\ bool$
  where $\_ \le$ **fail** | **fail** $\not\le$ **res** $\_$ | **res** $X \le$ **res** Y *iff* $X \subseteq Y$
fun **return** $::\, 'a \Rightarrow\, 'a\ nres$ where **return** $x \equiv$ **res** $\{x\}$
fun **bind** $::\, 'a\ nres \Rightarrow ('a \Rightarrow\, 'b\ nres) \Rightarrow\, 'b\ nres$
  where **bind fail** $f \equiv$ **fail** | **bind** (**res** X) $f \equiv SUP\ x \in X.\ f\,x$

The type $'a\,nres$ describes nondeterministic results, where **res** X describes the nondeterministic choice of an element from X, and **fail** describes a failed assertion. On nondeterministic results, we define the *refinement ordering* $\le$ by lifting the subset ordering, setting **fail** as top element. The intuitive meaning of $a \le b$ is that a *refines* b, i.e. results of a are also results of b. Note that the refinement ordering is a complete lattice with top element **fail** and bottom element **res** $\{\}$.

Intuitively, **return** x denotes the unique result x, and **bind** $m\,f$ denotes sequential composition: Select a result from m, and apply f to it.

Non-recursive programs can be expressed by these monad operations and Isabelle/HOL's **if** and **case**-combinators. Recursion is encoded by a fixed point combinator **rec** $::\ ('a \Rightarrow\, 'b\ nres) \Rightarrow\, 'a \Rightarrow\, 'b\ nres$, such that **rec** F is the greatest fixed point of the monotonic functor F wrt. the flat ordering of result sets with **fail** as the top element. If F is not monotonic, **rec** F is defined to be **fail**:

rec $F\,x \equiv$ **if** ($mono'\ F$) **then** ($flatf\_gfp\ F\,x$) **else fail**

Here, $mono'$ denotes monotonicity wrt. both the flat ordering and the refinement ordering. The reason is that for functors that are monotonic wrt. both orderings, the respective greatest fixed points coincide, which is useful to show proof rules for refinement.

Functors that only use the standard combinators described above are monotonic by construction. This is also exploited by the Partial Function Package [10], which allows convenient specification of recursive monadic functions.

Building on the combinators described above, the IRF also defines **while** and **foreach** loops to conveniently express tail recursion and folding over the elements of a finite set.

Example 4. Listing 1 displays the IRF formalization of a simple depth-first search algorithm that checks whether a directed graph, described by a (finite) set of edges E, has a path from source node s to target node t: With the tool support provided by the IRF, it is straightforward to prove this algorithm correct, and refine it to efficient functional code (cf. [13,17]).

4.2 Connection to Imperative/HOL

In this section, we describe how to refine a program specified in the nondeterminism monad of the IRF to a program specified in the heap-exception monad

Listing 1. Simple DFS algorithm formalized in the IRF

```
definition dfs :: ('v ×'v) set ⇒'v ⇒'v ⇒ bool nres where
  dfs E s t ≡ do {
    (_,r) ← rec (λdfs (V,v).
      if v ∈ V then return (V,False)
      else do {
        let V = insert v V;
        if v = t then return (V,True)
        else foreach ({v'. (v,v') ∈ E}) (λ(_,brk). ¬brk)
              (λv' (V,_). dfs (V,v')) (V,False)
      }
    ) ({},s);
    return r
  }
```

of Imperative/HOL. The main challenge is to refine abstract data to concrete data that may be stored on the heap and updated destructively.

At this point, we have a design choice: One option is to formalize a nondeterministic heap-exception monad, in which we encode an abstract program with a heap containing abstract data. In a second step, this program is refined to a deterministic program with concrete data structures. The other option is to omit the intermediate step, and directly relate abstract nondeterministic programs to concrete deterministic ones.

Due to limitations of the logic underlying Isabelle/HOL, we need a single HOL type that can encode all types we want to store on the heap. In Imperative/HOL, this type is chosen as $\mathbb{N}$, and thus only countable types can be stored on the heap. As long as we store concrete data structures, this is no real problem. However, abstract data types are in general not countable, nor does there exist a type in Isabelle/HOL that could encode all other types. This would lead to unnatural and clumsy restrictions on abstract data types, contradicting the goal of focusing the abstract proofs on algorithmic ideas rather than implementation details.

Thus, we opted for directly relating nondeterministic results with heap-modifying programs. We define the predicate *hnr* (short for *heap-nres refinement*) as follows:

$$hnr\ \Gamma\ c\ \Gamma'\ R\ m \equiv$$
$$m \neq \textbf{fail} \longrightarrow \langle \Gamma \rangle\ c\ \langle \lambda r.\ \Gamma' * (\exists x.\ R\ x\ r * \uparrow(\textbf{return}\ x \leq m)) \rangle_t$$

Intuitively, for an Imperative/HOL program c, $hnr\ \Gamma\ c\ \Gamma'\ R\ m$ states that on a heap described by assertion Γ, c returns a value that refines the nondeterministic result m wrt. the refinement assertion R. Additionally, the new heap contains Γ'.

In order to prove refinements, we derive a set of proof rules for the *hnr* predicate, including a frame rule, consequence rule, and rules relating the combinators of the heap monad with the combinators of the nondeterminism monad. For example, the consequence rule allows us to strengthen the precondition, weaken the postcondition, and refine the nondeterministic result:

$$[\![\Gamma_1 \Longrightarrow_A \Gamma_1'; \; hnr \; \Gamma_1' \; c \; \Gamma_2 \; R \; m; \; \Gamma_2 \Longrightarrow_A \Gamma_2'; \; m \le m']\!] \Longrightarrow hnr \; \Gamma_1 \; c \; \Gamma_2' \; R \; m'$$

For recursion, we get the following rule[4]:

assumes $\bigwedge cf \; af \; ax \; px.$ $[\![$
$\quad \bigwedge ax \; px. \; hnr \; (Rx \; ax \; px * \Gamma) \; (cf \; px) \; (\Gamma' \; ax \; px) \; Ry \; (af \; ax)]\!]$
$\quad \Longrightarrow hnr \; (Rx \; ax \; px * \Gamma) \; (Fc \; cf \; px) \; (\Gamma' \; ax \; px) \; Ry \; (Fa \; af \; ax)$
assumes $(\bigwedge x. \; mono\_Heap \; (\lambda f. \; Fc \; f \; x))$
assumes *precise Ry*
shows $hnr \; (Rx \; ax \; px * \Gamma) \; (heap.fixp\_fun \; Fc \; px) \; (\Gamma' \; ax \; px) \; Ry \; (\textbf{rec} \; Fa \; ax)$

Intuitively, we have to show that the concrete functor Fc refines the abstract functor Fa, assuming that the concrete recursive function cf refines the abstract one af. The argument of the call is refined by the refinement assertion Rx and the result is refined by Ry. Additionally, the heap may contain $\Gamma >$, and is transformed to $\Gamma' \; ax \; px$. Here, the ax and px that are attached to Γ' denote that the new heap may also depend on the argument to the recursive function.

Note that a refinement assertion needs not necessarily relate heap content to an abstract value. It can also relate a concrete non-heap assumes a Hoare triple. For a relation $R :: ('c \times 'a) \; set$ we define:

$$pure \; R \equiv (\lambda a \; c. \; \uparrow((c,a) \in R))$$

This allows us to mix imperative data structures with functional ones. For example, the refinement assertion *pure int_rel* describes the implementation of integer numbers by themselves, where $int\_rel \equiv Id::(int \times int) \; set$.

4.3   Automation

Using the rules for *hnr*, it is possible to manually prove refinement between an Imperative/HOL program and a program in the Isabelle Refinement Framework, provided they are structurally similar enough. However, this is usually a tedious and quite canonical task, as it essentially consists of manually rewriting the program from one monad to the other, thereby unfolding expressions into monad operations if they depend on the heap.

For this reason, we focused our work on automating this process: Given some hints which imperative data structures to use, we automatically synthesize the Imperative/HOL program and the refinement proof. The idea is similar to the Autoref tool [13], which automatically synthesizes efficient functional programs, and, indeed, we could reuse parts of its design for our tool.

In the rest of this section, we describe our approach to automatically synthesize imperative programs, focusing on the aspects that are different from the Autoref tool. The process of synthesizing consists of several consecutive phases: Identification of operations, monadifying, linearity analysis, translation, and cleaning up.

---

[4] Specified in Isabelle's *long goal format*, which is more readable for large propositions.

Identification of Operations. Given an abstract program in Isabelle/HOL, it is not always clear which abstract data types it uses. For example, maps are encoded as functions $'a \Rightarrow {'b}\,option$, and so are priority queues or actual functions. However, maps and priority queues are, also abstractly, quite different concepts. The purpose of this phase is to identify the abstract data types (e.g. maps and priority queues), and the operations on them. Technically, the identification is done by rewriting the operations to constants that are specific to the abstract data type. For example, $(f :: nat \Rightarrow nat\,option)\,x$ may be rewritten to $op\_map\_lookup\,f\,x$, provided that a heuristic identifies f as a map. If f is identified as a priority queue, the same expression would be rewritten to $op\_get\_prio\,f\,x$. The operation identification heuristic is already contained in the Autoref tool, and we slightly adapted it for our needs.

Monadifying. Once we have identified the operations, we flatten all expressions, such that each operation gets visible as a top-level computation in the monad. This transformation essentially fixes an evaluation order (which we choose to be left to right), and later allows us to translate the operations to heap-modifying operations in Imperative/HOL's heap monad.

Example 5. Consider the program $\mathbf{let}\,x = 1; \mathbf{return}\,\{x, x\}$.[5] Note that $\{x, x\}$ is syntactic sugar for $(insert\,x\,(insert\,x\,\{\}))$. A corresponding Imperative/HOL program might be:

$\mathbf{let}\ x = 1;\ s \leftarrow bv\_new;\ s \leftarrow bv\_ins\ x\ s;\ bv\_ins\ x\ s$

Note that the bv_new and bv_ins operations modify the heap, and thus have to be applied as monad operations and cannot be nested into a plain HOL expression. For this reason, the monadify phase flattens all expressions, and thus exposes all operations as monad operations. It transforms the above program to[6]:

$x \leftarrow \mathbf{return}\ 1;\ s \leftarrow \mathbf{return}\ \{\};\ s \leftarrow \mathbf{return}\ (insert\ x\ s);\ \mathbf{return}\ (insert\ x\ s)$

Note that operations that are not translated to heap-modifying operations will be folded again in the cleanup phase.

Linearity Analysis. In order to refine data to be contained on the heap, and destructively updated, we need to know whether the value of an operand may be destroyed. For this purpose, we perform a program analysis on the monadified program, which annotates each operand (which is a reference to a bound variable) to be *linear* or *nonlinear*. A linear operand is not used again, and can safely be destroyed by the operation, whereas a nonlinear operand needs to be preserved.

Example 6. Consider the program from Example 5. Linearity analysis adds the following annotations, where $\cdot^L$ means linear, and $\cdot^N$ means nonlinear:

$x \leftarrow \mathbf{return}\ 1;\ s \leftarrow \mathbf{return}\ \{\};\ s \leftarrow \mathbf{return}\ (insert\ x^N\ s^L);\ \mathbf{return}\ (insert\ x^L\ s^L)$

---

[5] Inserting x twice is redundant, but gives a nice example for our transformations.

[6] We applied α-conversion to give the newly created variables meaningful names.

That is, the insert operations may be translated to destructively update the set, while at least the first insert operation must preserve the inserted value.

Translation. Let a be the monadified and annotated program. We now synthesize a corresponding Imperative/HOL program. Assume the program a depends on the abstract parameters $a_1 \ldots a_n$, which are refined to concrete parameters $c_1 \ldots c_n$ by refinement assertions $R_1 \ldots R_n$. We start with a proof obligation of the form

$$hnr \ (R_1 \ a_1 \ c_1 \ * \ \ldots \ * \ R_n \ a_n \ c_n) \ ?c \ ?\Gamma' \ ?R \ a$$

Recall that ? indicates schematic variables, which are instantiated during resolution. We now repeatedly try to resolve with a set of syntax directed rules for the hnr predicate. There are rules for each combinator and each operator. If a rule would destroy an operand which is annotated as nonlinear, we synthesize code to copy the operand. For this, the user must have defined a copy operation for the operand's concrete data type.

Apart from hnr-predicates, the premises of the rules may contain frame inference and constraints on the refinement assertions. Frame inference is solved by a specialized tactic, which assumes that the frame consists of exactly one refinement assertion per variable. The rules are designed to preserve this invariant. If the content of a variable is destroyed, we still include a vacuous refinement assertion *invalid* for it, which is defined as $invalid \equiv \lambda_- \ _-. \ true$.

Apart from standard frame inference goals, which have the form

$$\Gamma \Longrightarrow_A R_1 \ a_1 \ c_1 \ * \ \ldots \ * \ R_n \ a_n \ c_n \ * \ ?F$$

we also have to solve goals of the form

$$R_1 \ a_1 \ c_1 \ * \ \ldots \ * \ R_n \ a_n \ c_n \ \vee \ R'_1 \ a_1 \ c_1 \ * \ \ldots \ * \ R'_n \ a_n \ c_n \Longrightarrow_A \ ?\Gamma$$

These goals occur when merging the different branches of if or case combinators, which may affect different data on the heap. Here, we keep refinement assertions with $R_i = R'_i$, and set the others to *invalid*.

Example 7. The rule for the if combinator is

assumes $P: \ \Gamma \Longrightarrow_A \Gamma_1 \ * \ pure \ bool\_rel \ a \ a'$
assumes $RT: \ a \Longrightarrow hnr \ (\Gamma_1 \ * \ pure \ bool\_rel \ a \ a') \ b' \ \Gamma_{2b} \ R \ b$
assumes $RE: \ \neg a \Longrightarrow hnr \ (\Gamma_1 \ * \ pure \ bool\_rel \ a \ a') \ c' \ \Gamma_{2c} \ R \ c$
assumes $MERGE: \ \Gamma_{2b} \vee_A \Gamma_{2c} \Longrightarrow_A \Gamma'$
shows $hnr \ \Gamma$ (**if** a' **then** b' **else** c') $\Gamma' \ R$ (**if** a **then** b **else** c)

Intuitively, it works as follows: We start with a heap described by the assertion Γ. First, the concrete value for the condition a is extracted by a frame rule (Premise P). Then, the *then* and *else* branches are translated (Premises RT and RE), producing new heaps described by the assertions Γ_{2b} and Γ_{2c}, respectively. Finally, these assertions are merged to form the assertion Γ' for the resulting heap after the if statement (Premise MERGE).

Another type of side conditions are constraints on the refinement assertions. For example, some rules require a refinement assertion to be precise (cf. Sect. 3). When those rules are applied, the corresponding refinement assertion may not be completely known, but (parts of) it may be schematic and only instantiated later. For this purpose, we keep track of all constraints during translation, and solve them as soon as the refinement assertion gets instantiated.

Using the resolution with rules for *hnr*, combined with frame inference and solving of constraints, we can automatically synthesize an Imperative/HOL program for a given abstract program. While there is only one rule for each combinator, there may be multiple rules for operators on abstract data types, corresponding to the different implementations. In the Autoref tool [13], we have defined some elaborate heuristic how to select the implementations. In our prototype implementation here we use a very simplistic strategy: Take the first implementation that matches the operation. By specifying the refinement assertions for the parameters of the algorithm, and declaring the implementations with specialized abstract types, this simplistic strategy already allows some control over the synthesized algorithm. In future work, we may adopt some more elaborate strategies for implementation selection.

Sometimes, functional data structures are more adequate than imperative ones, be it because they are accessed in a nonlinear fashion, or because we simply have no imperative implementation yet. Pure refinement assertions allow for mixing of imperative and functional data structures, and our tool can import rules from the Isabelle Collection Framework, thus making a large amount of functional data structures readily available.

Cleaning Up. After we have generated the imperative version of the program, we apply some rewriting rules to make it more readable. They undo the flattening of expressions performed in the monadify phase at those places where it was unnecessary, i.e. the heap is not modified. Technically, this is achieved by using Isabelle/HOL's simplifier with an adequate setup.

Example 8. Recall the DFS algorithm from Example 4. With a few (<10) lines of straightforward Isabelle text, our tool generates[7] the imperative algorithm displayed in Listing 2. From this, Imperative/HOL generates verified code in its target languages (currently OCaml, SML, Haskell, and Scala). Moreover, our tool proves the following refinement theorem:

hnr (*is_graph nat_rel E Ei* ∗ *pure nat_rel s si* ∗ *pure nat_rel t ti*)
   (*dfs_impl Ei si ti*)
   (*pure nat_rel t ti* ∗ *invalid s si* ∗ *is_graph nat_rel E Ei*)
   (*pure bool_rel*)
   (*dfs E s t*)

---

[7] Again, we applied α-conversion, to make the generated variable names more readable.

Listing 2. Imperative DFS algorithm generated by our tool.

```
dfs_impl Ei si ti ≡ do {
  V ← bv_new;
  (_,r) ← heap_rec (λdfs (V,v). do {
    visited ← bv_memb v V;
    if visited then return (V,False)
    else do {
      V ← bv_ins v V;
      if v = ti then return (V,True)
      else do {
        succ_list ← succi Ei v;
        imp_nfoldli succ_list (λ(_, brk). return (¬ brk))
          (λv (V,_). dfs (V,v)) (V,False)
      }
    }
  }) (V,si);
  return r
}
```

If we combine this with the correctness theorem of the abstract DFS algorithm
dfs, we immediately get the following theorem, stating total correctness of our
imperative algorithm:

corollary *dfs_impl_correct:*
  finite (reachable E s) $\Longrightarrow$
  $\langle is\_graph\ nat\_rel\ E\ Ei \rangle$
    dfs_impl Ei s t
  $\langle \lambda r.\ is\_graph\ nat\_rel\ E\ Ei * \uparrow(r \longleftrightarrow (s,t) \in E^*)\rangle_t.$

5   Case Studies

In this section, we present two case studies: We apply our method to a nested
depth-first search algorithm and Dijkstra's shortest paths algorithm. Both algorithms have already been formalized within the Isabelle Refinement Framework [6,22,23], and we were able to reuse the existing abstract algorithms and
correctness proofs unchanged. The resulting Imperative/HOL algorithms are
considerably more efficient than the original functional versions.

5.1   Nested Depth-First Search

For the CAVA model checker [6], we have verified various nested depth-first
search algorithms [26]. Here, we pick a version from the examples that come
with the Isabelle Collection Framework [11]. It contains an improvement by

Holzmann et al. [8], where the search already stops if the inner DFS finds a path back to a node on the stack of the outer DFS.

From the existing abstract formalization, it takes about 160 lines of mostly straightforward Isabelle text to arrive at the generated SML code and the corresponding correctness theorem, relating the imperative algorithm to its specification. The main part of the required Isabelle text consists of declaring parametricity rules for specific algebraic data types defined by the abstract formalization, and could be automated.

We compile the generated code with MLton [20] and benchmark it against the original functional refinement and an unverified implementation of the same algorithm in C++, taken from material accompanying [26]. The algorithm is run on state spaces extracted from the BEEM benchmark suite [24]: dining philosophers and Peterson's mutual exclusion algorithm. We have checked for valid properties only, such that the search has to explore the whole state space. The results are displayed in the table below:

Model	Property	#States	Fun	Fun*	Imp	Imp*	C++ (O3)	C++ (O0)
phils.4	ϕ_1	353668	975	75	70	63	48	66
phils.5		517789	1606	120	113	108	83	112
phils.4	$G(true)$	287578	740	59	53	46	40	54
phils.5		394010	1156	83	77	71	64	85
peterson.3	ϕ_2	58960	119	9	7	5	5	7
peterson.4		1120253	2476	184	142	110	111	158
peterson.3	$G(true)$	29289	55	4	3	2	3	4
peterson.4		576156	1314	88	70	55	54	78

where $\phi_1 = G(one_0 \implies one_0 \ W \ eat_0)$ and $\phi_2 = G(wait_0 \implies F(wait_0) \vee G(\neg ncs_0))$

The first column displays the name of the model, the second column the checked property, and the third column displays the number of states. The remaining columns show the time in ms required by the different implementations, on a 2.2 GHz i7 quadcore processor with 8 GiB of RAM. Fun denotes a purely functional implementation with red-black trees. Fun* denotes a purely functional implementation, relying on an unverified array implementation similar to Haskell's *Array.Diff*. Imp denotes the verified implementation generated by our tool, which uses array lists. Imp* denotes a verified implementation generated after hinting our tool to preinitialize the array lists to the correct size (which required 5 extra lines of Isabelle text), such that no array reallocation occurs during the search. Finally, the C++ columns denote the unverified C++ implementation, which uses arrays of fixed size. It was compiled using gcc 4.8.2 with (O3) and without (O0) optimizations.

The results are quite encouraging: Our tool generates code that is more than one order of magnitude faster than the purely functional code. We are also faster than the Fun*-implementation, which depends on an unverified component, and faster than the unoptimized C++ implementation. For the philosopher models, we come close to the optimized C++ implementation, and for the Peterson models, we even catch up.

5.2 Dijkstra's Shortest Paths Algorithm

We have performed a second case study, based on an existing formalization of Dijkstra's shortest paths algorithm [23]. The crucial data types in the existing formalization are a priority queue, a map from nodes to current paths and weights, and a map from nodes to outgoing edges that represents the graph. It took us about 130 lines of straightforward Isabelle text to set up our tool to produce an imperative version of Dijkstra's algorithm, using arrays for the maps. Currently, we do not have an imperative priority queue data structure, so we reused the existing functional one which is based on finger trees [7], demonstrating the capability of our tool to mix imperative and functional data structures. We benchmark our implementation (Imp) against the original functional version (Fun), and a reference implementation in Java (Java), taken from Sedgewick et al. [27]. The inputs are complete graphs with random weights and 1300 and 1500 nodes (cl1300, cl1500), as well as two examples from [27] (medium, large). The required times in ms are displayed in Fig. 1:

Name	Fun	Imp	Java
cl1300	278	167	28
cl1500	378	219	29
medium	2	2	3
large	45606	28861	1490

Fig. 1. Dijkstra benchmark

The results show that a significant speedup (factor 1.5–2) can be gained by replacing only some of the functional data structures by imperative ones. However, we are still one order of magnitude slower than the reference implementation in Java. Our profiling results indicate that most of the time in the Imperative/HOL implementation is spent to manage the finger tree-based priority queue, and we are currently formalizing an array-based min-heap — the same data structure as used in the Java implementation.

6 Conclusion

We have presented an Isabelle/HOL-based approach to automatically refine functional programs specified over abstract data types to imperative ones using heap-based data structures. Not only the program, but also the refinement proof is generated, such that we get imperative programs verified in Isabelle/HOL.

Our approach is based on the Isabelle Refinement Framework, for which many formalized algorithms already exist. These can now be refined to imperative implementations, without redoing their correctness proofs.

We have implemented a prototype tool, which we applied to generate a verified nested DFS algorithm, which is almost as efficient as an unverified implementation of the same algorithm in C++. Moreover, our approach can mix refinements to functional and imperative data structures, which we demonstrated by a refinement of Dijkstra's algorithm, where the priority queue is functional, but the graph representation and some maps are imperative. We gained a speedup of factor 1.5–2 wrt. the purely functional version, but are still an order of magnitude slower than an unverified implementation in Java.

Apart from extending the imperative collection framework by more data structures, future work includes improving the automation. Another interesting direction is to allow sharing of read-only data on the heap, which also would allow refinement of nested abstract data types, e.g. sets of sets to arrays of pointers to arrays. Fractional permissions [1] may be the right tool to achieve this.

6.1   Related Work

We are not aware of interactive theorem prover-based tools to automatically refine functional to imperative programs.

Separation logic has been implemented for various interactive theorem provers, e.g. [9,18,21,28]. The work closest to ours is probably the Ynot project [21]. They formalize a heap monad, a separation logic, and imperative data structures in Coq. Their code generator targets Haskell. However, we are not aware of any performance benchmarks.

For Isabelle/HOL, there is a second separation logic framework [9], which has been developed independently of ours. It can be instantiated to various heap models, while ours is specialized to Imperative/HOL. However, the provided automation is less powerful than ours.

The HOLFoot tool [28] implements a separation logic framework in HOL4. While it provides more powerful automation than our framework, its simplistic imperative language is less convenient for formalizing complex algorithms.

In Coq, various imperative OCaml programs, including Dijkstra's shortest paths algorithm, have been verified with *characteristic formulas* [5]. Apart from the genuine characteristic formula technique, the main difference to our work is that we use a top-down approach, refining an abstract algorithm down to executable code, while they use a bottom-up approach, starting with a translation of the OCaml code to characteristic formulas.

Acknowledgements. We thank Rene Meis for formalizing the basics of separation logic for Imperative/HOL. Moreover we thank Thomas Tuerk for interesting discussions about automation of separation logic.

References

1. Bornat, R., Calcagno, C., O'Hearn, P., Parkinson, M.: Permission accounting in separation logic. In: POPL, pp. 259–270. ACM (2005)
2. Bulwahn, L., Krauss, A., Haftmann, F., Erkök, L., Matthews, J.: Imperative functional programming with isabelle/HOL. In: Mohamed, O.A., Muñoz, C., Tahar, S. (eds.) TPHOLs 2008. LNCS, vol. 5170, pp. 134–149. Springer, Heidelberg (2008)
3. Calcagno, C., Distefano, D., O'Hearn, P., Yang, H.: Compositional shape analysis by means of bi-abduction. In: POPL 2009, pp. 289–300 (2009)

4. Calcagno, C., O'Hearn, P., Yang, H.: Local action and abstract separation logic. LICS **2007**, 366–378 (2007)
5. Charguéraud, A.: Characteristic formulae for the verification of imperative programs. In: ICFP, pp. 418–430. ACM (2011)
6. Esparza, J., Lammich, P., Neumann, R., Nipkow, T., Schimpf, A., Smaus, J.-G.: A fully verified executable LTL model checker. In: Sharygina, N., Veith, H. (eds.) CAV 2013. LNCS, vol. 8044, pp. 463–478. Springer, Heidelberg (2013)
7. Hinze, R., Paterson, R.: Finger trees: A simple general-purpose data structure. J. Funct. Program. **16**(2), 197–217 (2006)
8. Holzmann, G., Peled, D., Yannakakis, M.: On nested depth first search. In: SPIN. Discrete Mathematics and Theoretical Computer Science, vol. 32, pp. 23–32. American Mathematical Society (1996)
9. Klein, G., Kolanski, R., Boyton, A.: Mechanised separation algebra. In: Beringer, L., Felty, A. (eds.) ITP 2012. LNCS, vol. 7406, pp. 332–337. Springer, Heidelberg (2012)
10. Krauss, A.: Recursive definitions of monadic functions. In: Proceedings of PAR, vol. 43, pp. 1–13 (2010)
11. Lammich, P.: Collections framework. In: Archive of Formal Proofs, Dec 2009. http://afp.sf.net/entries/Collections.shtml. Formal proof development
12. Lammich, P.: Refinement for monadic programs. In: Archive of Formal Proofs (2012). http://afp.sf.net/entries/Refine_Monadic.shtml. Formal proof development
13. Lammich, P.: Automatic data refinement. In: Blazy, S., Paulin-Mohring, C., Pichardie, D. (eds.) ITP 2013. LNCS, vol. 7998, pp. 84–99. Springer, Heidelberg (2013)
14. Lammich, P.: Verified efficient implementation of gabow's strongly connected component algorithm. In: Klein, G., Gamboa, R. (eds.) ITP 2014. LNCS, vol. 8558, pp. 325–340. Springer, Heidelberg (2014)
15. Lammich, P., Lochbihler, A.: The isabelle collections framework. In: Kaufmann, M., Paulson, L.C. (eds.) ITP 2010. LNCS, vol. 6172, pp. 339–354. Springer, Heidelberg (2010)
16. Lammich, P., Meis, R.: A separation logic framework for imperative hol. In: Archive of Formal Proofs, Nov 2012. http://afp.sf.net/entries/Separation_Logic_Imperative_HOL.shtml. Formal proof development
17. Lammich, P., Tuerk, T.: Applying data refinement for monadic programs to hopcroft's algorithm. In: Beringer, L., Felty, A. (eds.) ITP 2012. LNCS, vol. 7406, pp. 166–182. Springer, Heidelberg (2012)
18. Marti, N., Affeldt, R.: A certified verifier for a fragment of separation logic. In: PPL-Workshop (2007)
19. Meis, R.: Integration von Separation Logic in das Imperative HOL-Framework. Master Thesis, WWU Münster (2011)
20. MLton Standard ML compiler. http://mlton.org/
21. Nanevski, A., Morrisett, G., Shinnar, A., Govereau, P., Birkedal, L.: Ynot: Reasoning with the awkward squad. In: ICFP (2008)
22. Neumann, R.: A framework for verified depth-first algorithms. In: Workshop on Automated Theory Exploration (ATX 2012), pp. 36–45 (2012)
23. Nordhoff, B., Lammich, P.: Formalization of Dijkstra's algorithm. In: Archive of Formal Proofs, Jan 2012. http://afp.sf.net/entries/Dijkstra_Shortest_Path.shtml. Formal proof development
24. Pelánek, R.: BEEM: benchmarks for explicit model checkers. In: Bošnački, D., Edelkamp, S. (eds.) SPIN 2007. LNCS, vol. 4595, pp. 263–267. Springer, Heidelberg (2007)

25. Reynolds, J.C.: Separation logic: A logic for shared mutable data structures. In: Proceedings of Logic in Computer Science (LICS), pp. 55–74. IEEE (2002)
26. Schwoon, S., Esparza, J.: A note on on-the-fly verification algorithms. In: Halbwachs, N., Zuck, L.D. (eds.) TACAS 2005. LNCS, vol. 3440, pp. 174–190. Springer, Heidelberg (2005)
27. Sedgewick, R., Wayne, K.: Algorithms, 4th edn. Addison-Wesley, Boston (2011)
28. Tuerk, T.: A separation logic framework for HOL. Technical report UCAM-CL-TR-799, University of Cambridge, Computer Laboratory, June 2011

Stream Fusion for Isabelle's Code Generator
Rough Diamond

Andreas Lochbihler$^{(\boxtimes)}$ and Alexandra Maximova

Institute of Information Security, Department of Computer Science,
ETH Zurich, Zürich, Switzerland
andreas.lochbihler@inf.ethz.ch

Abstract. Stream fusion eliminates intermediate lists in functional code. We formalise stream fusion for finite and coinductive lists in Isabelle/HOL and implement the transformation in the code preprocessor. Our initial results show that optimisations during code extraction can boost the performance of the generated code, but the transformation requires further engineering to be usable in practice.

1 Introduction

Over the last decade, code extraction spurred interest in writing executable specifications rather than abstract models in theorem provers [7,12]. For example, it is now possible to verify a conference management system in a prover and compile the model into a usable implementation [7]. Yet, satisfactory performance is hard to achieve. Existing work on efficiency [8,9] focuses on making efficient data structures available in the prover. The potential of optimisation during code extraction has been neglected so far.

Code extraction can boost efficiency in two ways. On the one hand, the generated code can use optimised libraries of the target language. To that end, one specifies manually how types and functions in the logic are mapped to the library. Such a mapping is unverified, i.e., the mapping and the libraries become part of the trusted code base. This is against the spirit of verification and should thus be avoided. On the other hand, extraction itself can transform and optimise the code. This seems sensible for three reasons. First, the transformations and the verification are carried out in the same formal framework. This ensures that they fit together. Second, the extractor can exploit the proven theorems (e.g., invariants, congruences) for optimisation. This knowledge gets lost during translation, as only the definitions are extracted. Thus, the target language compiler cannot exploit it. Third, the evaluation order and strictness requirements of the logic may be weaker than in the target language. This gives the extractor more freedom.

There is little benefit in re-implementing in the extractor all optimisations of the target language. Instead, one should focus on transformations that enable more optimisations in the target language. Fusion techniques [1,4,13] are a good candidate, as they transform a function designed for high-level proofs into one designed for optimisation.

© Springer International Publishing Switzerland 2015
C. Urban and X. Zhang (Eds.): ITP 2015, LNCS 9236, pp. 270–277, 2015.
DOI: 10.1007/978-3-319-22102-1_18

In this paper, we focus on stream fusion (see Sect. 2 for an introduction), as it is more powerful than other fusion techniques [1,3]. We have formalised stream fusion in Isabelle/HOL with its restrictive type system for finite and coinductive lists and implemented the transformation in the code generator (Sect. 3). It is designed such that fusion affects neither definitions nor proofs in applications. Our evaluation on micro-benchmarks shows that the transformation improves the run time of the generated code by 24 %–93 % (Sect. 4). Unfortunately, the transformation hardly triggers for applications at present. We have identified the obstacles and discuss how they could be overcome (Sect. 6).

The formalisation and implementation are available online [11].

2 Background on Stream Fusion

Stream fusion [1–3] transforms programs to enable optimisations. Consider, for example, the program $sum\text{-}odd\text{-}sq\ n\ =\ sum\ (map\ sq\ (filter\ odd\ [1..n]))$. The function $sum\text{-}odd\text{-}sq$ computes the sum of the squares (computed by sq) of all odd numbers up to n. The definition is designed for proving, as it composes functions like building blocks, so proofs can use the existing lemmas about them. Yet, if $sum\text{-}odd\text{-}sq$ is implemented as given, three lists are allocated at run-time: all numbers up to n, all odd numbers up to n, and their squares. In lazy languages, the lists are not allocated as a whole, but the cells still are. Ideally, no list would be needed at all. But as sum, map, and $filter$ are recursive, compilers are unlikely to inline and optimise them aggressively.

Stream fusion transforms $sum\text{-}odd\text{-}sq$ such that there is only one recursive function left. Then, subsequent optimisations like inlining and call specialisation can get rid of the intermediate allocations. To that end, it replaces a list of type α $list$ by a stream, which consists of a generator $g :: \sigma \Rightarrow (\alpha, \sigma)$ $step$ and a state $s :: \sigma$. Here, $step$ has three constructors: $Done$ indicates the end of the stream, $Yield\ x\ s$ produces the next element x and a new state s, and $Skip\ s$ represents a stuttering step. Lists and streams are linked via two functions $stream$ and $unstream$, which satisfy $unstream\ (stream, xs) = xs$.

$$stream\ [\,] = Done \qquad stream\ (x \cdot xs) = Yield\ x\ xs$$
$$unstream\ (g, s) = (case\ g\ s\ of\ Done \Rightarrow [\,] \quad |\ Skip\ s' \Rightarrow unstream\ (g, s')$$
$$|\ Yield\ x\ s' \Rightarrow x \cdot unstream\ (g, s'))$$

Functions on lists fall into three groups: producers such as $[\_ .. \_]$ return a list, consumers like sum take a list, and transformers (e.g., $filter$ and map) do both. Every such function f has a stream counterpart fS. For example, $filterS$ transforms a generator g into another generator $filterS\ P\ g$ as follows.

$$filterS\ P\ g\ s = (case\ g\ s\ of\ Done \Rightarrow Done \quad |\ Skip\ s' \Rightarrow Skip\ s'$$
$$|\ Yield\ x\ s' \Rightarrow if\ P\ x\ then\ Yield\ x\ s'\ else\ Skip\ s')$$

A fusion equation of the form $f\ xs = unstream\ (fS\ stream, xs)$ links f and fS, e.g., $filterP\ xs = unstream\ (filterS\ P\ stream, xs)$. Equations for producers and consumers omit *stream* and *unstream* as in $[n..m] = unstream ([\_..m]_S, n)$ and $sum \; xs = sumS \; stream \; xs$.

The stream fusion transformation operates in two steps. First, it replaces all list functions with stream functions using the fusion equations. This introduces conversions from streams to lists and back. Second, it eliminates adjacent conversions as in $unstream (f_1 \; stream, unstream (f_2 \; g, s))$, i.e., we get $unstream (f_1 (f_2 \; g), s)$. In the end, only functions on streams should be left. Coutts [1] identifies sufficient conditions for this to happen. Among others, only consumers of a stream may be recursive and transformers must pass *Skips* along unchanged. Then, later compiler stages can inline the functions and eliminate the step constructors and thereby the unnecessary allocations.

Note that the second step can change the type σ of the state for f_1. That is why streams actually have the existential type $\exists \sigma. \; (\sigma \Rightarrow (\alpha, \sigma) \; step) \times \sigma$. Thus, a consumer or transformer cannot examine the state type of the stream it takes. Hence, it can use the state only via the supplied generator, i.e., it behaves the same for all state types.

Transforming a function on lists to one on streams must currently be done manually. So, stream fusion requires that the programmer uses only library functions for which the setup has been provided. This restriction applies to other fusion techniques, too [4, 13].

3 Formalising and Performing Stream Fusion in Isabelle/HOL

Due to the restriction to pre-defined library functions, it is sensible to express fusion in HOL and prove it correct. Otherwise, the generated code must use those library functions of the target language, provided that they are available (we only know of list implementations in GHC [2,3]). Such adaptations exist partly, e.g., for the Haskell function *concatMap*, but they are unverified and error-prone, so it is better to avoid them.

Unfortunately, stream fusion cannot be formalised as is in HOL for two reasons. First, HOL does not have existential type quantifiers. Thus, the type variable σ cannot be hidden in the stream type. Consequently, the elimination of adjacent occurrences of *stream* and *unstream* cannot be expressed as an equality in HOL, because the state type changes. Neither would Coutts' induction proof over types with a logical relation [1] work, as HOL types are not syntactic. Second, stream fusion is designed for coinductive sequence types, as generators need not terminate. However, the formalisation should support finite lists, too, as they are the workhorse in Isabelle/HOL. Even coinductive lists [10], which may be infinite, pose a definitional challenge, as a generator might always return *Skip*, i.e., it refuses to decide whether the list ends. In a domain-theoretic setting, *unstream* could return undefined, but in HOL it must make a choice.

We avoid the first problem by changing the format of the fusion equations. In the former $f \; xs = unstream (fS \; stream, xs)$, we instantiate xs with $unstream (g, s)$ and eliminate the *stream-unstream* pair immediately in the equation. Thus, our format

$$f\ (unstream\ (g,s)) = unstream\ (fS\ g,s) \tag{1}$$

avoids *stream* entirely. Both formats are equivalent, as we get the standard equation back by setting $g = stream$ and $s = xs$ and rewriting with *unstream* (*stream*, xs) $= xs$. In our example, we have $[x..y] = unstream\ ([\_ .. y]_S, x)$ and *filter P* (*unstream* $(g,s))$ = *unstream* (*filterS P g,s*) and *sum* (*unstream* $(g,s))$ = *sumS g s*.

We address the second problem by defining two subtypes of generators. First, *terminating* generators that always reach *Done* after finitely many iterations. Second, *productive* generators whose iteration contains only finitely many consecutive *Skips*. The subtypes are defined using **typedef** and implemented via data refinement in the code generator [5]. For finite lists, we define *unstream* on terminating generators by well-founded recursion. Thus, well-founded induction is our proof principle for the fusion equations. For coinductive lists, we define *unstream* on arbitrary generators as a least fixpoint in the domain of functions on prefix-ordered lists [10], and proofs are by fixpoint induction. Thus, *unstream* interprets infinitely many consecutive *Skips* as the end of the list. Additionally, we lift *unstream* to the type of productive generators, as some functions have fusible implementations only for productive generators. For example, concatenation ++ of two coinductive lists is fusible only if the first list has a productive generator. Otherwise, the fusion equation *unstream* (g_1, s_1) ++ *unstream* (g_2, s_2) = *unstream* (*appendS* $g_1\ g_2\ s_2, s_1$) does not hold: for $g_1 = Skip$, the left hand side is *unstream* (g_2, s_2), but the right hand side equals $[\,]$, as *appendS* must pass *Skips* along.

We have defined stream versions for the fusible list functions in Isabelle/ HOL's list library, i.e., 4 producers, 17 transformers, and 13 consumers, and proved fusion equations for them. For *concatMap*, we also formalised the *flatten* operator from [3], which is easier to optimise. The consumers were the easiest, as they can be defined in terms of their list counterpart and *unstream*. The fusion equation and the recursive code equations were proved automatically from the definition. Producers and transformers are not recursive either, but the proofs of termination and the fusion equation require inductions. For coinductive lists, we have 3 producers, 10 transformers, and 7 consumers. When possible, they come in two versions for productive and arbitrary generators. Proofs are by induction on productivity and fixpoint induction, respectively.

The stream fusion transformation itself is implemented as a rewrite procedure in the code generator. Its preprocessor invokes the procedure on all subexpressions of the right-hand side of each code equation. The procedure tries to rewrite the given expression with the fusion equations. It succeeds only if there are no *unstream* functions left at the end; otherwise, the transformation is discarded for this invocation. Our format (1) for the fusion equations ensures that the rewriting terminates, as the *unstreams* are pushed from producers outwards through transformers to consumers. The check for left-over *unstreams* ensures that fusion transformation is complete, as only consumers can eliminate the *unstreams* that producers have introduced. It does not seem sensible to leave *unstreams* in the code, although other fusion systems do so [3,4]. In our setting, the target language compiler does not know that *stream* and *unstream* cancel out. Thus, the conversions would end up in the compiled code and might slow down the execution.

Our implementation is extensible. Users can register new *unstream* functions for other sequence types and new fusion equations for their constants. Overlapping fusion equations are tried in the order of registration. This allows us to use a specialised fusion equation for *flatten* when the inner generator does not depend on the outer's state.

4   Evaluation

To evaluate the potential of stream fusion in Isabelle/HOL, we applied it to three micro-benchmarks (enum, nested, merge) from [3]. They all consist of folding addition on integers over lists generated by *concatMap* and [_ .. _], i.e., they are designed to demonstrate the potential benefit of fusion. We generated Haskell and SML code with fusion enabled and disabled, and compiled it under GHC, PolyML, and mlton. Table 1 lists the run times averaged over ten runs (the parameter n has been set to 10 000 for enum and merge and to 1000 for nested). The measurements were performed on a 64-bit 2.4 GHz Intel i7-3630QM with 16 GB of RAM running Ubuntu Linux 12.04 LTS.

Surprisingly, the Haskell code with stream fusion enabled is slower than without. By looking at GHC's intermediate representation of the programs, we discovered that GHC does not eliminate all *step* constructors, because it does not specialise the consumer *foldlS* to the given combination of transformers and producers. Apparently, *foldlS* itself being recursive prevents the transformation. We manually applied the static argument transformation (SAT) to the generated consumer code such that the recursion occurs only in a nested function as shown in Fig. 1. Then, the specialisation happens and stream fusion enables run time improvements between 31 % and 42 % (row fusion + SAT in Table 1). Unfortunately, Isabelle's code generator cannot generate recursive subfunctions, although this can be expressed with local contexts in Isabelle/HOL itself.

The SML tests show that heavily optimising compilers like mlton (approx. 93 % faster) profit from stream fusion more than less optimising ones like PolyML (24 %–32 % faster). The manual SAT hardly affects PolyML and mlton as the differences are not statistically significant. Note that the folding in the test cases

Table 1. Run times in seconds averaged over ten runs; the relative standard deviation is <2.2 %.

Compiler	GHC 7.8.4 with -O3			PolyML 5.5.2			mlton 20100608		
Micro-benchmark	enum	nested	merge	enum	nested	merge	enum	nested	merge
No fusion	1.33	5.24	1.38	16.2	60.1	16.6	5.19	30.4	5.48
Fusion	1.53	5.61	1.48	12.3	41.1	12.3	.395	1.89	.392
Fusion + SAT	.918	3.05	.934	12.3	41.1	12.3	.388	1.90	.389

```
foldlS g f z s =                           foldlS g f = go
   (case generator g s of {                   where { go z s = (case generator g s of {
      Done – > z;                                 Done – > z;
      Skip a – > foldlS g f z a;                  Skip sa – > go z sa;
      Yield a sa – > let { za = f z a; }          Yield a sa – > let { za = f z a; }
         in Prelude.seq za (foldlS g f za sa); });    in Prelude.seq za (go za sa); })};
```

Fig. 1. Haskell code for *foldlS* as generated by Isabelle (left) and after manually applying SAT (right). The cast operator *generator* applies a terminating generator to a state.

ensures that everything gets evaluated eventually. Thus, the performance gains are due to saving allocations and enabling subsequent optimisations. In particular, the effects of laziness introduced by stream fusion can be neglected. Fusion replaces strict lists with streams a.k.a. lazy lists, i.e., it introduces laziness. This can result in huge savings, as we noted previously [9].

5   Related Work

Coutts [1] introduced stream fusion for Haskell and identified sufficient conditions for stream fusion being an optimisation. He calls our fusion equations "data abstraction properties" and proves some of them on paper, but his implementation uses the traditional format. He justifies eliminating *stream-unstream* pairs by induction over type syntax and invariants preserved by a fixed set of library functions.

Recently, Farmer et al. [3] showed that stream fusion outperforms other fusion techniques [4,13] when *concatMap* receives special treatment. Unlike in GHC's RULES system, the fusion equations necessary for that can be directly expressed in Isabelle.

Huffman [6] formalised stream fusion in Isabelle/HOLCF and proved fusion rules for seven functions on domain-theoretic lists. He focuses on proving Coutts' fusion equations correct, but does not implement the transformation itself. As HOLCF is incompatible with the code generator, we cannot use his work for our purposes.

Lammich's framework [8] transforms Isabelle/HOL programs such that they use efficient data structures. It assumes that the user has carefully written the program for efficiency. So, it does not attempt to eliminate any intermediate data structures.

6   Conclusion and Future Work

We have formalised stream fusion in Isabelle/HOL and implemented it in its code generator. Our initial results show that transformations performed during code extraction from theorem provers can make the compiled code much faster.

In fact, our implementation is just a first step. The transformation works well on code written with fusion in mind as in [9]. Yet, it hardly triggers in ordinary user-space programs. For example, the termination checker CeTA [12] generates 38 K lines of Haskell code, but stream fusion applies only once. Two main issues prevent performing stream fusion more widely. First, we perform fusion only when a single code equation contains the complete chain from producers via transformers to consumers. That is, if the calls to the producer and consumer occur in different functions, the preprocessor cannot see this and fusion is not applied. Control operators and *let* bindings break the chain, too. The first issue can be addressed by improving the implementation. Transformations like *let* floating and inlining of non-recursive functions can help to bring fusible functions together. Care is needed to ensure that sharing is preserved. Currently, Isabelle's code preprocessor does not support such global transformations. Moreover, for Haskell, support for local recursive functions is desirable. We leave this as future work.

Second, list functions are often defined recursively, even if they can be expressed with list combinators. Stream fusion ignores them, as there is no automatic conversion to streams. Yet, this restriction applies to all fusion implementations we know. At the moment, users must either prove the alternative definition in terms of combinators or define the counterparts on streams themselves. Fortunately, the existing definitions (and proofs) remain unchanged, as such additions are local.

Acknowledgements. We thank Joachim Breitner for helping with analysing the GHC compilation. He, Ralf Sasse, and David Basin helped to improve the presentation.

References

1. Coutts, D.: Stream fusion: practical shortcut fusion for coinductive sequence types. Ph.D. thesis, University of Oxford (2010)
2. Coutts, D., Leshchinskiy, R., Stewart, D.: Stream fusion: from lists to streams to nothing at all. In: ICFP 2007, pp. 315–326. ACM (2007)
3. Farmer, A., Höner zu Siederdissen, C., Gill, A.: The HERMIT in the stream. In: PEPM 2014, pp. 97–108. ACM (2014)
4. Gill, A., Launchbury, J., Peyton Jones, S.L.: A short cut to deforestation. In: FPCA 1993, pp. 223–232. ACM (1993)
5. Haftmann, F., Krauss, A., Kunčar, O., Nipkow, T.: Data refinement in Isabelle/HOL. In: Blazy, S., Paulin-Mohring, C., Pichardie, D. (eds.) ITP 2013. LNCS, vol. 7998, pp. 100–115. Springer, Heidelberg (2013)
6. Huffman, B.: Stream fusion. Archive of Formal Proofs, formal proof development (2009). http://afp.sf.net/entries/Stream-Fusion.shtml
7. Kanav, S., Lammich, P., Popescu, A.: A conference management system with verified document confidentiality. In: Biere, A., Bloem, R. (eds.) CAV 2014. LNCS, vol. 8559, pp. 167–183. Springer, Heidelberg (2014)
8. Lammich, P.: Automatic data refinement. In: Blazy, S., Paulin-Mohring, C., Pichardie, D. (eds.) ITP 2013. LNCS, vol. 7998, pp. 84–99. Springer, Heidelberg (2013)
9. Lochbihler, A.: Light-weight containers for Isabelle: efficient, extensible, nestable. In: Blazy, S., Paulin-Mohring, C., Pichardie, D. (eds.) ITP 2013. LNCS, vol. 7998, pp. 116–132. Springer, Heidelberg (2013)

10. Lochbihler, A., Hölzl, J.: Recursive functions on lazy lists via domains and topologies. In: Klein, G., Gamboa, R. (eds.) ITP 2014. LNCS, vol. 8558, pp. 341–357. Springer, Heidelberg (2014)

11. Lochbihler, A., Maximova, A.: Stream fusion in HOL with code generation. Archive of Formal Proofs, formal proof development (2014). http://afp.sf.net/entries/Stream_Fusion_Code.shtml

12. Sternagel, C., Thiemann, R.: Ceta 2.18. http://cl-informatik.uibk.ac.at/software/ceta/ (2014)

13. Svenningsson, J.: Shortcut fusion for accumulating parameters & zip-like functions. In: ICFP 2002, pp. 124–132. ACM (2002)

HOCore in Coq

Petar Maksimović[1,2] and Alan Schmitt[1 (✉)]

[1] Inria, Rennes, France
[2] Mathematical Institute of the Serbian Academy of Sciences and Arts,
Belgrade, Serbia
alan.schmitt@inria.fr

Abstract. We consider a recent publication on higher-order process calculi [12] and describe how its main results have been formalized in the Coq proof assistant. We highlight a number of important technical issues that we have uncovered in the original publication. We believe that these issues are not unique to the paper under consideration and require particular care to be avoided.

1   Introduction

Computer-aided verification has reached a point where it can be applied to state-of-the-art research domains, including compilers, programming languages, and mathematical theorems. Our goal is to contribute to this growing effort by formalizing a recent paper on process calculi.

The paper that we are examining is [12], by Lanese et al. It introduces HOCore—a minimal higher-order process calculus, which features input-prefixed processes, output processes, and parallel composition. Its syntax is as follows:

$$P ::= a(x).P \mid \overline{a}\langle P \rangle \mid P \parallel P \mid x \mid \mathbf{0}.$$

HOCore is minimal, in the sense that it features only the operators strictly necessary for higher-order communication. For one, there are no continuations following output messages. More importantly, there is no restriction operator, rendering all channels global and the dynamic creation of new channels impossible. The semantics of HOCore is presented in [12] in the form of a labeled transition system and it is shown that HOCore is a Turing-complete calculus. On the other hand, its observational equivalence is proven to be decidable. In fact, the main forms of strong bisimilarity considered in the literature (contextual equivalence, barbed congruence, higher-order bisimilarity, context bisimilarity, and normal bisimilarity [4,11,18,22]) all coincide in HOCore. Moreover, their synchronous and asynchronous versions coincide as well. Therefore, it is possible to decide that two processes are equivalent, yet it is impossible, in the general case, to decide whether or not they terminate. In addition, the authors give

This work has been partially supported by the ANR project 2010-BLAN-0305 PiCoq, as well as by the Serbian Ministry of Education, Science and Technological Development, through projects III44006 and ON174026.

C. Urban and X. Zhang (Eds.): ITP 2015, LNCS 9236, pp. 278–293, 2015.
DOI: 10.1007/978-3-319-22102-1_19

in [12] a sound and complete axiomatization of observational equivalence. Our formalization in Coq [13] addresses all of the results of [12] concerning decidability of observational equivalence, coincidence of synchronous bisimilarities, as well as the soundness and completeness of the axiomatization.

Contributions. We present a formalization of the HOCore calculus and its behavioral theory in Coq. Throughout the development, we have strived to preserve a high degree of visual and technical correspondence between, on the one hand, the formulations of theorems and their proofs in the original paper and, on the other hand, their Coq counterparts. We introduce in Sect. 2 the syntactic and semantic concepts of HOCore: processes, freshness of variables and channels, structural congruence, and the labeled transition system. In particular, we focus on modeling bound variables using the canonical approach of Pollack et al. [16].

Sections 3 and 4 contain the main contributions of this paper. In Sect. 3, we present the formalization of various strong bisimilarities used in [12]: higher-order, context, normal, open normal, and IO-bisimilarity, together with the formal proofs that these five bisimilarities coincide and that they are decidable. In Sect. 4, we present the axiomatization of HOCore and the formal proof of its soundness and completeness. In both of these sections, we detail the errors, inaccuracies, and implicit assumptions made in the pen-and-paper proofs, and show how to correct them. These issues are not unique to the paper under consideration, and require particular care to be avoided.

Section 5 is reserved for the overview of related work. Finally, we summarize what we have learned in Sect. 6. The complete formalization, available online,[1] contains approximately 4 thousands of lines of code (kloc) of specification and 22 kloc of proofs. It was developed intermittently over a period of three years by the authors, Simon Boulier, and Martín Escarrá. It relies on the TLC library—a general purpose Coq library equipped with many tactics for proof automation—developed by Arthur Charguéraud.[2] Due to lack of space, the details for some of the definitions and proofs can be found in the Appendix that is available online.

2   Formalizing HOCore

2.1   Syntax

HOCore is a minimal higher-order process calculus, based on the core of well-known calculi such as CHOCS [21], Plain CHOCS [23], and Higher-Order π-calculus [18–20]. The main syntactic categories of HOCore are channels (a, b, c), variables (x, y, z), and processes (P, Q, R). Channels and variables are atomic, whereas processes are defined inductively.

An input-prefixed process $a(x).P$ awaits to receive a process, e.g., Q, on the channel a; it then replaces all occurrences of the variable x in the process P

---

[1] http://www.irisa.fr/celtique/aschmitt/research/hocore/.

[2] http://www.chargueraud.org/softs/tlc/.

with Q, akin to λ-abstraction. An output process[3] $\overline{a}\langle Q \rangle$ emits the process Q on the channel a. Parallel composition allows senders and receivers to interact.

$$P,\ Q ::= \overline{a}\langle P \rangle \qquad\qquad \text{output process}$$
$$\mid a(x).P \qquad\qquad \text{input prefixed process}$$
$$\mid x \qquad\qquad \text{process variable}$$
$$\mid P \parallel Q \qquad\qquad \text{parallel composition}$$
$$\mid \mathbf{0} \qquad\qquad \text{empty process}$$

The only binding in HOCore occurs in input-prefixed processes. We denote the free variables of a process P by $\mathsf{fv}(P)$, and the bound ones by $\mathsf{bv}(P)$. In [12], processes are identified up to the renaming of bound variables. Processes that do not contain free variables are *closed* and those that do are *open*. If a variable x does not occur within a process P, we say that it is fresh with respect to that process, and denote this in the Coq development by $x \sharp P$. We denote lists with the symbol $\tilde{\ }$: a list of variables is, for instance, denoted by $\tilde{x}$. In addition, we define two notions of *size* for a process P: $s_{\parallel}(P)$, for which parallel compositions count; and $s(P)$, for which they do not. (see online Appendix)

Structural Congruence. The structural congruence relation ($\equiv$) is the smallest congruence relation on processes for which parallel composition is associative, commutative, and has $\mathbf{0}$ as the neutral element.

Canonical Representation of Terms. We first address our treatment of bound variables and α-conversion. We have opted for the *locally named* approach of Pollack et al. [16]. There, the set of variables is divided into two distinct subsets: free (global) variables, denoted by X, Y, Z, and bound (local) variables, denoted by x, y, z. The main idea is to dispense with α-equivalence by calculating the value of a local variable canonically at the time of its binding. For this calculation, we rely on a *height function* f that takes a global variable X and a term P within which X will be bound, and computes the value of the corresponding local variable x. We then replace every occurrence of X with x in P.

To illustrate, consider the process $a(x).(x \parallel x)$. It is α-equivalent to $a(y).(y \parallel y)$, for any local variable y. We wish to choose a canonical representative for this entire family of processes. To that end, we start from the process $X \parallel X$, where X is free, calculate $x = f_X(X \parallel X)$, and take this x to be our canonical representative. In the following, we write $a[X]_f.P$ for $a(f_X(P)).[f_X(P)/X]P$.

The height function needs to meet several criteria so that it does not, for one, return a local variable already used in a binding position around the global variable to be replaced, resulting in a capture. We use the criteria presented in [16] that guarantee its well-definedness. Although it is not strictly necessary to choose a particular height function f satisfying these criteria, it is important to demonstrate that it exists, which we do. (see online Appendix)

---

[3] We also refer to an output process as an output message or, simply, a message.

Adjusting the Syntax. The separation of variables into local and global ones is reflected in the syntax of processes

$$P,\ Q ::= a(x).P\ \mid\ \overline{a}\langle P\rangle\ \mid\ P\parallel Q\ \mid\ x\ \mid\ X\ \mid\ \mathbf{0}$$

which is encoded in Coq inductively as follows:

```
Inductive  process  :  Set  :=
  | Send      :  chan      ->  process  ->  process
  | Receive   :  chan      ->  lvar     ->  process  ->  process
  | Lvar      :  lvar      ->  process
  | Gvar      :  var       ->  process
  | Par       :  process   ->  process  ->  process
  | Nil       :  process.
```

Here, `chan`, `lvar`, and `var` are types representing channels, local variables, and global variables, respectively. We can now define substitution without any complications that would normally arise from the renaming of variables. We write $[Q/X]P$ for the substitution of the variable X with the process Q in P, and $[\widetilde{Q}/\widetilde{X}]P$ for the simultaneous substitution of distinct variables $\widetilde{X}$ with processes $\widetilde{Q}$ in P.

Well-Formed Processes. Given our height function f, we define *well-formed processes* as processes in which all local variables are bound[4] and computed using f. We refer to them as *wf-processes* and define the predicate `wf` that characterizes them inductively. Since all subsequent definitions, such as those of the labeled transition system, congruence, and bisimulations, consider wf-processes only, we define a dependent type corresponding to wf-processes, by pairing processes with the proof that they are well-formed.

```
Record wfprocess : Set := mkWfP { proc :> process ; wf_cond : wf proc}.
```

A `wfprocess` is a record with one constructor, `mkWfP`. This record contains two fields: a process, named `proc`, and a proof that `proc` is well-formed, named `wf_cond`. We use Coq's coercion mechanism to treat a `wfprocess` as a `process` when needed. This approach allows us to reason about wf-processes directly, without additional hypotheses in lemmas or definitions. Each time a wf-process is constructed, however, we are required to provide a proof of its well-formedness. To this end, we introduce well-formed counterparts for all process constructors; e.g., we express parallel composition of wf-processes in the following way:

---

[4] Here we have a slight overloading of terminology w.r.t. free and bound variables—while local variables are intended to model the bound variables of the object language, they can still appear free with respect to our adjusted syntax. For example, x is free in the process `Lvar x`, and we do not consider this process to be well-formed.

```
Definition wfp_Par (p q : wfprocess) :=
  mkWfP (Par p q) (WfPar p q (wf_cond p) (wf_cond q)).
```

where `WfPar` is used to construct well-formedness proofs for parallel processes:

```
| WfPar   : forall (p q : process), (wf p) -> (wf q) -> (wf (Par p q))
```

Our initial development did not use dependent types, but relied on additional well-formedness hypotheses for most lemmas instead. We have discovered that, in practice, it is more convenient to bundle processes and well-formedness together. We can, for instance, use Coq's built-in notion of reflexive relations without having to worry about relating non wf-processes. Also, we can define transitions and congruence directly on wf-processes, thus streamlining the proofs by removing a substantial amount of code otherwise required to show that processes obtained and constructed during the proofs are indeed well-formed.

2.2   Semantics

The original semantics of HOCore is expressed via a labeled transition system (LTS) applied to a calculus with α-conversion. It features the following transition types: internal transitions $P \xrightarrow{\tau} P'$, input transitions $P \xrightarrow{a(x)} P'$, where P receives on the channel a a process that is to replace a local variable x in the continuation P', and output transitions $P \xrightarrow{\overline{a}\langle P''\rangle} P'$, where P emits the process P'' on channel a and evolves to P'. This LTS is not satisfactory because input transitions mention the name of their bound variable. For one, the following transition has to be prevented to avoid name capture: $x \parallel a(x).x \xrightarrow{a(x)} x \parallel x$. To this end, side conditions are introduced and the entire semantics is defined up-to α-conversion to make sure terms do not get stuck because of the choice of a bound name. To simplify the formalization, we restate it using abstractions.

Abstractions and Agents. An *agent* A is either a process P or an abstraction F. An *abstraction* can be viewed as a λ-abstraction inside a context.

$$A ::= P \mid F \qquad F ::= (x).P \mid F \parallel P \mid P \parallel F$$

We call these abstractions *localized*, since the binder (x) does not move (the usual approach to abstractions involves "lifting" the binder so that the abstraction remains of the form $(x).P$). Our approach, similar to the one in [24], allows us to avoid recalculation of bound variables. The application of an abstraction to a process, denoted by $F \bullet Q$, follows the inductive definition of abstractions.

$$((x).P) \bullet Q = [Q/x]P \quad (F \parallel P) \bullet Q = (F \bullet Q) \parallel P \quad (P \parallel F) \bullet Q = P \parallel (F \bullet Q)$$

Removal Transitions. To simplify the definition of bisimulations that allow an occurrence of a free variable to be deleted from related processes, we also add transitions $X \xrightarrow{X} \mathbf{0}$, where a global variable dissolves into an empty process.

The Labeled Transition System. The LTS consists of the following seven rules.

Inp $a(x).P \xrightarrow{a} (x).P$ **Out** $\overline{a}\langle P\rangle \xrightarrow{\overline{a}\langle P\rangle} 0$ **Rem** $X \xrightarrow{X} 0$

Act1 If $P \xrightarrow{\alpha} A$, then $P \parallel Q \xrightarrow{\alpha} A \parallel Q$.

Act2 If $Q \xrightarrow{\alpha} A$, then $P \parallel Q \xrightarrow{\alpha} P \parallel A$.

Tau1 If $P \xrightarrow{\overline{a}\langle P''\rangle} P'$ and $Q \xrightarrow{a} F$, then $P \parallel Q \xrightarrow{\tau} P' \parallel (F \bullet P'')$.

Tau2 If $P \xrightarrow{a} F$ and $Q \xrightarrow{\overline{a}\langle P''\rangle} Q'$, then $P \parallel Q \xrightarrow{\tau} (F \bullet P'') \parallel Q'$.

In these rules, α denotes an arbitrary transition label. In Coq, we represent transitions inductively, providing a constructor for each rule. To illustrate, we present constructors corresponding to the rules OUT, IN, and REM.

```
| TrOut : forall a p, {{wfp_Send a p -- LabOut a p ->> AP wfp_Nil}}
| TrIn  : forall a X p, {{wfp_Abs a X p -- LabIn a ->> AA (AbsPure X p)}}
| TrRem : forall X, {{wfp_Gvar X -- LabRem X ->> AP wfp_Nil}}
```

3 Coincidence and Decidability of Bisimilarities

We now present the formalization of the decidability and coincidence theorems of HOCore. We emphasize that, in order to arrive at those results, it was necessary to prove a number of auxiliary lemmas about structural congruence, substitution, multiple substitution, and transitions that were implicitly considered true in the original paper. Some of these lemmas were purely mechanical, while others required a substantial effort. (see online Appendix)

We begin by presenting the five forms of strong bisimilarity commonly used in higher-order process calculi and describe their formalization in Coq. Following [12], we first define a number of basic bisimulation clauses. We show here only those adjusted during the formalization. (see online Appendix)

Definition 1. *A relation $\mathcal{R}$ on HOCore processes is:*

- *a variable relation, if when $P \mathcal{R} Q$ and $P \xrightarrow{X} P'$, there exists a process Q', such that $Q \xrightarrow{X} Q'$ and $P' \mathcal{R} Q'$.*
- *an input relation, if when $P \mathcal{R} Q$ and $P \xrightarrow{a} F$, there exists an abstraction F', such that $Q \xrightarrow{a} F'$, and for all global variables X fresh in P and Q, it holds that $(F \bullet X) \mathcal{R} (F' \bullet X)$.*
- *an input normal relation, if when $P \mathcal{R} Q$ and $P \xrightarrow{a} F$, there exists an abstraction F', such that $Q \xrightarrow{a} F'$, and for all channels m fresh in P and Q it holds that $(F \bullet \overline{m}\langle 0\rangle) \mathcal{R} (F' \bullet \overline{m}\langle 0\rangle)$.*
- *closed, if when $P \mathcal{R} Q$ and $P \xrightarrow{a} F$, then there exist Q' and F', such that $Q \xrightarrow{a} F'$, and for all closed R, it holds that $(F \bullet R) \mathcal{R} (F' \bullet R)$.*

```
Definition var_relation (R : RelWfP) : Prop :=
  forall (p q : wfprocess), (R p q) ->
    forall (X : var) (p' : wfprocess), {{p -- LabRem X ->> (AP p')}} ->
      exists (q' : wfprocess), {{q -- LabRem X ->> ( AP q')}} /\ (R p' q').
```

284   P. Maksimović and A. Schmitt

There are two main differences between these definitions and the ones presented in [12]. The first one, as previously mentioned, is the use of abstractions in place of variables in input transitions. The second difference is found in the definition of a *variable relation*; we use the transition **Rem**, while the one in [12] uses structural congruence. This change was motivated by the fact that we have chosen to define bisimulations semantically, in terms of processes capable of executing transitions. In that context, preserving the definition of a variable relation from [12] and involving structural congruence would be inconsistent. Moreover, working under the hypothesis that two processes are structurally congruent is very inconvenient, as their structure may be completely different.

Using these clauses, we define the five forms of strong bisimulation: higher-order ($\sim_{HO}$), context ($\sim_{CON}$), normal ($\sim_{NOR}$), input-output ($\sim_{IO}^{\circ}$), and open normal ($\sim_{NOR}^{\circ}$) bisimulations. We present here only the ones that differ from those in [12].

Definition 2. *A relation $\mathcal{R}$ on HOCore processes is:*

- *an input-output (IO) bisimulation if it is symmetric, an input relation, an output relation, and a variable relation;*
- *an open normal bisimulation if it is symmetric, a τ-relation, an input relation, an output normal relation, and a variable relation.*

```
Definition IO_bisimulation (R : RelWfP) : Prop :=
  (Symmetric R) /\ (in_relation R) /\ (out_relation R) /\ (var_relation R).
```

For each of these bisimulations, we also consider a corresponding *bisimilarity*, i.e., the union of all corresponding bisimulations, and each of these bisimilarities is proven to be a bisimulation itself. Bisimilarity can naturally be viewed as a co-inductive notion, but its co-inductive aspects can also be expressed in a set-theoretic way, as follows:

```
Definition IObis (p q : wfprocess) : Prop :=
  exists R, (IO_bisimulation R) /\ (R p q).
```

To compare the set-theoretic and the native Coq co-inductive versions of bisimilarity, we have also defined the latter and proven it equivalent to the set-theoretic one by using a variation of Park's principle [7]. We use these two definitions interchangeably, depending on which is more suited to a given proof. For instance, the co-inductive definition is more convenient for statements in which the candidate relations are simple and follow the formulation of the statement, as there is no need to supply the relation during the proof process. In the cases where the candidate relations are more complicated, however, we find it more natural and closer to the paper proofs to use the set-theoretic definition.

We also define the extensions of $\sim_{HO}$, $\sim_{CON}$, and $\sim_{NOR}$ to open processes by adding abstractions that bind all free variables, and denote these extensions by $\sim_{HO}^{\star}$, $\sim_{CON}^{\star}$, and $\sim_{NOR}^{\star}$, respectively. (see online Appendix)

Relations and Bisimilarities "up-to". An important notion used in our development is that of two processes P and Q being in a relation $\mathcal{R}$ *up to* another relation $\mathcal{R}'$: assuming the processes are related by $\mathcal{R}$ before a transition, they are related by $\mathcal{R}'\mathcal{R}\mathcal{R}'$ after the transition. We establish the following connection between bisimilarities and bisimilarities "up-to", illustrated here only for $\sim_{\mathrm{NOR}}^{\circ}$-bisimilarity, but holding for all five.

Lemma 1. *Let $\mathcal{R}'$ be an equivalence relation that is also an ONOR-bisimulation. Then, $p \sim_{\mathrm{NOR}}^{\circ} q$ if and only if $p \sim_{\mathrm{NOR}}^{\circ} q$ up to $\mathcal{R}'$.*

We mainly use this technique to reason up to structural congruence. We prove $\equiv$ is an equivalence relation, and that it is an $\sim_{\mathrm{NOR}}^{\circ}$-bisimulation. By Lemma 1, we may reason up to $\equiv$ to show that processes are $\sim_{\mathrm{NOR}}^{\circ}$-bisimilar. This allows us to consider small bisimulation candidates and use $\equiv$ after a transition to get back in the candidate.(see online Appendix)

Existential vs. Universal Quantification of Freshness. Definitions of input, input normal, and output normal relations all feature universal quantification on a fresh channel and/or variable. Our formalization has revealed that this condition leads to major problems in proving a number of statements, such as transitivity of $\sim_{\mathrm{NOR}}^{\circ}$-bisimilarity, Lemmas 1 and 3; these problems were not addressed in [12]. To correct this, we defined existentially-quantified $\sim_{\mathrm{IO}}^{\circ}$-, $\sim_{\mathrm{NOR}}$-, and $\sim_{\mathrm{NOR}}^{\circ}$-bisimilarities and proved they coincide with their universally-quantified counterparts. We could consider closing candidate relations under variable freshness instead, but this would complicate the formal proofs substantially. Much more importantly, existentially-quantified IO-bisimulation cannot be avoided in the formalization of the decidability procedure.

Decidability of IO-bisimilarity. The simplicity of IO-bisimilarity lets us show directly that not only it is a congruence, but that it is also decidable.

Lemma 2. *The following properties concerning $\sim_{\mathrm{IO}}^{\circ}$ hold:*

1. *$\sim_{\mathrm{IO}}^{\circ}$ is a congruence: if $P \sim_{\mathrm{IO}}^{\circ} P'$, then: (1) $a[X]_f.P \sim_{\mathrm{IO}}^{\circ} a[X]_f.P'$; (2) for all Q, $P \parallel Q \sim_{\mathrm{IO}}^{\circ} P' \parallel Q$; and (3) $\overline{a}\langle P \rangle \sim_{\mathrm{IO}}^{\circ} \overline{a}\langle P' \rangle$.*
2. *IO-bisimilarity and existential IO-bisimilarity coincide.*
3. *IO-bisimilarity and IO-bisimilarity up to $\equiv$ coincide.*
4. *$\sim_{\mathrm{IO}}^{\circ}$ is a τ-bisimulation.*
5. *$\sim_{\mathrm{IO}}^{\circ}$ is decidable.*

To show decidability, we specify a brute force algorithm that tests every possible transition for the two processes, up to a given depth. As IO-bisimilarity testing always results in smaller processes (there is no τ clause in this bisimilarity), we can show that this algorithm decides IO-bisimilarity, if the depth considered is large enough. This allows us to conclude:

```
Theorem IObis_constructive_decidable : forall p q, { p ≈ q} + { ~(p ≈ q)}.
```

We should note here that the brute force algorithm would not be applicable to the definition of IO-bisimilarity as stated in Definition 2, because we would need to consider universal quantification on the fresh variable X in the case of an input transition. However, since we have proven the equivalence of universally- and existentially-quantified IO-bisimilarities, it is sufficient to check bisimilarity for only one arbitrary fresh X. More importantly, we strongly emphasize that, although the TLC library is based on classical logic, the decidability procedure that we have formalized is fully constructive.

Coincidence of Bisimilarities. First, we address one of the two most technically challenging lemmas of this part of the development (Lemma 4.15 of [12]).

Lemma 3. *If $(m[X]_f.P') \parallel P \sim^{\circ}_{\text{NOR}} (m[X]_f.Q') \parallel Q$ for some fresh m, then $P \sim^{\circ}_{\text{NOR}} Q$ and $P' \sim^{\circ}_{\text{NOR}} Q'$.*

Proof. In order to show the first part of the lemma, we show that the relation

$$\mathcal{S} = \bigcup_{j=1}^{\infty} \left\{ (P, Q) \ : \ \left(\prod_{k \in 1..j} m_k[X_k]_f.P_k \right) \parallel P \sim^{\circ}_{\text{NOR}} \left(\prod_{k \in 1..j} m_k[X_k]_f.Q_k \right) \parallel Q \right\}$$

is an existential $\sim^{\circ}_{\text{NOR}}$-bisimulation, where $\{P_i\}_1^{\infty}$, $\{Q_i\}_1^{\infty}$ are arbitrary processes, $\{m_i\}_1^{\infty}$ are channels fresh with respect to P, Q, $\{P_i\}_1^{\infty}$, and $\{Q_i\}_1^{\infty}$, and each variable X_i is fresh with respect to P_i and Q_i. The proof proceeds as in [12], by examining possible transitions of P and showing these transitions are matched by Q. To this end, we study processes of the form $(\prod_{k \in 1..j} m_k[X_k]_f.P_k) \parallel P$ and prove several lemmas about their properties in relation to transitions and structural congruence. The original proof silently relies on the use of existentially- quantified $\sim^{\circ}_{\text{NOR}}$-bisimilarity and up-to structural congruence proof techniques for this bisimilarity (up-to techniques are proven, but only for IO-bisimilarity). These assumed results were not trivial to formally establish.

For the second part, we determined that it is unnecessary to formalize the informal procedure to consume $\sim^{\circ}_{\text{NOR}}$-bisimilar processes presented in [12]. Instead, it is sufficient to proceed by induction on $(s(P) + o(P))$, where $o(P)$ is the num- ber of outputs in P, and use the already proven first part, reducing the formal proof to a technical exercise. We believe that the formalization of this consump- tion procedure would have been very difficult and would require a substantial amount of time and effort. This illustrates the insights one can gain through formal proving when it comes to proof understanding and simplification.   □

We are now ready to state and prove the coincidence of bisimilarities.

Theorem 1. *The five strong bisimilarities coincide in HOCore.*

Proof. We prove the following implications, the direct corollary of which is our goal: (1) $\sim^{\circ}_{\text{IO}}$ implies $\sim^{\star}_{\text{HO}}$; (2) $\sim^{\circ}_{\text{HO}}$ implies $\sim^{\star}_{\text{CON}}$; (3) $\sim^{\star}_{\text{CON}}$ implies $\sim^{\star}_{\text{NOR}}$; (4) $\sim^{\star}_{\text{NOR}}$ implies $\sim^{\circ}_{\text{NOR}}$; (5) $\sim^{\circ}_{\text{NOR}}$ implies $\sim^{\circ}_{\text{IO}}$.

The only major auxiliary claim that needs to be proven here is that an open normal bisimulation also satisfies the output relation clause; this is an immediate consequence of Lemma 3. □

To conclude this section, we focus on the error discovered in the proof of the right-to-left direction of Lemma 4.14 of [12], stating that higher order bisimilarity implies IO-bisimilarity. Its proof was the most challenging to formalize in this part of the development and this formalization has led us to several important insights. First, we need the following auxiliary lemma.

Lemma 4. $P \sim_{\text{HO}}^{\star} Q$ if and only if *for all closed* $\widetilde{R}$, $[\widetilde{R}/\widetilde{X}]P \sim_{\text{HO}} [\widetilde{R}/\widetilde{X}]Q$, *where* $\widetilde{X} = \text{fv}(P) \cup \text{fv}(Q)$. *(Lemma 4.13 of [12])*.

While the proof of this claim takes only several lines on paper, its Coq version amounts to more than three hundred lines of code, requiring a number of auxiliary lemmas that appear evident in a pen-and-paper context. These lemmas mostly deal with permutations of elements inside a list and the treatment of channels in the "opening" of $\sim_{\text{HO}}$. We can now proceed to our desired claim.

Lemma 5. $\sim_{\text{HO}}^{\star}$ *implies* $\sim_{\text{IO}}^{\circ}$.

Discussion It suffices to prove that the relation $\mathcal{R} = \{(P, Q) \mid [\widetilde{M}/\widetilde{X}]P \sim_{\text{HO}}^{\star} [\widetilde{M}/\widetilde{X}]Q\}$ is an IO-bisimulation, where $\widetilde{X}$ and $\widetilde{M}$ are, respectively, lists of variables and messages carrying $\mathbf{0}$ on fresh channels. We show $\mathcal{R}$ is an input, an output, and a variable relation. The proofs of all three cases in [12] share the same structure that relies on the application of both directions of Lemma 4.

First, the unguarded variables[5] are substituted with processes of the form $\overline{m}\langle\mathbf{0}\rangle$. Then, the remaining free variables are substituted with arbitrary closed processes $\widetilde{R}$, using Lemma 4 left-to-right to show that the result is still in $\mathcal{R}$. Next, the processes do a transition, and it is proposed to apply Lemma 4 right-to-left to conclude. This last step is not justified, as we now explain. The main idea is to show that the bisimulation diagrams are closed under multiple substitution, i.e., that for all processes P and Q, all global variables X, and all closed processes $\widetilde{R}$, given $[\widetilde{R}/\widetilde{X}]P \sim_{\text{HO}} [\widetilde{R}/\widetilde{X}]Q$ and $[\widetilde{R}/\widetilde{X}]P \xrightarrow{\alpha} [\widetilde{R}/\widetilde{X}]P'$, there exists a Q' such that $[\widetilde{R}/\widetilde{X}]Q \xrightarrow{\alpha} [\widetilde{R}/\widetilde{X}]Q'$ and $[\widetilde{R}/\widetilde{X}]P' \sim_{\text{HO}} [\widetilde{R}/\widetilde{X}]Q'$. Then, the idea is to apply Lemma 4 right-to-left to obtain $P' \sim_{\text{HO}}^{\star} Q'$. This is accomplished by first showing that there exists an S, such that $[\widetilde{R}/\widetilde{X}]Q \xrightarrow{\alpha} S$ and $[\widetilde{R}/\widetilde{X}]P' \sim_{\text{HO}} S$, and then showing that there exists a Q', such that $S = [\widetilde{R}/\widetilde{X}]Q'$. The crucial mistake is that Q' here depends on $\widetilde{R}$, unless proven otherwise. Thus, the requirement for a unique Q' with a universal quantification on $\widetilde{R}$ of Lemma 4 right-to-left is not satisfied, and the lemma cannot be applied.

Our initial attempt at this proof followed this approach, which failed when Coq refused to let us generalize $\widetilde{R}$ since Q' depended on it. We thus proceeded differently, by first substituting every free variable with processes of the form $\overline{m}\langle\mathbf{0}\rangle$, using Lemma 4 left-to-right to show that we remain in the relation. We

---

[5] Variables that appear in an execution context, i.e., those not "guarded" by an input.

then applied several auxiliary lemmas to finish the proof. While the error in this proof was corrected relatively easily, its very existence in the final version of [12] is indicative of the need for a more formal treatment of proofs in this domain, a need further substantiated by the findings presented in the following Section.

4   Axiomatization of HOCore

In this section, we present a formal proof of the soundness and completeness of the axiomatization of HOCore and reflect on the errors discovered in the pen-and-paper proofs. First, we briefly outline the treatment of axiomatization in [12]. The authors begin by formulating a cancellation lemma for $\sim_{IO}^{\circ}$:

Lemma 6. *For all processes P, Q, R, if $P \parallel R \sim_{IO}^{\circ} Q \parallel R$, then also $P \sim_{IO}^{\circ} Q$.*

Next, they introduce the notion of a prime process and prime decomposition.

Definition 3. *A process P is prime if $P \not\sim_{IO}^{\circ} 0$ and $P \sim_{IO}^{\circ} P_1 \parallel P_2$ implies $P_1 \sim_{IO}^{\circ} 0$ or $P_2 \sim_{IO}^{\circ} 0$. When $P \sim_{IO}^{\circ} \prod_{i=1}^{n} P_i$, with each P_i prime, we say that $\prod_{i=1}^{n} P_i$ is a prime decomposition of P.*

The authors then prove that every process admits a prime decomposition unique up to bisimilarity and permutation of indices. Next, extended structural congruence $(\equiv_E,)$ is defined by adding to $\equiv$ the following distribution law:

$$a(x). \left(P \parallel \prod_{1}^{k-1} a(x).P \right) \equiv_E \prod_{1}^{k} a(x).P.$$

Next, a reduction relation $\rightsquigarrow$ and normal forms are defined as follows:

Definition 4. *We write $P \rightsquigarrow Q$ if there exist processes P' and Q', such that $P \equiv P'$, $Q' \equiv Q$, and Q' is obtained from P' by rewriting a subterm of P' using the distribution law left-to-right. A process P is in normal form if it cannot be further simplified in the system $\equiv_E$ using $\rightsquigarrow$.*

Finally, the following three claims conclude the axiomatization section of [12]:

Lemma 7. *If $a(x).P \sim_{IO}^{\circ} Q \parallel Q'$, with $Q, Q' \not\sim_{IO}^{\circ} 0$, then $a(x).P \sim_{IO}^{\circ} \prod_{1}^{k} a(x).R$, with $k > 1$ and $a(x).R$ in normal form.*

Lemma 8. *For all P, Q in normal form, if $P \sim_{IO}^{\circ} Q$, then $P \equiv Q$.*

Theorem 2. *$\equiv_E$ is a sound and complete axiomatization of $\sim_{IO}^{\circ}$ in HOCore.*

Our formalization of this section has led to the discovery of the following imprecisions and errors.

Cancellation Lemma. For the proof of Lemma 6 in [12], the authors attempt to re-use the proof from [14], that proceeds by induction on the sum of sizes of P, Q, and R. This, however, is not possible, as that proof was designed for a calculus with operators structurally different than the ones used in HOCore. Instead, we need to use the candidate relation $\mathcal{R}$, parameterized by a natural number n, such that $P \mathcal{R} Q$ at level n if and only if there exists a process R such that $s(P) + s(Q) + s(R) \leq n$ and $P \parallel R \sim_{\text{IO}}^{\circ} Q \parallel R$, and prove that this candidate relation is an IO-bisimulation by total induction on n.

"Deep" Prime Decomposition. In the proof of Lemma 7, the authors in [12] claim that every process is bisimilar to a parallel composition of a collection of prime processes in normal form. They prove this by taking the normal form of a process, performing a prime decomposition on this normal form, and concluding that all of the constitutents in this prime decomposition are both prime and in normal form. This, however, is not necessarily true, as there is no proof that taking the prime decomposition of a process preserves the fact that it is in normal form. In particular, the definition of primality considers only the top-level structure of the process: we can prove that every output message $\overline{a}\langle P \rangle$ is prime for all a and P, but we cannot obtain any information on the primality of P from this. On the other hand, normal forms are defined "in depth". Because of these discrepancies, the proof in [12] cannot be concluded. To remedy this, we need to define a "deep" version of primality that recursively requires of all sub-processes of P to be prime as well, together with the notion of "deep" prime decomposition, using which we can prove Lemma 7 properly.

Normal Forms and the Proof of Lemma 8. The proof of Lemma 8 in [12] exhibits several errors. It is performed by induction on $s(P)$, simultanously proving $\mathcal{A}$ and $\mathcal{C}$, where $\mathcal{A}$ is an auxiliary lemma stating that if P is an input-prefixed process in normal form, then it is prime, and $\mathcal{C}$ is the main claim. However, the induction must actually be performed on $\mathcal{A} \wedge (\mathcal{A} \rightarrow \mathcal{C})$, since $\mathcal{C}$ cannot be proven otherwise (the third case of item 2 on page 27 of [12] fails). Much more importantly, the seemingly trivial case when $P = a(x).P' \parallel \mathbf{0}$ is not covered by the proof, and it turned out that it cannot be covered at all with the proof structure as presented in [12]. As we were unable to immediately discover an alternate proof method, we decided to adjust the notion of normal form by disallowing empty processes in parallel composition. This meant that normal forms could no longer be defined using $\rightsquigarrow$, but had to be defined directly instead. Interestingly, we can say that normal forms are now "more normal" w.r.t. those in [12], as they are unique only up to commutativity and associativity of parallel composition. With these corrections, we were able to conclude the proof of Lemma 8.

5   Related Work

We first discuss alternative approaches to the treatment of binding and α-conversion, starting from *Nominal Isabelle* [10, 25, 27], where nominal datatypes are used to represent α-equivalence classes. This extension of Isabelle has been successfully applied in a number of formalizations of various calculi with binders [2,15,26], and taking advantage of it would probably reduce the size of our development. However, since one of our main results is the formalization of the decidability procedure for IO-bisimulation, and since stating decidability in Isabelle is not a trivial task, we have ultimately opted to use Coq.

The locally nameless approach [1,5] is the one most similar to the one we are using. There, variables are also split into two categories—local and global—but local variables are calculated by using de Bruijn indices instead of a height function. We find this approach to be both equally viable and equally demanding— freshness and well-formedness would both have to be defined, albeit differently, the proofs would retain their level of complexity, and the amount of code required would not be reduced substantially.

In the Higher-Order-Abstract-Syntax (HOAS), binders of the meta-language are used to encode the binding structure of the object language. In some settings, HOAS can streamline the development and provide an elegant solution for dealing with α-conversion, but in the case of process calculi it also brings certain drawbacks. As stated in [9], there are difficulties with HOAS in dealing with meta-theoretic issues concerning names, especially when it comes to the opening of processes, to the extent that certain meta-theoretic properties that involve substitution and freshness of names inside proofs and processes cannot be proven inside the framework and must be postulated instead.

In light of everything previously stated in this section and since HOCore does not have a restriction operator, we have decided to use the canonical locally named approach [16] for variables bound by the input operator. We have not yet considered more general approaches for name binding, such as [17]. In fact, our development may easily be adapted to other models for binders, as long as binders remain canonical (α-convertibility is equality).

We now turn to works related to the formalization of process calculi and their properties. To the best of the authors' knowledge, there have been no formalizations of the higher-order π-calculus so far. There are, however, many formalizations of the π-calculus in various proof assistants, including Coq, such as [8] (where the author uses de Bruijn indices) and [9] (where the authors use HOAS). The work closest to ours is the recent formalization in Isabelle of higher-order Psi-calculi [15], an extension of [2] to the higher-order setting. We have developed our own formalization because we wanted to stay as close to the paper proofs as possible, in particular when it came to the handling of higher-order and the definitions of the many bisimulations involved. Although we feel that translating HOCore and its bisimulations into a higher-order psi-calculus constitutes a very interesting problem, it is beyond the scope of this paper.

Preliminary versions of this paper have appeared in [3,6]. The former contained the initial results of the effort, which included the locally nameless technique for dealing with binders, the correction of the semantics, and the results on decidability. The latter added to this the new and more solid treatment of well-formed processes as records and the proof of the coincidence of bisimilarities.

This version rounds this formalization effort up with results on the soundness and completeness of the axiomatization, as well as the discovery and correction of a number of errors and imprecisions made in [12].

6   Lessons Learned and Conclusions

This formalization of HOCore in Coq has given us a deeper understanding of both the calculus itself and the level of precision required for proving properties of bisimulations in a higher-order setting. Moreover, it has led to the discovery of several major flaws in the proofs of [12]. First, there was the improper generalization of a hypothesis in the proof of Lemma 5. This flaw was due to a lapse in the tracking of the context inside which a property is derived, and it is in precisely such tracking that proof assistants excel. Next, the proof of Lemma 6 was incorrect, as it relied on a previously existing technique not applicable to higher-order process calculi. Finally, two notions crucial for the axiomatization of HOCore— prime decomposition and normal forms—were not defined correctly and did not take into account, respectively, the need for structural recursion in primality and the subtle impact of empty processes in parallel composition. These errors have all been corrected during the formalization and did not ultimately affect the correctness of the main results of [12], but we feel that it is their very presence in a peer-reviewed, state-of-the-art paper that strongly underlines the need for a more precise formal treatment of proofs in this domain.

We have also identified two missing results required by some lemmas. The first concerns existentially-quantified bisimulations: certain properties of bisimulations cannot be proven if the variables or channels featured in freshness conditions of bisimulation clauses are universally quantified. We solved this by introducing bisimulations with existential quantification and demonstrating that they are equivalent to the universally quantified ones. We have also discovered that existentially-quantified bisimulations are necessary for the formulation of the decidability procedure. The second missing result concerns transitivity of bisimulations, which is, in turn, required to prove that open normal bisimilarity and existential open normal bisimilarity coincide. This claim was not trivial to formalize; it required the use of up-to techniques, which were also left untreated for some bisimilarities in [12].

In the end, we were able to prove the decidability of IO-bisimilarity, the main coincidence theorem, and their supporting lemmas, as well as the results on the soundness and completeness of the axiomatization. We realized that some of the lemmas of the paper are actually not needed, such as Lemma 5. Additionally, we significantly simplified some proofs. For instance, the proof of Lemma 3 (Lemma 4.15 in [12]) does not require the consumption procedure described in [12]; induction on the combined measure $m(P)$ is sufficient and more elegant.

As a side effect of our precise treatment of bound variables, we have detected and corrected an error in the original semantics of HOCore, streamlining it in the process. Labels for input transitions no longer contain the variable to be substituted, but evolve to a localized abstraction instead. We have also introduced the deletion of a variable as a transition in the LTS, removing a special case in the definition of the bisimulations and avoiding the use of structural congruence.

In fact, many of the choices that were made during the formalization were guided by the intent to avoid structural congruence. Although it is very convenient to be able to freely change the structure of a process, doing so interferes considerably with local reasoning. There were cases, however, where we could not avoid it, in particular when proving properties of normal bisimulations. There, we have used "up-to structural congruence" techniques to keep the bisimulation candidates small enough. We have discovered that the discord between the rigid syntax of a formal process and the intuition that it models a "soup" where processes can move around freely and interact with each other made the formalizations of some of the proofs more difficult than anticipated.

We have also proved that the co-inductive and set-theoretic definitions of IO-bisimilarity are equivalent, and used them interchangeably, depending on the complexity of the candidate relation at hand.

As for further work, our immediate aim is to formalize the results presented in Sect. 5 of [12], pertaining to the correctness and completeness of IO-bisimilarity w.r.t. barbed congruence. At this point, we have the correctness proof, which, given Theorem 1, means that all of the five forms of bisimilarity in HOCore are correct w.r.t. barbed congruence. Afterwards, we plan to extend the formalization of HOCore with name restriction, before tackling more complex features such as passivation.

In conclusion, we hope that the experience described in this paper will serve as a motivational step towards a more systematic use of proof assistants in the domain of process calculi.

References

1. Aydemir, B., Charguéraud, A., Pierce, B.C., Pollack, R., Weirich, S.: Engineering formal metatheory. In: ACM SIGPLAN-SIGACT Symposium on Principles of Programming Languages (POPL), pp. 3–15. ACM, Jan 2008
2. Bengtson, J., Parrow, J.: Psi-calculi in Isabelle. In: Berghofer, S., Nipkow, T., Urban, C., Wenzel, M. (eds.) TPHOLs 2009. LNCS, vol. 5674, pp. 99–114. Springer, Heidelberg (2009)
3. Boulier, S., Schmitt, A.: Formalisation de HOCore en Coq. In: Actes des 23èmes Journées Francophones des Langages Applicatifs, Jan 2012
4. Cao, Z.: More on bisimulations for higher order π-calculus. In: Aceto, L., Ingólfsdóttir, A. (eds.) FOSSACS 2006. LNCS, vol. 3921, pp. 63–78. Springer, Heidelberg (2006)
5. Charguéraud, A.: The locally nameless representation. J. Autom. Reasoning, 1–46 (2011). doi:10.1007/s10817-011-9225-2
6. Escarrá, M., Maksimović, P., Schmitt, A.: HOCore in Coq. In: Actes des 26èmes Journées Francophones des Langages Applicatifs, Jan 2015
7. Gimenez, E.: A Tutorial on Recursive Types in Coq. Technical report No 0221 (1998)
8. Hirschkoff, D.: A full formalisation of pi-calculus theory in the calculus of constructions. In: Gunter, Elsa L., Felty, Amy P. (eds.) TPHOLs 1997. LNCS, vol. 1275, pp. 153–169. Springer, Heidelberg (1997)

9. Honsell, F., Miculan, M., Scagnetto, I.: pi-calculus in (co)inductive-type theory. Theoret. Comput. Sci. **253**(2), 239–285 (2000)
10. Huffman, B., Urban, C.: A new foundation for nominal Isabelle. In: Kaufmann, M., Paulson, L.C. (eds.) ITP 2010. LNCS, vol. 6172, pp. 35–50. Springer, Heidelberg (2010)
11. Jeffrey, A., Rathke, J.: Contextual equivalence for higher-order Pi-calculus revisited. Log. Meth. Comput. Sci. **1**(1), 1–22 (2005)
12. Lanese, I., Pérez, J.A., Sangiorgi, D., Schmitt, A.: On the expressiveness and decidability of higher-order process calculi. Inf. Comput. **209**(2), 198–226 (2011)
13. The Coq development team. Coq reference manual (2014). version. 8.4
14. Milner, R., Moller, F.: Unique decomposition of processes. Theor. Comput. Sci. **107**(2), 357–363 (1993)
15. Parrow, J., Borgström, J., Raabjerg, P., Åman Pohjola, J.: Higher-order psi-calculi. Math. Struct. Comput. Sci. FirstView 3, 1–37 2014
16. Pollack, R., Sato, M., Ricciotti, W.: A canonical locally named representation of binding. J. Autom. Reasoning, 1–23, May 2011. doi:10.1007/s10817-011-9229-y
17. Pouillard, N., Pottier, F.: A fresh look at programming with names and binders. In: Proceedings of the Fifteenth ACM SIGPLAN International Conference on Functional Programming (ICFP 2010), pp. 217–228, Sept 2010
18. Sangiorgi, D.: Expressing Mobility in Process Algebras: First-Order and Higher-Order Paradigms. Ph.D. thesis, Univ. of Edinburgh, Dept. of Comp. Sci. (1992)
19. Sangiorgi, D.: Bisimulation for higher-order process calculi. Inf. Comput. **131**(2), 141–178 (1996)
20. Sangiorgi, D.: π-calculus, internal mobility and agent-passing calculi. Theor. Comput. Sci. **167**(2), 235–274 (1996)
21. Thomsen, B.: A calculus of higher order communicating systems. In: Proceedings of POPL 1989, pp. 143–154. ACM Press (1989)
22. Thomsen, B.: Calculi for Higher Order Communicating Systems. Ph.D. thesis, Imperial College (1990)
23. Thomsen, B.: Plain CHOCS: A second generation calculus for higher order processes. Acta Inf. **30**(1), 1–59 (1993)
24. Tiu, A., Miller, D.: Proof search specifications of bisimulation and modal logics for the pi-calculus. ACM Trans. Comput. Logic (TOCL) **11**, 13:1–13:35 (2010)
25. Urban, C.: Nominal techniques in Isabelle/HOL. J. Autom. Reasoning **40**(4), 327–356 (2008)
26. Urban, C., Cheney, J., Berghofer, S.: Mechanizing the metatheory of LF. ACM Trans. Comput. Log. **12**(2), 15 (2011)
27. Urban, C., Kaliszyk, C.: General bindings and alpha-equivalence in nominal isabelle. In: Barthe, G. (ed.) ESOP 2011. LNCS, vol. 6602, pp. 480–500. Springer, Heidelberg (2011)

Affine Arithmetic and Applications to Real-Number Proving

Mariano M. Moscato[1][✉], César A. Muñoz[2], and Andrew P. Smith[1]

[1] National Institute of Aerospace, Hampton, VA 23666, USA
{mariano.moscato,andrew.smith}@nianet.org
[2] NASA Langley Research Center, Hampton, VA 23681, USA
cesar.a.munoz@nasa.gov

Abstract. Accuracy and correctness are central issues in numerical analysis. To address these issues, several self-validated computation methods have been proposed in the last fifty years. Their common goal is to provide rigorously correct enclosures for calculated values, sacrificing a measure of precision for correctness. Perhaps the most widely adopted enclosure method is interval arithmetic. Interval arithmetic performs well in a wide range of cases, but often produces excessively large overestimations, unless the domain is reduced in size, e.g., by subdivision. Many extensions of interval arithmetic have been developed in order to cope with this problem. Among them, affine arithmetic provides tighter estimations by taking into account linear correlations between operands. This paper presents a formalization of affine arithmetic, written in the Prototype Verification System (PVS), along with a formally verified branch-and-bound procedure implementing that model. This procedure and its correctness property enables the implementation of a PVS strategy for automatically computing upper and lower bounds of real-valued expressions that are provably correct up to a user-specified precision.

1 Introduction

Formal verification of safety-critical cyber-physical systems often requires proving formulas involving real-valued computations. At NASA, examples of such verification efforts include formally verified algorithms and operational concepts for the next generation of air traffic management systems [6,16,19]. Provably correct real-valued computations are also essential in areas such as analysis of floating point programs [1–3,7], verification of numerical algorithms [9,22], and in the formalization of mathematical results such as Kepler conjecture [8].

In general, the exact range of a nonlinear function of one or more variables over an interval domain cannot be determined in finite time. Enclosure methods are designed to provide sound intervals that are guaranteed to include, but may however overestimate, the true range of a nonlinear function over a bounded domain. More formally, given a function $f : \mathbb{R}^m \to \mathbb{R}$, an interval-valued function $F : \mathbb{IR}^m \to \mathbb{IR}$ is obtained, where $\mathbb{IR}$ denotes the set of closed non-empty real intervals, such that for all $\mathbf{V} \in \mathbb{IR}^m$

$$\mathbf{v} \in \mathbf{V} \implies f(\mathbf{v}) \in F(\mathbf{V}). \tag{1}$$

© Springer International Publishing Switzerland 2015
C. Urban and X. Zhang (Eds.): ITP 2015, LNCS 9236, pp. 294–309, 2015.
DOI: 10.1007/978-3-319-22102-1_20

Arithmetic may be performed on intervals, by providing definitions for elementary operators, logarithmic and trigonometric functions, and other real-valued functions that satisfy Formula (1). For example, if $\mathbf{a} = [\underline{a}, \overline{a}], \mathbf{b} = [\underline{b}, \overline{b}] \in \mathbb{IR}$, then $\mathbf{a} + \mathbf{b} = [\underline{a} + \underline{b}, \overline{a} + \overline{b}]$. Such definitions yield an enclosure method called *interval arithmetic* [14]. A natural interval extension E of any real expression e is then obtained by recursively replacing in e each constant by an interval containing the constant, each variable by its interval range, and each operator and function by their interval equivalents. Formalizations of interval arithmetic are available in several interactive theorem provers [4,13,21]. These formalizations also include proof strategies for performing provably correct real-valued computations.

While correct, enclosure methods often provide imprecise calculations of expressions involving real-valued functions due to the fact that approximation errors quickly accumulate. To mitigate this problem, enclosure methods often rely on the following property, which must be satisfied by any interval-valued function F. For all $\mathbf{U}, \mathbf{V} \in \mathbb{IR}^m$,

$$\mathbf{U} \subseteq \mathbf{V} \implies F(\mathbf{U}) \subseteq F(\mathbf{V}). \tag{2}$$

Formula (2) enables the use of domain subdivision techniques, whereby the starting domain is recursively subdivided into smaller box sub-domains, on which the enclosure methods provide suitable precision. *Branch and bound* is a recursive method for computing rigorous approximations that combines domain subdivision with pruning strategies. These strategies avoid unnecessary computations that do not improve already computed bounds. A formally verified branch and bound algorithm for generic enclosure methods is presented in [18].

It is well-known that enclosure methods such as interval arithmetic suffer from the *dependency problem*. This problem occurs when a real variable appears multiple times in an expression. In this case, large over-approximations may occur when each variable is treated as an independent interval. The dependency problem can be reduced by using enhanced data structures that, among other things, keep track of variable dependencies. In the case of polynomial and rational functions, a method based on *Bernstein polynomials* [12], provide better precision than interval arithmetic at the cost of increased computational time. Multivariate Bernstein polynomials and proof-producing strategies for rigorous approximation based on their properties are available in PVS [17].

The use of a particular enclosure method depends on a trade-off between precision and efficiency. At one extreme, interval arithmetic is computationally efficient but may produce imprecise bounds. At the other extreme, Bernstein polynomials offer precise bounds but they are computationally expensive. *Affine arithmetic* [5] is an enclosure method situated between these two extremes. By taking into account linear correlations between operands, affine arithmetic produces better estimates than interval arithmetic at a computational cost that is smaller than Bernstein polynomials.

This paper presents a deep embedding of affine arithmetic in the Prototype Verification System (PVS) [20] that includes addition, subtraction, multiplication, and power operation on variables. The embedding is used in an instantiation of a generic branch and bound algorithm [18] that yield a provably correct procedure for computing enclosures of polynomials with variables in interval domains. The formally verified branch and bound procedure is the foundation of a PVS proof strategy for mechanically and automatically finding lower and upper bounds for the minimum and maximum values of multivariate polynomials on a bounded domain.

The formal development presented in this paper is available as part of the NASA PVS Library.[1] All theorems presented in this paper are formally verified in PVS. For readability, standard mathematical notation is used throughout this paper. The reader is referred to the formal development for implementation details.

2   Affine Arithmetic

Affine arithmetic is a refinement of interval arithmetic that attempts to reduce the dependency problem by tracking linear dependencies between repeated variable instances and thereby retaining simple shape information. It is based on the idea that any real value a can be represented by an *affine form* $\widehat{a}$, defined as

$$\widehat{a} \stackrel{def}{=} a_{(0)} + \sum_{i=1}^{\infty} a_{(i)} \varepsilon_i, \tag{3}$$

where $\varepsilon_i \in [-1, 1]$, and $a_{(j)} \in \mathbb{R}$, with $j > 0$. It is assumed that the set of indices j such that $a_j \neq 0$ is finite. Henceforth, $\ell_{\widehat{a}}$ denotes the maximum element of that set or 0 if the set is empty. Each ε_i stands for an unknown error value introduced during the calculation of a. An affine form grows symmetrically around its *central value* $a_{(0)}$. Each $a_{(i)}$ represents the weight that the corresponding ε_i has on the overall uncertainty of a. The coefficients $a_{(i)}$ are called the *partial deviations* of the affine form.

There is a close relationship between affine forms and intervals. Given an affine form $\widehat{a}$ as in Formula (3), it is clear that the value a is included in the interval

$$[\widehat{a}] \stackrel{def}{=} \left[a_{(0)} - \sum_{i=1}^{\ell_{\widehat{a}}} |a_{(i)}|, a_{(0)} + \sum_{i=1}^{\ell_{\widehat{a}}} |a_{(i)}| \right]. \tag{4}$$

In fact, for every real value a' in $[\widehat{a}]$, there exists an assignment $\mathbf{N}$ of values from $[-1, 1]$ to each ε_i in $\widehat{a}$ such that $a' = \widehat{a}(\mathbf{N})$, where

$$\widehat{a}(\mathbf{N}) \stackrel{def}{=} a_{(0)} + \sum_{i=1}^{\ell_{\widehat{a}}} a_{(i)} \mathbf{N}(i). \tag{5}$$

As stated in [23], the semantics of affine arithmetic rely on the existence of a single $\mathbf{N}$ for which a' is equal to the ideal real value a. This property is called the *fundamental invariant* of affine arithmetic.

---

[1] http://shemesh.larc.nasa.gov/fm/ftp/larc/PVS-library/.

Conversely, any given interval $\mathcal{I} = [u, v]$ is equivalent to the affine form

$$\mathsf{V}_k^{\mathcal{I}} \stackrel{def}{=} \frac{v+u}{2} + \frac{v-u}{2}\varepsilon_k, \tag{6}$$

where all the partial deviations, except for the k-th coefficient, are equal to 0.

Arithmetic operations on affine forms can be defined. Operations that are affine combinations of their arguments are called *affine operations*. In contrast to affine operations, non-affine operations may require the addition of noise symbols that do not appear in their operands. For simplicity, variables are indexed by positive natural numbers. It is assumed that the number of variables in any expression is at most m. Thus, the first m noise symbols are reserved for variables. Using this convention, the affine form of variable x_m ranging over $\mathcal{I}$ is conveniently defined as $\mathsf{V}_m^{\mathcal{I}}$. An index $k > \max(m, \ell_{\widehat{a}})$ is referred to as a *fresh index* with respect to $\widehat{a}$.

Addition and multiplication of scalars, as unrestricted subtractions and additions, are affine operations. They are defined as shown next. Given an affine form $\widehat{a} = a_{(0)} + \sum_{i=1}^{\ell_{\widehat{a}}} a_{(i)}\varepsilon_i$ for a as in Formula (3) and $c \in \mathbb{R}$,

$$\begin{aligned}
\mathsf{add}_c(\widehat{a}) &\stackrel{def}{=} c + a_{(0)} + \sum_{i=1}^{\ell_{\widehat{a}}} a_{(i)}\varepsilon_i, \\
\mathsf{mul}_c(\widehat{a}) &\stackrel{def}{=} c \cdot a_{(0)} + \sum_{i=1}^{\ell_{\widehat{a}}} c \cdot a_{(i)}\varepsilon_i.
\end{aligned} \tag{7}$$

The negation operation is defined as $\mathsf{neg}(\widehat{a}) \stackrel{def}{=} \mathsf{mul}_{-1}(\widehat{a})$.

Given another affine form $\widehat{b} = b_{(0)} + \sum_{i=1}^{\ell_{\widehat{b}}} b_{(i)}\varepsilon_i$ for some real value b,

$$\mathsf{add}(\widehat{a}, \widehat{b}) \stackrel{def}{=} \left(a_{(0)} + b_{(0)}\right) + \sum_{i=1}^{\max(\ell_{\widehat{a}}, \ell_{\widehat{b}})} \left(a_{(i)} + b_{(i)}\right)\varepsilon_i. \tag{8}$$

The subtraction operation is defined as $\mathsf{sub}(\widehat{a}, \widehat{b}) \stackrel{def}{=} \mathsf{add}(\widehat{a}, \mathsf{neg}(\widehat{b}))$.

Whilst affine operations can be performed without the introduction of additional error terms, i.e., without any extra overestimation, a suitable rigorous affine approximation must be used for each non-affine operation. Next, definitions for multiplication and power operation on single variables (instead of arbitrary expressions) are presented. Multiplication of two affine forms is defined as

$$\mathsf{mul}_k(\widehat{a}, \widehat{b}) \stackrel{def}{=} a_{(0)}b_{(0)} + \sum_{i=1}^{\max(\ell_{\widehat{a}}, \ell_{\widehat{b}})} (a_{(0)}b_{(i)} + a_{(i)}b_{(0)})\varepsilon_i + \varepsilon_k \sum_{i=1}^{\ell_{\widehat{a}}} |a_{(i)}| \sum_{i=1}^{\ell_{\widehat{b}}} |b_{(i)}|, \tag{9}$$

where k is a fresh noise index with respect to $\widehat{a}$ and $\widehat{b}$, i.e., $k > \max(m, \ell_{\widehat{a}}, \ell_{\widehat{b}})$.

Even though the power operation can be implemented by reducing it to successive multiplications, the following definition gives a better performing alternative for the case where a single variable is raised to a power. Given an elementary affine form $\widehat{a} = a_{(0)} + a_{(m)}\varepsilon_m$ for a real value ranging over the interval $[\widehat{a}]$ and a collection of $n - 2$ fresh noise indices I with respect to $\widehat{a}$,

$$\mathsf{pow}_I(\widehat{a}, n) \stackrel{def}{=} \begin{cases} 1 & \text{if } n = 0, \\ a_{(0)}^n + n\, a_{(0)}^{n-1} a_{(m)}\varepsilon_m + \sum_{k=2}^{n} \binom{n}{k} a_{(0)}^{n-k} a_{(m)}^k \varepsilon_{I_{k-2}} & \text{otherwise.} \end{cases}$$

$$(10)$$

As in the case of interval arithmetic, the affine arithmetic operations satisfy containment properties. However, it is not generally true that $c \in [\widehat{a}]$ and $d \in [\widehat{b}]$ implies $c \circ d \in [\widehat{a} \circ \widehat{b}]$, for an arbitrary affine operation $\circ$. The correctness of the containment properties depends on a careful management of the noise symbols.

In general, the implementation of non-affine operations such as transcendental functions requires a series expansion with a rigorous error term. A Chebyshev approximation is well-suited and is sometimes used.

3   Formalization in PVS

In PVS, an affine form $\widehat{a}$, defined as in Formula (3), is represented by a record type that holds the central value $a_{(0)}$ and the list of coefficients $a_{(1)}, \ldots, a_{(n)}$. Since noise terms may be common to several affine forms, they are represented by an independent type Noise that denotes a mapping from positive natural numbers (the noise indices) to values in the interval $[-1, 1]$.

The following lemmas, which are proved in PVS, show the correctness of the affine arithmetic operations. In particular, they provide sufficient conditions for the containment properties to hold. For each lemma, the name of the corresponding PVS lemma is included in parentheses.

Lemma 1 (containment_interval). *Let* N *be a map of type* Noise, a *be a real number, and* $\widehat{a}$ *be an affine form of* a *over* N, *i.e.,* $\widehat{a}(N) = a$. *Then,* $a \in [\widehat{a}]$.

Proof. Using Formula (5), it has to be proved that $a = a_{(0)} + \sum_{i=1}^{\ell_{\widehat{a}}} a_{(i)}N(i) \in [a_{(0)} - \sum_{i=1}^{\ell_{\widehat{a}}} |a_{(i)}|, a_{(0)} + \sum_{i=1}^{\ell_{\widehat{a}}} |a_{(i)}|]$. A known property from interval arithmetic called containment_add (formally proved in the development interval_arith, part of the NASA PVS Library) states that $u + v \in [u_1 + v_1, u_2 + v_2]$, when $u \in [u_1, u_2]$ and $v \in [v_1, v_2]$. Then it suffices to show that $a_{(0)} \in [a_{(0)}, a_{(0)}]$, which is trivially true, and $\sum_{i=1}^{\ell_{\widehat{a}}} a_{(i)}N(i) \in [-\sum_{i=1}^{\ell_{\widehat{a}}} |a_{(i)}|, \sum_{i=1}^{\ell_{\widehat{a}}} |a_{(i)}|]$. This latter property can be proved by induction on $\ell_{\widehat{a}}$ and using containment_add.   □

Lemma 2 (containment_var). *Let* N *be a map of type* Noise, k *be a natural number,* $\mathcal{I}$ *be a real interval, and* $v \in \mathcal{I}$. *There exists a real number* $b \in [-1, 1]$ *such that* $V_k^{\mathcal{I}}(N \text{ with } [n \mapsto b]) = v$.

Proof. This lemma is proved by using the definition of $V_k^{\mathcal{I}}$ in Formula (6).   □

Lemma 3 (containment_aff_un). *Let* N *be a map of type* Noise, *and* a, c *be a pair of real numbers. Given* $\widehat{a} = a_{(0)} + \sum_{i=0}^{\ell_{\widehat{a}}} a_{(i)}\varepsilon_i$, *an affine form of* a *over* N,

$$\mathsf{neg}(\widehat{a})(N) = -a \quad and \quad \mathsf{add}_c(\widehat{a})(N) = c + a \quad and \quad \mathsf{mul}_c(\widehat{a})(N) = c \cdot a.$$

Proof. The equality $\mathsf{add}_c(\widehat{a})\,(\mathbf{N}) = c + a$ is a trivial consequence of Formulas (5) and (7). Meanwhile, $\mathsf{neg}(\widehat{a})\,(\mathbf{N}) = -a$ and $\mathsf{mul}_c(\widehat{a})\,(\mathbf{N}) = c \cdot a$ can be proved by induction on the length of the partial deviation of $\widehat{a}$ □

Lemma 4 (containment_aff_bin). *Let* $\mathbf{N}$ *be a map of type* `Noise` *and* a, b *be a pair of real numbers. Given* $\widehat{a} = a_{(0)} + \sum_{i=0}^{\ell_{\widehat{a}}} a_{(i)}\varepsilon_i$ *and* $\widehat{b} = b_{(0)} + \sum_{i=0}^{\ell_{\widehat{b}}} b_{(i)}\varepsilon_i$, *affine forms of* a *and* b, *resp., over* $\mathbf{N}$, *i.e.,* $\widehat{a}\,(\mathbf{N}) = a$ *and* $\widehat{b}\,(\mathbf{N}) = b$,

$$\mathsf{add}(\widehat{a}, \widehat{b})\,(\mathbf{N}) = a + b \quad and \quad \mathsf{sub}(\widehat{a}, \widehat{b})\,(\mathbf{N}) = a - b.$$

Proof. Both properties can be proved by induction on the sum of the lengths of the partial deviations of $\widehat{a}$ and $\widehat{b}$. □

Non-affine operations introduce new noise symbols. The following lemma states that fresh noise symbols can be soundly added to any affine representation.

Lemma 5 (eval_updb_no_idxs). *Let* $\mathbf{N}$ *be a map of type* `Noise`, a *be a real number, and* $\widehat{a}$ *an affine form of* a *over* $\mathbf{N}$, *i.e.,* $\widehat{a}\,(\mathbf{N}) = a$. *For any collection* $\{i_k\}_{k=1}^n$ *of fresh indices with respect to* $\widehat{a}$ *and* $b_1, \ldots, b_n$ *real numbers in* $[-1, 1]$,

$$\widehat{a}\,(\mathbf{N}) = \widehat{a}\,(\mathbf{N}\,with\,[i_1 \mapsto b_1, \ldots, i_n \mapsto b_n]).$$

Proof. The proof proceeds by induction on n. The interesting part is the base case ($n = 1$), which is proved by induction on $\ell_{\widehat{a}}$. That part of the proof relies on the fact that i_1 is a fresh index with respect to $\widehat{a}$. □

Henceforth, the notation $\mathbf{N}_p$, where p is a positive natural number, denotes the map that is equal to $\mathbf{N}$ in every index except in indices $i > p$, where $\mathbf{N}_p(i) = 0$.

Lemma 6 (containment_mul). *Let* $\mathbf{N}$ *be a map of type* `Noise` *and* a, b *be a pair of real numbers. Given* $\widehat{a} = a_{(0)} + \sum_{i=0}^{\ell_{\widehat{a}}} a_{(i)}\varepsilon_i$ *and* $\widehat{b} = b_{(0)} + \sum_{i=0}^{\ell_{\widehat{b}}} b_{(i)}\varepsilon_i$, *affine forms of* a *and* b *(resp.) over* $\mathbf{N}$, *for each index* $p > max(m, \ell_{\widehat{a}}, \ell_{\widehat{b}})$, *if*

$$\mathbf{N}(p) = \frac{\widehat{a}\,(\mathbf{N}_p)\,\widehat{b}\,(\mathbf{N}_p) - a_{(0)}b_{(0)} - \sum_{i=1}^{max(\ell_{\widehat{a}}, \ell_{\widehat{b}})} (a_{(0)}b_{(i)} + a_{(i)}b_{(0)})\mathbf{N}(i)}{\sum_{i=1}^{\ell_{\widehat{a}}} |a_{(i)}| \sum_{i=1}^{\ell_{\widehat{b}}} |b_{(i)}|},$$

then $\mathsf{mul}_p(\widehat{a}, \widehat{b})\,(\mathbf{N}) = a \cdot b$.

Proof. It can be shown that $\mathsf{mul}_p(\widehat{a}, \widehat{b})\,(\mathbf{N}) = \widehat{a}\,(\mathbf{N}_p) \cdot \widehat{b}\,(\mathbf{N}_p)$, by applying arithmetic manipulations and using Formulas (5) and (9) and the hypothesis on $\mathbf{N}(p)$. By Lemma 5, since $p > \ell_{\widehat{a}}$, $\widehat{a}\,(\mathbf{N}_p) = \widehat{a}\,(\mathbf{N})$. Also, if $\widehat{a}$ is an affine form of a over $\mathbf{N}$, $\widehat{a}\,(\mathbf{N}) = a$. Then, $\widehat{a}\,(\mathbf{N}_p) = a$. Similarly, $\widehat{b}\,(\mathbf{N}_p) = b$. □

Lemma 7 (containment_pow_var_ac). *Let* $\mathbf{N}$ *be a map of type* `Noise`, n *be a natural number,* a *be a real number, and* $\widehat{a} = a_{(0)} + a_{(l)}\varepsilon_l$ *an affine form of* a *over* $\mathbf{N}$. *Given* $\{i_k\}_{k=0}^{n-2}$, *a collection of fresh indices with respect to* $\widehat{a}$, *if*

$$\mathbf{N}(i_k) = \mathbf{N}(l)^{k+2} \quad for\ every\ k,\ with\ 0 \leq k \leq n$$

then $\mathsf{pow}_{\{i_k\}_{k=0}^{n-2}}(\widehat{a}, n)\,(\mathbf{N}) = a^n$.

Proof. The proof proceeds by separating cases according to the definition of $\mathsf{pow}_{\{i_k\}_{k=0}^{n-2}}(\widehat{a}, n)$. The case $n = 0$ is trivial. When $n > 0$, using Formulas (5) and (10) and the hypothesis on $\mathbf{N}(i_k)$, it can be shown that $\mathsf{pow}_{\{i_k\}_{k=0}^{n-2}}(\widehat{a}, n)\,(\mathbf{N})$ is the combinatorial expansion of $\widehat{a}\,(\mathbf{N})^k$. The hypothesis assuring $\widehat{a}\,(\mathbf{N}) = a$ can be used to conclude the proof. $\square$

In PVS, the fundamental property of affine arithmetic (Formula (2)) is proved on formal expressions containing constants, variables, addition, multiplication, and power operation on variables. Variables are indexed by positive natural numbers. A formal expression e represents a real number e by means of an evaluation function. More precisely, the PVS function eval_Γ from formal expressions into real numbers is recursively defined as follows, where Γ is a map from positive natural numbers, representing variable indices, into real values.

$$\mathsf{eval}_\Gamma(v_i) \stackrel{def}{=} \Gamma(i), \text{where } v_i \text{ represents the } i\text{-th variable,}$$

$$\mathsf{eval}_\Gamma(\mathsf{c}) \stackrel{def}{=} c, \text{where } \mathsf{c} \text{ represents the numerical constant } c,$$

$$\mathsf{eval}_\Gamma(-\mathsf{e}) \stackrel{def}{=} -\mathsf{eval}_\Gamma(\mathsf{e}),$$

$$\mathsf{eval}_\Gamma(\mathsf{e} + \mathsf{f}) \stackrel{def}{=} \mathsf{eval}_\Gamma(\mathsf{e}) + \mathsf{eval}_\Gamma(\mathsf{e}), \tag{11}$$

$$\mathsf{eval}_\Gamma(\mathsf{e} - \mathsf{f}) \stackrel{def}{=} \mathsf{eval}_\Gamma(\mathsf{e}) - \mathsf{eval}_\Gamma(\mathsf{e}),$$

$$\mathsf{eval}_\Gamma(\mathsf{e} \times \mathsf{f}) \stackrel{def}{=} \mathsf{eval}_\Gamma(\mathsf{e}) \cdot \mathsf{eval}_\Gamma(\mathsf{e}),$$

$$\mathsf{eval}_\Gamma(\mathsf{e}^n) \stackrel{def}{=} \mathsf{eval}_\Gamma(\mathsf{e})^n.$$

Algorithm 1, which is formally defined in PVS, recursively constructs an affine form of a formal expression e. It has as parameters the formal expression e containing at most m variables, a collection $\{\mathcal{I}_i\}_{i=1}^m$ of m intervals (one per variable), and a map that caches affine forms of sub-expressions of e. It returns a map of all sub-expressions of e, including itself, to affine forms. The cache map ensures that noise symbol indices are shared among common sub-expressions. In the algorithm, the notation $I_{[0\ldots k]}$ stands for a collection containing the first $k + 1$ indices in I, and $[a \ldots b]$ stands for the collection of consecutive indices from a to b.

The following theorem, which is proved in PVS, states the fundamental theorem of affine arithmetic.

Theorem 1. *Let e be a formal expression, $\{\mathcal{I}_i\}_{i=1}^m$ be a collection of intervals, Γ be map from variable indices in e to real numbers such that $\Gamma(i) \in \mathcal{I}_i$, $e = \mathsf{eval}_\Gamma(\mathsf{e})$, and $\widehat{e} = \mathtt{RE2AF}(\mathsf{e}, \{\mathcal{I}_j\}_{j=1}^m, \emptyset)(\mathsf{e})$. There exists a map $\mathbf{N}$ of type Noise such that $\widehat{e}\,(\mathbf{N}) = e$.*

Proof. The proof proceeds by structural induction on e of a more general statement, where the cache map may be non-empty. The proof of that statement uses the fact that every expression in the cache map is a sub-expression of e. Since this condition is encoded in the type of the parameter *cache*, it is guaranteed by the PVS type checker. The base cases are discharged with Lemmas 2–7. $\square$

```
1  RE2AF(e, {I_i}_{i=1}^m, cache)
2  │  if e is in cache then return cache;
3  │  else if e is the variable v_j then return cache with [e ↦ V_j^{I_j}];
4  │  else
5  │  │  do
6  │  │  │  switch e do
   │  │  │  │  // Affine Operations
7  │  │  │  │  case −e_1 return cache with [e ↦ neg(ê_1)];
8  │  │  │  │  case e_1 + k or k + e_1 return cache with [e ↦ add_k(ê_1)];
9  │  │  │  │  case e_1 × k or k × e_1 return cache with [e ↦ mul_k(ê_1)];
10 │  │  │  │  case e_1 + e_2 return cache with [e ↦ add(ê_1, ê_2)];
11 │  │  │  │  case e_1 − e_2 return cache with [e ↦ sub(ê_1, ê_2)];
   │  │  │  │  // Non-Affine Operations
12 │  │  │  │  case e_1 × e_2 return cache with [e ↦ mul_p(ê_1, ê_2)];
13 │  │  │  │  case v_i^k
14 │  │  │  │  │  if exists some (v_i^{k'} ↦ pow_I(V_j^{I_j}, k)) ∈ cache then
15 │  │  │  │  │  │  if k < k' then
16 │  │  │  │  │  │  │  return cache with [e ↦ pow_{I[0...k−2]}(V_j^{I_j}, k)];
17 │  │  │  │  │  │  else
18 │  │  │  │  │  │  │  return cache with [e ↦ pow_{I ∪[p...p+k−k']}(V_j^{I_j}, k)];
19 │  │  │  │  │  │  end
20 │  │  │  │  │  else return cache with [e ↦ pow_{[p...p+(k−2)]}(V_j^{I_j}, k)];
21 │  │  │  │  end
22 │  │  │  endsw
23 │  │  │  where k, k' are constants expressions and e_1, e_2 are non-constant
   │  │  │  expressions and I is a collection of noise indices
   │  │  │  and p > max(m, the greatest noise index in cache)
   │  │  │  and cache_1 = RE2AF(e_1,{I_i}_{i=1}^m,cache) and ê_1 = cache_1(e_1)
   │  │  │  and cache_2 = RE2AF(e_2,{I_i}_{i=1}^m,cache_1) and ê_2 = cache_2(e_2);
24 │  end
25 end
```

Algorithm 1. Construction of affine forms of all sub-expressions in e

4   Proof Strategy

The motivation for the formalization of affine arithmetic presented in this paper is not only to verify the correctness of its operations, but more importantly to develop a practical method for performing guaranteed computations inside a theorem prover. In particular, the following problem is considered.

Given a polynomial expression p with variables $x_1, \ldots, x_m$ ranging over real intervals $\{I_i\}_{i=1}^m$, respectively, and a positive natural number n (called *precision*) compute an interval enclosure $[u, v]$ for p, i.e., $p \in [u, v]$, that is provably correct up to the accuracy $\epsilon = 10^{-n}$, i.e., $v - \max(p) \leq \epsilon$ and $\min(p) - u \leq \epsilon$.

Using Algorithm 1 and Theorem 1, it is possible to construct an affine form $\hat{p}$ of any polynomial expression p. Lemma 1 guarantees that the interval $[\hat{p}]$, as defined in Formula (4), is a correct enclosure of p. This approach for computing correct polynomial enclosures can be easily automated in most theorem provers that support a soundness-preserving strategy language. However, this approach does not guarantee the quality of the enclosure.

As outlined in the introduction, a way to improve the quality of an enclosure consists in dividing the original range of the variables into smaller intervals and considering the union of all the enclosures computed on these smaller subdomains. This technique typically yields tighter range enclosures, albeit at computational cost.

The NASA PVS Library includes the formalization of a branch and bound algorithm that implements domain subdivision on a generic enclosure method [18]. The algorithm, namely simple_bandb, can be instantiated with concrete data types and heuristics for deciding the subdivision and pruning schemas. The instantiation presented here is similar to the one given in [18] using interval arithmetic. A simplified version of the signature of simple_bandb has as parameters a formal expression e, a domain box for the variables in e, an enclosure method evaluate, a subdivision schema subdivide, a function combine that combines results from recursive calls, and an upper bound maxd for the maximum recursion depth. The output of the algorithm is an interval indicating the maximum and the minimum of the values that the expression e takes over box and additional information regarding the performance of the algorithm such as number of subdivisions, maximum recursion depth, and precision of the solution.

Intervals are represented by the data type Interval. The parameter box is an element of type Box, which is a list of intervals. The abstract data type RealExpr is used to represent formal expressions such as e. All these types are available from the development interval_arith in the NASA PVS Library. The parameter *evaluate* corresponds to a generic enclosure method. In the case of affine arithmetic, that parameter corresponds to the following function.

$$\text{Eval}(e, box) \overset{def}{=} [\text{RE2AF}(e,\ box,\ \emptyset)(e)]. \tag{12}$$

The functions that correspond to the parameters subdivide and combine are defined as in [18]. The former takes as parameter a box and a natural number n, it returns two boxes, which only differ from the original box in the n-th interval. That interval is replaced in the first (resp. second) box by the lower (resp. upper) half of the n-th interval in the original box. The latter function is just the union of two intervals.

The soundness property of simple_bandb is expressed in terms of the following predicate.

$$\text{sound?}(\mathsf{e}, \text{box}, \mathcal{I}) \overset{def}{=} \forall \Gamma \in box\colon \text{eval}_\Gamma(e) \in \mathcal{I}. \tag{13}$$

Corollary 1 in [18] states that when $\mathcal{I}$ is the interval returned by simple_bandb applied to e, box, Eval, subdivide, combine, and maxd, three specific properties suffice to prove *sound?*(e, box, $\mathcal{I}$). Two of them are properties about the functions `subdivide` and `combine`. As they are the same as the interval arithmetic instantiation of the generic algorithm [18], the proofs of these properties are also the same. The remaining property is stated below.

$$\forall\, \text{box}, \text{e} : sound?(\text{e}, \text{box}, \text{Eval}(e, box)).$$

This property follows directly from Theorem 1, Formula (12), and Formula (13). The development in [18] includes a more sophisticated algorithm `bandb` that has some additional parameters. These parameters, which do not affect the soundness of the algorithm, enable the specification of a pruning heuristic and early termination conditions.

The formalization presented in [18] includes infrastructure for developing strategies via computational reflection using a provably correct instantiation of the generic branch and bound algorithm. In particular, it includes a function, written in the PVS strategy language, for constructing a formal expression e of type `RealExpr` representing a PVS arithmetic expression e of type `real` and an element `box` of type `Box` that contains of the interval ranges of the variables in e. Based on that infrastructure, the development presented in this paper includes a proof-producing PVS strategy `aff-numerical` that computes probably correct bounds of polynomial expressions up to a user specified precision.

In its more general form, the strategy `aff-numerical` has as parameter an arithmetic expression e. It adds to the current proof sequent the hypothesis $e \in \mathcal{I}$, where $\mathcal{I}$ is an enclosure computed by the affine arithmetic instantiation of `bandb` on default parameters. Optional parameters that can be specified by the user include desired precision, upper bound to the maximum depth, and strategy for selecting the variable direction for the subdivision schema.

Example 1. The left-hand side of Fig. 1 illustrates a PVS sequent involving the polynomial $P_1(x) = x^5 - 2x^3$, where the variable x is assumed to range over the open interval $(-1000, 0)$. In a sequent, the formulas appearing above the horizontal line are the *antecedent*, while the formulas appearing below the horizontal line are the *consequent*. The right-hand side of Fig. 1 illustrates the sequent after the application of the proof command

```
(aff-numerical "x^5-2*x^3" :precision 3 :maxdepth 14)
```

This command introduces a new formula to the antecedent, i.e., sequent formula $\{-1\}$. The new formula states that $P_1(x) \in [-999998000000000, 1.066]$, when $x \in (-1000, 0)$. The sequent can be easily discharged by unfolding the definition of `##`, which stands for inclusion of a real number in an interval. The optional parameters `:precision` and `:maxdepth` in the strategy `aff-numerical` are used by the branch and bound algorithm to stop the recursion. In this case, the algorithm stops when either the enclosure is guaranteed to be accurate up to 10^{-3} or when a maximum depth of 14 is reached. The branch and bound algorithm uses rational arithmetic to compute enclosures. Since the upper and lower bounds of these enclosures tend to have large denominators and numerators, the strategy computes another enclosure whose upper and lower bounds are decimal numbers. These numbers are the closest decimal numbers, for the given precision, to the rational numbers of the original enclosure.

```
                                  {-1}  x^5 - 2*x^3 ##
                                        [|-999998000000000, 1.066|]
  {-1}  x < 0                     {-2}  x < 0
  {-2}  x > -1000                 {-3}  x > -1000
     |-------                        |-------
  {1}    x^5 - 2*x^3 < 1.1        {1}    x^5 - 2*x^3 < 1.1
```

Fig. 1. Example of use of the strategy `aff-numerical`.

The strategy `aff-numerical` does not depend on external oracles since all the required theorems are proved within the PVS logic. Indeed, any particular instantiation of the strategy can be unfolded into a tree of proof commands that only includes PVS proof rules. The strategy does depend on the PVS ground evaluator [15], which is part of the PVS trusted code base, for the evaluation of the function branch and bound algorithm. It should be noted that while the soundness of the strategy depends on the correctness of the ground evaluator, the formal development presented in Sect. 3 does not. Furthermore, it is theoretically possible, although impractical, to replace every instance of the PVS ground evaluator in a proof by another strategy that only depends on deductive steps such as PVS's `grind`.

As part of the development presented in this paper, there is also available a strategy `aff-interval` that solves to a target enclosure or inequality as opposed to a target precision. For that kind of problem, `aff-interval` is more efficient than `aff-numerical`, since `aff-interval` takes advantage of early termination criteria, which are not available with `aff-numerical`.

5 Experimental Results

The objective of the experiments described in this section is to illustrate the performance of affine arithmetic compared to interval arithmetic using their PVS formalizations. The experiments use the strategies `numerical`, which is part of `interval_arith` in the NASA PVS Library, and `aff-numerical`, which is part of the development presented in this paper. The strategies share most of the strategy code and only differ in the enclosure method.

Both strategies were used to find enclosures of polynomials with different characteristics. The performance was measured not only in terms of the time consumed by each strategy in every case but also with respect to the quality of the results. These experiments were performed on a desktop PC with an Intel

Quad Core i5-4200U 1.60 GHz processor, 3.9 GiB of RAM, and 32-bit Ubuntu Linux.

The first part of this section presents the results obtained for polynomials in a single variable. The polynomials to be considered are $P_1 = x^5 - 2x^3$, from Example 1, and

$$P_2(x) = -\frac{10207769}{65536}x^{20} + \frac{3002285}{4096}x^{18} - \frac{95851899}{65536}x^{16} + \frac{6600165}{4096}x^{14} - \frac{35043645}{32768}x^{12}$$
$$+ \frac{1792791}{4096}x^{10} - \frac{3558555}{32768}x^8 + \frac{63063}{4096}x^6 - \frac{72765}{65536}x^4 + \frac{3969}{65536},$$

where $x \in (-1, 1)$. Turan's inequality for Legendre polynomials states that, for $x \in (-1, 1)$, $L_j(x)^2 > L_{j-1}(x)L_{j+1}(x)$ for all $x \in (-1, 1)$, where L_i stands for the i-th Legendre polynomial. The formula $P_2(x) > 0$ states Turan's inequality for $j = 10$.

The result of the comparison of the two enclosure methods using these examples is depicted in Fig. 2. The top graphic shows the magnitude of the overestimation produced by both strategies for each maximum subdivision depth (up to 43). In the bottom graphic, the y axis represents the time spent by the strategies for both examples and each depth.

The expected behavior of the affine arithmetic method — second-order convergence in overestimation with respect to depth, as compared to first-order for interval arithmetic — can be clearly seen in these graphs. For both P_1 and P_2 the rate of convergence for affine arithmetic method is significantly faster. Regarding P_1, even though the leftmost result is significantly worse than the one given by the interval arithmetic method, the convergence of the former is so fast that it reaches its best approximation at depth 16, while the latter needs a depth of 43 to reach an equivalent result. The difference in performance for P_2 is even sharper: only a depth of 4 is needed for the affine method to achieve its best result, which could not be matched with a depth below 20 for interval arithmetic.

For a given depth, the computation time for the affine method is always higher than the time for the interval method. Nevertheless, when considering the quality of the result, the former achieves a much better performance in the same amount of time.

The difference in the performance for P_2 is mostly due to the high level of dependency in the polynomial. In the corresponding affine form almost every sub-expression has noise terms shared with others in the expression. This sharing constrains the overall number of noise terms and tracks the dependencies to a considerable extent, allowing the affine method to reach much better results in less time.

Finally, both enclosure methods are tested on the following non-trivial multivariate polynomial.

$$P_3(x_0, x_1, x_2, x_3, x_4, x_5, x_6, x_7) = -x_0x_5^3 + 3x_0x_5x_6^2 - x_2x_6^3 + 3x_2x_6x_5^2 - x_1x_4^3$$
$$+ 3x_1x_4x_7^2 - x_3x_7^3 + 3x_3x_7x_4^2 - 0.9563453,$$

where $x_0 \in [-10, 40]$, $x_1 \in [40, 100]$, $x_2 \in [-70, -40]$, $x_3 \in [-70, 40]$, $x_4 \in [10, 20]$, $x_5 \in [-10, 20]$, $x_6 \in [-30, 110]$, and $x_7 \in [-110, -30]$. This polynomial

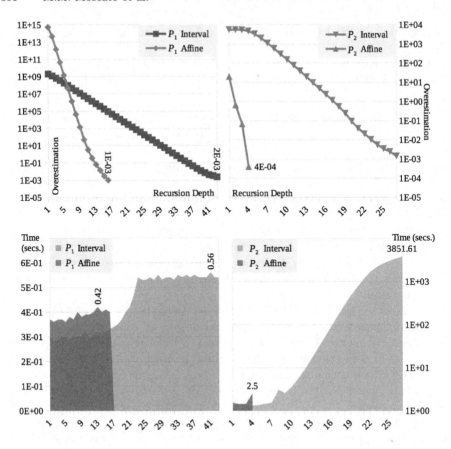

Fig. 2. Convergence rate for affine and interval arithmetic methods in P_1 and P_2

is taken from a database of demanding test problems for global optimization algorithms [24].

As shown in Table 1, both methods have similar results in this case. Despite starting with a better result (as in P_2) the interval method is overpassed by the affine method when the depth is set to 30. Nevertheless, the time taken by the latter is almost the double of the consumed by the former at this depth. The affine method is likely to perform better for smaller boxes or at even greater depth.

6   Conclusion and Further Work

The main contribution of this paper is a formalization of the affine arithmetic model [23] for self-validated numerical analysis. Although it has been performed in the Prototype Verification System (PVS), it does not depend on any specific feature of that system. The techniques presented in this paper could be implemented in any theorem prover with similar characteristics to PVS such as Coq,

Table 1. Data from P_3.

max depth	Interval result	time	Affine result	time
10	$[-895381 \times 10^2, 247381 \times 10^3]$	3.63 s.	$[-951004 \times 10^2, 247079 \times 10^3]$	9.2 s.
20	$[-824131 \times 10^2, 242342 \times 10^3]$	16.91 s.	$[-833045 \times 10^2, 236464 \times 10^3]$	55.96 s.
25	$[-800610 \times 10^2, 238942 \times 10^3]$	54.09 s.	$[-804002 \times 10^2, 233724 \times 10^3]$	150.87 s.
30	$[-794831 \times 10^2, 237518 \times 10^3]$	157.22 s.	$[-796329 \times 10^2, 233405 \times 10^3]$	281.45 s.
35	$[-793982 \times 10^2, 235733 \times 10^3]$	322.54 s.	$[-790525 \times 10^2, 232709 \times 10^3]$	530.89 s.
40	$[-791026 \times 10^2, 234318 \times 10^3]$	755.40 s.	$[-789961 \times 10^2, 232657 \times 10^3]$	963.66 s.
45	$[-790319 \times 10^2, 233936 \times 10^3]$	1524.40 s.	$[-789319 \times 10^2, 232480 \times 10^3]$	1627.00 s.

HOL, among others. Additionally, a proof-producing strategy for computing provably correct bounds of polynomials with variables over bounded domains was developed. This strategy relies heavily on the generic branch and bound algorithm introduced in [18]. The entire formalization, which is called `affine`, is available as part of the NASA PVS Library. The PVS formalization is organized into 11 theories, including the proofs of 193 properties and 483 non-trivial proof obligations automatically generated from typing conditions.

The performance of affine arithmetic is compared to interval arithmetic on some test cases using the PVS strategies developed for both enclosure methods. These experiments illustrate that, when dealing with problems with a high level of coupling between sub-expressions, the affine method performs significantly better than the interval method. The observed second-order rate of overestimation convergence for affine arithmetic accords with the theoretical result. In the presence of non-trivial functions over many variables, for a wide initial domain, it is possible that a large subdivision depth is necessary in order to realize this convergence.

Immler presents a formalization of affine arithmetic in Isabelle and uses it as part of developments intended to solve ordinary differential equations [9] and to calculate intersections between zonotopes and hyperplanes [10]. A minor difference with respect to the work presented in this paper is that formalization considers the multiplication inverse, which is not considered here, but it does not consider the power operation. The authors are not aware of any formalization of a subdivision technique or the development of proof-producing strategies using affine arithmetic as the one presented in the current paper. There are implementations of affine arithmetic in C and C++ [11], but a comparison with these tools would be also unfair since these are non-formal compiled codes, whereas the developed affine arithmetic strategy presented in this paper yields a formal proof.

The current approach only supports polynomial expressions. This support can easily be extended to a larger set of real-valued functions when affine forms for these functions are implemented. The performance of the proposed approach could be improved in several ways. The definition of Algorithm 1 uses simple data structures, which could be replaced by better-performing ones. Furthermore, the algorithm could also take advantage of some of the improvements proposed to the basic model of affine arithmetic. A comprehensive survey of such improvements can be found in [11]. Another well-known way to achieve better results is to combine both interval and affine methods by first applying the more efficient interval arithmetic approach and, after a certain subdivision depth, to take advantage of the better convergence rate of the affine algorithm.

References

1. Boldo, S., Clément, F., Filliâtre, J.C., Mayero, M., Melquiond, G., Weis, P.: Wave equation numerical resolution: A comprehensive mechanized proof of a C program. J. Autom. Reasoning **50**(4), 423–456 (2013). http://hal.inria.fr/hal-00649240/en/
2. Boldo, S., Clément, F., Filliâtre, J.C., Mayero, M., Melquiond, G., Weis, P.: Trusting computations: a mechanized proof from partial differential equations to actual program. Comput. Math. Appl. **68**(3), 325–352 (2014). http://www.sciencedirect.com/science/article/pii/S0898122114002636
3. Boldo, S., Marché, C.: Formal verification of numerical programs: From C annotated programs to mechanical proofs. Math. Comput. Sci. **5**, 377–393 (2011). http://dx.doi.org/10.1007/s11786-011-0099-9
4. Daumas, M., Lester, D., Muñoz, C.: Verified real number calculations: A library for interval arithmetic. IEEE Trans. Comput. **58**(2), 1–12 (2009)
5. de Figueiredo, L.H., Stolfi, J.: Affine arithmetic: Concepts and applications. Numer. Algorithms **37**(1–4), 147–158 (2004)
6. Galdino, A.L., Muñoz, C., Ayala-Rincón, M.: Formal verification of an optimal air traffic conflict resolution and recovery algorithm. In: Leivant, D., de Queiroz, R. (eds.) WoLLIC 2007. LNCS, vol. 4576, pp. 177–188. Springer, Heidelberg (2007)
7. Goodloe, A.E., Muñoz, C., Kirchner, F., Correnson, L.: Verification of numerical programs: From real numbers to floating point numbers. In: Brat, G., Rungta, N., Venet, A. (eds.) NFM 2013. LNCS, vol. 7871, pp. 441–446. Springer, Heidelberg (2013)
8. Hales, T., Adams, M., Bauer, G., Tat Dang, D., Harrison, J., Le Hoang, T., Kaliszyk, C., Magron, V., McLaughlin, S., Tat Nguyen, T., Quang Nguyen, T., Nipkow, T., Obua, S., Pleso, J., Rute, J., Solovyev, A., Hoai Thi Ta, A., Tran, T.N., Thi Trieu, D., Urban, J., Khac Vu, K., Zumkeller, R.: A formal proof of the Kepler conjecture. ArXiv e-prints, January 2015
9. Immler, F.: Formally verified computation of enclosures of solutions of ordinary differential equations. In: Badger, J.M., Rozier, K.Y. (eds.) NFM 2014. LNCS, vol. 8430, pp. 113–127. Springer, Heidelberg (2014). http://dx.doi.org/10.1007/978-3-319-06200-6
10. Immler, F.: A verified algorithm for geometric zonotope/hyperplane intersection. In: Proceedings of the 2015 Conference on Certified Programs and Proofs (CPP), pp. 129–136. ACM, New York (2015). http://doi.acm.org/10.1145/2676724.2693164
11. Kiel, S.: Yalaa: Yet another library for affine arithmetic. Reliable Comput. **16**, 114–129 (2012)
12. Lorentz, G.G.: Bernstein Polynomials, 2nd edn. Chelsea Publishing Company, New York (1986)
13. Melquiond, G.: Proving bounds on real-valued functions with computations. In: Armando, A., Baumgartner, P., Dowek, G. (eds.) IJCAR 2008. LNCS (LNAI), vol. 5195, pp. 2–17. Springer, Heidelberg (2008)

14. Moore, R.E., Kearfott, R.B., Cloud, M.J.: Introduction to Interval Analysis. SIAM, Philadelphia (2009)
15. Muñoz, C.: Rapid prototyping in PVS. Contractor Report NASA/CR-2003-212418, NASA, Langley Research Center, Hampton VA 23681-2199, USA (2003)
16. Muñoz, C., Carreño, V., Dowek, G., Butler, R.: Formal verification of conflict detection algorithms. Int. J. Softw. Tools Technol. Transf. 4(3), 371–380 (2003)
17. Muñoz, C., Narkawicz, A.: Formalization of a representation of Bernstein polynomials and applications to global optimization. J. Autom. Reasoning 51(2), 151–196 (2013). http://dx.doi.org/10.1007/s10817-012-9256-3
18. Narkawicz, A., Muñoz, C.: A formally verified generic branching algorithm for global optimization. In: Cohen, E., Rybalchenko, A. (eds.) VSTTE 2013. LNCS, vol. 8164, pp. 326–343. Springer, Heidelberg (2014)
19. Narkawicz, A., Muñoz, C., Dowek, G.: Provably correct conflict prevention bands algorithms. Sci. Comput. Program. 77(1–2), 1039–1057 (2012). http://dx.doi.org/10.1016/j.scico.2011.07.002
20. Owre, S., Rushby, J., Shankar, N.: PVS: A prototype verificationsystem. In: Kapur, D. (ed.) CADE 1992. LNCS, vol. 607, pp. 748–752. Springer, Heidelberg (1992)
21. Solovyev, A., Hales, T.C.: Formal verification of nonlinear inequalities with Taylor interval approximations. In: Brat, G., Rungta, N., Venet, A. (eds.) NFM 2013. LNCS, vol. 7871, pp. 383–397. Springer, Heidelberg (2013)
22. Solovyev, A., Hales, T.C.: Efficient formal verification of bounds of linear programs. In: Davenport, J.H., Farmer, W.M., Urban, J., Rabe, F. (eds.) MKM 2011 and Calculemus 2011. LNCS, vol. 6824, pp. 123–132. Springer, Heidelberg (2011)
23. Stolfi, J., Figueiredo, L.H.D.: Self-validated numerical methods and applications (1997)
24. Verschelde, J.: Algorithm 795: PHCpack: A general-purpose solver for polynomial systems by homotopy continuation. ACM Trans. Math. Softw. 25(2), 251–276 (1999)

Amortized Complexity Verified

Tobias Nipkow[✉]

Technische Universität München, Munich, Germany
nipkow@in.tum.de

Abstract. A framework for the analysis of the amortized complexity of (functional) data structures is formalized in Isabelle/HOL and applied to a number of standard examples and to three famous non-trivial ones: skew heaps, splay trees and splay heaps.

1 Introduction

Amortized complexity [3,14] of an algorithm averages the running times of a sequence of invocations of the algorithm. In this paper we formalize a simple framework for the analysis of the amortized complexity of functional programs and apply it to both the easy standard examples and the more challenging examples of skew heaps, splay trees and splay heaps. We have also analyzed pairing heaps [4] but cannot present them here for lack of space. All proofs are available online [9].

We are aiming for a particularly lightweight framework that supports proofs at a high level of abstraction. Therefore all algorithms are modeled as recursive functions in the logic. Because mathematical functions do not have a complexity they need to be accompanied by definitions of the intended running time that follow the recursive structure of the actual function definitions. Thus the user is free to choose the level of granularity of the complexity analysis. In our examples we simply count function calls.

Although one can view the artefacts that we analyze as functional programs, one can also view them as models or abstractions of imperative programs. For the amortized complexity we are only interested in the input-output behaviour and the running time complexity. As long as those are the same, it does not matter in what language the algorithm is implemented. In fact, the standard imperative implementations of all of our examples have the same complexity as the functional model. However, in a functional setting, amortized complexity reasoning may be invalid if the data structure under consideration is not used in a single-threaded manner [10].

1.1 Related Work

Hofmann and Jost [7] pioneered automatic type-based amortized analysis of heap usage of functional programs. This was later generalized in many directions, for example allowing multivariate polynomials [6]. Atkey [1] carries some of the

© Springer International Publishing Switzerland 2015
C. Urban and X. Zhang (Eds.): ITP 2015, LNCS 9236, pp. 310–324, 2015.
DOI: 10.1007/978-3-319-22102-1_21

ideas over to an imperative language with separation logic embedded in Coq. Charguéraud and Pottier [2] employ separation logic to verify the almost-linear amortized complexity of a Union-Find implementation in OCaml in Coq.

2   Lists and Trees

Lists are constructed from the empty list [] via the infix cons-operator ".", $|xs|$ is the length of xs, tl takes the tail and rev reverses a list.

Binary trees are defined as the data type $'a\ tree$ with two constructors: the empty tree or leaf $\langle\rangle$ and the node $\langle l,\ a,\ r\rangle$ with subtrees $l,\ r :: 'a\ tree$ and contents $a :: 'a$. The size of a tree is the number of its nodes:

$$|\langle\rangle| = 0 \qquad |\langle l,\ \_,\ r\rangle| = |l| + |r| + 1$$

For convenience there is also the modified size function $|t|_1 = |t| + 1$.

3   Amortized Analysis Formalized

We formalize the following scenario. We are given a number of operations that may take parameters and that update the state (some data structure) as a side effect. Our model is purely functional: the state is replaced by a new state with each invocation of an operation, rather than mutating the state. This makes no difference because we only analyze time, not space.

Our model of amortized analysis is a theory that is parameterized as follows (a *locale* in Isabelle-speak):

$'s$ is the type of state.
$'o$ is the type of operations.
$init :: 's$ is the initial state.
$nxt :: 'o \Rightarrow 's \Rightarrow 's$ is the next state function.
$inv :: 's \Rightarrow bool$ is an invariant on the state. We assume that the invariant
      holds initially ($inv\ init$) and that it is preserved by all operations
      ($inv\ s \Longrightarrow inv\ (nxt\ f\ s)$).
$t :: 'o \Rightarrow 's \Rightarrow real$ is the timing function: $t\ f\ s$ represents the time it takes to
      execute operation f in state s, i.e. $nxt\ f\ s$.

The effect of each operation f is modeled as a function $nxt\ f$ from state to state. Since functions are extensional, the execution time is modeled explicitly by function t. Alternatively one can instrument nxt with timing information and have it return a pair of a new state and the time the operation took. We have separated the computation of the result and the timing information into two functions because that is what one would typically do in a first step anyway to simplify the proofs. In particular this means that t need not be a closed form expression for the actual complexity. In all of our examples the definition of t will follow the (usually recursive) structure of the definition of nxt precisely. One could go one step further and derive t from an intensional formulation of nxt

automatically, but that is orthogonal to our aim in this paper, namely *amortized complexity*.

For the analysis of amortized complexity we formalize the *potential method*. That is, our theory has another parameter:

$\Phi :: {}'s \Rightarrow real$ is the *potential* of a state. We assume the potential is initially 0 ($\Phi\ init = 0$) and never becomes negative ($inv\ s \implies 0 \leq \Phi\ s$).

The potential of the state represents the savings that can pay for future restructurings of the data structure. Typically, the higher the potential, the more out of balance the data structure is. Note that the potential is just a means to an end, the analysis, but does not influence the actual operations.

Let us now analyze the complexity of a sequence of operations formalized as a function of type $nat \Rightarrow {}'o$. The sequence is infinite for convenience but of course we only analyze the execution of finite prefixes. For this purpose we define *state f n*, the state after the execution of the first n elements of f:

$$state :: (nat \Rightarrow {}'o) \Rightarrow nat \Rightarrow {}'s$$
$$state\ f\ 0 = init$$
$$state\ f\ (Suc\ n) = nxt\ (f\ n)\ (state\ f\ n)$$

Now we can define the amortized complexity of an operation as the actual time it takes plus the difference in potential:

$$a :: (nat \Rightarrow {}'o) \Rightarrow nat \Rightarrow real$$
$$a\ f\ i = t\ (f\ i)\ (state\ f\ i) + \Phi\ (state\ f\ (i + 1)) - \Phi\ (state\ f\ i)$$

By telescoping (i.e. induction) we obtain

$$\left(\textstyle\sum_{i<n} t\ (f\ i)\ (state\ f\ i)\right) = \left(\textstyle\sum_{i<n} a\ f\ i\right) + \Phi\ (state\ f\ 0) - \Phi\ (state\ f\ n)$$

where $\sum_{i<n} F\ i$ is the sum of all $F\ i$ with $i < n$. Because of the assumptions on Φ this implies that on average the amortized complexity is an upper bound of the real complexity:

$$\left(\textstyle\sum_{i<n} t\ (f\ i)\ (state\ f\ i)\right) \leq \left(\textstyle\sum_{i<n} a\ f\ i\right)$$

To complete our formalization we add one more parameter:

$U :: {}'o \Rightarrow {}'s \Rightarrow real$ is an explicit upper bound for the amortized complexity of each operation (in a certain state), i.e. we assume that $inv\ s \implies t\ f\ s + \Phi\ (nxt\ f\ s) - \Phi\ s \leq U\ f\ s$.

Thus we obtain that U is indeed an upper bound of the real complexity:

$$\left(\textstyle\sum_{i<n} t\ (f\ i)\ (state\ f\ i)\right) \leq \left(\textstyle\sum_{i<n} U\ (f\ i)\ (state\ f\ i)\right)$$

Instantiating this theory of amortized complexity means defining the parameters and proving the assumptions, in particular about U.

4   Easy Examples

Unless noted otherwise, the examples in this section come from a standard text-book [3].

4.1   Binary Counter

We start with the binary counter explained in the introduction. The state space $'s$ is just a list of booleans, starting with the least significant bit. There is just one parameterless operation "increment". Thus we can model type $'o$ with the unit type. The increment operation is defined recursively:

$incr\ [] = [\mathit{True}]$
$incr\ (\mathit{False} \cdot bs) = \mathit{True} \cdot bs$
$incr\ (\mathit{True} \cdot bs) = \mathit{False} \cdot incr\ bs$

In complete analogy the running time function for $incr$ is defined:

$t_{incr}\ [] = 1$
$t_{incr}\ (\mathit{False} \cdot bs) = 1$
$t_{incr}\ (\mathit{True} \cdot bs) = t_{incr}\ bs + 1$

Now we can instantiate the parameters of the amortized analysis theory:

$init = []$ $\qquad$ $nxt\ () = incr$ $\qquad$ $t\ () = t_{incr}$
$inv\ s = \mathit{True}$ $\qquad$ $\Phi\ s = |filter\ id\ s|$ $\qquad$ $U\ ()\ s = 2$

The key idea of the analysis is to define the potential of s as $|filter\ id\ s|$, the number of True bits in s. This makes sense because the higher the potential, the longer an increment may take (roughly speaking). Now it is easy to show that 2 is an upper bound for the amortized complexity: the requirement on U follows immediately from this lemma (which is proved by induction):

$t_{incr}\ bs + \Phi_{incr}\ (incr\ bs) - \Phi_{incr}\ bs = 2$

4.2   Stack with Multipop

The operations are

datatype $'a\ op_{stk} = Push\ 'a\ |\ Pop\ nat$

where $Pop\ n$ pops n elements off the stack:

$nxt_{stk}\ (Push\ x)\ xs = x \cdot xs$
$nxt_{stk}\ (Pop\ n)\ xs = drop\ n\ xs$

In complete analogy the running time function is defined:

$t_{stk}\ (Push\ x)\ xs = 1$
$t_{stk}\ (Pop\ n)\ xs = min\ n\ |xs|$

Now we can instantiate the parameters of the amortized analysis theory:

$init = []$ $\qquad$ $nxt = nxt_{stk}$ $\qquad$ $t = t_{stk}$ $\qquad$ $inv\ s = \mathit{True}$
$\Phi = length$ $\qquad$ $U\ f\ s = (\textbf{case}\ f\ \textbf{of}\ Push\ x \Rightarrow 2\ |\ Pop\ n \Rightarrow 0)$

The necessary proofs are all automatic.

4.3 Dynamic Tables

Dynamic tables are tables where elements are added and deleted and the table grows and shrinks accordingly. We ignore the actual elements because they are irrelevant for the complexity analysis. Therefore the operations

datatype $op_{tb} = Ins \mid Del$

do not have arguments. Similarly the state is merely a pair of natural numbers (n, l) that abstracts a table of size l with n elements. This is how the operations behave:

$nxt_{tb} \ Ins \ (n, \ l) = (n + 1, \ if \ n < l \ then \ l \ else \ if \ l = 0 \ then \ 1 \ else \ 2 * l)$
$nxt_{tb} \ Del \ (n, \ l) =$
$(n - 1, \ if \ n = 1 \ then \ 0 \ else \ if \ 4 * (n - 1) < l \ then \ l \ div \ 2 \ else \ l)$

If the table overflows upon insertion, its size is doubled. If a table is less than one quarter full after deletion, its size is halved. The transition from and to the empty table is treated specially.

This is the corresponding running time function:

$t_{tb} \ Ins \ (n, \ l) = (if \ n < l \ then \ 1 \ else \ n + 1)$
$t_{tb} \ Del \ (n, \ l) = (if \ n = 1 \ then \ 1 \ else \ if \ 4 * (n - 1) < l \ then \ n \ else \ 1)$

The running time for the cases where the table expands or shrinks is determined by the number of elements that need to be copied.

Now we can instantiate the parameters of the amortized analysis theory. We start with the system itself:

$init = (0, 0) \qquad nxt = nxt_{tb} \qquad t = t_{tb}$
$inv \ (n,l) = (if \ l = 0 \ then \ n = 0 \ else \ n \leq l \wedge l \leq 4 * n)$

This is the first time we have a non-trivial invariant. The potential is also more complicated than before:

$\Phi \ (n,l) = (if \ 2 * n < l \ then \ l \ / \ 2 - n \ else \ 2 * n - l)$

Now it is automatic to show the following amortized complexity:

$U \ f \ s = (case \ f \ of \ Ins \Rightarrow 3 \mid Del \Rightarrow 2)$

4.4 Queues

Queues have one operation for enqueueing a new item and one for dequeueing the oldest item:

datatype $'a \ op_q = Enq \ 'a \mid Deq$

We ignore accessing the oldest item because it is a constant time operation in our implementation.

The simplest possible implementation of functional queues (e.g. [10]) consists of two lists (stacks) $(xs, \ ys)$:

$$nxt_q \ (Enq \ x) \ (xs, \ ys) = (x \cdot xs, \ ys)$$
$$nxt_q \ Deq \ (xs, \ ys) = (if \ ys = [] \ then \ ([], \ tl \ (rev \ xs)) \ else \ (xs, \ tl \ ys))$$
$$t_q \ (Enq \ x) \ (xs, \ ys) = 1$$
$$t_q \ Deq \ (xs, \ ys) = (if \ ys = [] \ then \ |xs| \ else \ 0)$$

Note that the running time function counts only allocations of list cells and that it assumes rev is linear. Now we can instantiate the parameters of the amortized analysis theory to show that the average complexity of both Enq and Deq is 2. The necessary proofs are all automatic.

$$init = ([], []) \qquad nxt = nxt_q \qquad t = t_q \qquad inv \ s = True$$
$$\Phi \ (xs, ys) = |xs| \qquad U \ f \ s = (case \ f \ of \ Enq \ x \Rightarrow 2 \ | \ Deq \Rightarrow 0)$$

In the same manner I have also verified [9] a modified implementation where reversal happens already when $|xs| = |ys| + 1$; this improves the worst-case behaviour but (using $\Phi(xs, ys) = 2 * |xs|$) the amortized complexity of Enq increases to 3.

5 Skew Heaps

This section analyzes a beautifully simple data structure for priority queues: skew heaps [13]. Heaps are trees where the least element is at the root. We assume that the elements are linearly ordered. The central operation on skew heaps is $meld$, that merges two skew heaps and swaps children along the merge path:

$$meld \ h_1 \ h_2 =$$
$$(case \ h_1 \ of \ \langle \rangle \Rightarrow h_2$$
$$| \ \langle l_1, \ a_1, \ r_1 \rangle \Rightarrow$$
$$case \ h_2 \ of \ \langle \rangle \Rightarrow h_1$$
$$| \ \langle l_2, \ a_2, \ r_2 \rangle \Rightarrow$$
$$if \ a_1 \leq a_2 \ then \ \langle meld \ h_2 \ r_1, \ a_1, \ l_1 \rangle \ else \ \langle meld \ h_1 \ r_2, \ a_2, \ l_2 \rangle)$$

We consider the two operations of inserting an element and removing the minimal element:

datatype $'a \ op_{pq} = Insert \ 'a \ | \ Delmin$

They are implemented via $meld$ as follows:

$$nxt_{pq} \ (Insert \ a) \ h = meld \ \langle\langle\rangle, \ a, \ \langle\rangle\rangle \ h$$
$$nxt_{pq} \ Delmin \ h = del\_min \ h$$

$$del\_min \ \langle \rangle = \langle \rangle$$
$$del\_min \ \langle l, \ m, \ r \rangle = meld \ l \ r$$

For the functional correctness proofs see [9].

The analysis by Sleator and Tarjan is not ideal as a starting point for a formalization. Luckily there is a nice, precise functional account by Kaldewaij and Schoenmakers [8] that we will follow (although their $meld$ differs slightly from ours). Their cost measure counts the number of calls of $meld$, $Insert$ and $Delmin$:

$$t_{meld} \; \langle\rangle \; h = 1$$
$$t_{meld} \; h \; \langle\rangle = 1$$
$$t_{meld} \; \langle l_1, \; a_1, \; r_1\rangle \; \langle l_2, \; a_2, \; r_2\rangle =$$
$$(if \; a_1 \leq a_2 \; then \; t_{meld} \; \langle l_2, \; a_2, \; r_2\rangle \; r_1 \; else \; t_{meld} \; \langle l_1, \; a_1, \; r_1\rangle \; r_2) + 1$$

$$t_{pq} \; (Insert \; a) \; h = t_{meld} \; \langle\langle\rangle, \; a, \; \langle\rangle\rangle \; h + 1$$
$$t_{pq} \; Delmin \; h = (case \; h \; of \; \langle\rangle \Rightarrow 1 \mid \langle t_1, \; a, \; t_2\rangle \Rightarrow t_{meld} \; t_1 \; t_2 + 1)$$

Kaldewaij and Schoenmakers prove a tighter upper bound than Sleator and Tarjan, replacing the factor of 3 by 1.44. We are satisfied with verifying the bound by Sleator and Tarjan and work with the following simple potential function which is an instance of the one by Kaldewaij and Schoenmakers: it counts the number of "right heavy" nodes.

$$\Phi \; \langle\rangle = 0$$
$$\Phi \; \langle l, \; \_, \; r\rangle = \Phi \; l + \Phi \; r + (if \; |l| < |r| \; then \; 1 \; else \; 0)$$

To prove the amortized complexity of *meld* we need some further notions that capture the ideas of Sleator and Tarjan in a concise manner:

$$rheavy \; \langle l, \; \_, \; r\rangle = (|l| < |r|)$$

$$lpath \; \langle\rangle = []$$
$$lpath \; \langle l, \; a, \; r\rangle = \langle l, \; a, \; r\rangle \cdot lpath \; l$$

$$rpath \; \langle\rangle = []$$
$$rpath \; \langle l, \; a, \; r\rangle = \langle l, \; a, \; r\rangle \cdot rpath \; r$$

$$\Gamma \; h = |filter \; rheavy \; (lpath \; h)|$$
$$\Delta \; h = |filter \; (\lambda p. \; \neg \; rheavy \; p) \; (rpath \; h)|$$

Two easy inductive properties:

$$\Gamma \; h \leq \log_2 |h|_1 \quad (1) \qquad \Delta \; h \leq \log_2 |h|_1 \quad (2)$$

Now the desired logarithmic amortized complexity of *meld* follows:

$$t_{meld} \; t_1 \; t_2 + \Phi \; (meld \; t_1 \; t_2) - \Phi \; t_1 - \Phi \; t_2$$
$$\leq \Gamma \; (meld \; t_1 \; t_2) + \Delta \; t_1 + \Delta \; t_2 + 1 \qquad by \; induction \; on \; meld$$
$$\leq \log_2 |meld \; t_1 \; t_2|_1 + \log_2 |t_1|_1 + \log_2 |t_2|_1 + 1 \qquad by \; (1), \; (2)$$
$$= \log_2 (|t_1|_1 + |t_2|_1 - 1) + \log_2 |t_1|_1 + \log_2 |t_2|_1 + 1$$
$$\quad because \; |meld \; t_1 \; t_2| = |t_1| + |t_2|$$
$$\leq \log_2 (|t_1|_1 + |t_2|_1) + \log_2 |t_1|_1 + \log_2 |t_2|_1 + 1$$
$$\leq \log_2 (|t_1|_1 + |t_2|_1) + 2 * \log_2 (|t_1|_1 + |t_2|_1) + 1$$
$$\quad because \; \log_2 x + \log_2 y \leq 2 * \log_2 (x + y) \; if \; x, y \geq 1$$
$$= 3 * \log_2 (|t_1|_1 + |t_2|_1) + 1$$

Now it is easy to verify the following amortized complexity for *Insert* and *Delmin* by instantiating our standard theory with

$$U \; (Insert \; \_) \; h = 3 * \log_2 (|h|_1 + 2) + 2$$
$$U \; Delmin = 3 * \log_2 (|h|_1 + 2) + 4$$

Note that Isabelle supports implicit coercions, in particular from *nat* to *real*, that are inserted automatically [15].

6   Splay Trees

A splay tree [12] is a subtle self-adjusting binary search tree. It achieves its amortized logarithmic complexity by local rotations of subtrees along the access path. Its central operation is *splay* of type $'a \Rightarrow 'a\ tree \Rightarrow 'a\ tree$ that rotates the given element (of a linearly ordered type $'a$) to the root of the tree. Most presentations of *splay* confine themselves to this case where the given element is in the tree. If the given element is not in the tree, the last element found before a $\langle\rangle$ was met is rotated to the root. The complete definition is shown in Fig. 1.

Given splaying, searching for an element in the tree is trivial: you splay with the given element and check if it ends up at the root. For insertion and deletion, algorithm texts typically show pictures only. In contrast, we show the code only, in Figs. 2–3. To insert a, you splay with a to see if it is already there, and if it is not, you insert it at the top (which is the right place due to the previous splay action).

```
splay a ⟨⟩ = ⟨⟩
splay a ⟨cl, c, cr⟩ =
(if a = c then ⟨cl, c, cr⟩
 else if a < c
      then case cl of ⟨⟩ ⇒ ⟨cl, c, cr⟩
           | ⟨bl, b, br⟩ ⇒
               if a = b then ⟨bl, a, ⟨br, c, cr⟩⟩
               else if a < b
                    then if bl = ⟨⟩ then ⟨bl, b, ⟨br, c, cr⟩⟩
                         else case splay a bl of
                              ⟨al, a', ar⟩ ⇒ ⟨al, a', ⟨ar, b, ⟨br, c, cr⟩⟩⟩
                    else if br = ⟨⟩ then ⟨bl, b, ⟨br, c, cr⟩⟩
                         else case splay a br of
                              ⟨al, a', ar⟩ ⇒ ⟨⟨bl, b, al⟩, a', ⟨ar, c, cr⟩⟩
      else case cr of ⟨⟩ ⇒ ⟨cl, c, cr⟩
           | ⟨bl, b, br⟩ ⇒
               if a = b then ⟨⟨cl, c, bl⟩, a, br⟩
               else if a < b
                    then if bl = ⟨⟩ then ⟨⟨cl, c, bl⟩, b, br⟩
                         else case splay a bl of
                              ⟨al, a', ar⟩ ⇒ ⟨⟨cl, c, al⟩, a', ⟨ar, b, br⟩⟩
                    else if br = ⟨⟩ then ⟨⟨cl, c, bl⟩, b, br⟩
                         else case splay a br of
                              ⟨al, x, xa⟩ ⇒ ⟨⟨⟨cl, c, bl⟩, b, al⟩, x, xa⟩⟩
```

Fig. 1. Function *splay*

$insert_{st}$ a t =
(if $t = \langle\rangle$ then $\langle\langle\rangle, a, \langle\rangle\rangle$
 else case splay a t of
      $\langle l, a', r\rangle \Rightarrow$
        if $a = a'$ then $\langle l, a, r\rangle$
          else if $a < a'$ then $\langle l, a, \langle\langle\rangle, a', r\rangle\rangle$ else $\langle\langle l, a', \langle\rangle\rangle, a, r\rangle\rangle$

$delete_{st}$ a t =
(if $t = \langle\rangle$ then $\langle\rangle$
 else case splay a t of
      $\langle l, a', r\rangle \Rightarrow$
        if $a = a'$
          then if $l = \langle\rangle$ then r else case splay_max l of $\langle l', m, r'\rangle \Rightarrow \langle l', m, r\rangle$
          else $\langle l, a', r\rangle$)

Fig. 2. Functions $insert_{st}$ and $delete_{st}$

splay_max $\langle\rangle = \langle\rangle$
splay_max $\langle l, b, \langle\rangle\rangle = \langle l, b, \langle\rangle\rangle$
splay_max $\langle l, b, \langle rl, c, rr\rangle\rangle =$
(if $rr = \langle\rangle$ then $\langle\langle l, b, rl\rangle, c, \langle\rangle\rangle$
 else case splay_max rr of $\langle rrl, x, xa\rangle \Rightarrow \langle\langle\langle l, b, rl\rangle, c, rrl\rangle, x, xa\rangle$)

Fig. 3. Function splay_max

set_{tree} $\langle\rangle = \emptyset$
set_{tree} $\langle l, a, r\rangle = \{a\} \cup (set_{tree}\ l \cup set_{tree}\ r)$
bst $\langle\rangle = True$
bst $\langle l, a, r\rangle =$
($bst\ l \wedge bst\ r \wedge (\forall x \in set_{tree}\ l.\ x < a)\ \wedge (\forall x \in set_{tree}\ r.\ a < x))$

Fig. 4. Functions set_{tree} and bst

To delete a, you splay with a and if a ends up at the root, you replace it with the maximal element removed from the left subtree. The latter step is performed by splay_max that splays with the maximal element.

6.1 Functional Correctness

So far we had ignored functional correctness but for splay trees we actually need it in the verification of the complexity. To formulate functional correctness we

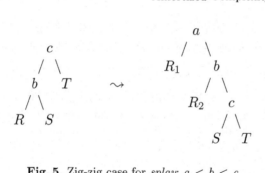

Fig. 5. Zig-zig case for *splay*: $a < b < c$

need the two auxiliary functions shown in Fig. 4. Function set_{tree} collects the elements in the tree, function *bst* checks if the tree is a *binary search tree* according to the linear ordering "$<$" on the elements. The key functional properties are that splaying does not change the contents of the tree (it merely reorganizes it) and that *bst* is an invariant of splaying:

$$set_{tree}\ (splay\ a\ t) = set_{tree}\ t$$
$$bst\ t \implies bst\ (splay\ a\ t)$$

Similar properties can be proved for insertion and deletion, e.g.,

$$bst\ t \implies set_{tree}\ (delete_{st}\ a\ t) = set_{tree}\ t - \{a\}$$

Now we present two amortized analyses: a simpler one that yields the bounds proved by Sleator and Tarjan [12] and a more complicated and precise one due to Schoenmakers [11].

6.2   Amortized Analysis

The timing functions are straightforward and not shown. Roughly speaking, they count only the number of splay steps: t_{splay} counts the number of calls of *splay*, t_{splay_max} counts the number of calls of *splay_max*; t_{delete} counts the time for both *splay* and *splay_max*.

The potential of a tree is defined as a sum of logarithms as follows:

$$\varphi\ t = \log_2 |t|_1$$
$$\Phi\ \langle\rangle = 0$$
$$\Phi\ \langle l,\ a,\ r\rangle = \Phi\ l + \Phi\ r + \varphi\ \langle l,\ a,\ r\rangle$$

The amortized complexity of splaying is defined as usual:

$$\mathcal{A}\ a\ t = t_{splay}\ a\ t + \Phi\ (splay\ a\ t) - \Phi\ t$$

Let *subtrees* yield the set of all subtrees of a tree:

$$subtrees\ \langle\rangle = \{\langle\rangle\}$$
$$subtrees\ \langle l,\ a,\ r\rangle = \{\langle l,\ a,\ r\rangle\} \cup (subtrees\ l \cup subtrees\ r)$$

The following logarithmic bound is proved by induction on t according to the recursion schema of *splay*: if *bst* t and $\langle l,\, a,\, r\rangle \in subtrees\ t$ then

$$\mathcal{A}\ a\ t \leq 3 * (\varphi\ t - \varphi\ \langle l,\, a,\, r\rangle) + 1 \tag{3}$$

Let us look at one case of the inductive proof in detail. We pick the so-called zig-zig case shown in Fig. 5. Subtrees with root x are called X on the left and X' on the right-hand side. Thus the figure depicts *splay* $a\ C = A'$ assuming the recursive call *splay* $a\ R = \langle R_1,\, a,\, R_2\rangle =: R'$.

$$
\begin{aligned}
\mathcal{A}\ a\ C &= \mathcal{A}\ a\ R + \varphi\ B' + \varphi\ C' - \varphi\ B - \varphi\ R' + 1 \\
&\leq 3 * (\varphi\ R - \varphi\ \langle l,\, a,\, r\rangle) + \varphi\ B' + \varphi\ C' - \varphi\ B - \varphi\ R' + 2 \\
&\quad \text{by ind.hyp.} \\
&= 2 * \varphi\ R + \varphi\ B' + \varphi\ C' - \varphi\ B - 3 * \varphi\ \langle l,\, a,\, r\rangle + 2 \\
&\quad \text{because } \varphi\ R = \varphi\ R' \\
&\leq \varphi\ R + \varphi\ B' + \varphi\ C' - 3 * \varphi\ \langle l,\, a,\, r\rangle + 2 \\
&\quad \text{because } \varphi\ B < \varphi\ R \\
&\leq \varphi\ B' + 2 * \varphi\ C - 3 * \varphi\ \langle l,\, a,\, r\rangle + 1 \\
&\quad \text{because } 1 + \log_2 x + \log_2 y < 2 * \log_2 (x + y) \text{ if } x, y > 0 \\
&\leq 3 * (\varphi\ C - \varphi\ \langle l,\, a,\, r\rangle) + 1 \qquad \text{because } \varphi\ B' \leq \varphi\ C
\end{aligned}
$$

This looks similar to the proof by Sleator and Tarjan but is different: they consider one double rotation whereas we argue about the whole input-output relationship; also our *log* argument is simpler.

From (3) we obtain in the worst case ($l = r = \langle\rangle$):

If *bst* t and $a \in set_{tree}\ t$ then $\mathcal{A}\ a\ t \leq 3 * (\varphi\ t - 1) + 1$.

In the literature the case $a \notin set_{tree}\ t$ is treated informally by stating that it can be reduced to $a' \in set_{tree}\ t$: one could have called *splay* with some $a' \in set_{tree}\ t$ instead of a and the behaviour would have been the same. Formally we prove by induction that if $t \neq \langle\rangle$ and *bst* t then

$$\exists a' \in set_{tree}\ t.\ splay\ a'\ t = splay\ a\ t \wedge t_{splay}\ a'\ t = t_{splay}\ a\ t$$

This gives us an upper bound for all binary search trees:

$$bst\ t \implies \mathcal{A}\ a\ t \leq 3 * \varphi\ t + 1 \tag{4}$$

The $\varphi\ t - 1$ was increased to $\varphi\ t$ because the former is negative if $t = \langle\rangle$.

We also need to determine the amortized complexity $\mathcal{A}m$ of *splay_max*

$$\mathcal{A}m\ t = t_{splay\_max}\ t + \Phi\ (splay\_max\ t) - \Phi\ t$$

A derivation similar to but simpler than the one for $\mathcal{A}$ yields the same upper bound: $bst\ t \implies \mathcal{A}m\ t \leq 3 * \varphi\ t + 1$.

Now we can apply our amortized analysis theory:

datatype $'a\ op_{st} = Splay\ 'a\ |\ Insert\ 'a\ |\ Delete\ 'a$

$$nxt_{st}\ (Splay\ a)\ t = splay\ a\ t \qquad t_{st}\ (Splay\ a)\ t = t_{splay}\ a\ t$$
$$nxt_{st}\ (Insert\ a)\ t = insert_{st}\ a\ t \qquad t_{st}\ (Insert\ a)\ t = t_{splay}\ a\ t$$
$$nxt_{st}\ (Delete\ a)\ t = delete_{st}\ a\ t \qquad t_{st}\ (Delete\ a)\ t = t_{delete}\ a\ t$$

$$init = \langle\rangle \qquad nxt = nxt_{st} \qquad t = t_{st} \qquad inv = bst \qquad \Phi = \Phi$$
$$U\ (Splay\ \_)\ t = 3 * \varphi\ t + 1$$
$$U\ (Insert\ \_)\ t = 4 * \varphi\ t + 2$$
$$U\ (Delete\ \_)\ t = 6 * \varphi\ t + 2$$

The fact that the given U is indeed a correct upper bound follows from the upper bounds for $\mathcal{A}$ and $\mathcal{A}m$; for $Insert$ and $Delete$ the proof needs more case distinctions and log-manipulations.

6.3   Improved Amortized Analysis

This subsection follows the work of Schoenmakers [11] (except that he confines himself to *splay*) who improves upon the constants in the above analysis. His analysis is parameterized by two constants $\alpha > 1$ and β subject to three constraints where all the variables are assumed to be ≥ 1:

$$(x + y) * (y + z)^\beta \leq (x + y)^\beta * (x + y + z)$$

$$\alpha * (l' + r') * (lr + r)^\beta * (lr + r' + r)^\beta$$
$$\leq (l' + r')^\beta * (l' + lr + r')^\beta * (l' + lr + r' + r)$$

$$\alpha * (l' + r') * (l' + ll)^\beta * (r' + r)^\beta$$
$$\leq (l' + r')^\beta * (l' + ll + r')^\beta * (l' + ll + r' + r)$$

The following upper bound is again proved by induction but this time with the help of the above constraints: if $bst\ t$ and $\langle l,\ a,\ r\rangle \in subtrees\ t$ then

$$\mathcal{A}\ a\ t \leq \log_\alpha\ (|t|_1\ /\ (|l|_1 + |r|_1)) + 1$$

From this we obtain the following main theorem just like before:

$$\mathcal{A}\ a\ t \leq \log_\alpha\ |t|_1 + 1$$

Now we instantiate the above abstract development with $\alpha = \sqrt[3]{4}$ and $\beta = 1/3$ (which includes proving the three constraints on α and β above) to obtain a bound for splaying that is only half as large as in (4):

$$bst\ t \implies \mathcal{A}_{34}\ a\ t \leq 3\ /\ 2 * \varphi\ t$$

The subscript $_{34}$ is our indication that we refer to the $\alpha = \sqrt[3]{4}$ and $\beta = 1/3$ instance. Schoenmakers additionally showed that this specific choice of α and β yields the minimal upper bound.

A similar but simpler development leads to the same bound for $\mathcal{A}m_{34}$ as for $\mathcal{A}_{34}$. Again we apply our amortized analysis theory to verify upper bounds for *Splay*, *Insert* and *Delete* that are also only half as large as before:

$$U \ (Splay \ \_) \ t = 3 \ / \ 2 * \varphi \ t + 1$$
$$U \ (Insert \ \_) \ t = 2 * \varphi \ t + 3 \ / \ 2$$
$$U \ (Delete \ \_) \ t = 3 * \varphi \ t + 2$$

The proofs in this subsection require a lot of highly nonlinear arithmetic. Only some of the polynomial inequalities can be automated with Harrison's sum-of-squares method [5].

7   Splay Heaps

Splay heaps are another self-adjusting data structure and were invented by Okasaki [10]. Splay heaps are organized internally like splay trees but they implement a priority queue interface. When inserting an element x into a splay heap, the splay heap is first partitioned (by rotations, like *splay*) into two trees, one $\leq x$ and one $> x$, and x becomes the new root:

$$insert \ x \ h = (\textbf{let} \ (l, \ r) = partition \ x \ h \ \textbf{in} \ \langle l, \ x, \ r \rangle)$$

$partition \ p \ \langle \rangle = (\langle \rangle, \ \langle \rangle)$
$partition \ p \ \langle al, \ a, \ ar \rangle =$
$(\textbf{if} \ a \leq p$
$\textbf{then} \ \textbf{case} \ ar \ \textbf{of} \ \langle \rangle \Rightarrow (\langle al, \ a, \ ar \rangle, \ \langle \rangle)$
$\quad | \ \langle bl, \ b, \ br \rangle \Rightarrow$
$\qquad \textbf{if} \ b \leq p \ \textbf{then} \ \textbf{let} \ (pl, \ y) = partition \ p \ br \ \textbf{in} \ (\langle \langle al, \ a, \ bl \rangle, \ b, \ pl \rangle, \ y)$
$\qquad \textbf{else} \ \textbf{let} \ (pl, \ pr) = partition \ p \ bl \ \textbf{in} \ (\langle al, \ a, \ pl \rangle, \ \langle pr, \ b, \ br \rangle)$
$\textbf{else} \ \textbf{case} \ al \ \textbf{of} \ \langle \rangle \Rightarrow (\langle \rangle, \ \langle al, \ a, \ ar \rangle)$
$\quad | \ \langle bl, \ b, \ br \rangle \Rightarrow$
$\qquad \textbf{if} \ b \leq p \ \textbf{then} \ \textbf{let} \ (pl, \ pr) = partition \ p \ br \ \textbf{in} \ (\langle bl, \ b, \ pl \rangle, \ \langle pr, \ a, \ ar \rangle)$
$\qquad \textbf{else} \ \textbf{let} \ (pl, \ pr) = partition \ p \ bl \ \textbf{in} \ (pl, \ \langle pr, \ b, \ \langle br, \ a, \ ar \rangle \rangle))$

Function *del_min* removes the minimal element and is similar to *splay_max*:

$del\_min \ \langle \rangle = \langle \rangle$
$del\_min \ \langle \langle \rangle, \ uu, \ r \rangle = r$
$del\_min \ \langle \langle ll, \ a, \ lr \rangle, \ b, \ r \rangle =$
$(\textbf{if} \ ll = \langle \rangle \ \textbf{then} \ \langle lr, \ b, \ r \rangle \ \textbf{else} \ \langle del\_min \ ll, \ a, \ \langle lr, \ b, \ r \rangle \rangle)$

In contrast to search trees, priority queues may contain elements multiple times. Therefore splay heaps satisfy the weaker invariant *bst*$_{eq}$:

$bst_{eq} \ \langle \rangle = True$
$bst_{eq} \ \langle l, \ a, \ r \rangle =$
$(bst_{eq} \ l \ \wedge \ bst_{eq} \ r \ \wedge \ (\forall x \in set_{tree} \ l. \ x \leq a) \ \wedge \ (\forall x \in set_{tree} \ r. \ a \leq x))$

This is an invariant for both *partition* and *del_min*:

If $bst_{eq} \ t$ and $partition \ p \ t = (l, \ r)$ then $bst_{eq} \ \langle l, \ p, \ r \rangle$.
If $bst_{eq} \ t$ then $bst_{eq} \ (del\_min \ t)$.

For the functional correctness proofs see [9].

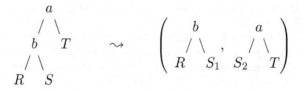

Fig. 6. Zig-zag case for *partition*: $b \leq p < a$

7.1 Amortized Analysis

Now we verify the amortized analysis due to Okasaki. The timing functions are straightforward and not shown: t_{part} and t_{dm} count the number of calls of *partition* and *del_min*. The potential of a tree is defined as for splay trees in Sect. 6.2. The following logarithmic bound of the amortized complexity $\mathcal{A}\ p\ t = t_{part}\ p\ t + \Phi\ l' + \Phi\ r' - \Phi\ t$ is proved by computation induction on *partition* t: if $bst_{eq}\ t$ and *partition* $p\ t = (l',\ r')$ then

$$\mathcal{A}\ p\ t \leq 2 * \varphi\ t + 1$$

Okasaki [10] shows the zig-zig case of the induction, I show the zig-zag case in Fig. 6. Subtrees with root x are called X on the left and X' on the right-hand side. Thus Fig. 6 depicts *partition* $p\ A = (B',\ A')$ assuming the recursive call *partition* $p\ S = (S_1,\ S_2)$.

$$
\begin{aligned}
\mathcal{A}\ p\ A &= \mathcal{A}\ p\ S + 1 + \varphi\ B' + \varphi\ A' - \varphi\ B - \varphi\ A \\
&\leq 2 * \varphi\ S + 2 + \varphi\ B' + \varphi\ A' - \varphi\ B - \varphi\ A \qquad \text{by ind.hyp.} \\
&= 2 + \varphi\ B' + \varphi\ A' \qquad \text{because } \varphi\ S < \varphi\ B \text{ and } \varphi\ S < \varphi\ A \\
&\leq 2 * \log_2 (|R|_1 + |S_1|_1 + |S_2|_1 + |T|_1 - 1) + 1 \\
&\qquad \text{because } 1 + \log_2 x + \log_2 y \leq 2 * \log_2 (x + y - 1) \text{ if } x, y \geq 2 \\
&= 2 * \varphi\ A + 1 \qquad \text{because } |S_1| + |S_2| = |S|
\end{aligned}
$$

The proof of the amortized complexity of *del_min* is similar to the one for *splay_max*: $t_{dm}\ t + \Phi\ (del\_min\ t) - \Phi\ t \leq 2 * \varphi\ t + 1$. Now it is routine to verify the following amortized complexities by instantiating our standard theory with $U\ (Insert\ \_)\ t = 3 * \log_2 (|t|_1 + 1) + 1$ and $U\ Delmin\ t = 2 * \varphi\ t + 1$.

Acknowledgement. Berry Schoenmakers patiently answered many questions about his work whenever I needed help.

References

1. Atkey, R.: Amortised resource analysis with separation logic. Logical Methods Comput. Sci. **7**(2), 33 (2011)
2. Charguéraud, A., Pottier, F.: Machine-checked verification of the correctness and amortized complexity of an efficient union-find implementation. In: Urban, C., Zhang, X. (ed.) ITP 2015. LNCS, vol. 9236, pp. 137–154. Springer, Heidelberg (2015)

3. Cormen, T.H., Leiserson, C.E., Rivest, R.L.: Introduction to Algorithms. MIT Press, New York (1990)
4. Fredman, M.L., Sedgewick, R., Sleator, D.D., Tarjan, R.E.: The pairing heap: A new form of self-adjusting heap. Algorithmica $1(1)$, 111–129 (1986)
5. Harrison, J.: Verifying nonlinear real formulas via sums of squares. In: Schneider, K., Brandt, J. (eds.) TPHOLs 2007. LNCS, vol. 4732, pp. 102–118. Springer, Heidelberg (2007)
6. Hoffmann, J., Aehlig, K., Hofmann, M.: Multivariate amortized resource analysis. ACM Trans. Program. Lang. Syst. $34(3)$, 14 (2012)
7. Hofmann, M., Jost, S.: Static prediction of heap space usage for first-order functional programs. In: Proceedings of the 30th ACM Symposium Principles of Programming Languages, pp. 185–197 (2003)
8. Kaldewaij, A., Schoenmakers, B.: The derivation of a tighter bound for top-down skew heaps. Inf. Process. Lett. **37**, 265–271 (1991)
9. Nipkow, T.: Amortized complexity verified. Archive of Formal Proofs (2014). http://afp.sf.net/entries/Amortized_Complexity.shtml. Formal proof development
10. Okasaki, C.: Purely Functional Data Structures. Cambridge University Press, Cambridge (1998)
11. Schoenmakers, B.: A systematic analysis of splaying. Inf. Process. Lett. **45**, 41–50 (1993)
12. Sleator, D.D., Tarjan, R.E.: Self-adjusting binary search trees. J. ACM $32(3)$, 652–686 (1985)
13. Sleator, D.D., Tarjan, R.E.: Self-adjusting heaps. SIAM J. Comput. $15(1)$, 52–69 (1986)
14. Tarjan, R.E.: Amortized complexity. SIAM J. Alg. Disc. Meth. $6(2)$, 306–318 (1985)
15. Traytel, D., Berghofer, S., Nipkow, T.: Extending hindley-milner type inference with coercive structural subtyping. In: Yang, H. (ed.) APLAS 2011. LNCS, vol. 7078, pp. 89–104. Springer, Heidelberg (2011)

Foundational Property-Based Testing

Zoe Paraskevopoulou[1,2], Cătălin Hriţcu[1(✉)], Maxime Dénès[1],
Leonidas Lampropoulos[3], and Benjamin C. Pierce[3]

[1] Inria Paris-Rocquencourt, Rocquencourt, France
catalin.hritcu@inria.fr
[2] ENS Cachan, Cachan, France
[3] University of Pennsylvania, Philadelphia, PA, USA

Abstract. Integrating property-based testing with a proof assistant creates an interesting opportunity: reusable or tricky testing code can be formally verified using the proof assistant itself. In this work we introduce a novel methodology for formally verified property-based testing and implement it as a foundational verification framework for QuickChick, a port of QuickCheck to Coq. Our framework enables one to verify that the executable testing code is testing the right Coq property. To make verification tractable, we provide a systematic way for reasoning about the set of outcomes a random data generator can produce with non-zero probability, while abstracting away from the actual probabilities. Our framework is firmly grounded in a fully verified implementation of QuickChick itself, using the same underlying verification methodology. We also apply this methodology to a complex case study on testing an information-flow control abstract machine, demonstrating that our verification methodology is modular and scalable and that it requires minimal changes to existing code.

1 Introduction

Property-based testing (PBT) allows programmers to capture informal conjectures about their code as executable specifications and to thoroughly test these conjectures on a large number of inputs, usually randomly generated. When a counterexample is found it is shrunk to a minimal one, which is displayed to the programmer. The original Haskell QuickCheck [12], the first popular PBT tool, has inspired ports to every mainstream programming language and led to a growing body of continuing research [11,15] and industrial interest [20]. PBT has also been integrated into proof assistants [5,8,13,14,21] as a way of reducing the cost of formal verification, finding bugs earlier in the verification process and decreasing the number of failed proof attempts: *Testing helps proving!* Motivated by these earlier successes, we have ported the QuickCheck framework to Coq, resulting in a prototype Coq plugin called QuickChick. With QuickChick, we hope to make testing a convenient aid during Coq proof development.

In this paper we explore a rather different way that testing and proving can cooperate in a proof assistant. Since our testing code (and most of QuickChick itself) is written in Coq, we can also formally verify this code using Coq. That

© Springer International Publishing Switzerland 2015
C. Urban and X. Zhang (Eds.): ITP 2015, LNCS 9236, pp. 325–343, 2015.
DOI: 10.1007/978-3-319-22102-1_22

is, *proving helps testing*! This verified-testing idea was first proposed a decade ago by Dybjer, Haiyan, and Takeyama [13,14,18] in the context of Agda/Alfa, but it has apparently received very little attention since then [4,7].

Why would one want verified testing? Because PBT is very rarely a push-button process. While frameworks such as QuickCheck provide generic infrastructure for writing testing code, it is normally up to the user to compose the QuickCheck combinators in creative ways to obtain effective testing code for the properties they care about. This testing code can be highly non-trivial, so mistakes are hard to avoid. Some types of mistakes are easily detected by the testing itself, while others are not: inadvertently testing a stronger property will usually fail with a counter-example that one can manually inspect, but testing a weaker or just a different property can succeed although the artifact under test is completely broken with respect to the property of interest. Thus, while PBT is effective at quickly finding bugs in formal artifacts, errors in testing code can conceal important bugs, instilling a false sense of confidence until late in the verification process and reducing the benefits of testing.

One response to this problem is providing more automation. QuickCheck uses type classes for this purpose, and other tools go much further—using, for instance, techniques inspired by functional logic programming and constraint solving [5,6,9,11,15,16]. While automation reduces user effort by handling easy but tedious and repetitive tasks, we are doubtful that the creative parts of writing effective testing code can be fully automated in general (any more than writing non-trivial programs in any other domain can); our experience shows that the parts that cannot be automated are usually tricky enough to also contain bugs [19]. Moreover, the more sophisticated the testing framework becomes, the higher the chances that it is going to contain bugs itself. Given the randomized nature of QuickCheck-style PBT, such bugs can go unnoticed for a long time.

Thus, both for tricky user code and for reusable framework code, verified testing may be an attractive solution. In particular, formal verification allows one to show that *non-trivial testing code is actually testing the intended property*. To make this process viable in practice, we need a modular and scalable way of reasoning about probabilistic testing code. Moreover, for high assurance, we desire a verification framework based on strong formal foundations, with precisely identified assumptions. More speculatively, such a foundational verification framework could serve as a target for certificate-producing metaprograms and external tools that produce testing code automatically (e.g., from inductive type specifications or boolean predicates).

Contributions. We introduce a novel methodology for formally verified PBT and implement it as a foundational verification framework for our QuickChick Coq plugin. To make verification tractable, we provide a systematic way for reasoning about the set of outcomes a random data generator can produce with non-zero probability, abstracting away from actual probabilities. This possibilistic abstraction is not only convenient in practice, but also very simple and intuitive. Beyond this abstraction, our framework is firmly grounded on a fully verified implementation of QuickChick itself. We are the first to successfully verify a QuickCheck-like library—significant validation for our verification methodology. We also describe a significant case study on testing an information-flow control abstract machine. These experimental results are encouraging, indicating that our verification methodology is modular and scalable, requiring minimal changes to existing code. Finally, porting QuickCheck to Coq is a useful side-contribution that is of independent interest, and we hope that we and others will build upon QuickChick in the future. Our verification framework relies on the SSReflect [17] extension to Coq, and is fully integrated into QuickChick, which is available under a permissive open source license at https://github.com/QuickChick.

Outline. Our verification framework is illustrated on an example in Sect. 2 and presented in full detail in Sect. 3. The information-flow case study is discussed in Sect. 4. We present related work in Sect. 5, before drawing conclusions and discussing future work in Sect. 6.

2 Example: Red-Black Trees

In this section we illustrate both the QuickChick plugin for Coq and our verification framework on a simple red-black tree example.[1] A red-black tree is a self-balancing binary search tree in which each non-leaf node stores a data item and is colored either Red or Black. We define the type of red-black trees of naturals in Coq as follows:

```
Inductive color := Red | Black.
Inductive tree := Leaf : tree | Node : color -> tree -> nat -> tree -> tree.
```

Inserting a new element into a red-black tree is non-trivial as it involves re-balancing to preserve the following invariants: (i) the root is black (ii) all leaves are black (iii) red nodes have two black children (iv) from any given node, all descendant paths to leaves have the same number of black nodes. (For simplicity, we ignore the binary-search-tree property and focus only on balancing here.) If we wanted to prove that an insert function of type nat -> tree -> tree preserves the red-black tree invariant, we could take inspiration from Appel [1] and express this invariant in declarative form:

```
Inductive is_redblack' : tree -> color -> nat -> Prop :=
| IsRB_leaf: forall c, is_redblack' Leaf c 0
| IsRB_r: forall n tl tr h, is_redblack' tl Red h -> is_redblack' tr Red h ->
                      is_redblack' (Node Red tl n tr) Black h
| IsRB_b: forall c n tl tr h, is_redblack' tl Black h -> is_redblack' tr Black h ->
                      is_redblack' (Node Black tl n tr) c (S h).
Definition is_redblack (t:tree) : Prop := exists h, is_redblack' t Red h.
```

The definition uses a helper inductive relation is_redblack', pronounced "is a red-black subtree," with three parameters: (i) a sub-tree (ii) the *color-context* c (i.e., the color of the parent node) (iii) the *black-height* h of the sub-tree (i.e., the number of black nodes in any path from the root of the sub-tree to a leaf). A leaf

---

[1] The complete code for this example is available at https://github.com/QuickChick/QuickChick/tree/master/examples/RedBlack.

is a well-formed red-black sub-tree in any color-context and has black-height o.
A node is a red-black sub-tree if both its child trees are red-black sub-trees and
if the color-context is black in case it has a red root. Moreover, the black-height
of the children must be equal, and we increase this height by 1 if the root is
black. Using this definition we might like to prove in Coq the following property
of insert:

```
Definition insert_preserves_redblack : Prop :=
  forall x s, is_redblack s -> is_redblack (insert x s).
```

Before starting a proof of this proposition we would like to quickly check that
we did not do any mistakes in the definition of insert or is_redblack. However,
the declarative definition of is_redblack is not well adapted to efficient testing.
Even if we were able to automatically give an executable interpretation to the
inductive definition of is_redblack' [2,23], we would still have to guess the existentially quantified black-height h, which would be highly inefficient. So in order
to effectively test the is_redblack invariant, we first manually devise an efficiently
executable version:

```
Definition is_redblack_bool (t : tree) : bool :=
  is_black_balanced t && has_no_red_red Red t.
```

We omit the definitions of the auxiliaries is_black_balanced and has_no_red_red. While
is_redblack_bool allows us to check whether a tree is red-black or not, in order
to test the invariant of insert using QuickChick, we also need a way to generate
random trees. We start by devising a generic tree *generator* using the QuickChick
combinators:

```
Definition genColor := elems [Red; Black].
Fixpoint genAnyTree_depth (d : nat) : G tree :=
  match d with
    | 0 => returnGen Leaf
    | S d' => freq [(1, returnGen Leaf);
                    (9, liftGen4 Node genColor (genAnyTree_depth d')
                                 arbitrary (genAnyTree_depth d'))]
  end.
Definition genAnyTree : G tree := sized genAnyTree_depth.
```

The genAnyTree_depth auxiliary generates trees of a given depth. If the depth is zero,
we generate a leaf using returnGen Leaf, otherwise we use the freq combinator to
choose whether to generate a leaf or to generate a color using genColor, a natural
number using arbitrary, the two sub-trees using recursive calls, and put everything
together using liftGen 4Node. The code illustrates several QuickChick combinators:
(i) elems chooses a color from a list of colors uniformly at random (ii) returnGen
always chooses the same thing, a Leaf in this case (iii) freq performs a biased
probabilistic choice choosing a generator from a list using user-provided weights
(in the example above we generate nodes 9 times more often than leafs) (iv)
liftGen4 takes a function of 4 arguments, here the Node constructor, and applies
it to the result of 4 other generators (v) arbitrary is a method of the Arbitrary
type class, which assigns default generators to frequently used types, in this case
the naturals (vi) sized takes a generator parameterized by a size, in this case
genAnyTree_depth, and produces a non-parameterized generator by iterating over

different sizes. Even this naive generator is easy to get wrong: our first take at it did not include the call to `freq` and was thus only generating full trees, which can cause testing to miss interesting bugs.

The final step for testing the `insert` function using this naive generator is combining the `genAnyTree` generator and the `is_redblack` boolean function into a *property checker*—i.e., the testing equivalent of `insert_preserves_redblack`:

```
Definition insert_preserves_redblack_checker (genTree : G tree) : Checker :=
  forAll arbitrary (fun n => forAll genTree (fun t =>
    is_redblack_bool t ==> is_redblack_bool (insert n t))).
```

This uses two checker combinators from QuickChick: (i) `forAll` produces data using a generator and passes it to another checker (ii) `c1 ==> c2` takes two checkers `c1` and `c2` and tests that `c2` does not fail when `c1` succeeds. The "`==>`" operator also remembers the percentage of inputs that do not satisfy the precondition `c1` and thus have to be discarded without checking the conclusion `c2`. In our running example `c1` and `c2` are two boolean expressions that are implicitly promoted to checkers. Now we have something we can test using the QuickChick plugin, using the `QuickChick` command:

```
QuickChick (insert_preserves_redblack_checker genAnyTree).

*** Gave up! Passed only 2415 tests
Discarded: 20000
```

We have a problem: Our naive generator for trees is very unlikely to generate red-black trees, so the premise of `insert_preserves_redblack_checker` is false and thus the property vacuously true 88 % of the time. The conclusion is actually tested infrequently, and if we collect statistics about the distribution of data on which it is tested, we see that the size of the trees that pass the very strong precondition is very small: about 95.3 % of the trees have 1 node, 4.2 % of them have 3 nodes, 0.4 % of them have 5 nodes, and only 0.03 % of them have 7 or 9 nodes. So we are not yet doing a good job at testing the property. While the generator above is very simple—it could probably even be generated automatically from the definition of `tree` [2, 24]—in order to effectively test the property we need to write a *property-based generator* that *only* produces red-black trees.

```
Program Fixpoint genRBTree_height (hc : nat*color) {wf wf_hc hc} : G tree :=
  match hc with
  | (0, Red) => returnGen Leaf
  | (0, Black) => oneOf [returnGen Leaf;
                    (do! n <- arbitrary; returnGen (Node Red Leaf n Leaf))]
  | (S h, Red) => liftGen4 Node (returnGen Black) (genRBTree_height (h, Black))
                      arbitrary (genRBTree_height (h, Black))
  | (S h, Black) => do! c' <- genColor;
                    let h' := match c' with Red => S h | Black => h end in
                    liftGen4 Node (returnGen c') (genRBTree_height (h', c'))
                        arbitrary (genRBTree_height (h', c')) end.

Definition genRBTree := bindGen arbitrary (fun h => genRBTree_height (h, Red)).
```

The `genRBTree_height` generator produces red-black trees of a given black-height and color-context. For black-height `0`, if the color-context is `Red` it returns a (black) leaf, and if the color-context is Black it uses the oneOf combinator to select between two generators: one that returns a leaf, and another that returns a Red node with leaf children and a random number. The latter uses do notation for bind ("do! n <- arbitrary;...") in the G randomness monad. For black-height larger than o and color-context Red we always generate a Black node (to prevent red-red conflicts) and generate the sub-trees recursively using a smaller black-height. Finally, for black-height larger than o and color-context Black we have the choice of generating a Red or a Black node. If we generate a Red node the recursive call is done using the same black-length. The function is shown terminating using a lexicographic ordering on the black-height and color-context.

With this new generator we can run 10000 tests on a laptop in less than 9 s, of which only 1 s is spent executing the tests. The the rest is spent extracting to OCaml and running the OCaml compiler (the extraction and compilation part could be significantly sped up; this time is also easily amortized for longer running tests):

```
QuickChick (insert_preserves_redblack_checker genRBTree).

+++ OK, passed 10000 tests
```

Moreover, none of the generated trees fails the precondition and the average size of the trees used for testing the conclusion is 940.7 nodes (compared to 1.1 nodes naively!)

In the process of testing, we have, however, written quite a bit of executable testing code—some of it non-trivial, like the generator for red-black trees. How do we know that this code is testing the declarative proposition we started with? Does our generator for red-black trees only produce red-black trees, and even more importantly can it in principle produce *all* red-black trees? Our foundational testing verification framework supports formal answers to these questions. In our framework the semantics of each generator is the set of values that have non-zero probability of being generated. Building on this, we assign a semantics to each checker expressing the logical proposition it tests, abstracting away from computational constraints like space and time as well as the precise probability distributions of the generators it uses. In concrete terms, a function semChecker assigns each Checker a Prop, and a function semGen assigns each generator of type G A a (non-computable) set of outcomes with Coq type A-> Prop.

```
semChecker : Checker -> Prop
semCheckable : forall (C : Type) '{Checkable C}, C -> Prop.
Definition set T := T -> Prop.
Definition set_eq {A} (m1 m2 : set A) := forall (a : A), m1 a <-> m2 a.
Infix "<-->" := set_eq (at level 70, no associativity) : set_scope.
semGen : forall A : Type, G A -> set A
semGenSize : forall A : Type, G A -> nat -> set A
```

Given these, we can prove that a checker c tests a declarative proposition P by showing that semChecker c is logically equivalent with P. Similarly, we can prove that a generator g produces the set of outcomes o by showing that the set semGen g is equal to o, using the extensional definition of set equality set_eq above. Returning to our red-black tree example we can prove the following top-level theorem:

```
Lemma insert_preserves_redblack_checker_correct :
  semChecker (insert_preserves_redblack_checker genRBTree)
  <-> insert_preserves_redblack .
```

The top-level structure of the checker and the declarative proposition is very similar in this case, and our framework provides lemmas about the semantics of `forAll` and "`==>`" that we can use to make the connection (`semCheckable` is just a variant of `semChecker` described further in Sect. 3.2):

```
Lemma semForAllSizeMonotonic {A C} '{Show A, Checkable C} (g : G A) (f : A -> C)
    '{SizeMonotonic _ g} '{forall a, SizeMonotonicChecker (checker (f a))} :
  (semChecker (forAll g f) <-> forall (a:A), a \in semGen g -> semCheckable (f a)).

Lemma semImplication {C} '{Checkable C} (c : C) (b : bool) :
  semChecker (b ==> c) <-> (b -> semCheckable c).

Lemma semCheckableBool (b : bool) : semCheckable b <-> b.
```

Using these generic lemmas, we reduce the original equivalence we want to show to the equivalence between `is_redblack` and `is_redblack_bool` (`reflect` is equivalence between a `Prop` and a `bool` in the SSReflect libraries).

```
Lemma is_redblackP t : reflect (is_redblack t) (is_redblack_bool t).
```

Moreover, we need to show that the generator for red-black trees is complete; i.e., they it can generate all red-black trees. We show this via a series of lemmas, including:

```
Lemma semColor : semGen genColor <--> [set : color].

Lemma semGenRBTreeHeight h c :
  semGen (genRBTree_height (h, c)) <--> [set t | is_redblack' t c h ].

Lemma semRBTree : semGen genRBTree <--> [set t | is_redblack t].
```

The proofs of these custom lemmas rely again on generic lemmas about the QuickChick combinators that they use. We list the generic lemmas that we used in this proof:

```
Lemma semReturn {A} (x : A) : semGen (returnGen x) <--> [set x].

Lemma semBindSizeMonotonic {A B} (g : G A) (f : A -> G B)
    '{Hg : SizeMonotonic _ g} '{Hf : forall a, SizeMonotonic (f a)} :
  semGen (bindGen g f) <--> \bigcup_(a in semGen g) semGen (f a).

Lemma semElems A (x : A) xs : semGen (elems (x ;; xs)) <--> x :: xs.

Lemma semOneOf A (g0 : G A) (gs : list (G A)) :
  semGen (oneOf (g0 ;; gs)) <--> \bigcup_(g in (g0 :: gs)) semGen g.
```

While the proof of the red-black tree generator still requires manual effort the user only needs to verify the code she wrote, relying on the precise high-level specifications of all combinators she uses (e.g., the lemmas above). Moreover, all proofs are in terms of propositions and sets, not probability distributions or low-level pseudo-randomness. The complete example is around 150 lines of proofs for 236 lines of definitions. While more aggressive automation (e.g., using SMT) could further reduce the effort in the future, we believe that verifying reusable or tricky testing code (like QuickChick itself or the IFC generators from Sect. 4) with our framework is already an interesting proposition.

3 Foundational Verification Framework

As the example above illustrates, the main advantage of using our verified testing framework is the ability to carry out abstract (possibilistic) correctness proofs of testing code with respect to the high-level specifications of the QuickChick combinators. But how do we know that those specifications are correct? And what exactly do we mean by "correct"? What does it mean that a property checker is testing the right proposition, or that a generator is in principle able to produce a certain outcome? To answer these questions with high confidence we have verified QuickChick all the way down to a small set of definitions and assumptions. At the base of our formal construction lies our possibilistic semantics of generators (Sect. 3.1) and checkers (Sect. 3.2), and an idealized interface for a splittable pseudorandom number generator (splittable PRNG, in Sect. 3.3). Our possibilistic abstraction allows us to completely avoid complex probabilistic reasoning at all levels, which greatly improves the scalability and ease of use of our methodology. On top of this we verify all the combinators of QuickChick, following the modular structure of the code (Sect. 3.4). We provide support for conveniently reasoning about sizes (Sect. 3.5) and about generators for functions (Sect. 3.6). Our proofs use a small library for reasoning about non-computable sets in point-free style (Sect. 3.7).

3.1 Set-of-Outcomes Semantics for Generators

In our framework, the semantics of a generator is the set of outcomes it can produce with non-zero probability. We chose this over a more precise abstraction involving probability distributions, because specifying and verifying probabilistic programs is significantly harder than nondeterministic ones. Our possibilistic semantics is simpler and easier to work with, allowing us to scale up our verification to realistic generators, while still being precise enough to find many bugs in them (Sect. 4). Moreover, the possibilistic semantics allows us to directly relate checkers to the declarative propositions they test (Sect. 3.2); in a probabilistic setting the obvious way to achieve this is by only looking at the support of the probability distributions, which would be equivalent to what we do, just more complex. Finally, with our set-of-outcomes semantics, set inclusion corresponds exactly to generator surjectivity from previous work on verified testing [13,14,18], while bringing significant improvements to proofs via point-free reasoning (Sect. 3.7) and allowing us to verify both soundness and completeness.

QuickChick generators are represented internally the same way as a reader monad with two parameters: a size and a random seed [12] (the bind of this monad splits the seed, which is further discussed in Sect. 3.3 and Sect. 3.4).

```
Inductive G (A:Type) : Type := MkGen : (nat -> RandomSeed -> A) -> G A.

Definition run {A : Type} (g : G A) := match g with MkGen f => f end.
```

Formally, the semantics of a generator g of type $G\ A$ is defined as the set of elements a of type A for which there exist a size s and a random seed r with $run\ g\ s\ r = a$.

```
Definition semGenSize {A : Type} (g : G A) (s : nat) : set A :=
  [set a : A | exists r, run g s r = a].
Definition semGen {A : Type} (g : G A) : set A :=
  [set a : A | exists s, a \in semGenSize g s].
```

We also define semGenSize, a variant of the semantics that assigns to a generator
the outcomes it can produce for a given size. Reasoning about sizes is discussed
in Sect. 3.5.

3.2 Possibilistic Semantics of Checkers

A property checker is an executable routine that expresses a property under test
so that is can be checked against a large number of randomly generated inputs.
The result of a test can be either successful, when the property holds for a given
input, or it may reveal a counterexample. Property checkers have type Checker and
are essentially generators of testing results.

In our framework the semantics of a checker is a Coq proposition. The proposition obtained from the semantics can then be proved logically equivalent to the
desired high-level proposition that we claim to test. More precisely, we map a
checker to a proposition that holds if and only if no counterexamples can possibly
be generated, i.e., when the property we are testing is always valid for the generators we are using. This can be expressed very naturally in our framework by
stating that all the results that belong to the set of outcomes of the checker are
successful (remember that checkers are generators), meaning that they do not
yield any counterexample. Analogously to generators, we also define semCheckerSize
that maps the checker to its semantics for a given size.

```
Definition semCheckerSize (c : Checker) (s : nat): Prop :=
  successful @: semGenSize c s \subset [set true].

Definition semChecker (c : Checker) : Prop := forall s, semCheckerSize c s.
```

Universally quantifying over all sizes in the definition of semChecker is a useful idealization. While in practice QuickChick uses an incomplete heuristic for trying out
different sizes in an efficient way, it would be very cumbersome and completely
unenlightening to reason formally about this heuristic. By deliberately abstracting away from this source of incompleteness in QuickChick, we obtain a cleaner
mathematical model. Despite this idealization, it is often not possible to completely abstract away from the sizes in our proofs, but we provide ways to make
reasoning about sizes convenient (Sect. 3.5).

In order to make writing checkers easier, QuickChick provides the type class
Checkable that defines checker, a coercion from a certain type (e.g., bool) to Checker. We
trivially give semantics to a type that is instance of Checkable with:

```
Definition semCheckableSize {A} '{Checkable A} (a : A) (s : nat) : Prop :=
  semCheckerSize (checker a) s.

Definition semCheckable {A} '{Checkable A} (a : A) : Prop := semChecker (checker a).
```

3.3 Splittable Pseudorandom Number Generator Interface

QuickChick's splittable PRNG is written in OCaml. The rest of QuickChick is written and verified in Coq and then extracted to OCaml. Testing happens outside of Coq for efficiency reasons. The random seed type and the low-level operations on it, such as splitting a random seed and generating booleans and numbers, are simply axioms in Coq. Our proofs also assume that the random seed type is inhabited and that the operations producing numbers from seeds respect the provided range. All these axioms would disappear if the splittable PRNG were implemented in Coq. One remaining axiom would stay though, since it is inherent to our idealized model of randomness:

```
Axiom randomSplit : RandomSeed -> RandomSeed * RandomSeed.
Axiom randomSplitAssumption :
  forall s1 s2 : RandomSeed, exists s, randomSplit s = (s1,s2).
```

This axiom says that the `randomSplit` function is surjective. This axiom has non-trivial models: the `RandomSeed` type could be instantiated with the naturals, infinite streams, infinite trees, etc. One can also easily show that, in all non-trivial models of this axiom, `RandomSeed` is an infinite type. In reality though, PRNGs work with finite seeds. Our axiom basically takes us from pseudorandomness to ideal mathematical randomness, as used in probability theory. This idealization seems unavoidable for formal reasoning and it would also be needed even if we did probabilistic as opposed to possibilistic reasoning. Conceptually, one can think of our framework as raising the level of discourse in two ways: (i) idealizing pseudo-randomness to probabilities (ii) abstracting probability distributions to their support sets. While the abstraction step could be formally justified (although we do not do this at the moment), the idealization step has to be taken on faith and intuition only. We believe that the possibilistic semantics from Sects. 3.1 and 3.2 and the axioms described here are simple and intuitive enough to be trusted; together with Coq they form the trusted computing base of our foundational verification framework.

3.4 Verified Testing Combinators

QuickChick provides numerous combinators for building generators and checkers. Using the semantics described above, we prove that each of these combinators satisfies a high-level declarative specification. We build our proofs following the modular organization of the QuickChick code (Fig. 1): only a few low-level generator combinators directly access the splittable PRNG interface and the concrete representation of generators. All the other combinators are built on top of the low-level

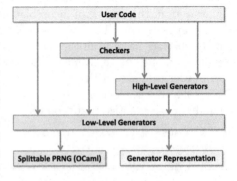

Fig. 1. QuickChick organization diagram

Table 1. Selected QuickChick combinators

Checker Combinators	`implication: forall {C : Type} '{Checkable C}, bool -> C -> Checker`
	`shrinking: forall {C A : Type} '{Checkable C},`
	`                (A -> list A) -> A -> (A -> C) -> Checker`
	`expectFailure: forall {C : Type} '{Checkable C}, C -> Checker`
	`forAll: forall {A C : Type} '{Show A, Checkable C},`
	`           G A -> (A -> C) -> Checker`
	`forAllShrink: forall {A C : Type} '{Show A, Checkable C},`
	`                 G A -> (A -> list A) -> (A -> C) -> Checker`

High-level Generator Combinators	`liftGen: forall {A B : Type}, (A -> B) -> G A -> G B`
	`sequenceGen: forall {A : Type}, list (G A) -> G (list A)`
	`foldGen: forall {A B : Type}, (A -> B -> G A) -> list B -> A -> G A`
	`oneof: forall {A : Type}, G A -> list (G A) -> G A`
	`frequency: forall {A : Type}, G A -> list (nat * G A) -> G A`
	`vectorOf: forall {A : Type}, nat -> G A -> G (list A)`
	`listOf: forall {A : Type}, G A -> G (list A)`
	`elements: forall {A : Type}, A -> list A -> G A`

Low-level Generator Combinators	`bindGen: forall {A B : Type}, G A -> (A -> G B) -> G B`
	`fmap: forall {A B : Type}, (A -> B) -> G A -> G B`
	`sized: forall {A: Type}, (nat -> G A) -> G A`
	`resize: forall {A: Type}, nat -> G A -> G A`
	`suchThatMaybe: forall {A : Type}, G A -> (A -> bool) -> G (option A)`
	`choose: forall {A : Type} '{ChoosableFromInterval A}, (A * A) -> G A`

generators. This modular organization is convenient for structuring our proofs all the way down. Table 1 illustrates an important part of the combinator library and how it is divided into low-level generators, high-level generators, and checkers.

The verification of low-level generators has to be done in a very concrete way that involves reasoning about random seeds. However, once we fully specify these generators in terms of their sets of outcomes, the proof of any generator that builds on top of them can be done in a fully compositional way that only depends on the set of outcomes specifications of the combinators used, abstracting away from the internal representation of generators, the implementation of the combinators, and the PRNG interface.

As we want the proofs to be structured in compositional way and only depend on the specifications and not the implementation of the combinators, we make the combinator implementations opaque for later proofs by enclosing them in a module that only exports their specifications. The size of the QuickChick framework (excluding examples) is around 2.4 kLOC of definitions and 2.0 kLOC of proofs.

3.5   Conveniently Reasoning About Sizes

QuickChick does not prescribe how the generators use the size parameter: some of them are actually *unsized* (they do not use their size parameter at all), while others are *sized* (they produce data depending on the size). For instance genColor from Sect. 2 is unsized—it always chooses uniformly at random between Red or Black—while genAnyTree and genRBTree are both sized. For sized generators, the precise size dependency can vary; indeed, there can be different notions of size for the same type: e.g., for genAnyTree size means depth, while for genRBTree it means black-height. Finally, some generators take the size parameter to mean the maximal size of the data they generate (e.g., the default generator for naturals, genAnyTree, genRBTree), while others take it to mean the exact size (e.g., sized (fun h => genRBTree_height (h, Red)) would be such a generator). Through our verification we discovered that unsized generators and sized generators using maximal size are easier to compose right since they allow stronger principles for compositional reasoning.

In Sect. 3.2 we defined the semantics of checkers by universally quantifying over all sizes, so one could naively expect that with this idealization there would be no need to unfold semGen and reason explicitly about the size parameter in terms of semGenSize in our generator proofs. Unfortunately, this is not always the case: low-level combinators taking several generators (or generator-producing functions) as arguments call all these arguments with the *same* size parameter (reader monad). For instance, bindGen g f returns a generator that given a size s and a seed r, splits r into (r1,r2), runs g on s and r1 in order to generate a value v, then applies f to v and runs the resulting generator with the same size s and with seed r2. It would be very tempting to try to give bindGen the following very intuitive specification, basically interpreting it as the bind of the nondetederminism monad:[2]

```
semGen (bindGen g f) <--> \bigcup_(a \in semGen g) semGen (f a).
```

This intuitive specification is, however, wrong in our setting. The set on the right-hand side contains elements that are generated from (f a) for some size parameter, whereas a is an element that has been generated from g with a different size parameter. This would allow us to prove the following generator complete

```
gAB = bindGen gA (fun a => bindGen gB (fun b => returnGen (a,b)))
```

with respect to [set : A * B] for any generators gA and gB for types A and B, even in the case when gA and gB are fixed-size generators, in which case gAB only produces pairs of equally-sized elements. In our setting, a correct specification of bindGen that works for arbitrary generators can only be given in terms of semGenSize, where the size parameter is also threaded through explicitly at the specification level:

```
Lemma semBindSize A B (g : G A) (f : A -> G B) (s : nat) :
  semGenSize (bindGen g f) s <--> \bigcup_(a in semGenSize g s) semGenSize (f a) s.
```

The two calls to semGenSize on the right-hand side are now passed the same size.

While in general we cannot avoid explicitly reasoning about the interaction between the ways composed generators use sizes, we can avoid it for two large classes of generators: unsized and size-monotonic generators. We call a generator

---

[2] Indeed, in a preliminary version of our framework the low-level generators were axiomatized instead of verified with respect to a semantics, and we took this specification as an axiom.

size-monotonic when increasing the size produces a larger (with respect to set inclusion) set of outcomes. Formally, these properties of generators are expressed by the following two type classes:

```
Class Unsized {A} (g : G A) := {
   unsized : forall s1 s2, semGenSize g s1 <--> semGenSize g s2 }.

Class SizeMonotonic {A} (g : G A) := {
   monotonic : forall s1 s2, s1 <= s2 -> semGenSize g s1 \subset semGenSize g s2 }.
```

The gAB generator above is in fact complete with respect to [set : A * B] if at least one of gA and gB is Unsized or if both gA and gB are SizeMonotonic. We can prove this conveniently using specialized specifications for bindGen from our library, such as the semBindSizeMonotonic lemma from Sect. 2 or the lemma below:

```
Lemma semBindUnsized1 {A B} (g : G A) (f : A -> G B) '{H : Unsized _ g}:
   semGen (bindGen g f) <--> \bigcup_(a in semGen g) semGen (f a).
```

Our library additionally provides generic type-class instances for proving automatically that generators are Unsized or SizeMonotonic. For instance Unsized generators are always SizeMonotonic and a bind is Unsized when both its parts are Unsized:

```
Declare Instance unsizedMonotonic {A} (g : G A) '{Unsized _ g} : SizeMonotonic g.

Declare Instance bindUnsized {A B} (g : G A) (f : A -> G B)
   '{Unsized _ g} '{forall x, Unsized (f x)} : Unsized (bindGen g f).
```

There is a similar situation for checkers, for instance the lemma providing a specification to forAll (internally just a bindGen) we used in Sect. 2 is only correct because of the preconditions that both the generator and the body of the forAll are SizeMonotonic.

3.6   Verified Generation of Functions

In QuickChick we emulate (and verify!) the original QuickCheck's approach to generating functions [10,12]. In order to generate a function f of type a->b we use a generator for type b, making sure that subsequent calls to f with the same argument use the same random seed. Upon function generation, QuickCheck captures a random seed within the returned closure. The closure calls a user-provided coarbitrary method that deterministically derives a new seed based on the captured seed and each argument to the function, and then passes this new seed to the generator for type b.

Conceptually, repeatedly splitting a random seed gives rise to an infinite binary tree of random seeds. Mapping arguments of type a to tree paths gives rise to a natural implementation of the coarbitrary method. For random generation to be correct, the set of all possible paths used for generation needs to be prefix-free: if any path is a subpath of another then the values that will be generated will be correlated.

To make our framework easier to use, we decompose verifying completeness of function generation into two parts. On the client side, the user need only provide an injective mapping from the function argument type to positives (binary positive integers) to leverage the guarantees of our framework. On the framework side, we made the split-tree explicit using lists of booleans as paths and proved a completeness theorem:

```
Theorem SplitPathCompleteness (l : list SplitPath) (f : SplitPath -> RandomSeed) :
   PrefixFree l -> exists (s : RandomSeed), forall p, In p l -> varySeed p s = f p.
```

Intuitively, given any finite prefix-free set of paths s and a function f from paths to seeds, there exists a random seed s such that following any path p from s in s's split-tree, we get f p. In addition, our framework provides a mapping from Coq's positives to a prefix-free subset of paths. Combining all of the above with the user-provided injective mapping to positives, the user can get strong correctness guarantees for function generation.

```
Theorem arbFunComplete '{CoArbitrary A, Arbitrary B} (max:A) (f:A-> B) (s:nat) :
   s = Pos.to_nat (coarbitrary max) -> (semGenSize arbitrary s <--> setT) ->
   exists seed, forall a, coarbLe a max -> run arbitrary s seed a = f a.
```

For generators for finite argument function types, the above is a full completeness proof, assuming the result type also admits a complete generator. For functions with infinite argument types we get a weaker notion of completeness: given any finite subset A of a and any function f : a->b, there exists a seed that generates a function f' that agrees with f in A. We have a paper proof (not yet formalized) extending this to a proof of completeness for arbitrary functions using transfinite induction and assuming the set of seeds is uncountable.

3.7   Reasoning About Non-computable Sets

Our framework provides convenient ways of reasoning about the set-of-outcomes semantics from Sect. 3.1. In particular, we favor as much as possible point-free reasoning by relating generator combinators to set operations. To this aim, we designed a small library for reasoning about non-computable sets that could be generally useful. A set A over type T is represented by a function P : T -> Prop such that P x expresses whether x belongs to A. On such sets, we defined and proved properties of (extensional) equality, inclusion, union and intersection, product sets, iterated union and the image of a set through a function. Interestingly enough, we did not need the set complement, which made it possible to avoid classical logic. Finally, in Coq's logic, extensional equality of predicates does not coincide with the primitive notion of equality. So in order to be able to rewrite with identities between sets (critical for point-free reasoning), we could have assumed some extensionality axioms. However, we decided to avoid this and instead used generalized rewriting [22], which extends the usual facilities for rewriting with primitive equality to more general relations.

4   Case Study: Testing Noninterference

We applied our methodology to verify the existing generators used in a complex testing infrastructure for information flow control machines [19]. The machines dynamically enforce noninterference: starting from any pair of indistinguishable states, any two executions result in final states are also indistinguishable. Instead of testing this end-to-end property directly we test a stronger single-step invariant proposed in [19]. Each generated machine state consists of instruction and data memories, a program counter, a stack and a set of registers. The generators we verified produce pairs of indistinguishable states according to a certain indistinguishability definition. The first state is generated arbitrarily and the second is produced by randomly varying the first in order to create an indistinguishable state.

We verified each of these generators with respect to a high-level specification. We proved soundness of the generation strategy, i.e. that any pair generated by the variation generators was indeed indistinguishable, thus state variation generation is sound with respect to indistinguishability. We also proved completeness of the generators with respect to a set of outcomes that is smaller than all possible indistinguishable states, precisely capturing the behavior of our generators. While generating all pairs of indistinguishable states seems good in theory, in practice it is more efficient to bound the size of the generated artifacts. However, the trade-off between completeness and efficiency needs to be considered carefully and our framework allowed us to understand and precisely characterize what additional constraints we enforce in our generation, revealing a number of bugs in the process. One of the trade-offs we had to precisely characterize in our specs is that we only generate instruction sequences of length 2, since we are only going to execute at most one instruction in a single step, but we must also allow the program counter to point to different instructions. This greatly improves efficiency since it is much cheaper to generate a couple than an entire sequence of reasonable instructions.

In some cases, we were not able to prove completeness with respect to the specification we had in mind when writing the generators. These cases revealed bugs in our generation that were not found during extensive prior testing and experiments. Some revealed hidden assumptions we had made while testing progressively more complex machines. For instance, in the simple stack-machine from [19], the label of the saved program counters on the stack was always decreasing. When porting the generators to more complex machines that invalidated this assumption, one should have restructured generation to reflect this. In our attempts to prove completeness this assumption surfaced and we were able to fix the problem, validate our fixes and see a noticeable improvement in bug-finding capabilities (some of the bugs we introduced on purpose in the IFC machine to evaluate testing were found much faster).

Other bugs we found were minor errors in the generation. For instance, when generating an indistinguishable atom from an existing one, most of the time we want to preserve the type of the atom (pointers to pointers, labels to labels, etc.) while varying the payload. This is necessary for example in cases where the atoms will be touched by the same instruction that expects a pointer at a location and finding something else there would raise a non-informative (for IFC) exception. On the other hand we did not *always* want to generate atoms of the same type, because some bugs might only be exposed in those cases. We had forgotten to vary the type in our generation which was revealed and fixed during proving. Fixing all these completeness bugs had little impact on generator efficiency, while giving us better testing.

In this case study we were able to verify existing code that was not written with verification in mind. For verifying ≈2000 lines of Coq code (of which around ≈1000 lines deal strictly with generation and indistinguishability and the other ≈1000 lines are transitively used definitions) our proofs required ≈2000 lines of code. We think this number could be further reduced in the future by taking better advantage of point-free reasoning and the non-computable sets library. With minimal changes to the existing code (e.g. fixing revealed bugs) we were able to structure our proofs in a compositional and modular way. We were able to locate incomplete generators with respect to our specification, reason about the exact sets of values that they can generate, and fix real problems in an already thoroughly tested code base.

5   Related Work

In their seminal work on combining testing and proving in dependent type theory [13,14], Dybjer et al., also introduce the idea of verifying generators and identify surjectivity (completeness) as the most important property to verify. They model generators in Agda/Alfa as functions transforming finite binary trees of naturals to elements of the domain, and prove from first principles the surjectivity of several generators similar in complexity to our red-black tree example generator from Sect. 2. They study a more complex propositional solver using a randomized Prolog-like search [14,18], but apparently only prove this generator correct informally, on paper [18, Appendix of Chap. 4]. The binary trees of naturals are randomly generated outside the system, and roughly correspond both to our seed and size. While Dybjer et al.'s idea of verifying generators is pioneering, we take this further and build a generic verification framework for PBT. By separating seeds and sizes, as already done in QuickCheck [12], we get much more control over the size of the data we can construct. While this makes formal verification a bit more difficult as we have to explicitly reason about sizes in our proofs, we support compositional size reasoning via type classes such as Unsized and SizeMonotonic (Sect. 3.5). Finally, our checkers do not have a fixed shape, but are also built and verified in a modular way.

In a previous attempt at bringing PBT to Coq, Wilson [24] created a simplified QuickCheck like tool for automatically generating test inputs for a small class of testable properties. His goal was to support dependently typed programming in Coq with both proof automation and testing support. In the same work, attempts are made to aid proof automation by disproving false generalizations using testing. However there is no support for writing generations in Coq and therefore also no way of proving interesting facts about generators. In addition, the generation is executed inside Coq which can lead to inefficiency issues without proper care. For example, as they report, a simple multiplication 200 x 200 takes them 0.35 s, while at the same time our framework can generate and test the insert property on around 400 red-black trees (Sect. 2).

A different approach at producing a formalized testing tool was taken in the context of FocalTest [7]. Their verification goal is different; they want to provide a fully verified constraint-based testing tool that automatically generates MC/DC compliant test suites from high-level specifications. They prove a translation from their high level ML-like language to their constraint solving language.

Isabelle provides significant support for testing, in particular via a push-button testing framework [5]. The current goals for QuickChick are different: we do not try to automatically generate test data satisfying complex invariant, but provide ways for the users to construct property-based generators. Both of these approaches have their advantages: the automatic generation of random test data in Isabelle is relatively easy to use for novices, while the approach taken by QuickChick gives the experienced user more control over how the data is generated. In the future, it would make sense to combine these approaches and obtain the best of both worlds.

A work perhaps closer related to ours, but still complementary, is the one by Brucker et al. [4], who in their HOL-TestGen framework also take a more foundational approach to testing methodologies, making certain assumptions explicit. Instead of using adequacy criteria like MC/DC [7], they provide feedback on "what remains to be proved" after testing. This is somewhat similar in concept to the notion of completeness of generators in our framework. However the tool's approach to generation is automatic small-scale exhaustive testing with no support for manual generators. Our experience is that randomized testing with large instances scales much better in practice. A more recent paper on HOL-TestGen [3] presents a complete case study and establishes a formal correspondence between the specifications of the program under test and the properties that will be tested after optimization.

6   Conclusion and Future Work

We introduce a novel methodology for formally verified PBT and implement it as a foundational verification framework for QuickChick, our Coq clone of Haskell QuickCheck. Our verification framework is firmly grounded in a verified implementation of QuickChick itself. This illustrates an interesting interaction between testing and proving in a proof assistant, showing that proving can help testing. This also reinforces the general idea that testing and proving are synergistic activities, and gives us hope that a virtuous cycle between testing and proving can be achieved in a theorem prover.

Future Work. Our framework reduces the effort of proving the correctness of testing code to a reasonable level, so verifying reusable or tricky code should already be an interesting proposition in many cases. The sets of outcomes abstraction also seems well-suited for more aggressive automation in the future (e.g., using an SMT solver).

Maybe more importantly, one should also strive to reduce the cost of effective testing in the first place. For instance, we are working on a property-based generator language in which programs can be interpreted both as boolean predicates and as generators for the same property. Other tools from the literature provide automation for testing [5,6,9,11,15,16], still, with very few exceptions [7], the code of these tools is fully trusted. While for some of these tools full formal verification might be too ambitious at the moment, having these tools produce certificates that can be checked in a foundational framework like ours seems well within reach.

Acknowledgments. We thank John Hughes for insightful discussions and the anonymous reviewers for their helpful comments. This work was supported by NSF award 1421243, *Random Testing for Language Design*.

References

1. Appel, A.W.: Efficient verified red-black trees, Manuscript (2011)
2. Berghofer, S., Bulwahn, L., Haftmann, F.: Turning inductive into equational specifications. In: Berghofer, S., Nipkow, T., Urban, C., Wenzel, M. (eds.) TPHOLs 2009. LNCS, vol. 5674, pp. 131–146. Springer, Heidelberg (2009)
3. Brucker, A.D., Brügger, L., Wolff, B.: Formal firewall conformance testing: an application of test and proof techniques. Softw. Test. Verification Reliab. **25**(1), 34–71 (2015)
4. Brucker, A.D., Wolff, B.: On theorem prover-based testing. Formal Aspects Comput. **25**(5), 683–721 (2013)
5. Bulwahn, L.: The new quickcheck for Isabelle. In: Hawblitzel, C., Miller, D. (eds.) CPP 2012. LNCS, vol. 7679, pp. 92–108. Springer, Heidelberg (2012)
6. Bulwahn, L.: Smart testing of functional programs in Isabelle. In: Bjørner, N., Voronkov, A. (eds.) LPAR-18 2012. LNCS, vol. 7180, pp. 153–167. Springer, Heidelberg (2012)
7. Carlier, M., Dubois, C., Gotlieb, A.: A first step in the design of a formally verified constraint-based testing tool: focaltest. In: Brucker, A.D., Julliand, J. (eds.) TAP 2012. LNCS, vol. 7305, pp. 35–50. Springer, Heidelberg (2012)
8. Chamarthi, H.R., Dillinger, P.C., Kaufmann, M., Manolios, P.: Integrating testing and interactive theorem proving. In: 10th International Workshop on the ACL2 Theorem Prover and its Applications. EPTCS, vol. 70, pp. 4–19 (2011)
9. Christiansen, J., Fischer, S.: EasyCheck — test data for free. In: Garrigue, J., Hermenegildo, M.V. (eds.) FLOPS 2008. LNCS, vol. 4989, pp. 322–336. Springer, Heidelberg (2008)
10. Claessen, K.: Shrinking and showing functions: (functional pearl). In: 5th ACM SIGPLAN Symposium on Haskell, pp. 73–80. ACM (2012)
11. Claessen, K., Duregård, J., Pałka, M.H.: Generating constrained random data with uniform distribution. In: Codish, M., Sumii, E. (eds.) FLOPS 2014. LNCS, vol. 8475, pp. 18–34. Springer, Heidelberg (2014)
12. Claessen, K., Hughes, J.: QuickCheck: a lightweight tool for random testing of Haskell programs. In: 5th ACM SIGPLAN International Conference on Functional Programming (ICFP), pp. 268–279. ACM (2000)

13. Dybjer, P., Haiyan, Q., Takeyama, M.: Combining testing and proving in dependent type theory. In: Basin, D., Wolff, B. (eds.) TPHOLs 2003. LNCS, vol. 2758, pp. 188–203. Springer, Heidelberg (2003)
14. Dybjer, P., Haiyan, Q., Takeyama, M.: Random generators for dependent types. In: Liu, Z., Araki, K. (eds.) ICTAC 2004. LNCS, vol. 3407, pp. 341–355. Springer, Heidelberg (2005)
15. Fetscher, B., Claessen, K., Pałka, M., Hughes, J., Findler, R.B.: Making random judgments: automatically generating well-typed terms from the definition of a type-system. In: Vitek, J. (ed.) ESOP 2015. LNCS, vol. 9032, pp. 383–405. Springer, Heidelberg (2015)
16. Fischer, S., Kuchen, H.: Systematic generation of glass-box test cases for functional logic programs. In: 9th International ACM SIGPLAN Conference on Principles and Practice of Declarative Programming (PPDP), pp. 63–74. ACM (2007)
17. Gonthier, G., Mahboubi, A.: An introduction to small scale reflection in Coq. J. Formalized Reasoning 3(2), 95–152 (2010)
18. Haiyan, Q.: Testing and Proving in Dependent Type Theory. Ph.D. thesis, Chalmers (2003)
19. Hriţcu, C., Hughes, J., Pierce, B.C., Spector-Zabusky, A., Vytiniotis, D., de Amorim, A.A., Lampropoulos, L.: Testing noninterference, quickly. In: 18th ACM SIGPLAN International Conference on Functional Programming (ICFP), pp. 455–468. ACM (2013)
20. Hughes, J.: QuickCheck testing for fun and profit. In: Hanus, M. (ed.) PADL 2007. LNCS, vol. 4354, pp. 1–32. Springer, Heidelberg (2007)
21. Owre, S.: Random testing in PVS. In: Workshop on Automated Formal Methods (2006)
22. Sozeau, M.: A new look at generalized rewriting in type theory. J. Formalized Reasoning 2(1), 41–62 (2009)
23. Tollitte, P.-N., Delahaye, D., Dubois, C.: Producing certified functional code from inductive specifications. In: Hawblitzel, C., Miller, D. (eds.) CPP 2012. LNCS, vol. 7679, pp. 76–91. Springer, Heidelberg (2012)
24. Wilson, S.: Supporting dependently typed functional programming with proof automation and testing. Ph.D. thesis, The University of Edinburgh, June 2011

A Linear First-Order Functional Intermediate Language for Verified Compilers

Sigurd Schneider[✉], Gert Smolka, and Sebastian Hack

Saarland University, Saarbrücken, Germany
sigurd.schneider@cs.uni-saarland.de

Abstract. We present the linear first-order intermediate language IL for verified compilers. IL is a functional language with calls to a nondeterministic environment. We give IL terms a second, imperative semantic interpretation and obtain a register transfer language. For the imperative interpretation we establish a notion of live variables. Based on live variables, we formulate a decidable property called coherence ensuring that the functional and the imperative interpretation of a term coincide. We formulate a register assignment algorithm for IL and prove its correctness. The algorithm translates a functional IL program into an equivalent imperative IL program. Correctness follows from the fact that the algorithm reaches a coherent program after consistently renaming local variables. We prove that the maximal number of live variables in the initial program bounds the number of different variables in the final coherent program. The entire development is formalized in Coq.

1 Introduction

We study the intermediate language IL for verified compilers. IL is a linear functional language with calls to a nondeterministic environment.

We are interested in translating IL to a register transfer language. To this end, we give IL terms a second, imperative interpretation called IL/I. IL/I interprets variable binding as assignment, and function application as *goto*, where parameter passing becomes parallel assignment.

For some IL terms, the functional interpretation coincides with the imperative interpretation. We call such terms *invariant*. We develop an efficiently decidable property we call *coherence* that is sufficient for invariance. To translate IL to IL/I, translating to the coherent subset of IL suffices, i.e. the entire translation can be done in the functional setting.

The notion of a live variable is central to the definition of coherence. Liveness analysis is a standard technique in compiler construction to over-approximate the set of variables the evaluation of a program depends on. Coherence is defined relative to the result of a liveness analysis.

Inspired by the correspondence between SSA [8] and functional programming [2,10], we formulate a register assignment algorithm [9] for IL and show that it realizes the translation to IL/I. For example, the algorithm translates program (a) in Fig. 1 to program (b). Correctness follows from two facts: First,

C. Urban and X. Zhang (Eds.): ITP 2015, LNCS 9236, pp. 344–358, 2015.
DOI: 10.1007/978-3-319-22102-1_23

```
1 let i = 1 in              1 i := 1;
2 fun f (j,p) =             2 fun f (i,n) =
3  let c = p <= m in        3  c := n <= m;
4  if c then                4  if c then
5    let k = p * j in       5    i := n * i;
6    let m = p + 1 in       6    n := n + 1;
7   f (k,m)                 7   f (i,n)
8  else                     8  else
9   j                       9   i
10 in f (i,n)               10 in f (i,n)
```

Fig. 1. Program (a) and (b) computing $F(n,m) := n * (n+1) * \ldots * m$

register assignment consistently renames program (a) such that the variable names correspond to program (b). Second, program (b) is coherent, hence let binding and imperative assignment behave equivalently. Parameter passing in IL/I can be eliminated by inserting parallel assignments [9]. In program (b), all parameters i, n can simply be removed, as they constitute self-assignments.

A key property of SSA-based register assignment is that the number of imperative registers required after register assignment is bounded by the maximal number of simultaneously live variables [9], which allows register assignment to be considered separate from spilling. We show that our algorithm provides the same bound on the number of different variable names in the resulting IL/I term.

1.1   Related Work

Correspondences between imperative and functional languages were investigated already by Landin [11]. The correspondence between SSA and functional programming is due to Appel [2] and Kelsey [10] and consists of a translation from SSA programs to functional programs in continuation passing style (CPS) [1,15]. Chakravarty et al. [6] reformulate SSA-based sparse conditional constant propagation on a functional language in administrative normal form (ANF) [16]. Our intermediate language IL is in ANF, and a sub-language (up to system calls) of the ANF language presented in Chakravarty et al. [6].

Two major compiler verification projects using SSA exist. CompCertSSA [3] integrates SSA-based optimization passes into CompCert [13]. VeLLVM [17,18] is an ongoing effort to verify the production compiler LLVM [12]. Both projects use imperative languages with ϕ-functions to enable SSA, and do not consider a functional intermediate language. As of yet, neither of the projects verifies register assignment in the SSA setting. In the non-SSA setting, a register allocation algorithm, which also deals with spilling, has been formally verified [5].

Beringer et al. [4] use a language with a functional and imperative interpretation for proof carrying code. They give a sufficient condition for the two semantics to coincide which they call Grail normal form (GNF). GNF requires functions to be closure converted, i.e. all variables a function body depends on must be parameters.

Chlipala [7] proves correctness for a compiler from Mini-ML to assembly including mutable references, but without system calls. Register assignment uses an interference graph constructed from liveness information. Chlipala restricts functions to take exactly one argument and requires the program to be closure converted prior to register assignment. This means liveness coincides with free variables and values shared or passed between functions reside in an (argument) tuple in the heap: Effectively, register assignment is function local. Chlipala does not prove bounds on the number of different variables used after register assignment and does not investigate the relationship to α-equivalence.

1.2   Contributions and Outline

– We formally define the functional intermediate language IL and its imperative interpretation, IL/I. We establish the notion of live variables via an inductive definition. We identify terms for which both semantic interpretations coincide via the decidable notion of coherence.
– Inspired by SSA-based register allocation, we formulate a register assignment algorithm for IL and prove that it realizes an equivalence preserving transformation to IL/I. We show the size of the maximal live set bounds the number of names after register assignment.
– All results in this paper have formal Coq proofs, and the development is available online (see Sect. 9). We omit proofs in the paper for space reasons. An extended version is available [**DBLP:journals/corr/SchneiderSH15**].

The paper is structured as follows: We introduce the functional language IL in Sect. 2 and the imperative language IL/I in Sect. 3. Program equivalence is defined in Sect. 4. We define invariance in Sect. 5, establish a notion of live variables in Sect. 6, and present coherence in Sect. 7. Register assignment is treated in Sect. 8.

2   IL

Values, Variables, and Expressions. We assume a set $\mathbb{V}$ of values and a function $\beta : \mathbb{V} \to \{0, 1\}$ that we use to simplify the semantic rule for the conditional. By convention, v ranges over $\mathbb{V}$. We use the countably-infinite alphabet $\mathcal{V}$ for names x, y, z of values, which we call *variables*.

We assume a type *Exp* of expressions. By convention, e ranges over *Exp*. Expressions are pure, their evaluation is deterministic and may fail, hence expression evaluation is a function $\llbracket \cdot \rrbracket : Exp \to (\mathcal{V} \to \mathbb{V}_\perp) \rightharpoonup \mathbb{V}_\perp$. Environments are of type $\mathcal{V} \to \mathbb{V}_\perp$ to track uninitialized variables. We assume a function $\text{fv} : Exp \to set\,\mathcal{V}$ such that for all environments V, V' that agree on $\text{fv}(e)$ we have $\llbracket e \rrbracket\, V = \llbracket e \rrbracket\, V'$. We lift $\llbracket \cdot \rrbracket$ pointwise to lists of expressions in a strict fashion: $\llbracket \bar{e} \rrbracket$ yields a list of values if none of the expressions in $\bar{e}$ failed, and $\perp$ otherwise.

$$\eta ::= e \mid \alpha \qquad\qquad\qquad \text{extended expression}$$

$$Term \ni s, t ::= \text{let } x = \eta \text{ in } s \qquad\qquad \text{variable binding}$$

$$\mid \text{if } e \text{ then } s \text{ else } t \qquad\qquad \text{conditional}$$

$$\mid e \qquad\qquad\qquad\qquad\qquad \text{value}$$

$$\mid \text{fun } f\,\overline{x} = s \text{ in } t \qquad\qquad \text{function definition}$$

$$\mid f\,\overline{e} \qquad\qquad\qquad\qquad\qquad \text{application}$$

Fig. 2. Syntax of IL

Syntax. IL is a functional language with a tail-call restriction and system calls. IL syntactically enforces a first-order discipline by using a separate alphabet $\mathcal{F}$ for names f, g of function type, which we call *labels*. IL uses a third alphabet $\mathcal{A}$ for names α which we call *actions*. The term $\text{let } x = \alpha \text{ in } \dots$ is like a system call α that non-deterministically returns a value. The formal development treats system calls with arguments. Their treatment is straightforward and omitted here for the sake of simplicity.

IL allows function definitions, but does not allow mutually recursive definitions. The syntax of IL is given in Fig. 2.

Semantics. The semantics of IL is given as small-step relation $\longrightarrow$ in Fig. 3. Note that the tail-call restriction ensures that no call stack is required. The reduction relation $\longrightarrow$ operates on **configurations** of the form (F, V, s) where s is the IL term to be evaluated. The semantics does not rely on substitution, but uses an environment $V : \mathcal{V} \to \mathbb{V}_{\perp}$ for variable definitions and a context F for function definitions. Transitions in $\longrightarrow$ are labeled with *events* ϕ. By convention, ψ ranges over events different from τ.

$$\mathcal{E} \ni \phi ::= \tau \mid v = \alpha$$

A **context** is a list of named definitions. A definition in a context may refer to previous definitions and itself. Notationally, we use contexts like functions: If a context F can be decomposed as $F_1; f : a; F_2$ where $f \notin dom\, F_2$, we write Ff for a and F^f for $F_1; f : a$. Otherwise, $Ff = \perp$. To ease presentation of partial functions, we treat $f : \perp$ as if f was not defined, i.e. $f \notin dom\,(f : \perp)$. We write $\emptyset$ for the empty context.

A **closure** is a tuple $(V, \overline{x}, s) \in \mathcal{C}$ consisting of an environment V, a parameter list $\overline{x}$, and a function body s. Since a function f in a context $F; f : \dots; F'$ can refer to function definitions in F (and to itself), the first-order restriction allows the closures to be non-recursive: function closures do not need to close under labels. An application $f\overline{e}$ causes the function context F to rewind to F^f, i.e. up to the definition of f (rule APP). In contrast to higher-order formulations, we do not define closures mutually recursively with the values of the language.

A **system call** $\text{let } x = \alpha \text{ in } s$ invokes a function α of the system, which is not assumed to be deterministic. This reflects in the rule EXTERN, which does not restrict the result value of the system call other than requiring that it is a

$$\text{OP} \quad \dfrac{[\![e]\!]\,V = v}{\begin{array}{l} F \mid V \qquad\qquad \mid \text{let } x = e \text{ in } s \\ \xrightarrow{\tau} F \mid V[x \mapsto v] \mid s \end{array}}$$

$$\text{COND} \quad \dfrac{[\![e]\!]\,V = v \qquad \beta(v) = i}{\begin{array}{l} F \mid V \mid \text{if } e \text{ then } s_0 \text{ else } s_1 \\ \xrightarrow{\tau} F \mid V \mid s_i \end{array}}$$

$$\text{EXTERN} \quad \dfrac{v \in \mathbb{V}}{F \mid V \mid \text{let } x = \alpha \text{ in } s \;\xrightarrow{v=\alpha}\; F \mid V[x \mapsto v] \mid s}$$

$$\text{LET} \quad \dfrac{}{\begin{array}{l} F \qquad\qquad\qquad \mid V \mid \text{fun } f\,\overline{x} = s \text{ in } t \\ \xrightarrow{\tau} F; f : (V, \overline{x}, s) \mid V \mid t \end{array}}$$

$$\text{APP} \quad \dfrac{[\![\overline{e}]\!]\,V = \overline{v} \qquad Ff = (V', \overline{x}, s)}{\begin{array}{l} F \qquad \mid V \qquad\qquad \mid f\,\overline{e} \\ \xrightarrow{\tau} F^f \mid V'[\overline{x} \mapsto \overline{v}] \mid s \end{array}}$$

Fig. 3. Semantics of IL

value. The semantic transition records the system call name α and the result value v in the event $v = \alpha$.

IL is **linear** in the sense that the execution of each term either passes control to a strict subterm, or applies a function that never returns. This ensures no run-time stack is required to manage continuations. WHILE, by contrast, uses sequentialization ; to manage a stack of continuations.

3   Imperative Interpretation of IL: IL/I

We are interested in a translation of IL to an imperative language that does not require function closures at run-time. We introduce a second semantic interpretation for IL which we call IL/I to investigate this translation. IL/I is an imperative language, where variable binding is interpreted as imperative assignment. Function application becomes a *goto*, and parameter passing is a parallel assignment to the parameter names. Closures are replaced by blocks $(\overline{x}, s) \in \mathcal{B}$ and blocks do not contain variable environments. Consequently, a called function can see all previous updates to variables. For example, the following two programs each return 5 in IL/I, but evaluate to 7 in IL:

```
1 let x = 7 in
2 fun f () = x in
3 let x = 5 in f ()
```

```
1 let x = 7 in
2 fun f () = x in
3 fun g x = f() in
4 let y = 5 in g y
```

To obtain the IL/I small-step relation $\longrightarrow_I$, we replace the rules F-LET and F-APP by the following rules:

$$\text{I-LET} \quad \dfrac{}{\begin{array}{l} L \qquad\qquad\qquad \mid V \mid \text{fun } f\,\overline{x} = s \text{ in } t \\ \xrightarrow{\tau}_I L; f : (\overline{x}, s) \mid V \mid t \end{array}}$$

$$\text{I-APP} \quad \dfrac{[\![\overline{e}]\!]\,V = \overline{v} \qquad Lf = (\overline{x}, s)}{\begin{array}{l} L \qquad \mid V \qquad\qquad \mid f\,\overline{e} \\ \xrightarrow{\tau}_I L^f \mid V[\overline{x} \mapsto \overline{v}] \mid s \end{array}}$$

4   Program Equivalence

To relate programs from different languages, we abstract from a configuration's internal behavior and only consider interactions with the environment (via system calls) and termination behavior. IL's reduction relation forms a labeled transition system (LTS) over configurations.

Definition 1. *A reduction system (RS) is a tuple* $(\Sigma, \mathcal{E}, \longrightarrow, \tau, res)$, *s.t.*

(1) $(\Sigma, \mathcal{E}, \longrightarrow)$ *is a LTS*   (3) $res : \Sigma \to \mathbb{V}_\perp$

(2) $\tau \in \mathcal{E}$ (4) $res\ \sigma = v \Rightarrow \sigma \longrightarrow$-terminal

An internally deterministic *reduction system (IDRS) additionally satisfies*

(5) $\sigma \xrightarrow{\phi} \sigma_1 \wedge \sigma \xrightarrow{\phi} \sigma_2 \Rightarrow \sigma_1 = \sigma_2$ action-deterministic

(6) $\sigma \xrightarrow{\phi} \sigma_1 \wedge \sigma \xrightarrow{\tau} \sigma_2 \Rightarrow \phi = \tau$ τ-deterministic

4.1   Partial Traces

We consider two configurations in an IDRS equivalent, if they produce the same partial traces. A partial trace π adheres to the following grammar:

$$\Pi \ni \pi ::= \epsilon \mid v \mid \perp \mid \psi\pi$$

We inductively define the relation $\triangleright \subseteq \Sigma \times \Pi$ such that $\sigma \triangleright \pi$ whenever σ produces the trace π. In the following, we write trace for partial trace.

Tr-Tau
$$\frac{\sigma \xrightarrow{\tau} \sigma' \qquad \sigma' \triangleright \pi}{\sigma \triangleright \pi}$$

Tr-End
$$\frac{}{\sigma \triangleright \epsilon}$$

Tr-Trm
$$\frac{\sigma \longrightarrow\text{-terminal}}{\sigma \triangleright res\ \sigma}$$

Tr-Evt
$$\frac{\sigma \xrightarrow{\psi} \sigma' \qquad \sigma' \triangleright \pi}{\sigma \triangleright \psi, \pi}$$

The traces a configuration produces are given as $\mathcal{P}\sigma = \{\pi \mid \sigma \triangleright \pi\}$.

Definition 2 (Trace Equivalence). $\sigma \simeq \sigma' :\Longleftrightarrow \mathcal{P}\sigma = \mathcal{P}\sigma'$

Lemma 1. σ *silently diverges if and only if* $\mathcal{P}\sigma = \{\epsilon\}$.

4.2   Bisimilarity

We give a sound and complete characterization of trace equivalence via bisimilarity. Bisimilarity enables coinduction as proof method for program equivalence, which is more concise than arguing about traces directly. We say a configuration σ is *ready* if the next step is a system call. We write $\sigma_2 \overset{R}{\rightsquigarrow} \sigma_1$ for $\forall \sigma_1', \sigma_1 \xrightarrow{\phi} \sigma_1' \Rightarrow \exists \sigma_2', \sigma_2 \xrightarrow{\phi} \sigma_2' \wedge \sigma_1' R \sigma_2'$. We write $\sigma \Downarrow w$ (where $w \in \mathbb{V}_\perp$) if $\sigma \longrightarrow^* \sigma'$ such that σ' is $\longrightarrow$-terminal and $res(\sigma') = w$.

Definition 3 (Bisimilarity). *Let* $(S, \mathcal{E}, \longrightarrow, res, \tau)$ *be an IDRS. Bisimilarity* $\sim\ \subseteq S \times S$ *is coinductively defined as the greatest relation closed under the following rules:*

BISIM-SILENT

$$\frac{\sigma_1 \longrightarrow^+ \sigma_1' \qquad \sigma_2 \longrightarrow^+ \sigma_2' \qquad \sigma_1' \sim \sigma_2'}{\sigma_1 \sim \sigma_2}$$

BISIM-TERM

$$\frac{\sigma_1 \Downarrow w \qquad \sigma_2 \Downarrow w}{\sigma_1 \sim \sigma_2}$$

BISIM-EXTERN

$$\frac{\sigma_1 \longrightarrow^* \sigma_1' \qquad \sigma_2 \longrightarrow^* \sigma_2' \qquad \sigma_1', \sigma_2' \text{ ready} \qquad \sigma_1' \overset{\sim}{\rightsquigarrow} \sigma_2' \qquad \sigma_2' \overset{\sim}{\rightsquigarrow} \sigma_1'}{\sigma_1 \sim \sigma_2}$$

BISIM-SILENT allows to match finitely many steps on both sides, as long as all transitions are silent. This makes sense for IDRS, but would not yield a meaningful definition otherwise. BISIM-SILENT ensures that every external transition of σ_1' is matched by the same external transition of σ_2', and vice versa. This ensures that if two programs are in relation, they react to every possible result value of the external call in a bisimilar way. The premises that σ_1', σ_2' are ready is there to simplify case distinctions by ensuring that the next event cannot be τ.

Theorem 1 (Soundness and Completeness). *Let* $(S, \mathcal{E}, \longrightarrow, res, \tau)$ *be an IDRS and* $\sigma, \sigma' \in S$. *Then:* $\sigma \sim \sigma' \iff \sigma \simeq \sigma'$

The semantics of IL and of IL/I each forms an IDRS. We define *res* such that $res(\sigma) = v$ if σ is of the form (F, V, e) and $\llbracket e \rrbracket\, V = v$. Otherwise, $res(\sigma) = \bot$. The definitions for IL/I are analogous. To relate configurations IL to IL/I, we form a reduction system on the sum $\Sigma_F + \Sigma_I$ of the configurations and lift $\longrightarrow$ and *res* accordingly. It is easy to see that the resulting reduction system is internally deterministic. If not clear from context, we use an index σ_F, σ_I to indicate which language a configuration belongs to.

5   Invariance

We call a term *invariant* if it has the same traces in both the functional and the imperative interpretation.

Definition 4 (Invariance). *A closed program* s *is invariant if*

$$\forall V, \ (\emptyset, V, s)_F \simeq (\emptyset, V, s)_I$$

Invariance is undecidable. We develop a syntactic, efficiently decidable criterion sufficient for invariance, which we call coherence. Coherence simplifies the translation between IL and IL/I.

Coherence is based on the observation that some IL programs do not really depend on information from the closure. Assume $Ff = (V', \overline{x}, s)$ and consider the following IL reduction according to rule APP:

$$(F, V, f\,\overline{e}) \quad \longrightarrow \quad (F^f, V'[\overline{x} \mapsto \overline{v}], s)$$

If V agrees with V' on all variables X that s depends on, then the configuration could have equivalently reduced to $(F^f, V[\overline{x} \mapsto \overline{v}], s)$. This reduction does not require the closure V' and is similar in spirit to the rule I-APP. Coherence is a syntactic criterion that ensures V and V' agree on a suitable set X at every function application. We proceed in two steps:

1. Section 6 introduces the notion of live variables, which identifies a set that contains all variables a program depends on.
2. Section 7 gives the inductive definition of coherence and shows that coherent programs are invariant.

6   Liveness

A variable x is *significant* to a program s and a context L, if there is an environment V and a value v such that $(L, V, s)_I \not\simeq (L, V[x \mapsto v], s)_I$. Significance is not decidable, as it is a non-trivial semantic property.

Liveness analysis is a standard technique in compiler construction to over-approximate the set of variables significant to the evaluation of an imperative program. While usual characterizations of live variables rely on data-flow equations [14], we define liveness inductively on the structure of IL's syntax. To the best of our knowledge, such an inductive definition is not in literature. The inductive definition factorizes the correctness aspect from the algorithmic aspect of liveness analysis.

We embed liveness information in the syntax of IL by introducing annotations for function definitions: The term $\mathsf{fun}\, f\,\overline{x} \,:\, X \,=\, s \,\mathsf{in}\, t$ is annotated with a set of variables X.

6.1   Inductive Definition of the Liveness Judgment

We define inductively the judgment **live**, which characterizes sound results of a liveness analysis.

$$\Lambda \vdash \mathbf{live}\, s : X \quad \text{where} \quad \begin{array}{rl} \Lambda : & context\,(set\,\mathcal{V}) \quad \text{liveness for functions} \\ X : & set\,\mathcal{V} \quad\quad\quad\;\; \text{live variables} \\ s : & Exp \quad\quad\quad\;\;\; \text{expression} \end{array}$$

The predicate $\Lambda \vdash \mathbf{live}\, s : X$ can be read as X *contains all variables significant to s in any context satisfying the assumptions Λ*. The context Λ records for every function f a set of variables X that we call the **globals** of f. Assuming $\overline{x}$ are the parameters of f, we will arrange things such that the set $X \cup \overline{x}$ contains all variables significant for the body of f, but never a parameter of f: $X \cap \overline{x} = \emptyset$. Throughout the paper, Λ is always a (partial) mapping from labels to globals, and X denotes a set of variables.

LIVE-OP
$$\frac{\mathrm{fv}(\eta) \subseteq X \qquad\qquad x \in X'}{X' \setminus \{x\} \subseteq X \qquad \Lambda \vdash \mathbf{live}\, s : X'}$$
$$\Lambda \vdash \mathbf{live}\, \mathbf{let}\, x = \eta\, \mathbf{in}\, s : X$$

LIVE-EXP
$$\frac{\mathrm{fv}(e) \subseteq X}{\Lambda \vdash \mathbf{live}\, e : X}$$

LIVE-APP
$$\frac{X_1 \subseteq X \qquad \mathrm{fv}(\bar{e}) \subseteq X}{\Lambda; f : X_1; \Lambda' \vdash \mathbf{live}\, f\, \bar{e} : X}$$

LIVE-COND
$$\frac{\mathrm{fv}(e) \subseteq X \qquad \Lambda \vdash \mathbf{live}\, s_1 : X_1}{X_1 \cup X_2 \subseteq X \qquad \Lambda \vdash \mathbf{live}\, s_2 : X_2}$$
$$\Lambda \vdash \mathbf{live}\, \mathbf{if}\, e\, \mathbf{then}\, s_1\, \mathbf{else}\, s_2 : X$$

LIVE-FUN
$$\frac{\Lambda; f : X_1 \vdash \mathbf{live}\, s_1 : X_1 \cup \bar{x} \qquad X_1 \cap \bar{x} = \emptyset}{\Lambda; f : X_1 \vdash \mathbf{live}\, s_2 : X_2 \qquad X_2 \subseteq X}$$
$$\Lambda \vdash \mathbf{live}\, \mathbf{fun}\, f\, \bar{x} : X_1 = s_1\, \mathbf{in}\, s_2 : X$$

Fig. 4. Liveness: An approximation of the significant variables for IL/I

Description of the Rules. LIVE-OP ensures that all variables free in η are live. Every live variable of the continuation s except x must be live at the assignment. We require x to be live in the continuation. LIVE-COND ensures that the live variables of a conditional at least contain the free variables of the condition, and the variables live in the consequence and alternative. LIVE-EXP ensures that for programs consisting of a single expression e at least the free variables of e are live. LIVE-APP ensures that the free variables of every argument are live, and that the globals X_1 of f are live at the call site. LIVE-FUN records the annotation X_1 as globals for f in Λ, ensures that $X_1 \cup \bar{x}$ is a large enough live set for the function body, and that X_1 does not contain parameters of f. The live variables X_2 of the continuation t must be live at the function definition (Fig. 4).

Theorem 2 (Liveness is Decidable). *For all Λ, X and annotated s, it is efficiently decidable whether $\Lambda \vdash \mathbf{live}\, s : X$ holds.*

The proof of Theorem 2 is constructive and yields an efficient, extractable decision procedure. The decision procedure recursively descends on the program structure, checking the conditions of the appropriate rule in every step.

6.2 Liveness Approximates Significance

We show that the live variables approximate the significant variables. We write $L \models \Lambda$ if a context L satisfies the assumptions Λ, and define:

LIVECTX1
$$\frac{L \models \Lambda \qquad X \cap \bar{x} = \emptyset \qquad \Lambda; f : X \vdash \mathbf{live}\, s : X \cup \bar{x}}{L; f : (\bar{x}, s) \models \Lambda; f : X}$$

LIVECTX2
$$\emptyset \models \emptyset$$

LiveCtx1 ensures that X does not contain parameters and that $X \cup \bar{x}$ is a large enough live set for the function body s under the context $\Lambda; f : X$.

We can now formally state the soundness of the live predicate. We prove that if $\Lambda \vdash \mathbf{live}\, s : X$, then X contains at least the significant variables of s in every context L that satisfies the assumptions Λ. We write $V =_X V'$ if V and V' agree on X, that is if $\forall x \in X, Vx = V'x$.

Theorem 3. *For every program* s, *if* $\Lambda \vdash$ ***live*** $s : X$ *and* $L \models \Lambda$ *and* $V =_X V'$, *then* $(L, V, s)_I \simeq (L, V', s)_I$.

7   Coherence

Coherence is a syntactic condition that ensures that a program is invariant. Coherence is defined relative to liveness information $\Lambda \vdash$ **live** $s : X$.

In the following programs, the set of globals of f is $\{x\}$. The program on the left is not invariant, while the program on the right is coherent.

```
1 let x = 7 in
2 fun f () : {x} = x in
3 let x = 5 in f ()
```

```
1 let x = 7 in
2 fun f () : {x} = x in
3 let y = 5 in f ()
```

In the program on the left in line 3, the value of x is 5 and disagrees with the value of x in the closure of f. In the program on the right, x was not redefined, hence both IL and IL/I will compute 7. We say a function f is *available* as long as none of f's globals were redefined. The inductive definition of coherence ensures only available functions are applied.

7.1   Inductive Predicate

The coherence judgment is of the form $\boxed{\Lambda \; \vdash \mathbf{coh}\, s}$, where s is an annotated program and Λ is similar to the context in the liveness judgment. We exploit that contexts realize a partial mapping, and maintain the invariant that Λ maps only *available* functions to their globals, and all other functions to $\bot$. The inductive definition given below ensures that only available functions are applied.

$$
\text{Coh-Op} \quad \frac{\lfloor \Lambda \rfloor_{\mathcal{V} \setminus \{x\}} \vdash \mathbf{coh}\, s}{\Lambda \vdash \mathbf{coh}\, \mathbf{let}\, x = \eta \,\mathbf{in}\, s} \qquad \text{Coh-Exp} \; \frac{}{\Lambda \vdash \mathbf{coh}\, e} \qquad \text{Coh-App} \; \frac{\Lambda f \neq \bot}{\Lambda \vdash \mathbf{coh}\, f\, \overline{y}}
$$

$$
\text{Coh-Cond} \quad \frac{\Lambda \vdash \mathbf{coh}\, s \qquad \Lambda \vdash \mathbf{coh}\, t}{\Lambda \vdash \mathbf{coh}\, \mathbf{if}\, x \,\mathbf{then}\, s \,\mathbf{else}\, t}
$$

$$
\text{Coh-Fun} \quad \frac{\Lambda; f : X \vdash \mathbf{coh}\, t \qquad \lfloor \Lambda; f : X \rfloor_x \vdash \mathbf{coh}\, s}{\Lambda \vdash \mathbf{coh}\, \mathbf{fun}\, f\, \overline{x}\, :\, X\, =\, s \,\mathbf{in}\, t}
$$

Description of the Rules. Coh-Op deals with binding a variable x. Every function that has x as a global (i.e. $x \in \Lambda f$) becomes unavailable, and must be removed from Λ. We write $\lfloor \Lambda \rfloor_X$ to remove all definitions from Λ that require more globals than X. Trivially, $\lfloor \Lambda \rfloor_{\mathcal{V}} = \Lambda$. To remove all definitions from Λ that use x as global, we use $\lfloor \Lambda \rfloor_{\mathcal{V} \setminus \{x\}}$.

Formally, the definition of $\lfloor \Lambda \rfloor_X$ exploits the list structure of contexts:

$$
\lfloor \emptyset \rfloor_X = \emptyset \qquad\qquad \lfloor \Lambda; f : X' \rfloor_X = \lfloor \Lambda \rfloor_X; f : X' \quad X' \subseteq X
$$

$$
\lfloor \Lambda; f : \bot \rfloor_X = \lfloor \Lambda \rfloor_X; f : \bot \qquad \lfloor \Lambda; f : X' \rfloor_X = \lfloor \Lambda \rfloor_X; f : \bot \quad X' \nsubseteq X
$$

COH-APP ensures only available functions can be applied, since Λ maps functions that are not available to $\bot$. COH-FUN deals with function definitions. When the definition of a function f is encountered, its globals X according to the annotation are recorded in Λ. In the function body s, only functions that require at most X as globals are available, so the context is restricted to $\lfloor\Lambda; f : X\rfloor_X$.

Theorem 4 (Coherence is Decidable). *For all Λ and annotated s, it is efficiently decidable whether $\Lambda \vdash \text{\textbf{coh}}\, s$ holds.*

7.2 Coherent Programs are Invariant

Given a configuration (F, V, t) such that $Ff = (V', \overline{x}, s)$, the **agreement invariant** describes a correspondence between the values of variables in the function closure V' and the environment V. If the closure of f is available, the closure environment V' agrees with the primary environment V on f's globals X: $V' =_X V$. We write $F, V \models \Lambda$ if $\forall f \in \text{dom}\, F \cap \text{dom}\, \Lambda$, $V' =_X V$ (where $\Lambda f = X$ and $Ff = (V', \overline{x}, s)$).

Function application continues evaluation with the function body from the closure. Assume $Ff = (V', \overline{x}, s)$ and consider the IL reduction:

$$(F, V, f\,\overline{e}) \longrightarrow (F^f, V'[\overline{x} \mapsto \overline{v}]a, s)$$

If coherence is to be preserved, s must be coherent under suitable assumptions. We say Λ approximates Λ' if whenever Λf is defined, it agrees with Λ' and define $\Lambda \preceq \Lambda' :\iff \forall f \in \text{dom}\, \Lambda,\ \Lambda f = \Lambda' f$. The **context coherence** predicate $\Lambda \vdash \text{\textbf{coh}}\, F$ ensures that all function bodies in closures are coherent. It is defined inductively on the context:

$$
\begin{array}{ccc}
& & \text{CoHC-Con} \\
& & \Lambda' \vdash \text{\textbf{live}}\, s : X \cup \overline{x} \quad \Lambda; f : X \preceq \Lambda' \\
\text{CoHC-Emp} & \text{CoHC-Bot} & \lfloor \Lambda; f : X \rfloor_X \vdash \text{\textbf{coh}}\, s \quad \Lambda \vdash \text{\textbf{coh}}\, F \\
\hline
\emptyset \vdash \text{\textbf{coh}}\, \emptyset & \dfrac{\Lambda \vdash \text{\textbf{coh}}\, F}{\Lambda; f : \bot \vdash \text{\textbf{coh}}\, F; f : b} & \Lambda, f : X \vdash \text{\textbf{coh}}\, F; f : (V, \overline{x}, s)
\end{array}
$$

COHC-CON encodes two requirements: First, the body of f must be coherent under the context restricted to the globals X of f (cf. COH-FUN). Second, $X \cup \overline{x}$ must suffice as live variables for the function body s under some assumptions Λ' such that $\Lambda; f : X$ approximates Λ'. Approximation ensures stability under restriction: $\Lambda \vdash \text{\textbf{coh}}\, F \Rightarrow \lfloor \Lambda \rfloor_X \vdash \text{\textbf{coh}}\, F$.

We define $strip(V, \overline{x}, s) = (\overline{x}, s)$ and lift $strip$ pointwise to contexts.

Theorem 5 (Coherence Implies Invariance). *Let $\Lambda \vdash \text{\textbf{coh}}\, s$ and $\Lambda \vdash \text{\textbf{coh}}\, F$ and $\Lambda' \vdash \text{\textbf{live}}\, s : X$ such that $\Lambda \preceq \Lambda'$. Then for all $V =_X V'$ such that $F, V \models \Lambda$, it holds $(F, V, s)_F \simeq (strip\, F, V', s)_I$.*

Theorem 5 reduces the problem of translating between IL/I and IL to the problem of establishing coherence. For the translation from IL to IL/I, it suffices to establish coherence while preserving IL semantics. Since SSA and functional programming correspond [2,10], the translation from IL/I to IL can be seen as SSA construction [8], and the translation from IL to IL/I, which we treat in the next section, as SSA destruction.

8   Translating from IL/F to IL/I via Coherence

The simplest method to establish coherence while preserving IL semantics is α-renaming the program apart. A renamed-apart program (for formal definition see [**DBLP:journals/corr/SchneiderSH15**]) is coherent, since every function is always available. The properties of α-conversion ensure semantic equivalence.

We present an algorithm that establishes coherence and uses no more different names than the maximal number of simultaneously live variables in the program. This algorithm corresponds to the assignment phase of SSA-based register allocation [9]. The algorithm requires a renamed-apart program as input to ensure that every consistent renaming can be expressed as a function from $\mathcal{V} \to \mathcal{V}$. We proceed in two steps:

1. We define the notion of *local injectivity* for a function $\rho : \mathcal{V} \to \mathcal{V}$. We show that renaming with a locally injective ρ yields an α-equivalent and coherent program $\rho\, s$.
2. We give an algorithm *rassign* and show that it constructs a locally injective ρ that uses the minimal number of different names.

We introduce **more liveness annotations** before every term in the syntax, i.e. wherever a term s appeared before, now a term $\langle X \rangle s$ appears that annotates s with the set X. From now on, s, t range over such annotated terms. We define the projection $[\langle X \rangle s] = X$. The annotation corresponds directly to the live set parameter X of the relation $\Lambda \vdash \mathbf{live}\, s : X$, hence it suffices to write $\Lambda \vdash \mathbf{live}\, s$ for annotated programs.

8.1   Local Injectivity

We define inductively a judgment $\rho \vdash \boldsymbol{inj}\, \boldsymbol{s}$ where $\rho : \mathcal{V} \to \mathcal{V}$ and s is an annotated program. We use the following notation for injectivity on X:

$$f \rightarrowtail X \; :\Longleftrightarrow \; \forall x\, y \in X,\; f\, x = f\, y \Longrightarrow x = y$$

The rules defining the judgement are given below and require ρ to be injective on every live set X annotating any subterm:

$$
\begin{array}{ccc}
\text{INJ-OP} & \text{INJ-VAL} & \text{INJ-APP} \\
\dfrac{\rho \rightarrowtail X \qquad \rho \vdash \boldsymbol{inj}\, s}{\rho \vdash \boldsymbol{inj}\, \langle X \rangle \operatorname{let} x = \eta \operatorname{in} s} & \dfrac{\rho \rightarrowtail X}{\rho \vdash \boldsymbol{inj}\, \langle X \rangle e} & \dfrac{\rho \rightarrowtail X}{\rho \vdash \boldsymbol{inj}\, \langle X \rangle f\, \overline{y}}
\end{array}
$$

$$
\begin{array}{cc}
\text{INJ-COND} & \text{INJ-FUN} \\
\dfrac{\rho \rightarrowtail X \qquad \rho \vdash \boldsymbol{inj}\, s \qquad \rho \vdash \boldsymbol{inj}\, t}{\rho \vdash \boldsymbol{inj}\, \langle X \rangle \operatorname{if} x \operatorname{then} s \operatorname{else} t} & \dfrac{\rho \rightarrowtail X \qquad \rho \vdash \boldsymbol{inj}\, s \qquad \rho \vdash \boldsymbol{inj}\, t}{\rho \vdash \boldsymbol{inj}\, \langle X \rangle \operatorname{fun} f\, \overline{x} : X_1 = s \operatorname{in} t}
\end{array}
$$

Let $\mathcal{V}_B(s)$ be the set of variables that occur in a binding position in s, and $\mathrm{fv}(s)$ be the set of free variables of s. For our theorems, several properties are required:

(1) The program must be without unreachable code, i.e. in every subterm $\mathbf{fun}\, f\, \overline{x} = s\, \mathbf{in}\, t$ it must be the case that f is applied in t.

(2) A variable in $\mathcal{V}_B(s)$ must not occur in a set of globals in Λ. We define $\Lambda \subseteq U :\Longleftrightarrow \forall f \in dom\, \Lambda,\ \Lambda f \subseteq U$.

(3) A variable in $\mathcal{V}_B(s)$ must not occur in the annotation $[s]$. We write $s \subseteq U$ if for every subterm t of s it holds that every $x \in [t]$ is either in U or bound at t in s.

For renamed-apart programs, these conditions ensure that the live set X in INJ-FUN always contains the globals X_1 of f (cf. LIVE-APP).

Theorem 6. *Let s be a renamed-apart program without unreachable code such that $\Lambda \vdash$ **live** s, $\Lambda \subseteq fv(s)$ and $s \subseteq fv(s)$. Then*

$$\rho \vdash \mathbf{inj}\, s \implies \rho\left(\lfloor \Lambda \rfloor_{[s]}\right) \vdash \mathbf{coh}\,(\rho\, s)$$

Theorem 6 states that the renamed program $\rho\, s$ is coherent under the assumptions $\rho\left(\lfloor \Lambda \rfloor_{[s]}\right)$, i.e. the point-wise image of $\lfloor \Lambda \rfloor_{[s]}$ under ρ.

Renaming with a locally injective renaming produces an α-equivalent program (for formal definition see [**DBLP:journals/corr/SchneiderSH15**]), and hence preserves program equivalence:

Theorem 7. *Let s be a renamed-apart program without unreachable code such that $\Lambda \vdash$ **live** s, $\Lambda \subseteq fv(s)$ and $s \subseteq fv(s)$. Let $\rho, d : \mathcal{V} \to \mathcal{V}$ such that ρ is the inverse of d on $fv(s)$. Then $\rho \vdash \mathbf{inj}\, s \implies \rho, d \vdash \rho\, s \sim_\alpha s$*

8.2 A Simple Register Assignment Algorithm

The algorithm rassign is parametrized by a function *fresh* $: set\, \mathcal{V} \to \mathcal{V}$ of which we require *fresh* $X \notin X$ for all finite sets of variables X. Based on *fresh*, we define a function *freshlist* $X\, n$ that yields a list of n pairwise-distinct variables such that $(freshlist\, X\, n) \cap X = \emptyset$. The SSA algorithm must process the program in an order compatible with the dominance order to work [9]. In our case it suffices to simply recurse on s as follows:

$$
\begin{aligned}
&rassign\, \rho\,(\langle X \rangle\, \mathbf{let}\, x = \eta\, \mathbf{in}\, s) &&= rassign\, (\rho[x \mapsto y])\, s \\
&\quad \text{where } y = fresh\, (\rho([s] \setminus \{x\})) \\
&rassign\, \rho\,(\langle X \rangle\, \mathbf{if}\, e\, \mathbf{then}\, s\, \mathbf{else}\, t) &&= rassign\, (rassign\, \rho\, s)\, t \\
&rassign\, \rho\,(\langle X \rangle\, e) &&= \rho \\
&rassign\, \rho\,(\langle X \rangle\, f\, \overline{e}) &&= \rho \\
&rassign\, \rho\,(\langle X \rangle\, \mathbf{fun}\, f\, \overline{x} : X' = s\, \mathbf{in}\, t) &&= rassign\, (rassign\, (\rho[\overline{x} \mapsto \overline{y}])\, s)\, t \\
&\quad \text{where } \overline{y} = freshlist\, (\rho([s] \setminus \overline{x}))\, |\overline{x}|
\end{aligned}
$$

We prove in Theorem 8 that the algorithm is correct for any choice of *fresh* and *freshlist*, as long as they satisfy the specifications above.

Theorem 8. *Let s be renamed-apart such that $\Lambda \vdash$ **live** s, $\Lambda \subseteq fv(s)$ and $s \subseteq fv(s)$. Let ρ be injective on $[s]$. Then: rassign $\rho\,s \vdash$ **inj** s.*

Our implementation of *fresh* implements the heuristic of simply choosing the smallest unused variable. Theorem 9 shows that for this choice of *fresh*, the largest live set determines the number of required names. We use $\mathcal{S}(k)$ to denote the set of the k smallest variables, and $\mathcal{V}_O(s)$ to denote the set of variables occurring (free or in a binding position) in s.

Theorem 9. *Assume fresh X yields a variable less or equal to $|X|$. Let s be renamed-apart such that $\Lambda \vdash$ **live** s, $\Lambda \subseteq fv(s)$ and $s \subseteq fv(s)$. Let k be the size of the largest set of live variables in s, and rassign $\rho\,s = \rho'$. If $\rho(fv(s)) \subseteq \mathcal{S}(n)$ then $\rho'(\mathcal{V}_O(s)) \subseteq \mathcal{S}(max\{n,k\})$.*

We prove a slightly generalized version of Theorem 9 by induction on s.

9 Formal Coq Development

Each theorem and lemma in this paper is proven as part of a larger Coq development, which is available online[1]. The development extracts to a simple compiler that, for instance, produces program (b) when given program (a) from the introduction as input.

The formalization uses De-Bruijn representation for labels, and named representation for variables. Notable differences to the paper presentation concern the treatment of annotations, the technical realization of the definition of liveness, and the inductive generalizations of Theorems 6–9.

10 Conclusion

We presented the functional intermediate language IL and developed the notion of coherence, which provides for a canonical and verified translation between functional and imperative programs. We formulate a register assignment algorithm by recursion on the structure of IL that achieves the same bound on the number of required registers as SSA-based register assignment. Coherence allowed us to justify correctness without directly arguing about program semantics by proving that the algorithm α-renames to a coherent program.

Acknowledgments. This research has been supported in part by a Google European Doctoral Fellowship granted to the first author.

---

[1] http://www.ps.uni-saarland.de/~sdschn/publications/lvc15.

References

1. Appel, A.W.: Compiling with Continuations. Cambridge University Press, Cambridge (1992)
2. Appel, A.W.: SSA is functional programming. In: SIGPLAN Notices, vol. 33, no. 4 (1998)
3. Barthe, G., Demange, D., Pichardie, D.: A formally verified SSA-based middleend. In: Seidl, H. (ed.) Programming Languages and Systems. LNCS, vol. 7211, pp. 47–66. Springer, Heidelberg (2012)
4. Beringer, L., MacKenzie, K., Stark, I.: Grail: a functional form for imperative mobile code. In: ENTCS, vol. 85, no. 1 (2003)
5. Blazy, S., Robillard, B., Appel, A.W.: Formal verification of coalescing graph-coloring register allocation. In: Gordon, A.D. (ed.) ESOP 2010. LNCS, vol. 6012, pp. 145–164. Springer, Heidelberg (2010)
6. Chakravarty, M.M.T., Keller, G., Zadarnowski, P.: A functional perspective on SSA optimisation algorithms. In: ENTCS, vol. 82, no. 2 (2003)
7. Chlipala, A.: A verified compiler for an impure functional language. In: POPL (2010)
8. Cytron, R., Ferrante, J., Rosen, B.K., Wegman, M.N., Zadeck, F.K.: Efficiently computing static single assignment form and the control dependence graph. In: TOPLAS, vol. 13, no. 4 (1991)
9. Hack, S., Grund, D., Goos, G.: Register allocation for programs in SSA-Form. In: CC (2006)
10. Kelsey, R.A.: A correspondence between continuation passing style and static single assignment form. In: SIGPLAN Notices, vol. 30, no. 3 (1995)
11. Landin, P.J.: Correspondence between ALGOL 60 and Church's Lambda-notation: part I. In: CACM, vol. 8, no. 2 (1965)
12. Lattner, C., Adve, V.S.: LLVM: a Compilation framework for lifelong program analysis and transformation. In: CGO (2004)
13. Leroy, X.: Formal verification of a realistic compiler. In: CACM, vol. 52, no. 7 (2009)
14. Nipkow, T., Klein, G.: Concrete Semantics: With Isabelle/HOL. Springer Publishing Company, Incorporated, Switzerland (2014)
15. Reynolds, J.C.: The discoveries of continuations. LSC $6(3-4)$, 23–247 (1993)
16. Sabry, A., Felleisen, M.: Reasoning about programs in continuation-passing style. In: LSC, vol. 6, no. (3-4) (1993)
17. Zhao, J., Nagarakatte, S., Martin, M.M.K., Zdancewic, S.: Formal verification of SSA-based Optimizations for LLVM. In: PLDI (2013)
18. Zhao, J., Nagarakatte, S., Martin, M.M.K., Zdancewic, S.: Formalizing LLVM intermediate representation for verified program transformations. In: POPL (2012)
19. Schneider, S., Smolka, G., Hack, S.: A first-order functional intermediate language for verified compilers. CoRR (2015). 1503.08665

Autosubst: Reasoning with de Bruijn Terms and Parallel Substitutions

Steven Schäfer[✉], Tobias Tebbi, and Gert Smolka

Saarland University, Saarbrücken, Germany
{schaefer,ttebbi,smolka}@ps.uni-saarland.de

Abstract. Reasoning about syntax with binders plays an essential role in the formalization of the metatheory of programming languages. While the intricacies of binders can be ignored in paper proofs, formalizations involving binders tend to be heavyweight. We present a discipline for syntax with binders based on de Bruijn terms and parallel substitutions, with a decision procedure covering all assumption-free equational substitution lemmas. The approach is implemented in the Coq library AUTOSUBST, which additionally derives substitution operations and proofs of substitution lemmas for custom term types. We demonstrate the effectiveness of the approach with several case studies, including part A of the POPLmark challenge.

1 Introduction

Proofs in the metatheory of programming languages and type systems are the kind of proofs that mathematicians hate for good reason: they are long, contain few essential insights, and have a lot of tedious but error-prone cases. Thus they appear to be an ideal target for computer-verification. Unfortunately, such efforts have often been hampered by the formalization of binders. On paper, it is consensus that issues of α-equivalence can be neglected with a clean conscience. For a machine-verifiable formalization, this is not an option.

We propose a style of reasoning about binders based on parallel de Bruijn substitutions [6]. A (parallel) substitution is a function mapping all variables to terms. The instantiation of a term s under a substitution σ implements capture-avoiding substitution, replacing all free variables simultaneously.

Parallel substitutions by themselves are already a useful tool. De Bruijn [6] used parallel substitutions to produce a simplified proof of the Church-Rosser theorem. Statements about typing judgements, such as weakening and substitutivity, can be generalized to manipulate the whole context at the same time [2,9]. When working with logical relations [8], we need parallel substitutions in the statement of the fundamental theorem. In all of these examples it is beneficial to quantify over parallel substitutions.

Parallel de Bruijn substitutions can be described in the σ-calculus of Abadi et al. [1], which is a first-order equational theory capturing the interaction of terms and substitutions. The σ-calculus is formulated as a confluent and terminating rewriting system. In addition, we recently showed that computing normal

© Springer International Publishing Switzerland 2015
C. Urban and X. Zhang (Eds.): ITP 2015, LNCS 9236, pp. 359–374, 2015.
DOI: 10.1007/978-3-319-22102-1_24

forms with the σ-calculus yields a decision procedure for equational substitution lemmas [14].

While none of the ingredients we use are novel, their combination yields a powerful and practical approach to syntactic theories. In particular, the application of the σ-calculus to prove substitution lemmas is novel.

There are a number of routine proofs required to adapt our approach to a given term language. We have implemented a Coq library AUTOSUBST [15] to automate this step. AUTOSUBST can automatically generate the substitution operations for a custom inductive type of terms and prove the corresponding substitution lemmas. AUTOSUBST offers tactics that implement the aforementioned normalization and decision procedure. AUTOSUBST also supports heterogeneous substitutions between multiple syntactic sorts, like terms with type binders in System F.

The paper is structured as follows. In Sect. 2, we present the approach behind AUTOSUBST using the untyped λ-calculus as a running example. We explain our handling of heterogeneous substitutions in Sect. 3. In Sect. 4, we report on our experiences using AUTOSUBST in practical formalizations.

1.1   Evaluation

We evaluate our approach, as well as AUTOSUBST, on a number of case studies.

- *Type Preservation of CC_ω*, a Martin-Löf style dependent type theory [11] with a predicative hierarchy of universes. This is a technically challenging development, which needs a number of auxiliary results (e.g., confluence of reduction). In total, the development takes 214 lines of specification and 233 lines of proof.
- *Normalization of System F*. The proof uses logical relations, which require some form of parallel substitutions in the fundamental theorem. We have mechanized both a proof for a call-by-value variant of System F, as well as a proof of strong normalization following Girard [8]. For the call-by-value case we require 99 lines of specification and 65 lines of proof. The strong normalization proof amounts to 153 lines of specification and 107 lines of proof.
- *Progress and Type Preservation of System F with Subtyping*. This is part A of the POPLmark challenge [4]. The whole development consists of 206 lines of specification and 240 lines of proof. This is less than half the length of existing solutions in Coq. The development has been kept fairly close to the paper proofs for easier comparison to other implementations. If we deviate from the paper proofs and omit well-formedness assumptions, we can shorten the proofs to 185 lines of specification and 157 lines of proof.

We report on these case studies in Sect. 4.

All case studies, as well as the source code of AUTOSUBST can be found at https://www.ps.uni-saarland.de/autosubst.

1.2   Related Work

There are a number of libraries and tools supporting proofs about syntax. We mention the ones that are available for Coq. What sets AUTOSUBST apart from the related work is our usage of parallel substitutions, which subsume single variable substitutions. This allows us to offer an automation tactic for solving substitution lemmas and simplifying terms involving substitutions based on a clearly defined equational theory. Additionally, AUTOSUBST is completely implemented within Coq and does not require external tools.

– CFGV [3] uses a generic type of context free grammars with variable binding. Custom term types are obtained by instantiating the generic construction. The library provides a number of general lemmas to work with CFGVs. The representation is named and explicitly deals with issues of α-equivalence.
– DBGen [12] is an external tool that generates de Bruijn substitution operations and proofs of corresponding lemmas. It includes support for syntax defined by mutual recursion. It generates a single-variable substitution operation.
– DBLib [13] is a Coq library for de Bruijn substitutions. It offers a generic type class interface and support for defining single-variable substitution operations in a uniform way. The substitution lemmas are solved by generic tactics. It offers tactics that nicely unfold the substitution operations. Also, DBLib offers some automation in the form of a hint database with frequently needed lemmas.
– GMeta [10] uses a generic term type and automatically generated isomorphisms between it and custom term types. It supports a de Bruijn and a locally nameless interface. Moreover, it has support for mutually recursive syntax and the corresponding heterogeneous substitutions. It supports single-variable substitutions, but lacks automation for substitution lemmas.
– Lambda Tamer [7] is a small library of general purpose automation tactics. It provides support for working with higher-order and dependently-typed abstract syntax.
– LNGen [5] is a generator for single-variable locally-nameless substitution operations and the corresponding lemmas for Coq. It is based on the specification syntax of the Ott tool [16].

2   From de Bruijn to Autosubst

In this section, we illustrate the de Bruijn representation as used in AUTOSUBST on the example of the untyped λ-calculus. We first recall the definition of de Bruijn terms and instantiation (Sect. 2.1) and argue that this yields a model of the σ-calculus [14]. We then demonstrate reasoning with de Bruijn terms at the example of the Takahashi, Tait, Martin-Löf proof of the Church-Rosser theorem (Sect. 2.2). In Sect. 2.3 we discuss the implementation of de Bruijn terms and substitutions in the interactive theorem prover Coq. This is automated in AUTOSUBST, and in Sect. 2.4 we present the same implementation using AUTOSUBST.

2.1   De Bruijn Representation and Substitution

In this section, we recall the de Bruijn representation of terms in the untyped λ-calculus and derive the corresponding definitions and lemmas.

The untyped λ-calculus is the archetype of a language with local definitions. A term of the untyped λ-calculus is either an application, a binder, or a variable. The idea behind a variable is that it is a reference — either to a binder within the same term or to an outside context. In de Bruijn representation we implement these references using natural numbers. The number n refers to the n-th enclosing binder, counting from 0.

We call the numbers by which we implement variables *de Bruijn indices*, or simply *indices*, and write them as x, y, z. We write *terms* $s, t \in \mathbb{T}$ as $s, t ::= x \mid s\,t \mid \lambda.\,s$. A *substitution* is a total function mapping indices to terms. A *renaming* is a substitution that replaces indices by indices. Note that we do not assume that renamings are bijective. The letters σ, τ, θ will denote substitutions, while ξ, ζ will stand for renamings.

A substitution can be seen as an infinite sequence $(s_0, s_1, s_2, \ldots)$ of terms. This view motivates the definition of a cons operation $s \cdot \sigma$.

$$(s \cdot \sigma)(x) := \begin{cases} s & \text{if } x = 0 \\ \sigma(x - 1) & \text{otherwise} \end{cases}$$

We also introduce notation for the *identity* and the *shift* substitution:

$$\mathsf{id}(x) := x \qquad\qquad \uparrow(x) := x + 1$$

Next, we define the *instantiation* and *composition* operations $s[\sigma]$ (read s under σ) and $\sigma \circ \tau$, which implement capture-avoiding substitution and composition of substitutions, respectively.

$$\begin{aligned} x[\sigma] &= \sigma(x) \\ (s\,t)[\sigma] &= (s[\sigma])\,(t[\sigma]) \qquad (\sigma \circ \tau)(x) = \sigma(x)[\tau] \\ (\lambda.\,s)[\sigma] &= \lambda.\,(s[\Uparrow\sigma]) \end{aligned}$$

In the third equation, we change the substitution using the operator $\Uparrow$ (pronounced "*up*"). This is necessary because λ is a binder and changes the interpretation of the indices in its scope. As we want to implement capture-avoiding substitution, we have to preserve the bound index 0 and increase all other indices to skip the additional binder. Combining both cases, we define $\Uparrow\sigma$ as

$$\Uparrow\sigma := 0 \cdot (\sigma \circ \uparrow)$$

This shows that the operation $\Uparrow$ does not need to be primitive.

This definition of instantiation is by mutual recursion between instantiation, $\Uparrow$ and composition. We have to argue termination to show that it is well-defined.

The mutual recursion can be broken up by considering the definition for renamings. For a renaming ξ, the definition of $\Uparrow\xi$ simplifies to a non-recursive one

$$(st)[\sigma] \equiv (s[\sigma])(t[\sigma]) \qquad\qquad\qquad \mathsf{id} \circ \sigma \equiv \sigma$$

$$(\lambda.\ s)[\sigma] \equiv \lambda.\ (s[0 \cdot \sigma \circ {\uparrow}]) \qquad\qquad \sigma \circ \mathsf{id} \equiv \sigma$$

$$0[s \cdot \sigma] \equiv s \qquad\qquad (\sigma \circ \tau) \circ \theta \equiv \sigma \circ (\tau \circ \theta)$$

$${\uparrow} \circ (s \cdot \sigma) \equiv \sigma \qquad\qquad (s \cdot \sigma) \circ \tau \equiv s[\tau] \cdot \sigma \circ \tau$$

$$s[\mathsf{id}] \equiv s \qquad\qquad s[\sigma][\tau] \equiv s[\sigma \circ \tau]$$

$$0[\sigma] \cdot ({\uparrow} \circ \sigma) \equiv \sigma \qquad\qquad 0 \cdot {\uparrow} \equiv \mathsf{id}$$

Fig. 1. The convergent rewriting system of the σ-calculus

(for a renaming ξ and a substitution σ, $\xi \circ \sigma$ is ordinary function composition) and thus the definition of $s[\xi]$ boils down to a structural recursion. Since ${\uparrow}$ is a renaming, we can now define ${\Uparrow}\sigma$ for general substitutions. Finally, we can define $s[\sigma]$ with a second structural recursion. Such a two-level definition of instantiation has been used by Adams in [2].

De Bruijn terms and parallel substitutions, together with the substitution operations (instantiation, composition, cons, shift, and id) form a model of the σ-calculus by Abadi et al. [1]. The σ-calculus is a calculus of explicit substitutions. Explicit substitutions were intended to analyze reduction and its implementation in a more fine-grained way. For us, it is interesting for two reasons.

First, the σ-calculus can express all substitutions necessary to describe reductions in the λ-calculus. For instance, β-reduction can be expressed using cons. In de Bruijn representation, a term $(\lambda.\ s)\,t$ reduces to the term s where the index 0 is replaced by t and every other index is decremented by 1, since we removed a binder. Using the substitution operations of the σ-calculus, this is expressed as $(\lambda.\ s)\,t \rhd s[t \cdot \mathsf{id}]$. We can express η-reduction as $(\lambda.\ s[{\uparrow}]\,0) \rhd s$. The bound index 0 cannot appear in the image of s under the shift substitution, which replaces an explicit side condition on the rule.

Second, the σ-calculus yields a useful decision procedure for equational substitution lemmas. The decision procedure is based on the rewriting system shown in Fig. 1. As shown in [14], it is also complete for the given model. Thus, we obtain a rewriting-based decision procedure for equations containing the operations defined in this section, the term constructors of the untyped λ-calculus, and universally quantified meta-variables for terms and substitutions.

This concludes the definitions needed for the untyped λ-calculus. It is interesting to note that the use of parallel substitutions goes back to de Bruijn [6]. Nevertheless, the use of single-index substitutions is widespread in the programming language community, although they are equally complicated to define, significantly less expressive, and their definition involves ad-hoc recursive "shift" functions. We believe that single-index substitutions are responsible for much of the dissatisfaction with de Bruijn terms. In contrast, the careful choice of substitution operations by Abadi et al. [1] makes this approach to formalizing syntax elegant.

2.2   Case Study: Confluence of Reduction

De Bruijn terms were originally introduced both as an implementation technique and as a way of simplifying paper proofs about the λ-calculus [6]. In this section we will outline a formal proof of the Church-Rosser theorem based on de Bruijn terms. There is almost no overhead in the definitions when compared to a presentation with named variables. We illustrate how parallel substitutions are used to obtain useful generalizations and how substitution lemmas can be shown using the σ-calculus.

The Church-Rosser theorem is equivalent to the statement that reduction is *confluent*, that is, if a term s reduces to t_1 and to t_2 in an arbitrary number of steps, then we can always find a common reduct of t_1 and t_2. The following proof is based on the work of Takahashi, Tait, and Martin-Löf [11,17].

Definition 1. Parallel reduction $s \rhd\!\!\!\rhd t$ *is defined by the following system of inference rules.*

$$\frac{s_1 \rhd\!\!\!\rhd s_2 \quad t_1 \rhd\!\!\!\rhd t_2}{(\lambda.\, s_1)\, t_1 \rhd\!\!\!\rhd s_2[t_2 \cdot id]} \qquad \frac{}{x \rhd\!\!\!\rhd x} \qquad \frac{s_1 \rhd\!\!\!\rhd s_2 \quad t_1 \rhd\!\!\!\rhd t_2}{s_1\, t_1 \rhd\!\!\!\rhd s_2\, t_2} \qquad \frac{s_1 \rhd\!\!\!\rhd s_2}{\lambda.\, s_1 \rhd\!\!\!\rhd \lambda.\, s_2}$$

Parallel reduction on substitutions is defined pointwise.

$$\sigma \rhd\!\!\!\rhd \tau := \forall x,\, \sigma(x) \rhd\!\!\!\rhd \tau(x)$$

Parallel reduction interpolates between ordinary reduction and many-step reduction, that is, $\rhd \subseteq \rhd\!\!\!\rhd \subseteq \rhd^*$. From this, we see that the confluence of parallel reduction implies the confluence of single-step reduction.

Parallel reduction may contract an arbitrary number of toplevel redexes in a single step. We consider a function ρ which performs a *maximal parallel reduction*.

$$\rho(x) = x$$
$$\rho(\lambda.\, s) = \lambda.\, \rho(s)$$
$$\rho((\lambda.\, s)\, t) = \rho(s)[\rho(t) \cdot id]$$
$$\rho(s\, t) = \rho(s)\, \rho(t) \qquad\qquad \text{if } s \text{ is not a } \lambda \text{ abstraction}$$

The result of applying ρ to a term s is maximal in the sense that we can always extend an arbitrary parallel reduction from s to ρs by contracting the remaining redices.

Lemma 1 (Triangle Property). *For terms s, t, if $s \rhd\!\!\!\rhd t$, then $t \rhd\!\!\!\rhd \rho(s)$.*

From this, the confluence of parallel reduction follows by a diagram chasing argument. The proof of Lemma 1 relies on a strong substitutivity property for parallel reductions. We need to show that if $s_1 \rhd\!\!\!\rhd s_2$ and $t_1 \rhd\!\!\!\rhd t_2$ then $s_1[t_1 \cdot id] \rhd\!\!\!\rhd s_2[t_2 \cdot id]$.

We cannot show this lemma by a direct induction, because the β-substitution $t_1 \cdot id$ is not stable when going under a binder. In the case of a $\lambda.\, s_1$, we would have to show by induction that $s_1[\Uparrow(t_1\cdot\mathsf{id})] \rhd s_2[\Uparrow(t_2\cdot\mathsf{id})]$. The minimal generalization that works is to consider all *single index substitutions*, that is, substitutions which replace an index k by a term s. In our notation, these substitutions can be written as $\Uparrow^k(s\cdot\mathsf{id})$. If we continue in this vein, we will notice that we also have to show a similar lemma for shift substitutions of the form $\Uparrow^k(\uparrow^l)$. A better strategy is to generalize the lemma over all substitutions.

Lemma 2 (Strong Substitutivity). *For all terms* $s \rhd t$ *and substitutions* $\sigma \rhd \tau$ *we have* $s[\sigma] \rhd t[\tau]$.

The proof proceeds by induction on the derivation of $s \rhd t$. In the case of the binder, we need to show that $\sigma \rhd \tau$ implies $\Uparrow\sigma \rhd \Uparrow\tau$. This statement in turn depends on a special case of the same lemma. In particular, we can show it by Lemma 2 specialized to renamings. Intuitively, this is because the proof has to follow the inductive structure of the instantiation operation.

The other interesting case in the proof of Lemma 2 is the case for β-reduction, where we have to show a substitution lemma.

$$s[\Uparrow\tau][t[\tau]\cdot\mathsf{id}] = s[t\cdot\mathsf{id}][\tau]$$

This equation holds as a consequence of the axioms of the σ-calculus. In particular we can show it completely mechanically by rewriting both sides of the equation to the normal form $s[t[\tau]\cdot\tau]$.

2.3   Realization in Coq

In Coq we can define the terms of the untyped λ-calculus as an inductive type.

```
Inductive term : Type :=
 | Var (x : nat)
 | App (s t : term)
 | Lam (s : term).
```

The cons operation can be defined generically for functions $\mathbb{N} \to X$ for every type X. We introduce the notation s .: σ to stand for $s\cdot\sigma$. The identity substitution id will be written as `ids` and corresponds to the variable constructor. By post-composing with the identity substitution, we can lift arbitrary renamings (functions ξ : nat $\to$ nat) to substitutions. Since there is no standard notation for forward composition, we introduce the notation f >>> g for forward composition of functions. For readability, we also introduce notation for the coercion of renamings into substitutions, ren ξ := ξ >>> ids. The shift renaming $\uparrow$ is written (+1).

As mentioned before, we implement the instantiation operation by first specializing to renamings.

```
Fixpoint rename (ξ : nat → nat) (s : term) : term :=
  match s with
   | Var x ⇒ Var (ξ x)
   | App s t ⇒ App (rename ξ s) (rename ξ t)
   | Lam s ⇒ Lam (rename (0 .: ξ >>> (+1)))
  end.
```

Using `rename`, we now define up σ := ids 0 .: σ >>> rename (+1). Finally, using up, we define the full instantiation operation on terms.

```
Fixpoint inst (σ : nat → term) (s : term) : term :=
  match s with
  | Var x ⇒ σ x
  | App s t ⇒ App (inst σ s) (inst σ t)
  | Lam s ⇒ Lam (inst (up σ) s)
  end.
```

With instantiation, we define substitution composition $\sigma \gg \tau$ as σ >>> inst τ. We write s.[σ] for $s[\sigma]$.

To complete the definition of instantiation we need to show that `rename` is a special case of `inst`. Concretely we must have `rename` ξ s = `inst` (ren ξ) s for all renamings ξ and terms s. The proof proceeds by induction on s and allows us to forget about `rename` in the remainder.

In order to show that the definitions above yield a model of the σ-calculus, we have to show every equation in Fig. 1. Note however, that the only definitions that are specific to the term language at hand are the identity substitution, and the definition of instantiation. In order to show that our definitions yield a model of the σ-calculus, we only need to know that instantiation and the identity substitution behave correctly. For this, it suffices to establish

$$\mathsf{id}(x)[\sigma] = \sigma(x)$$
$$s[\mathsf{id}] = s$$
$$s[\sigma][\tau] = s[\sigma \circ \tau]$$

It is easy to check that given these three equations all the other equations in Fig. 1 follow without any other assumptions about id or instantiation.

In the particular case of the untyped λ-calculus, the first equation $\mathsf{id}(x)[\sigma] = \sigma(x)$ holds by definition, while the second follows by a straightforward term induction. The third equation is more interesting, since the proof has to follow the inductive structure of the instantiation operation.

Formally, the proof proceeds in three steps. First we show $s[\xi][\tau] = s[\xi \circ \tau]$, by induction on s. Using this result we can show $s[\sigma][\xi] = s[\sigma \circ \xi]$, again by induction on s. We finally show the full equation, with another term induction. This proof boils down to showing that $\Uparrow\sigma \circ \Uparrow\tau = \Uparrow(\sigma \circ \tau)$, which depends on both specializations.

2.4   Realization in Autosubst

The library AUTOSUBST generalizes and automates the constructions from the previous sections for arbitrary term languages. A term language is specified as an inductive type with annotations.

```
subst_comp s σ τ : s.[σ].[τ] = s.[σ >> τ]
subst_id s        : s.[ids] = s
id_subst x σ      : (ids x).[σ] = σ x
rename_subst ξ s  : rename ξ s = s.[ren ξ]
```

Fig. 2. Substitution Lemmas in `SubstLemmas`

For the untyped λ-calculus, we define the type of terms as follows.

```
Inductive term : Type :=
| Var (x : var)
| App (s t : term)
| Lam (s : {bind term}).
```

Every term language must contain exactly one constructor for de Bruijn indices, i.e., a constructor which takes a single argument of type `var`. As before, this constructor serves as the identity substitution. The type `var` is an alias for `nat`.

Additionally, all binders must be marked. In this example, `Lam` is the only binder. It introduces a new index into the scope of its argument `s`. The annotation `{bind term}` is definitionally equal to `term`.

Using this definition of terms, we can derive the instantiation operation by defining an instance of the `Subst` type class using the tactic `derive`. This is comparable to the usage of deriving-clauses in the programming language Haskell. In addition, we need to define instances for the two auxiliary type classes `Ids` and `Rename`, defining the identity substitution and the instantiation of renamings.

```
Instance Ids_term    : Ids term.    derive. Defined.
Instance Rename_term : Rename term. derive. Defined.
Instance Subst_term  : Subst term.  derive. Defined.
```

Next, we derive the substitution lemmas from Fig. 1 by generating an instance of the `SubstLemmas` type class.

```
Instance SubstLemmas_term : SubstLemmas term. derive. Qed.
```

This class contains the lemmas depicted in Fig. 2.

These are all the necessary definitions to start using the library. In proofs involving terms and substitutions we can now benefit from the `autosubst` and `asimpl` tactics. The `autosubst` tactic is a decision procedure for equations between substitution expressions, that is, terms or substitutions build with the substitution operations. The tactic `asimpl` normalizes substitution expressions which appear in the goal. There are also focused variants, `asimpl in H` and `asimpl in *`, for simplifying a specific assumption `H` or all assumptions as well as the goal.

In order to use these tactics effectively, it is sometimes necessary to produce equational goals by hand. For instance, consider the following constructor for β-reduction.

```
step_beta s t : step (App (Lam s) t) s.[t .: ids]
```

This constructor can only be applied if the goal contains a term which is *syntactically* of the shape `s.[t .: ids]`. In order to apply the constructor to a goal of the form `step (App (Lam s) t) u`, we need a lemma with an equational hypothesis.

```
step_ebeta s u t : u = s.[t .: ids] → step (App (Lam s) t) u
```

We can then apply `step_ebeta` and show the subgoal using `autosubst`. Alternatively, we could have used `step_ebeta` directly in the definition of the reduction relation. With this design choice, we can show simple properties like substitutivity of reduction with an almost trivial proof script.

```
Lemma step_subst s t :
  step s t → ∀σ, step s.[σ] t.[σ].
Proof.
  induction 1; constructor; subst; autosubst.
Qed.
```

The tactic `autosubst` solves the crucial substitution lemma

$$\texttt{s.[up } \sigma\texttt{].[t.[}\sigma\texttt{] .: ids] = s.[t .: ids].[}\sigma\texttt{]}$$

Similar considerations apply to every lemma statement involving specific substitutions.

Internally, AUTOSUBST differs from a straightforward implementation of the σ-calculus in several ways. The σ-calculus contains a number of equations which are specific to the untyped λ-calculus. If we wanted to use the same approach for a new term language, we would need to extend the rewriting system with equations specific to the new term language. Instead, we construct a definition of instantiation with the appropriate simplification behavior. The automation tactics use a combination of rewriting and term simplification.

Internally, instantiation is defined in terms of renaming. We hide this fact by treating `up` as if it was an additional primitive. The automation tactics work by unfolding these definitions and rewriting substitution expressions to an internal normal form. In the case of `asimpl`, we then fold the expressions back into a more readable format.

AUTOSUBST is completely written in Coq. We synthesize the substitution operations using the meta-programming capabilities of Ltac.

3   Heterogeneous Substitutions

So far, we have only considered single-sorted syntactic theories. However, many practical programming languages distinguish at least between terms and types. In such a setting, we may have more than one substitution operations on terms, which may interact with one another.

As before, we shall introduce the necessary machinery with a concrete example. In this section we consider a two-sorted presentation of System F.

$$A, B ::= X \mid A \rightarrow B \mid \forall. A$$
$$s, t ::= x \mid s\,t \mid \lambda A.\,s \mid \Lambda.\,s \mid s\,A$$

Substituting types in types works as before. Term substitutions are complicated by the fact that terms contain type binders.

$$x[\sigma] = \sigma(x)$$
$$(s\,t)[\sigma] = s[\sigma]\,t[\sigma] \qquad (s\,A)[\sigma] = s[\sigma]\,A$$
$$(\lambda A.\,s)[\sigma] = \lambda A.\,s[\Uparrow\sigma] \qquad (\Lambda.\,s)[\sigma] = \Lambda.\,s[\sigma \bullet \uparrow]$$

Upon traversing a type binder, we need to increment all type variables in σ. In order to substitute types in terms, we introduce heterogeneous instantiation and composition operations ($s[\![\theta]\!]$ and $\sigma \bullet \theta$). To avoid confusion with term substitutions, we will write θ, θ' for type substitutions.

$$x[\![\theta]\!] = x \qquad (s\,A)[\![\theta]\!] = s[\![\theta]\!]\,A[\theta]$$
$$(s\,t)[\![\theta]\!] = s[\![\theta]\!]\,t[\![\theta]\!] \qquad (\Lambda.\,s)[\![\theta]\!] = \Lambda.\,s[\![\Uparrow\theta]\!]$$
$$(\lambda A.\,s)[\![\theta]\!] = \lambda A[\theta].\,s[\![\theta]\!] \qquad (\sigma \bullet \theta)(x) = \sigma(x)[\![\theta]\!]$$

We need a number of lemmas about heterogeneous instantiation in order to work with it using equational reasoning. The situation is analogous to the case of ordinary instantiation, expect that heterogeneous substitutions have to be invariant in the image of the identity substitution on terms.

$$\mathsf{id}(x)[\![\theta]\!] = \mathsf{id}(x)$$
$$s[\![\mathsf{id}]\!] = s$$
$$s[\![\theta]\!][\![\theta]\!]' = s[\![\theta \circ \theta']\!]$$

Note that the identity substitution in the first equation is the identity substitution on terms, while in the second equation, it refers to the identity substitution on types.

Furthermore, in order to show the substitution lemmas for terms, we need to know something about the interaction between the two kinds of substitutions. The fact to take care of is that terms may contain types, but types do not contain terms. Thus, we can push a type substitution under a term substitution, so long as we take care that the type substitution was applied in the image of the term substitution.

$$s[\sigma][\![\theta]\!] = s[\![\theta]\!][\sigma \bullet \theta]$$

4   Case Studies

We have used AUTOSUBST to mechanize proofs about the metatheory of System F, System F$_{<:}$, and CC$_\omega$. In each case study we define an inductive type of terms and derive the substitution operations using AUTOSUBST. In the case of System F and System F$_{<:}$, this uses AUTOSUBST's support for heterogeneous substitutions as presented in Sect. 3.

For the proofs, we follow the same general strategy as presented in Sect. 2.2. Whenever we encounter a lemma concerning a specific substitution, we try to find a generalization over all substitutions. For instance, in proofs of type preservation, we must show the admissibility of β-substitution for the typing relation. We generalize and show the admissibility of every "well-typed" substitution.

$$\frac{\Gamma, A \vdash s : B \qquad \Gamma \vdash t : A}{\Gamma \vdash s[t \cdot \mathrm{id}] : B} \quad \rightsquigarrow \quad \frac{\Gamma \vdash s : A \qquad \sigma : \Delta \to \Gamma}{\Delta \vdash s[\sigma] : A}$$

where $\sigma : \Delta \to \Gamma$ means that every $\sigma(x)$ is a term of type Γ_x in context Δ. However, the details of this definition depend on the typing relation. In the literature, such well-typed substitutions are known as *context morphisms* [9].

Context morphisms are more powerful than the β-substitution lemma. In particular, we obtain weakening by showing $\uparrow : \Gamma, A \to \Gamma$ and the admissibility of β by showing that if $\Gamma \vdash s : A$, then $s \cdot \mathrm{id} : \Gamma \to \Gamma, A$.

The proof of a context morphism lemma always follows the inductive structure of the instantiation operation. We first show the lemma for renamings. Then using the lemma for renamings, we show that $\Uparrow$ corresponds to context extension, that is, $\Uparrow\sigma : \Delta, A \to \Gamma, A$, whenever $\sigma : \Delta \to \Gamma$. From this, the full context morphism lemma follows.

In every case study, we show all equational substitution lemmas automatically with AUTOSUBST. With the correct definitions and lemma statements, most proofs are straightforward.

4.1   Type Preservation for CC_ω

Type preservation of CC_ω boils down to a number of inversion lemmas for typing derivations, as well as the admissibility of substitution. The former depend on the Church-Rosser theorem, whose proof proceeds exactly as illustrated in Sect. 2.2.

The main difference between a normal paper presentation of CC_ω and the de Bruijn presentation is in the case of the variable rule. Context lookup is needed to implement the variable rule.

$$\frac{(x, A) \in \Gamma}{\Gamma \vdash x : A} \quad \rightsquigarrow \quad \frac{x < |\Gamma|}{\Gamma \vdash x : \Gamma_x}$$

In the non-dependent case, we can implement Γ_x by list lookup. In this case, however, the context is dependent. For dependent contexts Γ, we need to ensure that Γ_x is a valid type in Γ. In Coq, we totalize this function by adding a case for the empty context.

$$(\Gamma, A)_0 = A[\uparrow]$$
$$(\Gamma, A)_{x+1} = \Gamma_x[\uparrow]$$

The context morphism lemma for CC_ω takes the form

$$\frac{\Gamma \vdash s : A \qquad \sigma : \Delta \to \Gamma}{\Delta \vdash s[\sigma] : A[\sigma]} \qquad \sigma : \Delta \to \Gamma := \forall x < |\Gamma|, \Delta \vdash \sigma(x) : \Gamma_x[\sigma]$$

Apart from the obvious instances for weakening and β, we also obtain a context conversion lemma as a corollary. If we have two types $A \equiv B$, then id $: \Gamma, B \to \Gamma, A$. Instantiating the context morphism lemma yields

$$\frac{\Gamma, A \vdash s : C \qquad A \equiv B}{\Gamma, B \vdash s : C}$$

4.2 Strong Normalization for System F

We follow Girard's proof of strong normalization [8]. The proof consists of defining a type-indexed family of sets of terms $L(A)$ such that $s \in L(A)$ implies that s is strongly normalizing. We then show that every term of type A is contained in $L(A)$, which implies the strong normalization result. The main technical difficulty with this proof is that we need to generalize the latter statement. Instead of showing that $\Gamma \vdash s : A$ implies $s \in L(A)$, we need to show that $s[\sigma] \in L(A)$ for all substitutions σ such that $\sigma(x) \in L(\Gamma_x)$. This generalization is standard and extensively documented in the literature.[1] In the proof, we have some occasion to generalize lemmas over all substitutions, but otherwise there are no surprises.

We still need to show how to define System F using de Bruijn terms.

In the named presentation of System F, we have two contexts, a context Δ for type variables and a context Γ for term variables. The rule for type abstraction is usually rendered with a side condition.

$$\frac{\Delta, X; \Gamma \vdash s : A}{\Delta; \Gamma \vdash \Lambda X. s : \forall X. A} \quad (X \notin \mathrm{FV}(\Gamma))$$

For de Bruijn terms, we can drop the context Δ, since it only asserts that at most a certain number of type variables are free. This constraint is orthogonal to the typing relation. The freshness assumption can in turn be implemented similar to η-reduction. Note that we do not abstract over an arbitrary variable X, but rather over the index 0. If Γ is an arbitrary context, then 0 is not free in the context $\Gamma[\uparrow]$, where we instantiate every type in Γ under $\uparrow$. This leads to the following concise definition.

$$\frac{\Gamma[\uparrow] \vdash s : A}{\Gamma \vdash \Lambda. s : \forall. A}$$

All other inference rules are straightforward.

4.3 POPLmark Challenge

The POPLmark [4] is a benchmark to "measur[e] progress [...] in mechanizing the metatheory of programming languages". We solve part A, which concerns progress and preservation of System $F_{<:}$.

---

[1] There are a large number of details which are swept under the carpet in this short overview. The formalization tells the whole story.

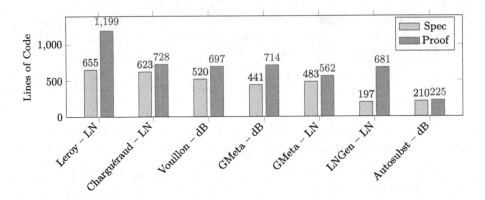

Fig. 3. Comparison of Coq solutions to part A of the POPLmark challenge

The typing relation of System $F_{<:}$ carries two contexts: A context Δ with subtyping assumptions for type variables and a context Γ with typing assumptions for terms. The context Δ is dependent, while Γ is non-dependent.

The proof of type preservation decomposes into a sequence of context morphism lemmas. In part 1A of the POPLmark challenge, we show the context morphism lemma for subtyping. For part 2A, we prove similar context morphism lemmas for the typing relation. These proofs are all simple and regular. The other lemmas are all very close to the informal proof.

In Fig. 3, we compare lines of code using `coqwc`, excluding parts that the authors marked as reusable libraries.

- Xavier Leroy's solution[2] is a self-contained development using the locally nameless representation.
- Arthur Charguéraud's solution[3] also uses the locally nameless representation, but uses a small library to reason about the locally nameless representation.
- Jérôme Vouillon[4] uses de Bruijn indices in a self-contained development.
- There are two solutions using the GMeta [10] library. One using locally nameless and one using de Bruijn indices.
- There is a solution[5] using the LNGen [5] library, which generates substitution operations for locally nameless terms and proofs for the corresponding infrastructure lemmas.

5   Conclusion

There is an abundance of proposals for the formalization of binders: named syntax with α-equivalence and freshness-assumptions, de Bruijn indices, the locally

2 http://www.seas.upenn.edu/~plclub/poplmark/leroy.html.
3 http://www.chargueraud.org/softs/ln/.
4 http://www.seas.upenn.edu/~plclub/poplmark/vouillon.html.
5 http://www.cis.upenn.edu/~sweirich/papers/lngen/.

nameless representation, higher-order abstract syntax, and nominal logic to name just a few contenders. The POPLmark Challenge [4] has been set up as a benchmark for these approaches and to encourage further development of solutions that work and scale.

However, the right method to choose depends on a lot of factors, the most important of which is probably the logic or theorem prover that one is working with. For example, higher-order abstract syntax and nominal logic require special support from the meta-logic in most variations. On the other hand, pure first-order approaches like named syntax, de Bruijn indices and the locally nameless representation work in most systems.

Another important distinction is the distance to paper proofs. Here, approaches with named contexts like named terms and locally nameless are closer, at the price of having to deal with issues of freshness manually. Our approach is rather distant from the usual paper proofs, which tend to use single-variable substitutions. In this respect, it is more distant than the more common approach using single-variable de Bruijn substitutions. On the other hand, our approach removes most of the technicalities of single-variable de Bruijn substitutions. Our decision procedure relies crucially on the fact that we use parallel substitutions. We believe that the elegance of this pure and parallel de Bruijn style is worth the paradigm shift.

5.1 Future Work

We identified the following shortcomings with the current implementation of AUTOSUBST, which we would like to address in future work.

- The current approach to heterogeneous substitutions is rather ad-hoc. It is not extensible to mutually recursive types and it is lacking a completeness result.
- AUTOSUBST lacks support for binders with variable arity. This is needed to formalize pattern matching, as used in part B of the POPLmark challenge.
- AUTOSUBST generates definitions using Ltac, which is rather fragile. Proper error reporting is currently not available, and in some cases it is necessary to manually inspect the generated code for errors. It seems to be almost impossible to build an Ltac script that is reliable in corner cases, due to the unpredictable nature of the Ltac semantics. Additionally, AUTOSUBST cannot handle mutually recursive term types. These problems could be solved by writing the code generator in an external programming language, or by extending the metaprogramming capabilities of Coq.

References

1. Abadi, M., Cardelli, L., Curien, P.L., Lévy, J.J.: Explicit substitutions. J. Funct. Program. **1**(4), 375–416 (1991)
2. Adams, R.: Formalized metatheory with terms represented by an indexed family of types. In: Filliâtre, J.-C., Paulin-Mohring, C., Werner, B. (eds.) TYPES 2004. LNCS, vol. 3839, pp. 1–16. Springer, Heidelberg (2006)

3. Anand, A., Rahli, V.: A generic approach to proofs about substitution. In: Proceedings of the 2014 International Workshop on Logical Frameworks and Meta-languages: Theory and Practice, p. 5. ACM (2014)
4. Aydemir, B.E., et al.: Mechanized metatheory for the masses: the POPLMARK challenge. In: Hurd, J., Melham, T. (eds.) TPHOLs 2005. LNCS, vol. 3603, pp. 50–65. Springer, Heidelberg (2005)
5. Aydemir, B.E., Weirich, S.: LNgen: Tool support for locally nameless representations. Technical report, University of Pennsylvania (2010)
6. de Bruijn, N.G.: Lambda calculus notation with nameless dummies, a tool for automatic formula manipulation, with application to the Church-Rosser theorem. Indagationes Mathematicae (Proceedings) **75**(5), 381–392 (1972)
7. Chlipala, A.: Parametric higher-order abstract syntax for mechanized semantics. In: ACM Sigplan Notices, vol. 43, pp. 143–156. ACM (2008)
8. Girard, J.Y., Taylor, P., Lafont, Y.: Proofs and Types, vol. 7. Cambridge University Press, Cambridge (1989)
9. Goguen, H., McKinna, J.: Candidates for substitution. LFCS report series - Laboratory for Foundations of Computer Science ECS LFCS (1997)
10. Lee, G., Oliveira, B.C.D.S., Cho, S., Yi, K.: GMETA: a generic formal metatheory framework for first-order representations. In: Seidl, H. (ed.) ESOP 2012. LNCS, vol. 7211, pp. 436–455. Springer, Heidelberg (2012)
11. Martin-Löf, P.: An intuitionistic theory of types. Twenty-five Years Constructive Type Theory **36**, 127–172 (1998)
12. Polonowski, E.: Automatically generated infrastructure for de bruijn syntaxes. In: Blazy, S., Paulin-Mohring, C., Pichardie, D. (eds.) ITP 2013. LNCS, vol. 7998, pp. 402–417. Springer, Heidelberg (2013)
13. Pottier, F.: DBLIB, a Coq library for dealing with binding using de Bruijn indices, Dec 2013. https://github.com/fpottier/dblib
14. Schäfer, S., Smolka, G., Tebbi, T.: Completeness and Decidability of de Bruijn Substitution Algebra in Coq. In: Proceedings of the 2015 Conference on Certified Programs and Proofs, CPP 2015, pp. 67–73. ACM, New York, Jan 2015
15. Schäfer, S., Tebbi, T.: Autosubst: Automation for de Bruijn syntax and substitution in Coq, August 2014. www.ps.uni-saarland.de/autosubst
16. Sewell, P., Nardelli, F.Z., Owens, S., Peskine, G., Ridge, T., Sarkar, S., Strniša, R.: Ott: Effective tool support for the working semanticist. J. Funct. Program. **20**(1), 71 (2010)
17. Takahashi, M.: Parallel reductions in λ-calculus. Inf. Comput. **118**(1), 120–127 (1995)

ModuRes: A Coq Library for Modular Reasoning About Concurrent Higher-Order Imperative Programming Languages

Filip Sieczkowski[(✉)], Aleš Bizjak, and Lars Birkedal

Department of Computer Science, Aarhus University, Aarhus, Denmark
{filips,abizjak,birkedal}@cs.au.dk

Abstract. It is well-known that it is challenging to build semantic models of type systems or logics for reasoning about concurrent higher-order imperative programming languages. One of the key challenges is that such semantic models often involve constructing solutions to certain kinds of recursive domain equations, which in practice has been a barrier to formalization efforts. Here we present the ModuRes Coq library, which provides an easy way to solve such equations. We show how the library can be used to construct models of type systems and logics for reasoning about concurrent higher-order imperative programming languages.

1 Introduction

In recent years we have seen a marked progress in techniques for reasoning about higher-order, effectful programming languages. However, many of the resulting models and logics have eschewed formal verification. The main reason seems to be the rising complexity of these theories, as well as the increasing use of sophisticated mathematical structures, which impose a substantial barrier to entry for potential formalization efforts. One of the crucial features that is difficult to model is circularity, in its various shapes. Its prototypical form stems from higher-order store, where we can store functions and references as well as first-order data, but circularities can arise in many other ways. One that is commonly encountered in program logics for concurrency, for instance, is shared invariants, which can be accessed by any one thread of computation — provided that the access preserves the invariant. Since the state that a particular invariant describes could well refer to other invariants, a circularity arises. This circularity is quite similar in nature to the one that arises when one attempts to interpret types of higher-order references: the semantic description of the store has to include the interpretation of the type for each location — but the interpretation of the type itself depends on the description of the store.

In this paper we present the ModuRes Coq library[1], which provides an easy way to solve these kinds of circularities, and help use the solutions to build models of programming languages and program logics.

---

[1] Available at http://cs.au.dk/~birke/modures/tutorial.

C. Urban and X. Zhang (Eds.): ITP 2015, LNCS 9236, pp. 375–390, 2015.
DOI: 10.1007/978-3-319-22102-1_25

Example: Consider a simply-typed lambda-calculus extended with ML-style higher-order references. As discussed in [1,2,9,11,17] it is natural to try to give the semantic interpretation of its types in the following (flawed) way:

$$\mathsf{type} = \mathsf{world} \to \mathcal{P}(\mathsf{value}) \qquad\qquad \mathsf{world} = \mathsf{loc} \rightharpoonup_{\mathsf{fin}} \mathsf{type}. \qquad (1)$$

The idea is that the world associates with each location its semantic type, and the semantic types use the world to give the interpretation to the reference types: the interpretation of a reference type should only contain locations that are associated with the appropriate type by the world. However, the recursive dependency between type and world is not well-founded and thus a simple set-theoretic solution to such an equation does not exist in general (for a more detailed treatment of this circularity see e.g. [9]).

Various solutions to this problem have been proposed, using tools like explicit step-indexing [1,2], Hobor *et al.*'s indirection theory, which also comes with an associated Coq formalization [17], and ultrametric spaces [9,11], among others. In our library, we formalize a variant of the ultrametric-space approach described by Birkedal et al. [9–11], that not only gives us a general way of solving the recursive domain equations, but puts it in a larger context that allows us to easily utilize it to build higher-order logics with recursively defined predicates. In the following, we explain some of the basics of our approach, and why we believe that it forms a low-boilerplate and powerful tool, and present how it can be used to model and solve equations like (1). We also report on a recent successful application of the ModuRes Coq library to formalize a state-of-the-art program logic for a concurrent higher-order imperative programming language [20].

2   COFE in Coq

As mentioned above, the recursive domain equations that often arise when modeling sophisticated type systems and logics for advanced programming languages, in general do not have a solution using standard sets and functions. In the ModuRes library we use the approach of Birkedal *et al.* [10] and move from the category **Set** of sets and functions to categories enriched over certain kinds of ultrametric spaces. For formalization purposes, it is important that this move is not too cumbersome in practical use. The ModuRes library and this paper therefore makes use of an alternative, simpler, presentation of the necessary ultrametric spaces, known as complete ordered families of equivalences, or COFEs [16]. A tutorial presentation of COFEs and their application to models of higher-order logics can be found in Birkedal and Bizjak [8]. In this section we focus on our formalization in Coq, and introduce the types that are necessary to understand the statement and solution of recursive domain equations.

2.1   Using Type Classes: Types with Equality

COFEs will be represented in Coq as types enriched with some additional information. To make it easier to work with such enriched types, we make use of

Coq's type classes [23]. In this subsection we explain the general approach we follow, by first considering how to represent types with an equality provided by the user (known in the Coq standard library as a setoid).

We begin by calling to mind the setoid definition from the standard library:

Class Setoid A := {
    equiv : relation A ;
    setoid_equiv :> Equivalence equiv }.

This definition means that for any type T for which we can find an instance of the Setoid class, we have a relation, and we know that this relation is an equivalence relation. The type class system allows us to build many generic constructions on setoids, such as products and subset types. Consider, for instance the product of two setoids, with the usual, pointwise equality. The following snippet gives the definition, with the proof that the relation is indeed an equivalence elided (recall == is the standard library infix notation for equiv).

Context '{eqU : Setoid U} '{eqV : Setoid V}.
Definition prod_equiv (p1 p2 : U * V) :=
    fst p1 == fst p2 /\ snd p1 == snd p2.
Global Program Instance prod_setoid : Setoid (U * V) :=
    Build_Setoid (equiv := prod_equiv) _.
Next Obligation. ... **Qed**.

The principal strength of the type class mechanism as applied as above is in the lightweight usage: any type T * U can be considered as a setoid as long as we can find a proof that both arguments have a setoid structure too.

We can build many general constructions on types of equality, but one of particular importance is the function space on equality types. As with many other mathematical structures, the general functions between the underlying types are too loose: we need a space of "good" functions. In the case of setoids, the appropriate notion of goodness is, unsurprisingly, preservation of equality. We can define equality-preserving maps as follows:

Record morphism S T '{eqS : Setoid S} '{eqT : Setoid T} :=
    mkMorph {
        morph :> U -> V;
        morph_resp : Proper (equiv ==> equiv) morph}.
Infix " -=>" := morphism (at level 45, right associativity).

Several things are worth noting here. Firstly, an equality preserving function is actually a *record* that carries with itself the proof of that property. Note that while we could declare it a typeclass, this would not give us much, since finding a proof of preservation of equality is beyond the type-directed proof search that the type class mechanism employs. Secondly, we are using the standard library Proper class, which expresses the general notion of a relation being preserved by a function symbol — in our case, that the function morph preserves setoid equality, i.e., takes equiv arguments to equiv results. Finally, in contrast to the general function spaces, we can define the setoid structure for the equality preserving maps, as illustrated by the following snippet:

Context '{eqT : Setoid T} '{eqU : Setoid U}.
Definition morph_equiv (f g : T −=> U) := forall t, f t == g t.
Global Program Instance morph_setoid : Setoid (T −=> U) :=
   Build_Setoid (equiv := morph_equiv) _.
Next Obligation. ... **Qed**.

This is the gist of the pattern of enriching the types that we follow throughout the library. It is reminiscent and indeed inspired by the approach of Spitters and van der Weegen [24]; we discuss the relationship in more detail in Sect. 5. The library provides many additional constructions on setoids, as well as useful definitions and lemmas. However, the most important aspect is that we provide a lightweight way of enriching the types, particularly the function spaces.

2.2   Complete Ordered Families of Equivalences

After seeing the general pattern we use to enrich types, we can move to defining the actual COFEs. Conceptually, ordered families of equivalences are an extension of setoids: where the setoid provides one notion of equality for a given type, an ordered family of equivalences (OFE) provides a whole family of equivalences that *approximate* equality. Formally, we define them as follows:

Class ofe T {eqT : Setoid T} :=
   { dist : nat −> T −> T −> Prop;
     dist_morph n :> Proper (equiv ==> equiv ==> iff) (dist n);
     dist_refl : forall x y, (forall n, dist n x y) <−> x == y;
     dist_sym n :> Symmetric (dist n);
     dist_trans n :> Transitive (dist n);
     dist_mono : forall n x y, dist (S n) x y −> dist n x y;
     dist_bound : forall x y, dist 0 x y}.

As we can see, in addition to a setoid structure, an ofe has a relation for each natural number. These are *ordered*, in the sense that $dist_{mono}$ implies: a relation at level n is included in all smaller levels. Furthermore, the relation is trivial at level 0, and the family's limit is the setoid equality — i.e., two elements of an OFE are considered equal iff they are approximately equal at all levels, as enforced by $dist_{refl}$. The somewhat cryptic statement of $dist_{morph}$ ensures that the approximate equality respects the setoid equality, i.e., that one can freely substitute equal arguments of dist. We use the following notation for the approximate equality:

Notation "x_'='_n_'='_y" := (dist n x y).

For the OFEs to serve their purpose, we need one extra property: completeness of all Cauchy chains. Cauchy chains are sequences (or chains) whose elements get arbitrarily close together. An OFE T is *complete* (a **COFE**) if all Cauchy chains *converge* to an element of T: for any level of approximation n there is a tail of the chain whose elements are all n-equal to the limit. Formally

Definition chain (T : **Type**) := nat −> T.
Class cchain '{ofeT : ofe T} (c : chain T) :=
  chain_cauchy : forall n i j (HLei : n <= i) (HLej : n <= j), (c i) = n = (c j).
Definition converges (c : chain T) (m : T) :=
  forall n, exists k, forall i (HLe : k <= i), m = n = (c i)
Class cofe T '{ofeT : ofe T} :=
  { compl : forall c {cc : cchain c}, T;
    conv_cauchy : forall c {cc : cchain c}, converges c (compl c)}.

Constructions on COFEs. Much like in the case of types with equality, we can provide various standard constructions on COFEs. For most standard constructions, such as products and indexed products, or subset types, it suffices to define the approximation pointwise.[2] However, we can also define another simple, but important space: step-indexed propositions, or, *uniform predicates*. The (slightly simplified) definition is as follows:

Record UPred T :=
  mkUPred {p :> nat −> T −> Prop;
          uni_pred : forall n m t (HLe : m <= n), p n t −> p m t}.

As we can see, these are families of predicates over some type T, such that the predicates are *downwards-closed* in the natural number component. This makes it easy to define non-trivial approximate equality on them:

Definition up_dist {T} n (p q : UPred T) :=
  forall m t, m < n −> (p m t <−> q m t).

Note that we require the propositions to be equivalent only for *strictly smaller* indices: this ensures that any two uniform predicates are equal at level 0.

Non-expansive Maps. For function spaces between COFEs, we can proceed in a manner similar to the equality types. Here, the appropriate notion of a function space consists of the non-expansive functions, i.e., the equality preserving maps that also preserve all the approximations:

Record ofe_morphism T U '{oT : ofe T} '{oU : ofe U} :=
  mkUMorph { ofe_morph :> T −=> U;
               ofe_morph_nonexp n : Proper (dist n ==> dist n) met_morph}.
Infix " $\rightarrow_{ne}$ " := ofe_morphism (at level 45, right associativity).

    Following the pattern set by the equality types, we can show that this notion of function space between COFEs itself forms a COFE with the usual application and abstraction properties. In other words, the category with objects COFEs and with morphisms non-expansive functions is cartesian closed.

---

[2] Although in the case of subset types an extra condition is required to make sure the subset is complete.

2.3 Contractiveness and Fixed Points

A *contractive* function is a non-expansive function for which the level of approximate equality between the results of function application not only persists, but actually increases. Formally, we can define it as

Class contractive (f : T $\rightarrow_{ne}$ U) := contr n : Proper (dist n ==> dist (S n)) f.

Observe, that if we have a contractive endofunction f on some space T, the results of iterating f on any two elements get progressively more and more equal — any two elements of T are 0-equal, so the results of applying f to them are 1-equal, and so on. This results in a version of Banach's fixed-point theorem for contractive maps on COFEs, i.e., we get the following properties:

Definition fixp (f : T $\rightarrow_{ne}$ T) {HC : contractive f} (x : T) : T := ...
Lemma fixp_eq f x {HC : contractive f} : fixp f x == f (fixp f x).
Lemma fixp_unique f x y {HC : contractive f} : fixp f x == fixp f y.

The term fixp constructs the fixed point of a contractive function f by iterating f starting at x and taking the limit (hence the need for completeness). The lemma fixp_eq expresses that fixp f x is indeed a fixed point of f for any x and lemma fixp_unique shows that the starting point of the iteration is irrelevant.

When using the library to build models of type systems and logics, fixed-point properties allow us to interpret recursively defined types and predicates.

3 Solving the Recursive Domain Equation

Following Birkedal *et al.* [10], the ModuRes library provides solutions to recursive domain equations in categories enriched over COFEs. Here, for simplicity of presentation, we just present our approach to recursive domain equations for the concrete category of COFEs,[3] which suffices for many applications.

We first describe the interface of the input to the part of the library solving recursive domain equations, then describe the interface of the provided solution, and finally describe how one may use the solution.

3.1 Interface of the Recursive Domain Equation

The ModuRes library provides solutions to recursive domain equations in the category of COFEs. A recursive domain equation must be specified by a suitable functor F on the category of COFEs. To accommodate mixed-variance recursive domain equations, the functor F must be a bifunctor, such that positive and negative occurrences of the recursion variable are split apart. In the example from the Introduction, we were seeking a solution to the equation

$$T \simeq (\mathsf{Loc} \rightarrow_{fin} T) \rightarrow \mathsf{Pred}(\mathsf{Val})$$

---

[3] Since the category of COFEs itself is cartesian closed it is indeed enriched over itself, and hence it is a special case of the general approach provided by the library.

In this case, there are no positive occurrences, and just one negative one. Thus, in this case, our functor F can be defined by

$$F(T^-, T^+) = (\mathsf{Loc} \rightarrow_{\mathsf{fin}} T^-) \rightarrow_{\mathsf{ne}} \mathsf{UPred}(\mathsf{Val}).$$

Note that since the result needs to be a COFE, we need to use non-expansive function spaces ($\rightarrow_{\mathsf{ne}}$) and the step-indexed predicate space, UPred.[4]

The key parts of the interface to a recursive domain equation are:

Module Type SimplInput.

   ...

   Parameter F : COFE −> COFE −> COFE.
   Parameter FArr : BiFMap F.
   Parameter FFun : BiFunctor F.
   Parameter F_ne : unit $\rightarrow_{\mathsf{ne}}$ F unit unit.
End SimplInput.

The interface consists of several components. The function F works on type COFE — a record that packs a type together with a proof that it is indeed a COFE. We have to enforce that F is not just any function on COFEs, but a *functor*. To this end, we need to provide a definition of how it transforms non-expansive functions on its arguments into non-expansive functions on the result: a requirement expressed by the BiFMap typeclass. Moreover, the result has to be contravariant in the first argument, and covariant in the second, as shown in the following formulation, specialized to the category of COFEs

Class BiFMap (F : COFE −> COFE −> COFE) :=
  fmorph : forall $\{t\ t'\ u\ u'\}$, $(t' \rightarrow_{\mathsf{ne}} t) * (u \rightarrow_{\mathsf{ne}} u') \rightarrow_{\mathsf{ne}} (F\ t\ u \rightarrow_{\mathsf{ne}} F\ t'\ u')$.

In our running example, this means that for any spaces $T_1^-, T_1^+, T_2^-, T_2^+$, and any functions $f : T_2^- \rightarrow_{\mathsf{ne}} T_1^-, g : T_1^+ \rightarrow_{\mathsf{ne}} T_2^+$ we need to build – in a non-expansive way – a function of type

$$((\mathsf{Loc} \rightarrow_{\mathsf{fin}} T_1^-) \rightarrow_{\mathsf{ne}} \mathsf{UPred}(\mathsf{Val})) \rightarrow_{\mathsf{ne}} (\mathsf{Loc} \rightarrow_{\mathsf{fin}} T_2^-) \rightarrow_{\mathsf{ne}} \mathsf{UPred}(\mathsf{Val}).$$

As is usual in such cases, there is only one sensible definition: we define FArr as

$$\mathsf{FArr}\ (f : T_2^- \rightarrow_{\mathsf{ne}} T_1^-, g : T_1^+ \rightarrow_{\mathsf{ne}} T_2^+)\ (P : (\mathsf{Loc} \rightarrow_{\mathsf{fin}} T_1^-) \rightarrow_{\mathsf{ne}} \mathsf{UPred}(\mathsf{Val}))$$
$$(w : \mathsf{Loc} \rightarrow_{\mathsf{fin}} T_2^-) = P(\mathsf{map}\ f\ w),$$

i.e., use the function f to map over the finite map that describes the world, and finish the definition by using the predicate we took as an argument. Of course, we also need to check that this definition is non-expansive in all of the arguments, which is a simple excercise.

To prove that F actually forms a functor, we need one more check: the definition of FArr has to preserve compositions of non-expansive maps, as well as the

---

[4] For the actual application to modeling ML-style reference types, the functions should not only be non-expansive, but also monotone wrt. a suitable extension ordering on the worlds. The ModuRes library includes support for such, but we omit that here.

identity function. This is expressed using the BiFunctor typeclass. Again, these
conditions typically amount to simple proofs.

The final ingredient in the interface is that we should provide a proof that
the type produced by F, if it is applied to the singleton space, is inhabited.
Obviously, this is easy to achieve for our running example: any constant uniform
predicate is a good candidate, and we have the freedom to pick any one.

3.2   Interface of the Solution

Now that we know how to represent a recursive domain equation, we can look at
what our general solution provides. The signature looks as follows (here, again,
specialised to the type of COFEs, rather than more general enriched categories).

Module Type SolutionType(InputF : SimplInput).
  Import InputF.
  Axiom TInf : COFE.
  Axiom Fold : ▷ (F TInf TInf) $\to_{ne}$ TInf.
  Axiom Unfold : TInf $\to_{ne}$ ▷ (F TInf TInf).
  Axiom FU_id : Fold ∘ Unfold == id TInf.
  Axiom UF_id : Unfold ∘ Fold == id (▷ (F TInf TInf)).
  . . .
End SolutionType.

First of all, the solution provides a COFE by the name of TInf. This is the
solution to the recursive domain equation, an abstract complete, ordered family
of equivalences. Since we don't know its definition (the definition is not provided
as part of the interface) — and, indeed, we do not need to know it — we need
some other ways to use it. This is provided by the two dual functions: Fold
and Unfold, that convert between TInf and the input functor that defines the
recursive domain equation, F.

The Unfold function takes an object of type TInf to (F TInf TInf): but there
is a twist, namely, the ▷ operator, usually called "later". This operator acts on
the distances of the spaces that is its argument, bringing them "one step down".
That is, if we have $m_1 = n + 1 = m_2$ in some space M, we only have $m_1 = n = m_2$
in the space ▷ M. This has consequences for the Unfold function: assume we have
some elements t_1, t_2 of our solution TInf, and that $t_1 = n + 1 = t_2$. In such a
case, we can learn something about the structure of these elements by applying
Unfold (remember that we know the definition of F), but we can only establish
that Unfold $t_1 = n$ = Unfold t_2. We will see in the following section that while
this affects the way we use the solution, it does not give rise to any problems.

While Unfold provides us with a way of extracting information from the
solution, Fold does the converse: it gives us a way to convert an object of type
(F TInf TInf) into a TInf. Similarly to Unfold, there is a later operator, although
in this case it allows us to *increase* the level of approximation, since it appears
on the argument type.

Finally, the solution provides two equations that describe the effects of Fold
and Unfold. These effects are simple: if one follows Fold with an Unfold, the

resulting function is equivalent to an identity; similarly if we follow unfolding with folding. This provides us with the only way to eliminate the calls to Unfold and Fold that can appear in the definitions that use the solution.

3.3 Using the Solution

As we can see, apart for the later operators appearing in some places, the solution interface is remarkably simple. However, this leads to the question of how we can use it in practice. To address this question, in this section we take the solution that arises from our running example, and show some key cases of a simple unary logical relation that uses the solution as a foundation.

In Sect. 3.1 we have defined the functor F that describes the recursive domain equation that arises from the presence of reference types. The solution of this equation has given us a type TInf. The final types we need to define now are the semantic types that we will use to interpret the syntax (the type T), and the useful helper type of worlds W.

Definition T := F TInf TInf.
Definition W := Loc $\rightharpoonup_{fin}$ TInf.

The easiest way to define a logical relation in Coq is by first defining semantic operators that will be used to interpret the programming language type constructors. Particularly interesting is the semantic operator for interpreting reference types. Naturally, the operator, call it sref, should map semantic types to semantic types: sref : T $\rightarrow_{ne}$ T. The idea is that if R is a semantic type interpreting a programming language type τ then sref(R) should interpret the programming language type ref τ, and thus, loosely speaking, sref(R) should, given a world w consist of those locations in the world whose type matches R. Now, if we unroll the definitions of T and F, we can see that there are a lot of side conditions about non-expansiveness that we will need to prove when defining sref. Thus, it is easiest to start by building the actual predicate. This can be done roughly as follows (explanation follows below):

Inductive refR : T $->$ W $->$ nat $->$ val $->$ Prop :=
| RRef (w : W) (R : T) (Rw : TInf) l n (HLup : w l = Some Rw)
    (HEqn : R = n = Unfold Rw) : refR R w n (vLoc l).

The next step is to show that refR is in reality a uniform predicate, and that the remaining function spaces are non-expansive. We elide these straightforward proofs, and instead examine the definition itself.

Let us look at the arguments of refR. The first, R, is the semantic type that models the argument of the reference type. The following, w, is the world followed by the step index of the uniform predicate, and the value that inhabits our type. Obviously, this value should be a location, but what makes a location a proper value of a reference type? The answer lies in the assumptions: HLup ensures that the location is defined in the world, and gives us a type Rw. However, Rw is of the type TInf, and R is at type T. This is where the conversion functions from the solution interface come in: we can use Unfold to cast Rw to type T (with the equalities brought "one step down"), which allows us to compare it to R with the HEqn assumption: we claim that R is indistinguishable from Rw at the given approximation level.

The other place in the definition of the logical relation that needs to use the world in a nontrivial way is the interpretation of computations. To interpret computations we need to take a heap that *matches the world* and, in addition to ensuring that the resulting value is in the interpretation of the type, we need to check that the resulting heap also matches the world (allowing for its evolution). What does it mean for a heap to match the world? Clearly, all the locations in the world should be defined in the heap. Additionally, the values stored in the heap should belong to the semantic types stored in the world. This gives us the following definition:

Definition interpR $(w : W)$ $(n : nat)$ $(h : heap) :=$
    forall l R $(HLu : w\ k = Some\ R)$, exists v, $h\ k = Some\ v$ $/\backslash$ Unfold R w n v.

Like in the definition of refR, we use the Unfold function to map the element stored in the world back to type T, in this case in order to apply it to arguments.

One may be concerned that we only use the Unfold function in these definitions: clearly there should be places in the logical relations argument where we need to use the Fold function. The answer is, that Fold is used in the proof of compatibility of the allocation rule. This is the one rule that forces us to extend the world, and to do this we need to come up with an object of type TInf — which we can only procure by using Fold. This is also the only place where we need to use the proof that Fold and Unfold compose to identity, to ensure that the resulting heap matches the extended world.

3.4 Summary

In summary, we have now seen how a user of the ModuRes library can obtain a solution to a recursive domain equation in the category of COFEs by providing a suitable bifunctor. The library then provides a solution with associated isomorphisms, which the user of the library can use to build, e.g., a logical relations interpretation of programming language reference types.

The user of the ModuRes library does not need to understand how the solution is constructed. The construction in Coq follows the proof in Birkedal *et al.* [10].

4   Building Models of Higher-Order Logics

In this section we explain how the ModuRes library can be used to build models of higher-order logics for reasoning about concurrent higher-order imperative programs. Concretely, we describe the core part of the model of Iris, a recently published state-of-the-art program logic [20]. For reasons of space, we cannot explain in detail here *why* the core part of Iris is defined as it is; for that we refer the reader to *loc. cit.* Instead, we aim at showing how the library supports working with a model of higher-order logic using a recursively defined space of truth values.

To reason about mutable state, Iris follows earlier work on separation logic where a variety of structures with which models of propositions can be built have been proposed [12,13,15,22], see [19, Chap. 7] for a recent review. Iris settles on a simple yet expressive choice to model propositions as suitable subsets of a partial commutative monoid. In the simplest case, the partial commutative monoid can be the one of heaps, as used in classical separation logic, but in general it is useful to allow for much more elaborate partial commutative monoids [20]. To reason about concurrent programs, Iris propositions are indexed over named invariants, which are used to describe how some shared state may evolve. There has been a lot of research on developing a rich language for describing invariants (*aka* protocols); one of the observations of the work on Iris is that it suffices to describe an invariant by a predicate itself, when the predicate is over a suitably rich notion of partial commutative monoid. In other words, the invariants in Iris are described by worlds, which map numbers (names of invariants) to propositions, and a proposition is a map from worlds to (a suitable) powerset of a partial commutative monoid M. Thus, to model Iris propositions using the ModuRes library, we use as input the following bifunctor:

$$F(P^-, P^+) = (\mathbb{N} \rightharpoonup_{fin} P^-) \to_{ne,mon} \mathsf{UPred}(M),$$

Note that the functor is very similar to the one for modeling ML reference types. Here $\mathsf{UPred}(M)$ is the COFE object consisting of uniform predicates over the partial commutative monoid M, and $\to_{ne,mon}$ denotes the set of non-expansive and monotone functions. For monotonicity, the order on $(\mathbb{N} \rightharpoonup_{fin} P^-)$ is given by the extension ordering (inclusion of graphs of finite functions) and the order on $\mathsf{UPred}(M)$ is just the ordinary subset ordering. Using the library, we obtain a solution, which we call $\mathsf{PreProp}$. Then we define

$$\mathsf{Wld} = \mathbb{N} \rightharpoonup_{fin} \mathsf{PreProp} \qquad \mathsf{Props} = \mathsf{Wld} \to_{ne,mon} \mathsf{UPred}(M).$$

The type Props serves as the semantic space of Iris propositions. Recall that the interface of the solution gives us an isomorphism between $\mathsf{PreProp}$ and $\triangleright\mathsf{Props}$ — a fact that is not necessary to show Props form a general separation logic, but crucial when we want to make use of the invariants in the logic-specific development of Iris.

Let us take stock. What we have achieved so far is to define the propositions (the object of truth values) of Iris, as a complete ordered family of equivalences, and we obtained it by solving a recursive domain equation. We want to use this to get a model of higher-order separation logic. In Iris, types are modeled by complete ordered families of equivalences, terms by non-expansive functions, and the type of propositions is modeled by Props. We can show on paper (see the tutorial [8]) that this is an example of a so-called BI-hyperdoctrine, a categorical notion of model of higher-order separation logic [7]. In the ModuRes library we do not provide a general formalization of BI-hyperdoctrines. Instead, we have formalized a particular class of BI-hyperdoctrines, namely those where types and terms are modeled by COFEs and non-expansive functions and propositions are modeled by a preordered COFE, the preorder modeling logical entailment. We now describe how that is done, and later we return to the instantiation to Iris.

Let T be a preordered COFE — think of T as the object of truth values. We write ⊑ for the ordering relation. Logical connectives will be described by the laws they need to satisfy. Most of the connectives are relatively easy to model: the usual binary connectives of separation logic are maps of type T →ne T →ne T that satisfy certain laws. For instance, implication is defined by the two axioms

```
and_impl : forall P Q R, and P Q ⊑ R <-> P ⊑ impl Q R;
impl_pord :> Proper (pord --> pord ++> pord) impl;
```

The former of these establishes the adjoint correspondence between conjunction and implication, and the latter ensures that implication is contravariant in its first argument and covariant in its second argument, with respect to the entailment relation (pord denotes the pre-order for which ⊑ is the infix notation).

This leaves us with the issue of quantifiers. Since the library provides the notion of BI-hyperdoctrine that is modeled by complete ordered families of equivalences, the quantifiers should work only for functions between COFEs — indeed, only for functions that are nonexpansive. This gives rise to the following types of the quantifiers:

```
all : forall {U} '{cofeU : cofe U}, (U →ne T) →ne T;
xist : forall {U} '{cofeU : cofe U}, (U →ne T) →ne T;
```

As with the other connectives, we require certain laws to hold. For example, below are the laws for the existential quantifier. The first law expresses that existential quantification is left adjoint to reindexing, while the second law expresses functoriality of existential quantification.

```
xist_L U '{cU : cofe U} :
    forall (P : U →ne T) Q, (forall u, P u ⊑ Q) <-> xist P ⊑ Q;
xist_pord U '{cU : cofe U} (P Q : U →ne T) :
    (forall u, P u ⊑ Q u) -> xist P ⊑ xist Q
```

Now that we have a description of what a model of higher-order separation logic is, we can proceed to show that if we let T = Props, then all the required properties hold. With the help of the ModuRes library we can establish this automatically, and in a modular fashion. Firstly, the library contains a fact that UPred(M), the space of uniform predicates over a partial commutative monoid M, forms a model of higher-order separation logic. Moreover, it also shows that the set of monotone and non-expansive maps from a preordered COFE, such as Wld, to any structure that models a higher-order separation logic itself models higher-order separation logic. Thus, it follows easily — and, thanks to the use of typeclasses, automatically — that the space of Props also satisfies the rules of the logic. In the development process, this allows us to focus on the application-specific parts of the logic, rather than on the general BI structure.

Recursive Predicates. To reason about recursive programs, Iris includes guarded recursively defined predicates, inspired by earlier work of Appel et. al. [3]. In the Coq formalization they are modeled roughly as follows. First, we show that there is an endo-function on UPred(M), also denoted ▷ and also called "later" (though it is different from the later functor on COFEs we described earlier). It is defined by

Definition laterF (p : nat −> T −> Prop) n t :=
  match n **with** O => True
                | S n => p n t
  end.

This later operator extends pointwise to Props. Now if we have a non-expansive function $\varphi : (U \to_{ne} \text{Props}) \to_{ne} (U \to_{ne} \text{Props})$, then if we compose it with the later operator on $U \to_{ne}$ Props we can show that the resulting function is contractive, and hence we get a fixed-point, as described in Sect. 2.3.

The combination of higher-order separation logic with guarded recursively defined predicates makes it possible to give abstract specifications of layered and recursive program modules [25]. Indeed, it is to facilitate this in full generality that we model the types and terms of Iris by COFEs and non-expansive functions, rather than simply by Coq types and Coq functions.

In the formalization of Iris, we also use this facility to define the meaning of Hoare triples via a fixed-point of a contractive function.

We have explained how some of the key features of Iris are modeled using the ModuRes library. The actual formalization of Iris also includes a treatment of the specific program logic rules that Iris includes. They are modeled and proved sound in our formalization of Iris, but we do not discuss that part here, since that part only involves features of the library that we have already discussed.

5   Related Work

Benton *et al.* [6] and Huffman [18] provide libraries for constructing solutions to recursive domain equations using classical domain theory. In recent work on program logics the spaces involved are naturally equipped with a metric structure and the functions needed are suitably non-expansive, a fact used extensively in the ModuRes library. In contrast solutions using domain theory do not appear to provide the necessary properties for modeling such higher-order program logics.

The line of work that is possibly the closest related to the ModuRes library is Hobor *et al.*'s theory of indirection [17] and their associated Mechanized Semantic Library in Coq [4]. It provides an *approximate* step-indexed solution of a recursive domain equation expressed as a functor in the category of sets. The explicit aim of indirection theory is to obtain the approximations in a *simple* setting, and it manages to do so: in the Mechanized Semantic Library, the approximate solution is represented as a usual Coq type. Thus, one can use standard Coq function spaces and standard higher-order quantification, over Coq types.

As argued in detail in [9], the ultrametric — or, equivalently, COFE — treatment that the ModuRes library implements subsumes indirection theory. What's more, it is not necessarily limited to step-indexing over operational models, but can also be readily used over classical domain-theoretic denotational semantics. Generally, it also seems easier to enrich the constructions with additional structure, such as the preorder in Sect. 4, or build recursive objects and predicates through the use of Banach's fixed-point operator.

Another important aspect is pointed out in the recent work by Svendsen and Birkedal [25], where they explore program logics for reasoning about layered and recursive abstractions. To this end, they need to quantify over higher-order predicates, and use them within definitions of other, recursive, predicates. In general this can lead to problems with well-definedness. However, since in their setup all maps are non-expansive by default, the recursive predicate can be defined without problems. In contrast, if the model admitted other functions, the question would be far from certain, and probably require explicit restrictions to non-expansive maps in certain places, and more frequent reasoning explicitly in the model. Thus to model higher-order logics with guarded recursion in general, functions should be modeled by non-expansive functions (rather than all functions). This is the approach taken by the ModuRes library. See Sect. 4 (Exercise 4.17) and Sect. 5 (Corollary 5.22) of the tutorial [8] for more details.

Throughout the development, we intended the ModuRes library to provide a readable and easy to grasp set of tools. This would not be possible without some recent developments in proof engineering. Throughout the development we heavily use the type class feature of Coq, due to Sozeau and Oury [23] to abstract from the particular kind of enriched structure we are working in and present a simple and meaningful context to the user. In setting up the pattern for the hierarchy of classes that forms the backbone of the development — the types with equality, OFEs and COFEs — we follow closely Spitters and van der Weegen's work [24], which inspired our attempt to set up this hierarchy in a clean and readable fashion. We deviate somewhat from their pattern, however, in that we package the operations together with the proofs (see dist in the definition of ofe in Sect. 2), where their approach would separate them from the propositional content. This works because our hierarchy does not contain problematic diamond inheritance patterns, and we found that it improves performance.

6   Conclusions and Future Work

A question one could ask of any development in this line of work is, how easy the library is to adopt. Our early experiences seem encouraging in this regard. An intern, who recently worked with us on applying the library, was able to pick it up with little trouble. Moreover, in the development and formalization of Iris [20] this was also our experience. We believe that the abstract presentation of the solution of the recursive domain equation, and the structured construction of the spaces that the library supports will allow this experience to scale. However, since the authors were involved in both of these projects, it is not a certainty. To make the library more accessible to others, we have started developing a tutorial for the ModuRes library. While still a work in progress, we believe it can already be helpful for potential users of the library. It can be found, along with the library, online at http://cs.au.dk/~birke/modures/tutorial. It features, in particular, a detailed tutorial treatment of the running example we used in this paper, a logical relations interpretation of an ML-like language with general references.

Outside the tutorial context, we see the future of the ModuRes library as serving in the development of formalized models of programming languages and program logics. It would be of particular interest to try to extend the simpler models currently used by tools that attempt to verify programs *within* Coq, such as Bedrock [14] or Charge! [5], in order to enhance the expressive power of their respective logics. The proof-engineering effort required to achieve this in an efficient and scalable is non-trivial, however. We hope that the recent progress by Malecha and Bengtson on reflective tactics can help in this regard [21].

In conclusion, we have presented a library that provides a powerful, low-boilerplate set of tools that can serve to build models of type systems and logics for reasoning about concurrent higher-order imperative programing languages. The structures and properties provided by the library help the user focus on the issues specific to the model being constructed, while the well-definedness of the model and some of the common patterns can be handled automatically. Thus, we believe that this line of work can significantly lower the barrier to entry when it comes to formalizing the complex models of programming languages of today.

Acknowledgements. The formalization of the general solution of the recursive domain equation is inspired by an earlier, unpublished development by Varming and Birkedal. While both the proof engineering methods used and the scope of the ModuRes library differ significantly from this earlier effort, some of the setup is borrowed from that. Yannick Zakowski was the first user of the library, providing important feedback, as well as a formalization of the example used in Sect. 3. We thank the anonymous reviewers for their comments.

This research was supported in part by the ModuRes Sapere Aude Advanced Grant from The Danish Council for Independent Research for the Natural Sciences (FNU).

References

1. Ahmed, A.: Semantics of Types for Mutable State. Ph.D. thesis, Princeton University (2004)
2. Ahmed, A., Appel, A.W., Virga, R.: A stratified semantics of general references embeddable in higher-order logic. In: LICS (2002)
3. Appel, A., Melliès, P.-A., Richards, C., Vouillon, J.: A very modal model of a modern, major, general type system. In: POPL (2007)
4. Appel, A.W., Dockins, R., Hobor, A.: (2009). http://vst.cs.princeton.edu/msl/
5. Bengtson, J., Jensen, J.B., Birkedal, L.: Charge! - a framework for higher-order separation logic in Coq. In: Beringer, L., Felty, A. (eds.) ITP 2012. LNCS, vol. 7406, pp. 315–331. Springer, Heidelberg (2012)
6. Benton, N., Kennedy, A., Varming, C.: Some domain theory and denotational semantics in Coq. In: Berghofer, S., Nipkow, T., Urban, C., Wenzel, M. (eds.) TPHOLs 2009. LNCS, vol. 5674, pp. 115–130. Springer, Heidelberg (2009)

7. Biering, B., Birkedal, L., Torp-Smith, N.: BI hyperdoctrines and higher-order separation logic. In: Sagiv, M. (ed.) ESOP 2005. LNCS, vol. 3444, pp. 233–247. Springer, Heidelberg (2005)
8. Birkedal, L., Bizjak, A.: A taste of categorical logic - tutorial notes (2014). http://cs.au.dk/~birke/modures/tutorial/categorical-logic-tutorial-notes.pdf
9. Birkedal, L., Reus, B., Schwinghammer, J., Støvring, K., Thamsborg, J., Yang, H.: Step-indexed kripke models over recursive worlds. In: POPL (2011)
10. Birkedal, L., Støvring, K., Thamsborg, J.: The category-theoretic solution of recursive metric-space equations. Theor. Comput. Sci. **411**(47), 4102–4122 (2010)
11. Birkedal, L., Støvring, K., Thamsborg, J.: Realizability semantics of parametric polymorphism, general references, and recursive types. In: de Alfaro, L. (ed.) FOSSACS 2009. LNCS, vol. 5504, pp. 456–470. Springer, Heidelberg (2009)
12. Brotherston, J., Villard, J.: Parametric completeness for separation theories. In: POPL (2014)
13. Calcagno, C., O'Hearn, P.W., Yang, H.: Local action and abstract separation logic. In: LICS (2007)
14. Chlipala, A.: The bedrock structured programming system: combining generative metaprogramming and hoare logic in an extensible program verifier. In: ICFP (2013)
15. Dockins, R., Hobor, A., Appel, A.W.: A fresh look at separation algebras and share accounting. In: Hu, Z. (ed.) APLAS 2009. LNCS, vol. 5904, pp. 161–177. Springer, Heidelberg (2009)
16. Di Gianantonio, P., Miculan, M.: A unifying approach to recursive and co-recursive definitions. In: Geuvers, H., Wiedijk, F. (eds.) TYPES 2002. LNCS, vol. 2646, pp. 148–161. Springer, Heidelberg (2003)
17. Hobor, A., Dockins, R., Appel, A.: A theory of indirection via approximation. In: POPL (2010)
18. Huffman, B.: A purely definitional universal domain. In: Berghofer, S., Nipkow, T., Urban, C., Wenzel, M. (eds.) TPHOLs 2009. LNCS, vol. 5674, pp. 260–275. Springer, Heidelberg (2009)
19. Jensen, J.B.: Enabling Concise and Modular Specifications in Separation Logic. Ph.D. thesis, IT University of Copenhagen (2014)
20. Jung, R., Swasey, D., Sieczkowski, F., Svendsen, K., Turon, A., Birkedal, L., Dreyer, D.: Iris: monoids and invariants as an orthogonal basis for concurrent reasoning. In: POPL (2015)
21. Malecha, G., Bengtson, J.: Rtac – a reflective tactic language for Coq (2015) (Submitted for publication)
22. Pottier, F.: Syntactic soundness proof of a type-and-capability system with hidden state. JFP **23**(1), 38–144 (2013)
23. Sozeau, M., Oury, N.: First-class type classes. In: Mohamed, O.A., Muñoz, C., Tahar, S. (eds.) TPHOLs 2008. LNCS, vol. 5170, pp. 278–293. Springer, Heidelberg (2008)
24. Spitters, B., van der Weegen, E.: Type classes for mathematics in type theory. Math. Struct. Comput. Sci. **21**(4), 795–825 (2011)
25. Svendsen, K., Birkedal, L.: Impredicative concurrent abstract predicates. In: Shao, Z. (ed.) ESOP 2014 (ETAPS). LNCS, vol. 8410, pp. 149–168. Springer, Heidelberg (2014)

Transfinite Constructions in Classical Type Theory

Gert Smolka$^{(\boxtimes)}$, Steven Schäfer, and Christian Doczkal

Saarland University, Saarbrücken, Germany
{smolka,schaefer,doczkal}@ps.uni-saarland.de

Abstract. We study a transfinite construction we call tower construction in classical type theory. The construction is inductive and applies to partially ordered types. It yields the set of all points reachable from a starting point with an increasing successor function and a family of admissible suprema. Based on the construction, we obtain type-theoretic versions of the theorems of Zermelo (well-orderings), Hausdorff (maximal chains), and Bourbaki and Witt (fixed points). The development is formalized in Coq assuming excluded middle.

1 Introduction

We are interested in type-theoretic versions of set-theoretic theorems involving transfinite constructions. Here are three prominent examples we will consider in this paper:

- *Zermelo 1904* [12,13]. Every set with a choice function can be well-ordered.
- *Hausdorff 1914* [5]. Every poset with a choice function has a maximal chain.
- *Bourbaki-Witt 1951* [2,9,11]. Every increasing function on a chain-complete poset has a fixed point.

All three results can be obtained in type theory with a transfinite construction we call tower construction. We start with a partial order $\leq$ on a type X and a *successor function* $f : X \to X$ such that $x \leq fx$ for all x. We also assume a subset-closed family of *admissible sets* (i.e., unary predicates on X) and a *join function* $\sqcup$ that yields the suprema of admissible sets. Now for each point a in X we inductively define a set Σ_a we call the tower for a:

$$\frac{}{a \in \Sigma_a} \qquad \frac{x \in \Sigma_a}{fx \in \Sigma_a} \qquad \frac{p \subseteq \Sigma_a \quad p \text{ admissible and inhabited}}{\sqcup p \in \Sigma_a}$$

For every tower we show the following:

1. Σ_a is well-ordered by $\leq$.
2. If Σ_a is admissible, then $\sqcup \Sigma_a$ is the greatest element of Σ_a.
3. $x \in \Sigma_a$ is a fixed point of f iff x is the greatest element of Σ_a.

© Springer International Publishing Switzerland 2015
C. Urban and X. Zhang (Eds.): ITP 2015, LNCS 9236, pp. 391–404, 2015.
DOI: 10.1007/978-3-319-22102-1_26

The proofs make frequent use of excluded middle but no other assumptions are needed.

The tower construction formalizes Cantor's idea of transfinite iteration in type theory. No notion of ordinals is used. In axiomatic set theory, an instance of the tower construction appears first in 1908 in dualized form in Zermelo's second proof [13] of the well-ordering theorem. The general form of the tower construction in set theory was identified by Bourbaki [2] in 1949. Bourbaki [2] defines Σ_a as the intersection of all sets closed under the rules of our inductive definition and proves the three results stated above. Felscher [3] discusses the tower construction in set theory and gives many historical references. Felscher reports that the tower construction was already studied in 1909 by Hessenberg [6] in almost general form (set inclusion as order and union as supremum).

The Bourbaki-Witt theorem stated at the beginning of this section does not appear in Bourbaki [2]. The theorem is, however, an immediate consequence of the results Bourbaki [2] shows for towers (see above). As admissible sets we take chains; Result 1 tells us that Σ_a is a chain; hence it follows with Results 2 and 3 that $\sqcup\Sigma$ is a fixed point of f. In fact, the argument gives us a generalized version of the Bourbaki-Witt fixed point theorem that requires suprema only for well-ordered subsets and that asserts the existence of a fixed point above any given point. Explicit statements of the Bourbaki-Witt fixpoint theorem appear in Witt [11] and Lang [9]. In contrast to Bourbaki [2], who like us sees the tower construction as the main object of interest, Witt and Lang see the fixed point theorem as the main result and hide the tower construction in the proof of the theorem.

The theorems of Zermelo and Hausdorff mentioned above can also be obtained with the tower construction (Bourbaki [2] argues the case for Zermelo's theorem and Zorn's lemma, a better known variant of Hausdorff's theorem). For both theorems, we start with a base type B and take for X the type of all sets over B. We assume that the sets over B are extensional. As ordering we take set inclusion and as admissible sets we take all families of sets over B. For Zermelo's theorem, the successor function adds an element as determined by the given choice function. The well-ordering of the tower $\Sigma_\emptyset$ for the empty set then induces a well-ordering of B. For Hausdorff's theorem, we define the successor function as the function that based on the given choice function adds an element such that a chain is obtained if this is possible. Then $\sqcup\Sigma_\emptyset$ is a maximal chain.

The present paper is organized as follows. We first study a specialized tower construction that suffices for the proofs of Zermelo's and Hausdorff's theorem. We present the specialized tower construction in addition to the general construction since it comes with simpler proofs. The specialized tower construction operates on sets and uses a successor function that adds at most one element. We obtain proofs of Hausdorff's and Zermelo's theorem. From Zermelo's theorem we obtain a result asserting the existence of well-ordered extensions of well-founded relations. We then restart in a more abstract setting and study the general tower construction. We prove that towers are well-ordered and obtain the Bourbaki-Witt fixed point theorem.

The development of this paper is formalized in Coq assuming excluded middle and can be found at https://www.ps.uni-saarland.de/extras/itp15. The development profits much from Coq's support for inductive definitions. The material in this paper is a perfect candidate for formalization since it is abstract and pretty formal anyway. The interactive support Coq provides for the construction and verification of the often technical proofs turned out to be beneficial. The fact that Coq comes without built-in assumptions like choice and extensionality also proved to be beneficial since this way we could easily identify the minimal assumptions needed for the development.

We mention some related work in type theory. A type-theoretic proof of Zermelo's theorem based on Zorn's lemma appears in Isabelle's standard library [8] (including well-ordered extensions of well-founded relations). Ilik [7] presents a type-theoretic proof of Zermelo's theorem formalized in AgdaLight following Zermelo's 1904 proof. Bauer and Lumsdaine [1] study the Bourbaki-Witt fixed point principle in an intuitionistic setting.

2   Sets as Unary Predicates

Assumption 1. *We assume excluded middle throughout the paper.*

Let X be a type. A *set over* X is a unary predicate on X. We write

$$set\ X\ :=\ X \to Prop$$

for the type of sets over X and use familiar notations for sets:

$$x \in p := px \qquad\qquad \{x \mid s\} := \lambda x.s$$
$$p \subseteq q := \forall x \in p.\ x \in q \qquad\qquad \{x\} := \lambda z.\ z = x$$
$$p \subset q := p \subseteq q \land \exists x \in q.\ x \notin p \qquad X := \lambda x.\top \quad \emptyset := \lambda x.\bot$$
$$p \cap q := \{x \mid x \in p \land x \in q\} \qquad p \cup q := \{x \mid x \in p \lor x \in q\}$$
$$p \setminus q := \{x \mid x \in p \land x \notin q\} \qquad \neg p := \{x \mid x \notin p\}$$

We call a set *inhabited* if it has at least one member, and *unique* if it has at most one member. A *singleton* is a set that has exactly one member. We call a set *empty* if it has no member. We call two sets p and q *comparable* if either $p \subseteq q$ or $q \subseteq p$.

We call a type X *extensional* if two sets over X are equal whenever they have the same elements. If we assume excluded middle, sets over an extensional type are very much like the familiar mathematical sets.

Fact 2. *Let p and q be sets over an extensional type X. Then:*

1. *If $p \subseteq q$ and $q \subseteq p$, then $p = q$.*
2. *$p \subseteq q$ iff $p \subset q$ or $p = q$.*
3. *$p \subset q$ iff $p \subseteq q$ and $p \neq q$.*
4. *$\neg\neg p = p$, $\neg(p \cap q) = \neg p \cup \neg q$, $\neg(p \cup q) = \neg p \cap \neg q$, and $p \setminus q = p \cap \neg q$.*

By a *family over* X we mean a set over $\text{set } X$. We define *intersection* and *union* of families as one would expect:

$$\bigcap F := \{\, x \mid \forall p \in F.\ x \in p \,\} \qquad \bigcup F := \{\, x \mid \exists p \in F.\ x \in p \,\}$$

Note that the members of a family are ordered by inclusion.

3   Orderings and Choice Functions

Let X be a type.

Let R be a binary predicate on X and p be a set over X. We define

$$LRp := \{\, x \in p \mid \forall y \in p.\ Rxy \,\}$$

and speak of the *set of least elements for R and p*.

Let p be a set over X. A *partial ordering of p* is a binary predicate $\leq$ on X satisfying the following properties for all $x, y, z \in p$:

- *Reflexivity:* $x \leq x$.
- *Antisymmetry:* If $x \leq y$ and $y \leq x$, then $x = y$.
- *Transitivity:* If $x \leq y$ and $y \leq z$, then $x \leq z$.

A *linear ordering of p* is a partial ordering $\leq$ of p such that for all $x, y \in p$ either $x \leq y$ or $y \leq x$. A *well-ordering of p* is a partial ordering $\leq$ of p such that every inhabited subset of p has a least element (i.e., $L(\leq)q$ is inhabited whenever $q \subseteq p$ is inhabited). Note that every well-ordering of p is a linear ordering of p.

A *partial ordering of X* is a partial ordering of $\lambda x. \top$. *Linear orderings of X* and *well-orderings of X* are defined analogously.

We define the notation $x < y := x \leq y \wedge x \neq y$.

Fact 3 (Inclusion and Families). *Let F be a family over an extensional type X. Then inclusion $\lambda pq.\ p \subseteq q$ is a partial ordering of F. Moreover:*

1. *Inclusion is a linear ordering of F iff two members of F are always comparable.*
2. *Inclusion is a well-ordering of F iff $\bigcap G \in G$ for every inhabited $G \subseteq F$.*

A *choice function for X* is a function $\gamma : \text{set } X \to \text{set } X$ such that $\gamma p \subseteq p$ for every p and γp is a singleton whenever p is inhabited. Our definition of a choice function is such that no description operator is needed to obtain a choice function from a well-ordering.

Fact 4. *If $\leq$ is a well-ordering of X, then $L(\leq)$ is a choice function for X.*

4   Special Towers

An *extension function* for a type X is a function $\eta : \text{set } X \to \text{set } X$ such that $\eta p \subseteq \neg p$ for all p. An extension function η is called *unique* if ηp is unique for every p.

Assumption 5. *Let* X *be an extensional type and* η *be a unique extension function for* X.

We define $p^+ := p \cup \eta p$ and speak of *adjunction*. We inductively define a family Σ over X:

$$\frac{F \subseteq \Sigma}{\bigcup F \in \Sigma} \qquad\qquad \frac{p \in \Sigma}{p^+ \in \Sigma}$$

We refer to the definition of Σ as *specialized tower construction* and call the family Σ *tower* and the elements of Σ *stages*. We have $\bigcup \emptyset = \emptyset \in \Sigma$. Note that $\emptyset$ is the least stage and $\bigcup \Sigma$ is the greatest stage. Also note that Σ is a minimal family over X that is closed under union and adjunction.

Fact 6. $\eta(\bigcup \Sigma) = \emptyset$.

Proof. Suppose $x \in \eta(\bigcup \Sigma)$. Since $(\bigcup \Sigma)^+ \subseteq \bigcup \Sigma$, we have $x \in \bigcup \Sigma$. Contradiction since η is an extension function. □

We show that Σ is linearly ordered by inclusion. We base the proof on two lemmas.

Lemma 7. *Let a set* p *and a family* F *be given such that* p *is comparable with every member of* F. *Then* p *is comparable with* $\bigcup F$.

Proof. Suppose $\bigcup F \not\subseteq p$. Then there exists $q \in F$ such that $q \not\subseteq p$. Thus $p \subseteq q$ by assumption. Hence $p \subseteq \bigcup F$. □

Lemma 8. *Let* $p \subseteq q^+$ *and* $q \subseteq p^+$. *Then* p^+ *and* q^+ *are comparable.*

Proof. The claim is obvious if $\eta p \subseteq q$ or $\eta q \subseteq p$. Otherwise, we have $x \in \eta p \setminus q$ and $y \in \eta q \setminus p$. We show $p = q$, which yields the claim.

Let $z \in p$. Then $z \in q^+$. Suppose $z \in \eta q$. Then $z = y$ by uniqueness of ηq. Contradiction since $y \notin p$.

The other direction is analogous. □

Fact 9 (Linearity). Σ *is linearly ordered by inclusion.*

Proof. Let p and q be in Σ. We show by nested induction on $p \in \Sigma$ and $q \in \Sigma$ that p and q are comparable. There are four cases. The cases where p or q is a union follow with Lemma 7. The remaining case where $p = p_1^+$ and $q = q_1^+$ follows with Lemma 8 by exploiting the inductive hypotheses for the pair p_1 and q and the pair q_1 and p. □

Fact 10 (Inclusion). *Let* $p, q \in \Sigma$ *and* $p \subset q$. *Then* $p^+ \subseteq q$.

Proof. By Linearity we have $p^+ \subseteq q$ or $q \subseteq p^+$. The first case is trivial. For the second case we have $p \subset q \subseteq p^+ = p \cup \eta p$. Since ηp is unique, we have $q = p^+$. The claim follows. □

Fact 11 (Greatest Element). *Let $p^+ = p \in \Sigma$. Then p is the greatest element of Σ.*

Proof. Let $q \in \Sigma$. We show $q \subseteq p$ by induction on $q \in \Sigma$.

Union. Let $F \subseteq \Sigma$. We show $\bigcup F \subseteq p$. Let $r \in F$. We show $r \subseteq p$. The claim follows by the inductive hypothesis.

Adjunction. Let $q \in \Sigma$. We show $q^+ \subseteq p$. We have $q \subseteq p$ by the inductive hypothesis. If $q = p$, $q^+ = p^+ = p$ by the assumption. If $q \neq p$, we have $q^+ \subseteq p$ by Fact 10. □

Fact 12. *Let $p^+ = p \in \Sigma$. Then $p = \bigcup \Sigma$.*

Proof. We know that $\bigcup \Sigma$ is the greatest element of Σ. Since greatest elements are unique, we know $p = \bigcup \Sigma$ by Fact 11. □

Lemma 13. *Let $F \subseteq \Sigma$ be inhabited. Then $\bigcap F \in F$.*

Proof. By contradiction. Suppose $\bigcap F \notin F$. Then $\bigcap F \subset q$ whenever $q \in F$. We define

$$p := \bigcup \{ q \in \Sigma \mid q \subseteq \bigcap F \}$$

We have $p \subseteq \bigcap F$. Since $\bigcap F \notin F$, we have $p \subset q$ whenever $q \in F$. Thus $p^+ \subseteq \bigcap F$ by Fact 10. Hence $p^+ \subseteq p$ by the definition of p. Thus $p^+ = p$. Let $q \in F$. We have $p \subseteq \bigcap F \subseteq q \subseteq \bigcup \Sigma = p$ by Fact 12. Thus $\bigcap \Sigma = q \in F$. Contradiction. □

Theorem 1 (Well-Ordering). *Σ is well-ordered by inclusion.*

Proof. Follows with Lemma 13 and Fact 3. □

Fact 14 (Intersection). *Σ is closed under intersections.*

Proof. Follows with Lemma 13 and Fact 15. □

We call an extension function η *exhaustive* if ηp is inhabited whenever $\neg p$ is inhabited.

Fact 15. *If η is exhaustive, then $\bigcup \Sigma = X \in \Sigma$.*

Proof. Follows with Fact 6. □

5   Hausdorff's Theorem

Assumption 16. *Let R be a binary predicate on X and γ be a choice function for X.*

We call a set p over X a *chain* if for all $x, y \in p$ either Rxy or Ryx. We call a family over X *linear* if it is linearly ordered by inclusion.

Fact 17. *The union of a linear family of chains is a chain.*

We show that the tower construction gives us a maximal chain. We choose the extension function $\eta p := \gamma(\lambda x.\ x \notin p \wedge chain\ (p \cup \{x\}))$. Clearly, η is unique. Moreover, all stages are chains since the rules of the tower construction yield chains when applied to chains. That the union rule yields a chain follows from Fact 17 and the linearity of Σ. We have $\eta(\bigcup \Sigma) = \emptyset$ for the greatest stage. By the definition of η it now follows that $\bigcup \Sigma$ is a maximal chain. This completes our proof of Hausdorff's theorem.

Theorem 2 (Existence of Maximal Chains). *Let X be an extensional type with a choice function. Then every binary predicate on X has a maximal chain.*

Proof. Follows from the development above. □

Hausdorff's theorem can be strengthened so that one obtains a maximal chain extending a given chain. For the proof one uses an extension function that gives preference to the elements of the given chain. If q is the given chain, the extension function $\lambda p.\ \gamma(\lambda x.\ x \notin p \wedge chain\ (p \cup \{x\}) \wedge (x \in q \vee q \subseteq p))$ does the job.

We remark that Zorn's lemma is a straightforward consequence of Hausdorff's theorem.

6   Zermelo's Theorem

We can obtain a well-ordering of $\bigcup \Sigma$ by constructing an injective embedding of $\bigcup \Sigma$ into Σ. This gives us a well-ordering of X if the extension function is exhaustive. Since a choice function γ for X gives us a unique and exhaustive extension function $\lambda p.\ \gamma(\neg p)$, we have arrived at a proof of Zermelo's theorem.

We define the *stage for x* as the greatest stage not containing x:

$$\overline{x} := \bigcup \{ q \in \Sigma \mid x \notin q \}$$

Fact 18. *Let $x \in \bigcup \Sigma$. Then $\eta \overline{x} = \{x\}$.*

Proof. Since η is unique it suffices to show $x \in \eta \overline{x}$. Suppose $x \notin \eta \overline{x}$. Then $\overline{x}^+ \subseteq \overline{x}$ since $\overline{x}^+$ is a stage not containing x. Thus $\overline{x} = \bigcup \Sigma$ by Fact 12. Contradiction since $x \notin \overline{x}$. □

Fact 19 (Injectivity). *Let $x, y \in \bigcup \Sigma$ and $\overline{x} = \overline{y}$. Then $x = y$.*

Proof. Follows with Fact 18 and the uniqueness of η since $\eta \overline{x} = \eta \overline{y}$. □

Theorem 3. $x \leq_\eta y := \overline{x} \subseteq \overline{y}$ *is a well-ordering of* $\bigcup \Sigma$.

Proof. Follows with Fact 19 and Theorem 1. □

We prove some properties of the well-ordering we have obtained for $\bigcup \Sigma$.

Fact 20. *Let $x \in p \in \Sigma$. Then $\overline{x} \subset p$.*

Proof. By Linearity it suffices to show that $p \subseteq \overline{x}$ is contradictory, which is the case since $x \in p$ and $x \notin \overline{x}$. $\square$

Fact 21. *Let* $x \in \bigcup \Sigma$. *Then* $\overline{x} = \{\, z \in \bigcup \Sigma \mid \overline{z} \subset \overline{x} \,\}$.

Proof. Let $z \in \overline{x}$. We have $\overline{z} \subset \overline{x}$ by Fact 20. Let $z \in \bigcup \Sigma$ and $\overline{z} \subset \overline{x}$. We show $z \in \overline{x}$. We have $z \in \eta \overline{z} \subseteq \overline{x}$ by Facts 10 and 18. $\square$

Let $\leq$ be a partial ordering of X. We define the *segment for x and $\leq$* as $S_x := \{\, z \mid z < x \,\}$. We say that a well-ordering $\leq$ *agrees* with a choice function γ if $\gamma(\neg S_x) = \{x\}$ for every x.

Theorem 4 (Existence of Well-Orderings). *Let X be an extensional type and γ be a choice function for X. Then there exists a well-ordering of X that agrees with γ.*

Proof. Let $\eta p := \gamma(\neg x)$. Then η is a unique and exhaustive extension function for X. We have $X = \bigcup \Sigma$ by Fact 15. Thus $\leq_\eta$ is a well-ordering of X by Theorem 3. Let S_x^η be the segment for x and $\leq_\eta$. We have $S_x^\eta = \overline{x}$ by Fact 21. Moreover, $\leq_\eta$ agrees with γ since $\gamma(\neg S_x^\eta) = \eta(S_x^\eta) = \eta \overline{x} = \{x\}$ by Facts 18 and 21. $\square$

One can show that there exists at most one well-ordering of X that agrees with a given choice function. Thus our construction yields the unique well-ordering that agrees with the given choice function. This is also true for the well-orderings obtained with Zermelo's proofs [12,13].

We show that with Theorem 4 we can get well-ordered extensions of well-founded predicates. Let R be a binary predicate on X and p be a set over X. We define

$$ MRp := \{\, x \in p \mid \forall y \in p.\ Ryx \to y = x \,\} $$

and call R *well-founded* if MRp is inhabited whenever p is inhabited. Note that $\lambda xy.\bot$ is well-founded, and that every well-ordering is well-founded. We say that a binary predicate $\leq$ *extends* R if $x \leq y$ whenever Rxy.

Corollary 1 (Existence of Well-Ordered Extensions). *Let X be an extensional type with a choice function. Then every well-founded predicate on X can be extended into a well-ordering of X.*

Proof. Let R be a well-founded predicate on X. From the given choice function we obtain a choice function γ such that $\gamma p \subseteq MRp$ for all sets p over X. This is possible since R is well-founded. By Theorem 4 we have a well-ordering $\leq$ of X that agrees with γ. Let Ryx. We show $y \leq x$ by contradiction. Suppose not $y \leq x$. By Linearity and excluded middle we have $x < y$. Since $\leq$ agrees with γ, we have $x \in \gamma(\neg S_x) \subseteq MR(\neg S_x)$. Since $y \in \neg S_x$ and Ryx, we have $y = x$. Contradiction. $\square$

We return to the tower construction and show that Σ contains exactly the lower sets of the well-ordering $\leq_\eta$ of $\bigcup \Sigma$.

Fact 22. *Let $p, q \in \Sigma$.*

1. If $p \subset q$, then $\eta p \neq \eta q$.
2. If $\eta p = \eta q$, then $p = q$.

Proof. Claim 2 follows with excluded middle and Linearity from Claim 1. To show Claim 1, let $p \subset q$. Then $\eta p \subseteq q$ by Fact 10. Suppose $\eta p = \eta q$. Then $\eta q = \emptyset$ and thus $\eta p = \emptyset$. Hence $p = \bigcup \Sigma$ by Fact 12. Contradiction since $p \subset q \in \Sigma$. $\square$

Fact 23. $p \in \Sigma$ *if and only if $p = \bigcup \Sigma$ or $p = \overline{x}$ for some $x \in \bigcup \Sigma$.*

Proof. The direction from right to left is obvious. For the other direction assume $p \in \Sigma$ and $p \neq \bigcup \Sigma$. By Fact 12 we have some $x \in \eta p \in \bigcup \Sigma$. By Fact 22 we have $p = \overline{x}$ if $\eta p = \eta \overline{x}$. Follows by Fact 18 and the uniqueness of η. $\square$

Bourbaki [2] shows that Zermelo's theorem can be elegantly obtained from Zorn's lemma. A similar proof appears in Halmos [4]. Both proofs can be based equally well on Hausdorff's theorem.

7   General Towers

Our results for the specialized tower construction are not fully satisfactory. Intuition tells us that Σ should be well-ordered for every extension function, not just unique extension functions. However, in the proofs we have seen, uniqueness of the extension function is crucial. The search for a general proof led us to a generalized tower construction where the initial stage of a tower can be chosen freely (so far, the initial stage was always $\emptyset$). Now the lemmas and proofs can talk about all final segments of a tower, not just the full tower. This generality is needed for the inductive proofs to go through.

We make another abstraction step, which puts us in the setting of the Bourbaki-Witt fixed point theorem. Instead of taking sets as stages, we now consider towers whose stages are abstract points of some partially ordered type X. We assume an increasing function $f : X \to X$ to account for adjunction, and a subset-closed family S of sets over X having suprema to account for union. Given a starting point a, the tower for a is obtained by closing under f and suprema from S. We regain the concrete tower construction by choosing set inclusion as ordering and the family of all families of sets for S.

To understand the general tower construction, it is best to forget the specialized tower construction and start from the intuitions underlying the general construction. We drop all assumptions made so far except for excluded middle.

Assumption 24. *Let X be a type and $\leq$ be a partial order on X. Moreover, let f be a function $X \to X$, S be a family over X, and $\sqcup$ be a function set $X \to X$ such that:*

1. For all x : $x \leq fx$ (f is increasing*).*
2. For all $p \subseteq q$: $p \in S$ if $q \in S$ (S is subset-closed*).*
3. For all $p \in S$: $\sqcup p \leq x \leftrightarrow \forall z \in p.\ z \leq x$ ($\sqcup$ yields suprema on S*).*

We call fx the successor *of x, $\sqcup p$ the* join *of p, and the sets in S* admissible.

We think of a tuple $(X, \leq, \sqcup, S)$ satisfying the above assumptions as an *S-complete partial order* having suprema for all admissible sets.

Definition 1. *We inductively define a binary predicate $\rhd$ we call* reachability:

$$\frac{}{x \rhd x} \qquad \frac{x \rhd y}{x \rhd fy} \qquad \frac{p \in S \qquad p \text{ inhabited} \qquad \forall y \in p.\ x \rhd y}{x \rhd \sqcup p}$$

Informally, $x \rhd y$ means that x can reach y by transfinite iteration of f. We define some helpful notations:

$$\Sigma_x := \{ y \mid x \rhd y \} \qquad\qquad\qquad\qquad \textit{tower for } x$$
$$x \bowtie y := x \rhd y \lor y \rhd x \qquad\qquad\qquad \textit{x and y connected}$$
$$x \rhd p := p \in S \land p \text{ inhabited} \land \forall y \in p.\ x \rhd y$$

We will show that every tower is well-ordered by $\leq$ and that the predicates $\leq$ and $\rhd$ agree on towers. With the notation $x \rhd p$ we can write the join rule more compactly:

$$\frac{x \rhd p}{x \rhd \sqcup p}$$

Fact 25. *If $x \in p \in S$, then $x \leq \sqcup p$.*

Fact 26. *If $x \rhd y$, then $x \leq y$.*

Proof. By induction on $x \rhd y$ exploiting the assumption that f is increasing. □

Lemma 27 (Strong Join Rule). *The following rule is admissible for $\rhd$.*

$$\frac{p \in S \qquad p \text{ inhabited} \qquad \forall y \in p\ \exists z \in p.\ y \leq z \land x \rhd z}{x \rhd \sqcup p}$$

Proof. We have $\sqcup p = \sqcup(p \cap \Sigma_x)$ by the assumptions, Fact 25, and the subset-closedness of S. We also have $x \rhd p \cap \Sigma_x$. Thus the conclusion follows with the join rule. □

Fact 28 (Successor). *If $x \rhd y$, then either $x = y$ or $fx \rhd y$.*

Proof. By induction on $x \rhd y$. There is a case for every rule. The case for the first rule is trivial.

Successor. Let $x \rhd y$. We show $x = fy$ or $fx \rhd fy$. By the inductive hypotheses we have $x = y$ or $fx \rhd y$. The claim follows.

Join. Let $x \rhd p$. We show $x = \sqcup p$ or $fx \rhd \sqcup p$. Case analysis with excluded middle.

1. p unique. We have $\sqcup p = y$ for some $y \in p$ since p is inhabited. By the inductive hypothesis for $x \rhd y$ we have $x = y$ or $fx \rhd y$. The claim follows.
2. p not unique. We show $fx \rhd \sqcup p$ by Lemma 27. Let $y \in p$. We need some $z \in p$ such that $y \leq z$ and $x \rhd z$. By the inductive hypothesis for $x \rhd y$ we have two cases.
   (a) $fx \rhd y$. The claim follows with $z := y$.
   (b) $x = y$. We need some $z \in p$ such that $x \leq z$ and $fx \rhd z$. Since p is not unique, we have $z \in p$ different from x. By the inductive hypothesis for $x \rhd z$ we have $fx \rhd z$. The claim follows with $z := z$. $\qquad\square$

Lemma 29. *Let $x \bowtie \sqcup q$ for all $x \in p$. Then either $\sqcup p \leq \sqcup q$ or $\sqcup q \rhd \sqcup p$.*

Proof. Case analysis using excluded middle.

1. $\sqcup q \rhd x$ for some $x \in p$. We show $\sqcup q \rhd \sqcup p$ with Lemma 27. Let $y \in p$. We need some $z \in p$ such that $y \leq z$ and $\sqcup q \rhd z$. By assumption we have $y \bowtie \sqcup q$. If $y \rhd \sqcup q$, the claim follows with $z := x$ since $y \leq x$ by Fact 26. If $\sqcup q \rhd y$, the claim follows with $z := y$.
2. $\sqcup q \rhd x$ for no $x \in p$. We show $\sqcup p \leq \sqcup q$. Let $y \in p$. We show $x \leq \sqcup q$. By assumption we have $x \bowtie \sqcup q$. If $x \rhd \sqcup q$, the claim follows by Fact 26. If $\sqcup q \rhd x$, we have a contradiction. $\qquad\square$

Lemma 30. *Let $a \rhd x$ and $a \rhd y$. Then $x \bowtie y$.*

Proof. By nested induction on $a \rhd x$ and $a \rhd y$. The cases where $x = a$ or $y = a$ are trivial. The cases were $x = fx'$ or $y = fy'$ are straightforward with Fact 28. Now only the case where x and y are both joins remains.

We have $a \rhd \sqcup p$, $a \rhd \sqcup q$, $a \rhd p$, and $a \rhd q$. The inductive hypotheses give us $x \bowtie \sqcup q$ for all $x \in p$ and $x \bowtie \sqcup p$ for all $x \in q$. By Lemma 29 we have $\sqcup p \leq \sqcup q$ or $\sqcup q \rhd \sqcup p$ and also $\sqcup q \leq \sqcup p$ or $\sqcup p \rhd \sqcup q$. Thus $\sqcup p \bowtie \sqcup q$ by antisymmetry of $\leq$. $\qquad\square$

Theorem 5 (Coincidence and Linearity). *Let $x, y \in \Sigma_a$. Then:*

1. $x \leq y$ iff $x \rhd y$.
2. $x \leq y$ or $y \leq x$.

Proof. Follows with Facts 26 and Lemma 30. $\qquad\square$

Lemma 31. *Let $a \rhd b = fb$. Then $x \rhd b$ whenever $a \rhd x$.*

Proof. We show $x \rhd b$ induction on $a \rhd x$. For the first rule the claim is trivial.

Successor. Let $a \rhd x$. We show $fx \rhd b$. We have $x \rhd b$ by the inductive hypothesis. The claim follows with Fact 28.

Join. Let $a \rhd p$. We show $\sqcup p \rhd b$. By Theorem 5 it suffices to show $\sqcup p \leq b$. Let $x \in p$. We show $x \leq b$. Follows by the inductive hypothesis. $\qquad\square$

Theorem 6 (Fixed Point).

1. $fx = x$ and $x \in \Sigma_a$ iff x is the greatest element of Σ_a.
2. If $\Sigma_a \in S$, then $\sqcup \Sigma_a$ is the greatest element of Σ_a.

Proof. Claim (1) follows with Lemma 31 and Fact 26. For Claim (2) let $\Sigma_a \in S$. Then $a \triangleright \Sigma_a$. Thus $a \triangleright \sqcup \Sigma_a$ and $\sqcup \Sigma_a \in \Sigma_a$. The claim follows with Fact 25.   □

Fact 32 (Successor). *Let $x, y \in \Sigma_a$ and $x \leq y \leq fx$. Then either $x = y$ or $y = fx$.*

Proof. By Coincidence we have $x \triangleright y$. By Fact 28 we have $x = y$ or $fx \triangleright y$. If $x = y$, we are done. Let $fx \triangleright y$. By Coincidence we have $fx \leq y$. Thus $y = fx$ by the assumption and antisymmetry of $\leq$.   □

Fact 33 (Join). *Let $x \in \Sigma_a$, $p \subseteq \Sigma_a$, $p \in S$, and $x < \sqcup p$. Then there exists $y \in p$ such that $x < y$.*

Proof. By contradiction. Suppose $x \not< y$ for all $y \in p$. By Linearity we have $y \leq x$ for all $y \in p$. Thus $\sqcup p \leq x$ and therefore $x < x$. Contradiction.   □

8   Well-Ordering of General Towers

We already know that every tower Σ_a is linearly ordered by $\leq$ (Theorem 5). We now show that every tower is well-ordered by $\leq$. To do so, we establish an induction principle for towers, from which we obtain the existence of least elements. We use an inductive definition to establish the induction principle.

Definition 2. *For every a in X we inductively define a set I_a:*

$$\frac{x \in \Sigma_a \qquad \forall y \in \Sigma_a.\ y < x \to y \in I_a}{x \in I_a}$$

Lemma 34 (Induction). $\Sigma_a \subseteq I_a$ *and* $I_a \subseteq \Sigma_a$.

Proof. $I_a \subseteq \Sigma_a$ is obvious from the definition of I_a. For the other direction, let $x \in \Sigma_a$. We show $x \in I_a$ by induction on $a \triangleright x$. We have three cases. If $x = a$, the claim is obvious.

Sucessor. Let $a \triangleright x$. We show $fx \in I_a$. Let $y \in \Sigma_a$ such that $y < fx$. We show $y \in I_a$. We have $x \in I_a$ by the inductive hypothesis. By Linearity and excluded middle we have three cases. If $x < y$, we have a contradiction by Fact 32. If $x = y$, the claim follows from $x \in I_a$. If $y < x$, the claim follows by inversion of $x \in I_a$.

Join. Let $a \triangleright p$. We show $\sqcup p \in I_a$. Let $y \in \Sigma_a$ such that $y < \sqcup p$. We show $y \in I_a$. Since $y < \sqcup p$, we have $z \in p$ such that $y < z$ by Fact 33. By the inductive hypothesis we have $z \in I_a$. The claim follows by inversion.   □

Theorem 7 (Well-Ordering). Σ_a *is well-ordered by* $\leq$.

Proof. Let $x \in p \subseteq \Sigma_a$. Then $x \in I_a$ by Lemma 34. We show by induction on $x \in I_a$ that $L(\leq)p$ is inhabited. If $x \in L(\leq)p$, we are done. Otherwise, we have $y \in p$ such that $y < x$ by Linearity (Theorem 5). The claim follows by the inductive hypothesis. □

We can now prove a generalized version of the Bourbaki-Witt theorem.

Theorem 8 (Bourbaki-Witt, Generalized). *Let X be a type, $\leq$ be a partial order of X, and $\sqcup$ be a function set $X \to X$ such that $\sqcup p$ is the supremum of p whenever p is well-ordered by $\leq$. Let f be an increasing function $X \to X$ and a be an element of X. Then f has a fixed point above a.*

Proof. Let S be the family of all sets over X that are well-ordered by $\leq$. Then all assumptions made so far are satisfied. By Theorem 7 we know that Σ_a is well-ordered by $\leq$. Thus $\Sigma_a \in S$. The claim follows with Theorem 6. □

The general tower construction can be instantiated so that it yields the special tower construction considered in the first part of the paper. Based on this instantiation, the theorems of Hausdorff and Zermelo can be shown as before. For the generalized version of Hausdorff's theorem (extension of a given chain), the fact that the general construction yields a tower for every starting point provides for a simpler proof.

9 Final Remarks

We have studied the tower construction and some of its applications in classical type theory (i.e., excluded middle and impredicative universe of propositions). The tower construction is a transfinite construction from set theory used in the proofs of the theorems of Zermelo, Hausdorff, and Bourbaki and Witt. The general form of the tower construction in set theory was identified by Bourbaki [2] in 1949.

Translating the tower construction and the mentioned results from set theory to classical type theory is not difficult. There is no need for an axiomatized type of sets. The sets used in the set-theoretic presentation can be expressed as types and as predicates (both forms are needed). The tower construction can be naturally expressed with an inductive definition.

We have studied a specialized and a general form of the tower construction, both expressed with an inductive definition. The specialized form enjoys straightforward proofs and suffices for the proofs of the theorems of Zermelo and Hausdorff. The general form applies to a partially ordered type and is needed for the proof of the Bourbaki-Witt fixed point theorem. Our proofs of the properties of the tower construction are different from the proofs in the set-theoretic literature in that they make substantial use of induction. The proofs in the set-theoretic literature have less structure and use proof by contradiction in place of induction (with the exception of Felscher [3]).

There are two prominent unbounded towers in axiomatic set theory: The class of ordinals and the cumulative hierarchy. The cumulative hierarchy is usually obtained with transfinite recursion on ordinals. If we consider an axiomatized type of sets in type theory, the general tower construction of this paper yields a direct construction of the cumulative hierarchy (i.e., a construction not using ordinals). Such a direct construction of the cumulative hierarchy will be the subject of a forthcoming paper.[1]

We have obtained Hausdorff's and Zermelo's theorem under the assumption that the underlying base type is extensional. This assumption was made for simplicity and can be dropped if one works with extensional choice functions and equivalence of sets.

Acknowledgement. It was Chad E. Brown who got us interested in the topic of this paper when in February 2014 he came up with a surprisingly small Coq formalization of Zermelo's second proof of the well-ordering theorem using an inductive definition for the least Θ-chain. In May 2015, Frédéric Blanqui told us about the papers of Felscher and Hessenberg in Tallinn.

References

1. Bauer, A., Lumsdaine, P.L.: On the Bourbaki-Witt principle in toposes. Math. Proc. Camb. Philos. Soc. **155**, 87–99 (2013)
2. Bourbaki, N.: Sur le théorème de Zorn. Archiv der Mathematik **2**(6), 434–437 (1949)
3. Felscher, W.: Doppelte Hülleninduktion und ein Satz von Hessenberg und Bourbaki. Archiv der Mathematik **13**(1), 160–165 (1962)
4. Halmos, P.R.: Naive Set Theory. Springer, New York (1960)
5. Hausdorff, F.: Grundzüge der Mengenlehre. Viet, Leipzig (1914). English translation appeared with AMS Chelsea Publishing
6. Hessenberg, G.: Kettentheorie und Wahlordnung. Journal für die reine und Angewandte Mathematik **135**, 81–133 (1909)
7. Ilik, D.: Zermelo's well-ordering theorem in type theory. In: Altenkirch, T., McBride, C. (eds.) TYPES 2006. LNCS, vol. 4502, pp. 175–187. Springer, Heidelberg (2007)
8. Isabelle/HOL: a generic proof assistant (2015). http://isabelle.in.tum.de
9. Lang, S.: Algebra. Graduate Texts in Mathematics, 3rd edn. Springer, New York (2002)
10. van Heijenoort, J.: From Frege to Gödel: a source book in mathematical logic. Harvard University Press, Cambridge (1967)
11. Witt, E.: Beweisstudien zum Satz von M. Zorn. Mathematische Nachrichten **4**(1–6), 434–438 (1950)
12. Zermelo, E.: Beweis, daß jede Menge wohlgeordnet werden kann. (Aus einem an Herrn Hilbert gerichteten Briefe). Mathematische Annalen **59**, 514–516 (1904). English translation in [10]
13. Zermelo, E.: Neuer Beweis für die Möglichkeit einer Wohlordnung. Mathematische Annalen **65**, 107–128 (1908). English translation in [10]

---

[1] Our initial motivation for studying the general tower construction was the direct construction of the cumulative hierarchy in axiomatic set theory. Only after finishing the proofs for the general tower construction in type theory, we discovered Bourbaki's marvelous presentation [2] of the tower construction in set theory.

A Mechanized Theory of Regular Trees in Dependent Type Theory

Régis Spadotti$^{(\boxtimes)}$

Institut de Recherche en Informatique de Toulouse (IRIT),
Université de Toulouse, Toulouse, France
`Regis.Spadotti@irit.fr`

Abstract. Regular trees are potentially infinite trees such that the set of distinct subtrees is finite. This paper describes an implementation of regular trees in dependent type theory. Regular trees are semantically characterised as a coinductive type and syntactically as a nested datatype. We show how tree transducers can be used to define regular tree homomorphisms by guarded corecursion. Finally, we give examples on how transducers are used to obtain decidability of properties over regular trees. All results have been mechanized in the proof assistant Coq.

1 Introduction

Infinite trees arise in the formalization of the semantics of programming languages or process algebras. For instance, they can be used to give semantics to *while loops* in imperative programming languages where the loops are interpreted as their potentially infinite unfolding. In addition, infinite trees can be used to model diverging or never ending computations.

More abstractly, infinite trees can be used to model cyclic structures. Such circular structures are found, for instance, in cyclic lists, cyclic binary trees [16], iso-recursive types in functional programming algebras [12], or labeled transition systems. However, in practice, it is often the case that these infinite trees are finitely defined in the sense that they are obtained from a *finite cyclic structure* and thus embed a regularity property, *e.g.* the infinite paths of finitely branching trees are *ultimately periodic.*

Coinductive types provide a natural framework to define and reason about these infinite objects. However, the support of coinductive types in proof assistants based on dependent type theory (such as COQ or AGDA) is less mature than for their counterparts, namely inductive types, and is still subject to active research [2,10].

Consequently, reasoning with coinductive types in such proof assistants can be sometimes quite challenging, especially when coinduction is mixed with induction [7,8,15].

The motivation for this work is to provide a mechanized theory of a subset of infinite trees called regular trees in the proof assistant COQ. In particular, this library could be used within COQ as an implementation of regular coinductive

© Springer International Publishing Switzerland 2015
C. Urban and X. Zhang (Eds.): ITP 2015, LNCS 9236, pp. 405–420, 2015.
DOI: 10.1007/978-3-319-22102-1_27

types or it could be used outside of COQ through extraction since the whole development is constructive and axiom-free.

Contributions. In this paper, we make the following contributions[1]:

- We define a characterization of regular trees as a subset of a coinductive type and give a syntax for regular trees as a nested datatype.
- We show that top-down tree transducers may be used to define, by means of guarded corecursion, regular tree homomorphisms.
- Finally, we provide an embedding of a fragment of Computation Tree Logic (CTL [9]) for regular trees and show how, in combination with top-down tree transducers [19], it can be used to decide some properties over regular trees.

Notations. We introduce briefly the type theory we are working with along with various notations. Even though all of the results contained in this article are fully mechanized in COQ, we choose to present this work with slightly different notations, yet very close to the ones used within the implementation, in an effort to improve readability. In particular, we use mixfix notations to avoid to declare notations and some form of dependent pattern-matching in order to hide match annotations.

Dependent product is written $(a : A) \rightarrow B\ a$ while the dependent sum is denoted $(a : A) \times B\ a$ or $\sum_{a:A} B\ a$. When no dependence is involved, we shall drop the formal parameter and simply write $A \rightarrow B$ and $A \times B$ for functions and product of types respectively. Type annotations may be omitted when they can be inferred from the context. The coproduct of two types is written $A \uplus B$, propositional equality (also known as Leibniz equality) of two terms t and s of type A is noted $t =_A s$ or simply $t = s$. The cumulative hierarchy of types is written $\mathcal{U}_i$ but we omit the level and simply write $\mathcal{U}$.

Inductive (resp. coinductive) datatypes are introduced with the keyword **inductive** (resp. **coinductive**). Indexed datatypes representing propositions may rather be introduced with inference rules where single (resp. double) line rules denote inductive (resp. coinductive) definitions. Note that coinductive datatypes are introduced in term of constructors rather than destructors. As a result, we assume that the names of the arguments of the constructors define (implicitly) the destructors.

Finally, under the Curry-Howard isomorphism propositions are interpreted as types but we also use the quantifiers $\forall$, $\exists$ and connectors $\wedge$, $\vee$ (in place of dependent product, dependent sum, and coproduct) in order to emphasize the propositional nature of a statement.

2   Regular Trees

In this section, we aim at defining formally the type of regular trees. Intuitively, a regular tree is a potentially infinite tree having the property that there exist only *finitely many distinct* subtrees [6].

---

[1] COQ files available at http://www.irit.fr/~Regis.Spadotti/regular-tree/.

2.1 Signatures and Trees

Our goal is to reason about trees abstractly, thus we introduce a notion of signature [1] (ranked alphabet) from which the type of trees is derived:

record Sig : $\mathcal{U}$
    constructor $-\lhd-$
      | Op   :) $\mathcal{U}$
      | Dec-Op : $\forall\,(o_1\,o_2 : \mathsf{Op})$, Dec $(o_1 = o_2)$
      | Ar   : $\mathsf{Op} \to \mathbb{N}$

The first field Op describes the set of *operators* or *function symbols* while the field Ar assigns *an arity* to each operator. Note that there is a coercion from Sig to $\mathcal{U}$, thus we write $o : \Sigma$ instead of $o : \Sigma.\mathsf{Op}$ whenever the signature Σ is used in a context where a type is expected. In addition, we assume that there is a coercion from $\mathbb{N}$ to $\mathcal{U}$ through Fin, where Fin n is the type representing the set of $\{0, \ldots, n-1\}$. The second field Dec-Op requires the equality over the type of operators to be decidable, where $\mathsf{Dec}(T)$ means $T \uplus \neg T$. Each signature induces a functor on $\mathcal{U}$, called the extension of a signature:

ext : Sig $\to \mathcal{U} \to \mathcal{U}$
ext S X $\equiv$ $\sum_{(o:S)}$ Vec X $(S.\mathsf{Ar}\,o)$

When T is a type, we write $\Sigma\,T$ to mean ext $\Sigma\,T$.

Given a signature Σ, the type of finite (resp. potentially infinite) trees is obtained as follows:

inductive $\mu\,(\Sigma : \mathsf{Sig}) : \mathcal{U}$
  | $-\lhd-$: $(o : \Sigma) \to$ Vec $(\mu\,\Sigma)\,(\Sigma.\mathsf{Ar}\,o) \to \mu\,\Sigma$

coinductive $\nu\,(\Sigma : \mathsf{Sig}) : \mathcal{U}$
  | $-\lhd-$: $(\mathsf{root} : \Sigma) \to (\mathsf{br} : \mathsf{Fin}\,(\Sigma.\mathsf{Ar}\,o) \to \nu\,\Sigma) \to \nu\,\Sigma$

Informally, the type $\mu\,\Sigma$ (resp. $\nu\,\Sigma$) represents the least (resp. greatest) fixpoint of the extension functor induced by Σ. Note that we use two different representations for the type of subtrees, namely a vector (Vec, defined as a bounded list) and a finite map.

The choice of the vector is motivated by the fact that we want to remain "compatible" with the propositional equality. If we used the finite map representation, proving two trees to be equal would require function extensionality, which is not available within the type theory we are working with.

However, the representation with a finite map for the coinductive type is due to a limitation of COQ since no proper support for nesting inductive and coinductive types is provided. From now on, we shall write coT Σ instead of $\nu\,\Sigma$ to denote the type of potentially infinite trees over the signature Σ.

2.2 Equality Between Infinite Trees

Leibniz equality is not adequate to identify two infinite trees, mainly because the identity type is defined inductively whereas the type of trees is defined coinductively. Thus, we need a more suitable definition to assess equality between two trees, also known as bisimilarity, this time defined *coinductively*:

Definition 1 (Bisimilarity). $\boxed{-\sim-\ \{\Sigma\} : \mathsf{coT}\ \Sigma \to \mathsf{coT}\ \Sigma \to \mathcal{U}}$

$$\sim\text{-intro}\ \frac{(\mathsf{root\text{-}=}\ :\ \mathsf{root}\ t_1 = \mathsf{root}\ t_2) \qquad (\mathsf{br\text{-}}\sim\ :\ \forall\ k,\ \mathsf{br}\ t_1\ k \sim \mathsf{br}\ t_2\ (\mathsf{root\text{-}=}_*\ k))}{t_1 \sim t_2}$$

Because there is a dependence between a function symbol and its arity, we have to transport the equality root-= in the type of k in order to obtain the kth subtree of t_2. This is written as $\mathsf{root\text{-}=}_* : \Sigma.\mathsf{Ar}\ (\mathsf{root}\ t_1) \to \Sigma.\mathsf{Ar}\ (\mathsf{root}\ t_2)^2$.

An alternative definition of bisimilarity is to consider the limit of a chain of equalities of tree approximations.

Definition 2 (Bisim. depth). $\boxed{-\sim_-\ -\ \{\Sigma\} : \mathsf{coT}\ \Sigma \to \mathbb{N} \to \mathsf{coT}\ \Sigma \to \mathcal{U}}$

$$\sim_0\text{-intro}\ \frac{}{t_1 \sim_0 t_2}$$

$$\sim_n\text{-intro}\ \frac{(\mathsf{root\text{-}=}\ :\ t_1 = t_2) \qquad (\mathsf{br\text{-}}\sim\ :\ \forall\ k,\ \mathsf{br}\ t_1\ k\ \sim_n\ \mathsf{br}\ t_2\ (\mathsf{root\text{-}=}_*\ k))}{t_1 \sim_{1+n} t_2}$$

The relationship between these two definitions is established in the next section.

2.3 Coalgebra-Based Approach

It is interesting to view the coinductive type as a (weakly) final coalgebra over a suitable functor in order to circumvent the syntactic guard condition. In our case, we define the coiterator for the type of $\mathsf{coT}\ \Sigma$ as follows:

```
coiterator {Σ} {X} : (X → Σ X) → X → coT Σ
coiterator  outₓ  x  ≡  π₁ (outₓ x)  ◁  coiterator ∘ π₂ (outₓ x)
```

As a result, in order to define a tree over a signature Σ, it suffices to give a type X a Σ-coalgebra structure.

The same principle can be applied to the type witnessing the bisimilarity of two trees $-\sim-$ by proving that it is actually the greatest bisimulation relation:

Lemma 1 (Greatest bisimulation). *Let a signature Σ. For any binary relation $-\mathcal{R}-$ over $\mathsf{coT}\ \Sigma$ such that:*

$$\forall\ s\ t,\ s\ \mathcal{R}\ t \to \sum_{e\,:\,\mathsf{root}\ s\,=\,\mathsf{root}\ t} (\forall\ k,\ (\mathsf{br}\ t_1\ k)\ \mathcal{R}\ (\mathsf{br}\ t_2\ (e_*\ k)))$$

we have $\mathcal{R} \subseteq\ \sim$, i.e., $\forall\ s\ t,\ s\ \mathcal{R}\ t \to s \sim t$.

A binary relation satisfying the condition of the above lemma is called a *bisimulation relation* over $\mathsf{coT}\ \Sigma$. Finally, we have yet another proof method at our disposal, this time based on induction, thanks to the following result:

---

[2] rew-in in COQ.

Lemma 2. *The binary relation defined as*

$$s \, \mathcal{R} \, t := (\varphi : \forall n, \, s \sim_n t)$$

is a bisimulation relation.

The proof of Lemma 2 relies on the fact that equality over the type of operators ($\Sigma.\mathsf{Op}$) is decidable and thus satisfies the *Uniqueness of Identity Proofs* principle. Essentially, this means that the choice of the proof witnessing the equality between the roots is irrelevant. However, without such principle the definition of $-\mathcal{R}-$ is not strong enough to complete the proof and would require an additional hypothesis such as:

$$\forall \, n, \, \mathsf{trunc} \, (\varphi \, (1 + n)) = \varphi \, n$$

where the function trunc has the following type: $s \sim_{1+n} t \to s \sim_n t$.

2.4 Reasoning Modulo Bisimilarity

Contrary to Leibniz equality, the bisimilarity relation is not provably substitutive. This means that it is not possible to substitute bisimilar trees in an arbitrary context unless there is a proof that the context respects bisimilarity. As a result, most of the development containing coinductive trees has to be carried modulo this bisimilarity relation. Without proper quotient types, this is usually achieved by considering types as setoids, *i.e.*, as a pair $(T, - \approx_T -)$ where $\approx_T$ is an equivalence relation over T, and functions and predicates as setoid morphisms. Within the COQ development, setoids and setoid rewriting is achieved by means of type classes. This has the benefit of leaving much of the details implicit while relying on instance resolution to fill in the missing parts. We follow the same approach in the remaining sections of this paper.

In particular, given a type A we can always form the setoid $(A, - = -)$. Given a signature Σ, the setoid over $\mathsf{coT} \, \Sigma$ is defined as the pair $(\mathsf{coT} \, \Sigma, - \sim -)$. Finally, given a setoid $(A, - \approx_A -)$ and family of types $B : A \to \mathcal{U}$ such that B respects the equivalence relation $\approx_A$, the setoid over the type $S := \sum_{(a:A)} B \, a$ is defined as $(S, - \approx_S -)$ where $s \approx_S s' := \pi_1(s) \approx_A \pi_1(s')$.

2.5 A Formal Definition of the Type of Regular Trees

Informally, a tree t is said to be regular when there are only *finitely many distinct* subtrees. In order to translate this statement into a formal definition, we define the type of subtrees and a characterization of finite type.

Definition 3 (Subtree). $\boxed{\, - \preccurlyeq - \, \{\Sigma\} : \mathsf{coT} \, \Sigma \to \mathsf{coT} \, \Sigma \to \mathcal{U} \,}$

$$\preccurlyeq\text{-refl} \, \frac{t_1 \sim t_2}{t_1 \preccurlyeq t_2} \qquad \preccurlyeq\text{-br} \, \frac{k : \Sigma.\mathsf{Ar} \, (\mathsf{root} \, t_2) \quad t_1 \preccurlyeq \mathsf{br} \, t_2 \, k}{t_1 \preccurlyeq t_2}$$

Obviously, the subtree relation respects bisimilarity and is provably transitive. A proof of $t' \preccurlyeq t$ is thus a *finite* path from t to t'. For a given tree t, we note $[- \preccurlyeq t]$ to mean $\sum_{t'}(t' \preccurlyeq t)$.

Definition 4 (Finite type). *Let* $(A, - \approx -)$ *be a setoid. The type* A *is said to be finite if there exists a surjection from a finite type to* A. *More formally, there exist:*

1. *a natural number* ucard
2. *a map* to-index $: A \to$ Fin ucard
3. *a map* from-index $:$ Fin ucard $\to A$

such that for all $(a : A)$, from-index (to-index a) $\approx a$.

There exist others definition of finite type [18]. Some of them require the underlying equality over the type A to be decidable, which, in general, is not the case for coinductive types.

Definition 5 (Regular tree). *Let* Σ *be a signature. A tree* $(\chi : \mathsf{coT}\ \Sigma)$ *is said to be regular if the type* $[- \preccurlyeq \chi]$ *is finite.*
We note Reg_Σ *the type of regular trees over* Σ.

3   A Syntax for Regular Trees

In this section, we aim at defining a syntax for regular trees, also know as *rational expressions* [6]. We proceed as follows.

First, we describe the syntax of regular trees as an inductively-defined heterogeneous datatype, called cyclic terms [16]. Then, we give an interpretation of cyclic terms as a coinductive type corresponding to their infinite unfolding.

Finally, we prove soundness and completeness of the interpretation of cyclic terms in the sense that each cyclic term yields a regular tree and conversely.

3.1   Cyclic Terms

The syntax is defined as an inductive datatype. As a result, we have to find a way to *implicitly* encode cycles. The key idea is to represent a cycle through a binder [16]. The binder will act as a marker for the beginning of the cycle while the bound variable will represent a back-pointer to it, thus creating an implicit cycle. We define the syntax of cyclic terms for a given signature Σ as a nested datatype with De Bruijn index [3]:

```
inductive C (Σ : Sig) (n : ℕ) : 𝒰
  var        : Fin n → C Σ n
  −◁−        : (o : Σ) → Vec (C Σ n) (Σ.Ar o) → C Σ n
  rec − ◁ −  : (o : Σ) → Vec (C Σ (1 + n)) (Σ.Ar o) → C Σ n
```

The definition of C Σ is similar to the definition of the term algebra over Σ (see Sect. 2). The main difference is the presence of two additional constructors, namely var and rec $-\lhd-$, that are used to represent bound variables and binders respectively.

The constructor introducing the binder, namely rec $-\lhd-$, contains some form of redundancy to guarantee that each binder rec is guarded by a function symbol. This will ensure that the interpretation function is total. The type of closed terms $C\ \Sigma\ 0$ is abbreviated to $C\ \Sigma$.

Finally, the type of cyclic terms carries the structure of a monad. As such, we note parallel substitution as

$$- \ggeq - : C\ \Sigma\ n \to (\text{Fin } n \to C\ \Sigma\ m) \to C\ \Sigma\ m$$

while substitution in a rec-bound term with a newly introduced free variable is written as

$$-[* := -] : C\ \Sigma\ (1+n) \to C\ \Sigma\ n \to C\ \Sigma\ n.$$

3.2   Semantics of Cyclic Terms

Cyclic terms over a signature Σ will be interpreted as infinite trees coT Σ. First, we introduce the type of closures in order to ensure that each free variable has a definition:

inductive Closure $\{\Sigma : \text{Sig}\} : \mathbb{N} \to \mathcal{U}$
| [] : Closure Σ 0
| $- :: - : \forall\ \{n\},\ C\ \Sigma\ n \to$ Closure $\Sigma\ n \to$ Closure $\Sigma\ (1+n)$

The type Closure is defined in such a way as to ensure that each free variable may only refer to terms that have been previously introduced within the closure. As a result, this ensures that each free variable is defined by a closed term:

var-def $\{\Sigma\}\ \{n\}$: Closure $\Sigma\ n \to$ Fin $n \to C\ \Sigma$
var-def $(t :: \rho)$ zero $\equiv$ $t \ggeq$ var-def ρ
var-def $(t :: \rho)$ (suc i) $\equiv$ var-def $\rho\ i$

Note that the function var-def terminates since the number n of free variables decreases in-between each recursive call. Given a term and a closure, we get a closed term as follows:

close-term $\{\Sigma\}\ \{n\}$: Closure $\Sigma\ n \to C\ \Sigma\ n \to C\ \Sigma$ zero
close-term ρ t $\equiv$ $t \ggeq$ var-def ρ

Now, we give two interpretation functions of cyclic terms as infinite trees based on two different implementation strategies. The first one is based on substitution and operates on closed terms:

$[\![-]\!]\ \{\Sigma\}\ \{n\}$: $C\ \Sigma \to$ coT Σ
$[\![\ o \lhd os\ ]\!]$ $\equiv$ $o \lhd \lambda i \cdot [\![\ os[i]\ ]\!]$
$[\![\ \text{rec } o \lhd os\ ]\!]$ $\equiv$ $o \lhd \lambda i \cdot [\![\ os[i]\ [\ * := \ \text{rec } o \lhd os\ ]\ ]\!]$

Note that the case for var is omitted since we are working with closed terms.

The second interpretation function is defined for terms that may contain free variables while carrying their definitions within a closure. However, we have to deal with one technicality in defining the function by guarded corecursion. When the term is a free variable, its definition has to be looked up within the closure in order to extract a function symbol in order to ensure productivity. Moreover, note that, nothing prevents the definition of a free variable to be again a free variable. Consequently, the process of looking up the definition of the variable has to be repeated until a term in guarded form (*i.e.* guarded by $- \vartriangleleft -$) is obtained. Though, this process is guaranteed to terminate eventually since closures are inductively defined terms. As a result, we use the coiterator mapping Σ-coalgebras into infinite trees to define the interpretation function. The carrier of the Σ-coalgebra consists in the pairs of terms in guarded form associated with a closure:

$$\mathcal{G} : \mathsf{Sig} \to \mathcal{U}$$
$$\mathcal{G} \ \Sigma \ \equiv \ (n : \mathbb{N}) \times (o : \Sigma) \times \mathsf{Vec} \ (\mathcal{C} \ \Sigma \ n) \ (\Sigma.\mathsf{Ar} \ o) \times \mathsf{Closure} \ \Sigma \ n$$

while the morphism is defined as to inductively compute the next guarded term, by first looking up the definition of a free variable in the closure:

$$\mathsf{lookup} \ \{\Sigma\} \ \{n\} : \mathsf{Fin} \ n \to \mathsf{Closure} \ \Sigma \ n \to \mathcal{G} \ \Sigma$$
$$
\begin{array}{llll}
\mathsf{lookup} & \mathsf{zero} & (\mathsf{var} \ k :: \rho) & \equiv \ \mathsf{lookup} \ k \ \rho \\
\mathsf{lookup} & \mathsf{zero} & (o \vartriangleleft os :: \rho) & \equiv \ (\_ , o , os , \rho) \\
\mathsf{lookup} & \mathsf{zero} & (\mathsf{rec} \ o \vartriangleleft os :: \rho) & \equiv \ (\_ , o , os , \mathsf{rec} \ o \vartriangleleft os :: \rho) \\
\mathsf{lookup} & (\mathsf{suc} \ i) & (\_ :: \rho) & \equiv \ \mathsf{lookup} \ i \ \rho
\end{array}
$$

Finally, any term associated with a closure can be turned into guarded form:

$$\mathsf{to}\mathcal{G} \ \{\Sigma\} \ \{n\} : \mathsf{C} \ \Sigma \ n \to \mathsf{Closure} \ \Sigma \ n \to \mathcal{G} \ \Sigma$$
$$
\begin{array}{llll}
\mathsf{to}\mathcal{G} & (\mathsf{var} \ k) & \rho & \equiv \ \mathsf{lookup} \ k \ \rho \\
\mathsf{to}\mathcal{G} & (o \vartriangleleft os) & \rho & \equiv \ (\_ , o , os , \rho) \\
\mathsf{to}\mathcal{G} & (\mathsf{rec} \ o \vartriangleleft os) & \rho & \equiv \ (\_ , o , os , \mathsf{rec} \ o \vartriangleleft os :: \rho)
\end{array}
$$

The semantics of cyclic term as an infinite tree is thus obtained as follows:

$$\mathcal{G}\text{-coalg} \ \{\Sigma\} : \mathcal{G} \ \Sigma \to \Sigma \ (\mathcal{G} \ \Sigma)$$
$$\mathcal{G}\text{-coalg} \ (m , o , os , \rho) \ \equiv \ (o , \lambda \ i \cdot \mathsf{to}\mathcal{G} \ (os[i]) \ \rho)$$

$$[\![-]\!]_- \ \{\Sigma\} \ \{n\} : \mathsf{C} \ \Sigma \ n \to \mathsf{Closure} \ \Sigma \ n \to \mathsf{coT} \ \Sigma$$
$$[\![t]\!]_\rho \ \equiv \ \mathsf{coiterator} \ \mathcal{G}\text{-coalg} \ (\mathsf{to}\mathcal{G} \ t \ \rho)$$

Theorem 1 (Equivalence of semantics). *Given a signature Σ, a cyclic term t with n free variables and a closure ρ then*

$$[\![\ \mathsf{close-term} \ t \ \rho \ ]\!] \sim [\![\ t \ ]\!]_\rho$$

3.3   From Cyclic Terms to Regular Trees

So far, we have defined two equivalent functions mapping cyclic terms to infinite trees but we have yet to show that these infinite trees are actually regular. It is interesting to remark that the following definitions and theorems are a generalization (to an arbitrary signature) of [5], which deals with the problem of giving a complete axiomatization of the subtyping relation of iso-recursive types. First, we specify, in the form of a structural operational semantics, the subset of terms from which a cyclic term may unfold to:

Definition 6 (Immediate subterm/Subterm modulo unfolding).

$$\sqsubseteq_0\text{-}\vartriangleleft \frac{k : \Sigma.\text{Ar } o}{o \vartriangleleft os \sqsubseteq_0 os[k]} \qquad \sqsubseteq_0\text{-rec} \frac{k : \Sigma.\text{Ar } o}{\text{rec } o \vartriangleleft os \sqsubseteq_0 os[k] \, [\, * := \text{rec } o \vartriangleleft os \,]}$$

We note $-\sqsubseteq-$ *the reflexive-transitive closure of* $-\sqsubseteq_0-$.

In order to show that the type $[\, t \sqsubseteq - \,]$ is finite, we show that it is included in a finite set t^* computed as follows:

$$
\begin{aligned}
-^* \, \{\Sigma\} \, \{n\} &: \mathsf{C} \, \Sigma \, n \to \text{list } (\mathsf{C} \, \Sigma \, n) \\
(\text{var } x)^* &\equiv [\, \text{var } x \,] \\
(o \vartriangleleft os)^* &\equiv (o \vartriangleleft os) :: \bigcup \text{map}^{\text{Vec}} \, -^* \, os \\
(\text{rec } o \vartriangleleft os)^* &\equiv (\text{rec } o \vartriangleleft os) :: \text{map}^{\text{list}} \, [\, * := \text{rec } o \vartriangleleft os \,] \, \big(\bigcup \text{map}^{\text{Vec}} \, -^* \, os\big)
\end{aligned}
$$

Theorem 2. *For any cyclic term t, we have:*

1. *for all u such that $t \sqsubseteq u$ then $u \in t^*$,*
2. *for all $u \in t^*$ then $t \sqsubseteq u$.*

hence the type $[\, t \sqsubseteq - \,]$ is finite.

Theorem 3 (Soundness). *For all closed cyclic term t, the tree $[\![\, t \,]\!]$ is regular.*

Proof. We construct a surjection from $[\, t \sqsubseteq - \,]$ to $[\, - \preccurlyeq t \,]$ and conclude with Theorem 2. $\qquad\qquad\square$

3.4   From Regular Trees to Cyclic Terms

In Sect. 3.3, we showed that each cyclic term interpreted as a potentially infinite tree yields a regular tree. We are now interested in the converse, *i.e.*, to define a function:

$$[\![\, - \,]\!]^{\text{inv}} : \text{Reg } \Sigma \to \mathsf{C} \, \Sigma$$

By definition, regular trees embed a finite structure and thus an iteration principle. To define $[\![\, - \,]\!]^{\text{inv}}$, we introduce an auxiliary function that operates over an arbitrary subterm of a fixed regular tree that we call χ. Intuitively, to construct the cyclic term from the regular tree we have to store the subterms that have already been processed. We know that this will eventually terminate since the set of subterms of a regular tree is finite. However, since equality of infinite trees is not decidable, we are faced with the issue of testing whether a subterm has already been processed. One solution is to exploit the fact that the type $[\, - \preccurlyeq \chi \,]$ is finite, which means by definition of finite, that we know how to index subterms of χ. Thus, instead of storing the subterms we store their indexes, for which equality is decidable. As a result, given a finite type A, we can define a function

$$- \in^? - : \forall \, (a : A) \, (l : \text{list } (\text{Fin ucard}_A)), \left(\sum_{i \, : \, \text{Fin (len } l)} \overline{l(i)} \approx_A a\right) \uplus (\text{to-index}_A \, t \notin l).$$

such that, given an element a of A and a list l, returns either the position of a in the list l, if the index of a is in the list, or a proof that the index of a is not in the list. The notation $\overline{l(i)}$ denotes an element a' such from-index$_A$ $(l(i)) = a'$.

$$[\![-]\!]_\chi^{inv} : \forall t,\ t \preccurlyeq \chi \to \lceil l \rceil \to \mathsf{Env}\ \Sigma\ (\mathsf{len}\ l) \to \mathsf{C}\ \Sigma\ (\mathsf{len}\ l)$$
$$[\![\ t\ ]\!]_\chi^{inv}\ p\ B\ \Gamma \equiv$$
$$\quad \textbf{match}\ t\ \in^i?\ l\ \textbf{with}$$
$$\quad\ |\ \mathsf{yes}\ (i\ ,\ \_)\ \Rightarrow\ \Gamma(i)$$
$$\quad\ |\ \mathsf{no}\ nl\quad \Rightarrow\ \mathsf{rec}\ (\mathsf{root}\ t) \vartriangleleft i \mapsto [\![\ \mathsf{br}\ t\ i\ ]\!]_\chi^{inv}\ (nl\ \lceil :: \rceil B)\ \_\ (\Gamma \oplus \mathsf{var}\ 0)$$

The termination of this recursive function $[\![-]\!]_\chi^{inv}$ is witnessed by the term B whose type is given by

$$\lceil - \rceil\ \{A\} : \mathsf{list}\ A \to \mathcal{U}$$
$$\lceil - \rceil\ \equiv\ \mathsf{Acc}\ (\lambda\ l\ l'\ \cdot \exists\ x,\ x \notin l \wedge l' = x :: l)$$

where $\mathsf{Acc}\ \mathcal{R}$ defines the accessibility predicate over a binary relation $\mathcal{R}$. A list l is said to be accessible, written $\lceil l \rceil$, if the process of adding elements which are not yet in l eventually terminates, *i.e.*, there exists a bound. Given a proof $(nl : x \notin l)$ and a term $(B : \lceil l \rceil)$ we can form the term $(np\ \lceil :: \rceil\ B : \lceil x :: l \rceil)$, obtained by inversion of B, which results in a structurally smaller term than B. Finally, the inverse function $[\![-]\!]^{inv}$ is obtained by starting at the root of χ:

$$[\![-]\!]^{inv}\ \{\Sigma\} : \mathsf{Reg}\ \Sigma \to \mathsf{C}\ \Sigma$$
$$[\![\ \chi\ ]\!]^{inv}\ \equiv\ [\![\ \chi\ ]\!]_\chi^{inv}\ \mathsf{init}\ \preccurlyeq\text{-refl}\ \mathsf{empty}$$

where init is a proof that the empty list of indexes is accessible and empty denotes the empty environment.

Theorem 4 (Completeness). *For all regular tree χ, there exists a cyclic term $\overline{\chi}$ such that $\chi \sim [\![\ \overline{\chi}\ ]\!]$.*

Proof. $\overline{\chi}$ is defined by $[\![\ \chi\ ]\!]^{inv}$. $\square$

4   Defining Regular Tree Homomorphisms

A regular tree homomorphism between an input signature Σ and an output signature Δ is a morphism $\varphi : \mathsf{coT}\ \Sigma \to \mathsf{coT}\ \Delta$ preserving regularity. As an example, consider the two following mutually corecursive functions where $\Sigma = \Delta = \mathsf{Stream}\ A$:

$\varphi_1 : \mathsf{Stream}\ A \to \mathsf{Stream}\ A$ $\qquad$ $\varphi_2 : \mathsf{Stream}\ A \to \mathsf{Stream}\ A$
$\varphi_1\ (x :: xs)\ \equiv\ f_1(x) :: \varphi_2\ xs$ $\qquad$ $\varphi_2\ (x :: xs)\ \equiv\ f_2(x) :: \varphi_1\ xs$

To show that both φ_1 and φ_2 preserves regularity amounts to construct a surjection between the set of subtrees of a regular input stream σ and the set of subtrees of φ_i.

On the other hand, let's first observe through an other example, a tree morphism that does not preserve regularity. For $i \in \mathbb{N}$, we define the family of corecursive functions φ_i as follows:

ψ_i : Stream $\mathbb{N} \to$ Stream $\mathbb{N}$
$\psi_i \ (x :: xs) \ \equiv \ f_i(x) :: \psi_{i+1} \ xs$

If we define, for all x, $f_i(x) = i$, then, for any input stream σ, the resulting stream $\psi_0(\sigma) = \omega$ is not regular.

The difference between the two examples is that ψ is defined using an *infinite family of corecursive equations*. In order to capture this difference, we use a shallow embedding of the formalism of tree transducers [19]. Then, we show that tree transducers induce regular tree homomorphisms.

4.1 Top-Down Tree Transducers

Definition 7 (Top-down tree transducer). *Given a finite type Q, an input signature Σ and an output signature Δ, the type of top-down tree transducer is defined as:*

tdtt : $\mathcal{U} \to Sig \to Sig \to \mathcal{U}$
tdtt $Q \ \Sigma \ \Delta \ \equiv \ Q \to (o : \Sigma) \to \Delta^+(Q \times \Sigma.\mathsf{Ar}\ o)$

The type Q is called *the state space* of the transducer and is used to represent a set of *mutually corecursive definitions*. The definition of tdtt describes a set of rewrite rules: for each state q and function symbol $(o : \Sigma)$ we specify a tree δ to substitute for o. The leafs of δ are pairs composed of the next state and a variable indexing a subtree of o. The tree δ is called the *right-hand side* of the rewrite rule. Finally, for a type X, the type $\Delta^*(X)$ represents the free monad over the signature Δ, inductively defined from the two constructors: var : $X \to \Delta^*(X)$ and $\downarrow_\Delta : \Delta^+(X) \to \Delta^*(X)$.

Definition 8 (Induced natural transformation). *Given a tree transducer $(\tau :$ tdtt $Q \ \Sigma \ \Delta)$ the natural transformation induced by τ over the functors $Q \times \Sigma -$ and $\Delta^+ (Q \times -)$ is defined as:*

$\eta : \forall \{\alpha\}, \ Q \times \Sigma \ \alpha \to \Delta^+ (Q \times \alpha)$
$\eta \ (q \, , \, t) \ \equiv \ \mathsf{map}^{\Delta^+(Q \times -)} \ (\pi_2 \ t) \ (\tau \ q \ (\pi_1 \ t))$

Definition 9 (Induced tree morphism). *A tree transducer $(\tau :$ tdtt $Q \ \Sigma \ \Delta)$ induces a tree morphism defined by coiteration, as follows:*

$\langle\!\langle - \rangle\!\rangle :$ tdtt $Q \ \Sigma \ \Delta \to Q \to \mathsf{coT} \ \Sigma \to \mathsf{coT} \ \Delta$
$\langle\!\langle \tau \rangle\!\rangle \ q \ t \ \equiv \ \mathsf{coiterator} \ \left(\mathsf{fold}^{\Delta^*} \ \eta' \ (\mathsf{map}^\Delta \ \downarrow_\Delta) \right) \ (\mathsf{var} \ (\ q \, , \, t \ ))$
$\quad \textbf{where} \ \ \eta' \ \equiv \ \eta(\tau) \circ \mathsf{map}^{Q \times -} \ \mathsf{out}_\Sigma$

where fold^{Δ^*} *is the non-dependent eliminator of the free monad, and* $\mathsf{out}_\Sigma :$ $\mathsf{coT} \ \Sigma \to \Sigma \ (\mathsf{coT} \ \Sigma)$ *is the destructor for the type* $\mathsf{coT} \ \Sigma$. *Figure 1 illustrates how the tree transducer operates on an input tree.*

Theorem 5. *Given a tree transducer $(\tau :$ tdtt $Q \ \Sigma \ \Delta)$, a state q and a regular tree χ, then $\langle\!\langle \tau \rangle\!\rangle \ q \ \chi$ is regular.*

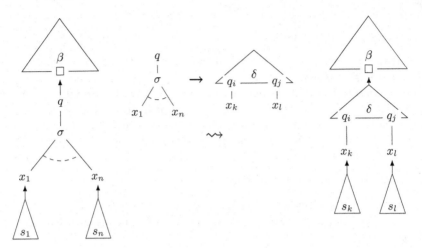

Fig. 1. A derivation step induced by τ.

4.2 2-Ary Tree Transducers

We modify the type of top-down tree transducer defined previously to operate on two trees rather than one.

Definition 10 (2-ary top-down tree transducer). *Given a finite type Q, two input signatures Σ and Δ and an output signature Γ the type of 2-ary top-down tree transducer is defined as:*

$$\mathsf{tdtt}_2 : \mathcal{U} \to Sig \to Sig \to \mathcal{U}$$
$$\mathsf{tdtt}_2 \ Q \ \Sigma \ \Delta \ \Gamma \equiv$$
$$Q \to (\sigma : \Sigma) \to (\delta : \Delta) \to \Gamma^+(Q \times (\Sigma.\mathsf{Ar}\ \sigma \times \Delta.\mathsf{Ar}\ \delta \uplus \Sigma.\mathsf{Ar}\ \sigma \uplus \Delta.\mathsf{Ar}\ \delta))$$

We introduce a function lifting a product of trees to a tree over a product of signatures. Consequently, this means that the induced tree morphism of a 2-ary transducer can be defined as the induced tree morphism of a (unary) transducer. First, we define the product of signature as follows:

$$- \otimes - : Sig \to Sig \to Sig$$
$$\Sigma \ \otimes \ \Delta \equiv \{ \ \ \mathsf{Op} \qquad := \ \Sigma \times \Delta$$
$$; \ \mathsf{Ar}\ (\sigma \ , \ \delta) \ := \ \Sigma.\mathsf{Ar}\ \sigma \times \Delta.\mathsf{Ar}\ \delta + \Sigma.\mathsf{Ar}\ \sigma + \Delta.\mathsf{Ar}\ \delta \ \}$$

while the lifting of a product of trees to a tree over the product of signature is defined as:

$$- \otimes - \{\Sigma\} \{\Delta\} : \mathsf{coT}\ \Sigma \to \mathsf{coT}\ \Delta \to \mathsf{coT}\ (\Sigma \otimes \Delta)$$
$$t_1 \ \otimes \ t_2 \equiv (\mathsf{root}\ t_1 \ , \ \mathsf{root}\ t_2) \lhd \lambda\, i \cdot \mathbf{match}\ [\, i\,]\ \mathbf{with}$$
$$\qquad | \ \iota_1\,(\iota_1\,(i\ ,\ j)) \Rightarrow \mathsf{br}\ t_1\ i \otimes \mathsf{br}\ t_2\ j$$
$$\qquad | \ \iota_1\,(\iota_2\,i) \Rightarrow \mathsf{br}\ t_1\ i \otimes t_2$$
$$\qquad | \ \iota_2\,j \Rightarrow t_1 \otimes \mathsf{br}\ t_2\ j$$

where the function $[\,-\,]$ converts a type $\mathsf{Fin}(a \times b + c + d)$ to the type $\mathsf{Fin}(a) \times \mathsf{Fin}(b) \uplus \mathsf{Fin}(c) \uplus \mathsf{Fin}(d)$. It has an inverse that we write $[\,-\,]^{\mathsf{inv}}$.

Definition 11 (Induced tree morphism). *A transducer* $(\tau : \mathsf{tdtt}_2 \, Q \, \varSigma \, \varDelta)$
induces a tree morphism defined by coiteration, as follows:

$$\langle\!\langle - \rangle\!\rangle_2 : \mathsf{tdtt}_2 \, Q \, \varSigma \, \varDelta \, \varGamma \to \mathsf{coT} \, \varSigma \to \mathsf{coT} \, \varDelta \to \mathsf{coT} \, \varGamma$$
$$\langle\!\langle \tau \rangle\!\rangle_2 \quad q \quad t_1 \quad t_2 \quad \equiv \quad \langle\!\langle \tau' \rangle\!\rangle \, q \, (t_1 \otimes t_2)$$
$$\textbf{where} \quad \tau' \quad q \quad (\sigma \,, \delta) \quad \equiv \quad \mathsf{map}^{\varGamma + (Q \times -)} \, [-]^{\mathsf{inv}} \, (\tau \, q \, \sigma \, \delta)$$

Theorem 6. *Given a tree transducer* $(\tau : \mathsf{tdtt}_2 \, Q \, \varSigma \, \varDelta \, \varGamma)$, *a state* q *and two regular trees* χ_1 *and* χ_2, *then* $\langle\!\langle \tau \rangle\!\rangle_2 \, q \, \chi_1 \, \chi_2$ *is regular.*

4.3 Induced Tree Morphism on Cyclic Terms

So far, we defined the type of top-down tree transducers and showed that the tree morphisms induced by tree transducers preserve the regularity of trees. Consequently, we can lift the induced tree morphism to operate on cyclic terms since we know how to obtain a cyclic term back from a regular tree:

$$\langle\!\langle - \rangle\!\rangle^{\mathsf{C}} : \mathsf{tdtt} \, Q \, \varSigma \, \varDelta \to Q \to \mathsf{C} \, \varSigma \to \mathsf{C} \, \varDelta$$
$$\langle\!\langle \tau \rangle\!\rangle^{\mathsf{C}} \quad q \quad \equiv \quad [\![-]\!]^{\mathsf{inv}} \circ \langle\!\langle \tau \rangle\!\rangle \, q \circ [\![-]\!]$$

In particular, the induced tree morphism $\langle\!\langle - \rangle\!\rangle^{\mathsf{C}}$ satisfies the following equation:

$$\langle\!\langle \tau \rangle\!\rangle^{\mathsf{C}} \, q \, (\mathsf{rec} \, o \lhd os) \approx \langle\!\langle \tau \rangle\!\rangle^{\mathsf{C}} \, q \, (o \lhd os \, [\, * := \mathsf{rec} \, o \lhd os \,]) \qquad (*)$$

Essentially, this means that $\langle\!\langle \tau \rangle\!\rangle^{\mathsf{C}}$ respects the semantics of cyclic terms in the sense that cycles ought to be indistinguishable from their unfolding. As a result, tree transducers give us a way to define functions on cyclic terms corecursively: a termination problem (*i.e.* the term $o \lhd os \, [\, * := \mathsf{rec} \, o \lhd os \,]$ is not a subterm of $\mathsf{rec} \, o \lhd os$) is reduced to productivity problem.

Example. To illustrate the approach consider the problem of defining the synchronous parallel product of processes for a fragment of process algebra such as CSP [12]. The syntax Proc of processes is defined inductively as

$$P, Q ::= \mathsf{STOP} \mid \mathsf{SKIP} \mid a \to P \mid P \,\square\, Q \mid \mu X \cdot P \mid X$$

where SKIP (resp. STOP) denotes the successful (resp. unsuccessful) terminating process. The process $a \to P$ accepts the letter a and then behaves as P. The non-deterministic choice between P and Q is noted $P \,\square\, Q$. Finally, $\mu X \cdot P$ represents a recursively defined process.

We call $\varSigma^{\mathsf{Proc}}$ the signature of CSP process consisting of the constructors STOP, SKIP and $\square$. Thus the type $\mathsf{C} \, \varSigma^{\mathsf{Proc}}$ (defined in Sect. 3.1) can be regarded as the subset of guarded closed terms of Proc. The synchronous parallel product is axiomatized in CSP as follows:

STOP	$\parallel$	$\_$	$= \mathsf{STOP}$
$\_$	$\parallel$	STOP	$= \mathsf{STOP}$
$(a \to P)$	$\parallel$	$(b \to Q)$	$= \textbf{if } a \overset{?}{=} b \textbf{ then } a \to (P \parallel Q) \textbf{ else STOP}$
$(P \,\square\, Q)$	$\parallel$	R	$= (P \parallel R) \,\square\, (Q \parallel R)$
R	$\parallel$	$(P \,\square\, Q)$	$= (R \parallel P) \,\square\, (R \parallel P)$
SKIP	$\parallel$	P	$= P$
P	$\parallel$	SKIP	$= P$

These axioms can also be read as a recursive definition of the parallel operator, provided that this definition can extended to the μ constructor. However, unfolding of μ could lead to a non structural recursive definition. Instead of trying to find a termination criterion, we can reinterpret the axioms as a set of rewrite rules. These rules are clearly expressible as a 2-ary tree transducer. By exploiting the induced tree morphism $\langle\!\langle\, -\, \rangle\!\rangle^{\mathsf{C}}$, we can derive a recursive function on the syntax. Moreover, the behavior for recursive processes conforms to the specification given by Equation $(*)$.

5   Decidability Results

We define a fragment of Computation Tree Logic [9] CTL^- over infinite trees $\mathsf{coT}\ \Sigma$. The syntax is defined inductively as follows:

$$\Phi ::= \top \mid P \mid \neg\, \Phi \mid \Phi \,\wedge\, \Phi \mid \mathbf{AG}\ \Phi \mid \mathbf{AX}\ \Phi$$

where P denotes predicates over infinite trees. The semantics of CTL^- is given as:

$$
\begin{aligned}
[\![\, - \,]\!] &: \mathsf{CTL}^- \to \mathsf{coT}\ \Sigma \to \mathcal{U} \\
[\![\quad \top \quad]\!] &\equiv \mathsf{const}\ \top \\
[\![\quad P \quad]\!] &\equiv P \\
[\![\quad \neg\, \Phi \quad]\!] &\equiv \neg \circ [\![\, \Phi \,]\!] \\
[\![\quad \Phi_1 \,\wedge\, \Phi_2 \quad]\!] &\equiv [\![\, \Phi_1 \,]\!] \wedge [\![\, \Phi_2 \,]\!] \\
[\![\quad \mathbf{AG}\ \Phi \quad]\!] &\equiv \mathsf{AG}\ [\![\, \Phi \,]\!] \\
[\![\quad \mathbf{AX}\ \Phi \quad]\!] &\equiv \mathsf{AX}\ [\![\, \Phi \,]\!]
\end{aligned}
$$

where $\mathsf{AX}\ P\ t := \forall\, k,\ P\ (\mathsf{br}\ t\ k)$ and $\mathsf{AG}\ P := \nu\ Z \cdot P \wedge \mathsf{AX}\ Z$ (coinductively defined). The set of predicates of a formula Φ is noted $\mathsf{Preds}(\Phi)$.

Theorem 7 (Model-checking). *Given a regular tree $\chi : \mathsf{coT}\ \Sigma$, a formula of CTL^-, if all predicates in $\mathsf{Preds}(\Phi)$ are decidable then $[\![\, \Phi \,]\!]\ \chi$ is decidable.*

5.1   Decidability of Bisimilarity for Regular Trees

We give a proof of the decidability of equality (bisimilarity) between regularity trees. The proof is based on both the encoding of CTL^- and the usage of top-down tree transducers.

$$
\begin{aligned}
\tau_\sim &: \mathsf{coT}\ \Sigma \to \mathsf{coT}\ \Sigma \to \mathsf{coT}\ \Sigma^\perp \\
\tau_\sim\ t_1\ t_2 &\equiv \mathbf{match}\ \mathsf{root}\ t_1 \overset{?}{=} \mathsf{root}\ t_2\ \mathbf{with} \\
&\quad\mid\ \mathsf{yes}\ e \Rightarrow \iota_1\ (t_1, t_2, e) \lhd \lambda\,k \cdot \tau_\sim\ (\mathsf{br}\ t_1\ k)\ (\mathsf{br}\ t_2\ (e_* k)) \\
&\quad\mid\ \mathsf{no}\ \_ \Rightarrow \perp
\end{aligned}
$$

Lemma 3 (Bisimilarity encoding). *Given two infinite trees χ_1 and χ_2, we have*

$$\chi_1 \sim \chi_2 \Leftrightarrow [\![\, \mathbf{AG}\ \checkmark \,]\!]\ (\tau_\sim\ \chi_1\ \chi_2)$$

where $\checkmark$ is a state predicate that is true when the root symbol of a tree is not $\perp$.

Theorem 8 (Decidability of bisimilarity). *Given two regular trees χ_1 and χ_2, it is decidable whether χ_1 is bisimilar to χ_2.*

6   Conclusion

In this paper, we described the implementation of a mechanized theory of regular trees. Regular trees are characterized both coinductively, as a subset of a coinductive type satisfying a regularity property, and inductively, as a cyclic term. We showed that top-down tree transducers induce maps which preserve regularity of trees. Moreover, we gave a proof of the decidability of bisimilarity of regular trees through a reduction to a model-checking problem.

The choice to represent cyclic structures as nested datatypes is motivated in work of [16]. They show, by considering increasingly more general signatures, that such datatypes carry both an algebra and coalgebra structures. However, they do not address the problem of defining regular tree homomorphisms.

Close to our work, in the context of mechanizing regular trees, is [8], dealing with the problem of subtyping iso-recursive type. This work exploits the ability to define in AGDA mixed induction/coinduction which is not available in COQ.

Another example, is the work of [13], where they describe an implementation of regular coinductive types in the context of the OCAML programming language. We could investigate whether some part of their implementation could be obtained through extraction of our COQ library. Alternatively, we could consider bringing some of the extensions defined in [13] by means of a COQ plugin.

As future work, we would like to extend the expressiveness of top-down tree transducers while still preserving the regularity property of the induced tree morphism. Examples of such tree transducers include extended top-down tree transducers [4] or macro tree transducers [11].

Finally, we could consider more expressive logics over regular trees such as CTL* [17], or even the modal μ-calculus [14] and use these logics to derive decidability results on regular trees.

Acknowledgements. I am grateful to Jean-Paul Bodeveix and Mamoun Filali for helpful discussions about this work and the anonymous referees for suggesting improvements and clarifications for this paper.

References

1. Abbott, M., Altenkirch, T., Ghani, N.: Containers: Constructing strictly positive types. Theor. Comput. Sci. **342**(1), 3–27 (2005)
2. Abel, A., Pientka, B., Thibodeau, D., Setzer, A.: Copatterns: programming infinite structures by observations. In: The 40th Annual ACM SIGPLAN-SIGACT Symposium on Principles of Programming Languages, POPL 2013, Rome, Italy, January 23–25, 2013, pp. 27–38 (2013)
3. Altenkirch, T., Reus, B.: Monadic presentations of lambda terms using generalized inductive types. In: Flum, J., Rodríguez-Artalejo, M. (eds.) CSL 1999. LNCS, vol. 1683, pp. 453–468. Springer, Heidelberg (1999)
4. Arnold, A., Dauchet, M.: Bi-transductions de forêts. In: ICALP, pp. 74–86 (1976)
5. Brandt, M., Henglein, F.: Coinductive axiomatization of recursive type equality and subtyping. Fundam. Inform. **33**(4), 309–338 (1998)

6. Courcelle, B.: Fundamental properties of infinite trees. Theor. Comput. Sci. **25**, 95–169 (1983)
7. Danielsson, N.A.: Beating the productivity checker using embedded languages. In: Proceedings Workshop on Partiality and Recursion in Interactive Theorem Provers, PAR 2010, Edinburgh, UK, 15th July 2010, pp. 29–48 (2010)
8. Danielsson, N.A., Altenkirch, T.: Subtyping, declaratively. In: Bolduc, C., Desharnais, J., Ktari, B. (eds.) MPC 2010. LNCS, vol. 6120, pp. 100–118. Springer, Heidelberg (2010)
9. Emerson, E.A., Halpern, J.Y.: Decision procedures and expressiveness in the temporal logic of branching time. In: Proceedings of the Fourteenth Annual ACM Symposium on Theory of Computing, STOC 1982, pp. 169–180. ACM, New York (1982)
10. Endrullis, J., Hendriks, D., Bodin, M.: Circular coinduction in coq using bisimulation-up-to techniques. In: Blazy, S., Paulin-Mohring, C., Pichardie, D. (eds.) ITP 2013. LNCS, vol. 7998, pp. 354–369. Springer, Heidelberg (2013)
11. Engelfriet, J., Vogler, H.: Macro tree transducers. J. Comput. Syst. Sci. **31**(1), 71–146 (1985)
12. Hoare, C.A.R.: Communicating sequential processes. Commun. ACM **21**(8), 666–677 (1978)
13. Jeannin, J.-B., Kozen, D., Silva, A.: CoCaml: Programming with coinductive types. Technical report. http://hdl.handle.net/1813/30798, Computing and Information Science, Cornell University, December 2012. Fundamenta Informaticae, to appear
14. Kozen, D.: Results on the propositional mu-calculus. Theor. Comput. Sci. **27**, 333–354 (1983)
15. Nakata, K., Uustalu, T., Bezem, M.: A proof pearl with the fan theorem and bar induction - walking through infinite trees with mixed induction. In: Yang, H. (ed.) APLAS 2011. LNCS, vol. 7078, pp. 353–368. Springer, Heidelberg (2011)
16. Uustalu, T., Ghani, N., Hamana, M., Vene, V.: Representing cyclic structures as nested datatypes. In: Proceedings of 7th Trends in Functional Programming, pp. 173–188. Intellect (2006)
17. Pnueli, A.: The temporal logic of programs. In: 18th Annual Symposium on Foundations of Computer Science, Providence, Rhode Island, USA, 31 October–1 November 1977, pp. 46–57 (1977)
18. Spiwack, A., Coquand, T.: Constructively finite? In: Pardo, L.L., Ibáñez, A.R., García, J.R. (eds.) Contribuciones científicas en honor de Mirian Andrés Gómez, pp. 217–230. Universidad de La Rioja (2010)
19. Thatcher, J.W.: Generalized sequential machine maps. J. Comput. Syst. Sci. **4**(4), 339–367 (1970)

Deriving Comparators and Show Functions in Isabelle/HOL

Christian Sternagel and René Thiemann[✉]

Institute of Computer Science, University of Innsbruck, Innsbruck, Austria
{christian.sternagel,rene.thiemann}@uibk.ac.at

Abstract. We present an Isabelle/HOL development that allows for the automatic generation of certain operations for user-defined datatypes. Since the operations are defined within the logic, they are applicable for code generation. Triggered by the demand to provide readable error messages as well as to access efficient data structures like sorted trees in generated code, we provide show functions that compute the string representation of a given value, comparators that yield linear orders, and hash functions. Moreover, large parts of the employed machinery should be reusable for other operations like read functions, etc.

In contrast to similar mechanisms, like Haskell's "deriving," we do not only generate definitions, but also prove some desired properties, e.g., that a comparator indeed orders values linearly. This is achieved by a collection of tactics that discharge the corresponding proof obligations automatically.

1 Introduction

Before shedding light on *how* things are handled internally, let us have a look at *what* the new mechanism does by means of an example.

As reasonably simple datatypes consider lists and rose trees

datatype α *list* = *Nil* | *Cons* α (α *list*)   datatype α *tree* = *Tree* α (α *tree list*)

where both datatypes store content of type α. Typical operations that are required on specific lists or trees include the following: determine which of two values is smaller, e.g., for sorting; turning a value into a string, e.g., for printing; computing a hash code for a value, e.g., for efficient indexing; etc.

With our development, obtaining such functionality for trees—assuming that it is already available for lists—is as easy as issuing

derive *compare tree*   derive *show tree*   derive *hashable tree*

which may be read as "derive a compare function for trees, then derive a show function for trees, and finally derive a hash function for trees." Afterwards, we can easily handle sets of trees or dictionaries where trees are used as keys in code generation [3]: comparisons or hash codes are required to invoke the efficient algorithms from Isabelle's collections framework [6] and container library [7].

© Springer International Publishing Switzerland 2015
C. Urban and X. Zhang (Eds.): ITP 2015, LNCS 9236, pp. 421–437, 2015.
DOI: 10.1007/978-3-319-22102-1_28

From the deriving mechanism we get the following functions at our disposal where *order* is a type which consists of the three elements *Eq*, *Lt*, and *Gt*:

compare :: (α::*compare*) *tree* $\Rightarrow$ α *tree* $\Rightarrow$ *order*
show :: (α::*show*) *tree* $\Rightarrow$ *string*
hashcode :: (α::*hashable*) *tree* $\Rightarrow$ *hashcode*

Here, the annotation α::c denotes that type variable α has to be in type class c, i.e., trees are comparable ("showable", "hashable") if their node contents are.

This is exactly what one would expect from similar mechanisms like Haskell's `deriving` or Scala's *case classes* (which support automatic definitions of equality checks, show functions, and hash functions).

However, we are in the formal setting of the proof assistant Isabelle/HOL [9], and can thus go one step further and in addition to automatic function definitions also provide automatic proofs of some properties that these functions have to satisfy: since HOL is a logic of total functions, totality is obligatory; for comparators we guarantee that they implement a linear order (see Sect. 3); and for show functions that they adhere to the *show law* (see Sect. 4).

Overview. After presenting some preliminaries and related work in Sect. 2, we first present our two main applications: comparators are the topic of Sect. 3 and show functions are discussed in Sect. 4. While we also support the generation of hash functions (without proving any properties about them), we do not discuss them in the remainder, since this would give no further insight.

Afterwards we explain the main parts of the process to generate class instances. Since this is mostly generic, we will present each part with only one of the classes as a leading example. In general the process is divided into the following steps:

1. First, in Sect. 5, we show how to define the desired operations as recursive functions. To this end, we illustrate a two-level construction principle that guarantees totality.
2. In Sect. 6, we further illustrate how the defining equations of operations are post-processed for code generation with the aim of improved efficiency.
3. After that, in Sect. 7, we discuss how to generate proof obligations for the desired properties and use induction along the recursive structure of a datatype to discharge them. Once these properties are proved it is straightforward to derive class instances.

After the explanation of the deriving mechanism, we continue in Sect. 8 and illustrate how the new infrastructure for comparators can be integrated into existing Isabelle/HOL formalizations. Finally, we conclude in Sect. 9.

Our formalization is part of the development version of the archive of formal proofs (AFP). Instructions on how to access our formalization as well as details on our experiments are provided at: http://cl-informatik.uibk.ac.at/software/ceta/experiments/deriving

2   Preliminaries and Related Work

Let us start with two remarks about notation: [] and # are syntactic sugar for the list constructors *Nil* and *Cons*, and we use both notations freely; function composition is denoted ∘.

Our work is built on top of Isabelle/HOL's new datatype package [1,15], which thus had a strong impact on the specifics of our implementation. Therefore, some more details might be helpful. The new datatype package is based on the notion of *bounded natural functors* (BNFs). A BNF is a type constructor equipped with a map function, set functions (one for each type parameter; collecting all elements of that type which are part of a given value), and a cardinality bound (which is however irrelevant for our purposes). Moreover, BNFs are closed under composition as well as least and greatest fixpoints. At the lowest level, BNF-based (co)datatypes correspond to fixpoint equations, where multiple curried constructors are modeled by disjoint sums (+) of products (×). Finite lists, for example, are obtained as least fixpoint of the equation $\beta = unit + \alpha \times \beta$.

While in principle this might provide opportunity for generic programming *a la* Magalhães et al. [8]—which makes a sum-of-products representation available to the user—there is the severe problem that we not only need to define the generic functions, but also have to prove properties about them. For this reason, we do not work on the low-level representation, but instead access BNFs via its high-level interface, e.g., we utilize the primrec command for specifying primitive recursive functions, and heavily depend on the induction theorems that are generated by the datatype package. A further problem in realizing the approach of [8] is the lack of multi-parameter type classes in Isabelle/HOL. In the following, whenever we mention "primitive recursion," what we actually mean is the specific version of primitive recursion provided by primrec.

Given a type constructor κ with n type parameters $\alpha_1, \ldots, \alpha_n$—written $(\alpha_1, \ldots, \alpha_n)\ \kappa$—the corresponding map function is written map_κ and the set functions $set_\kappa^1, \ldots, set_\kappa^m$, where the superscript is dropped in case of a single type parameter.[1] In the following we will use "datatype" synonymously with "BNF" but restrict ourselves to BNFs obtained as least fixpoints. Moreover, we often use types and type constructors synonymously.

In general we consider arbitrary datatypes of the form

$$\text{datatype } (\alpha_1, \ldots, \alpha_n)\ \kappa = C_1\ \tau_{11} \ldots \tau_{1k_1} \mid \ldots \mid C_m\ \tau_{m1} \ldots \tau_{mk_m} \qquad (1)$$

where each τ_{ij} may only consist of the type variables $\alpha_1, \ldots, \alpha_n$, previously defined types, and $(\alpha_1, \ldots, \alpha_n)$ (for which the type parameters $\alpha_1, \ldots, \alpha_n$ may not be instantiated differently; a standard restriction in Isabelle/HOL). In general several mutually recursive datatypes may be defined simultaneously. For simplicity's sake we restrict to the case of a single datatype in our presentation.

---

[1] A further technicality—allowing for things like *phantom types*—is the separation between "used" and "unused" type parameters. To simplify matters, we consider all variables to be "used" in the remainder, even though our implementation also supports "unused" ones.

Related Work. Our work is inspired by the deriving mechanism of Haskell [10, Chap. 10], which was later extended by Hinze and Peyton Jones [4] as well as Magalhães et al. [8]. Similar functionality is also provided by Scala's[2] case classes and Janestreet's `comparelib`[3] for OCaml. To the best of our knowledge there is no previous work on a deriving mechanism that also provides formally verified guarantees about its generated functions (apart from our previous work [12,14] on top of the, now old, datatype package that was never described in detail).

However, the basics of generic programming in theorem provers—automatically deriving functions for arbitrary datatypes, but without proofs—where already investigated by Slind and Hurd [11] (thanks to the anonymous referees for pointing us to that work). Not surprisingly, our basic constructions bear a strong resemblance to that work, where the relationship will be addressed in more detail in Sect. 5.

3   Linear Orders and Comparators

Several efficient data structures and algorithms rely on a linear order of the underlying element type. For example, to uniquely represent sets by lists, they have to be sorted; dictionaries often require a linear order on keys, etc. Hence, in order to use these algorithms within generated code we require linear orders for the corresponding element types.

There are at least two alternative approaches when representing linear orders for some type α. The first one is to provide one (or both) of the orders $<$ or $\leq$, and the second one is to demand a comparator of type $\alpha \Rightarrow \alpha \Rightarrow order$. In the following, we favor the approach using comparators for several reasons:

The first one is related to simplicity. When constructing comparators, only one function has to be synthesized, whereas linear orders in Isabelle/HOL require both of $<$ and $\leq$. Of course we could just synthesize one of those, say $<$, and then define the other using the built-in equality, i.e., $x \leq y$ iff $x < y \vee x = y$. But then even a single invocation of $\leq$ might result in two comparisons.

Concerning efficiency, for some algorithms, one has to invoke two comparisons of elements of type α, where a comparator only needs one. For example, when traversing a binary search tree, it may require two comparisons to figure out, whether we have to go on left, or right, or whether we are already at the right node. Similarly, also when generating linear orders for complex types, we might want to define the lexicographic order on pairs, where we again may need to perform two comparisons between the first entries of the pairs. In contrast, for both examples we get all the required information from one invocation of the comparator. This may lead to an exponential difference when comparing two tree-shaped values. Despite these benefits of comparators we do not want to hide that there are also some disadvantages. For example for numeral types, where $<$ and $\leq$ can be seen as built-in functions, there might be some overhead when computing the full comparison result of a comparator (which needs two

---

[2] http://scala-lang.org.

[3] https://github.com/janestreet/comparelib.

comparisons), where a single invocation of $<$ might have been sufficient as in a sorting algorithm.

The last and perhaps most important reason for preferring comparators is the fact that using comparators as a new *dedicated* class for sorting, etc., one does not interfere with the remaining formalization. The problem here is, that Isabelle's type class for orders allows to specify exactly *one* order $<$. As an example, in IsaFoR [13] we defined $<$ on positions (of terms) as the standard prefix order, which is the natural choice for a large part of the whole formalization, except for sorting. Still we can invoke a sorting algorithm via the comparator for positions, since the orders within the classes "comparator" and "order" may differ.

In order to formalize comparators in Isabelle/HOL, we started by defining a predicate *is-cmp* :: α *comparator* $\Rightarrow$ *bool* that demands three crucial properties for symmetry, equality, and transitivity.

$$invert\text{-}order\ (c\ x\ y) = c\ y\ x$$
$$c\ x\ y = Eq \Longrightarrow x = y \qquad\qquad (2)$$
$$c\ x\ y = Lt \Longrightarrow c\ y\ z = Lt \Longrightarrow c\ x\ z = Lt$$

Here, α *comparator* is a type abbreviation for $\alpha \Rightarrow \alpha \Rightarrow$ *order*, and *invert-order* :: *order* $\Rightarrow$ *order* swaps Lt with Gt.

We further provide definitions to switch between comparators and linear orders, which eases the integration of our results in the existing Isabelle/HOL infrastructure which primarily works on linear orders, when it comes to sorting, etc., cf. Sect. 8.

$$comparator\text{-}of\ x\ y = (\text{if } x < y \text{ then } Lt \text{ else if } x = y \text{ then } Eq \text{ else } Gt)$$
$$le\text{-}of\text{-}comp\ c\ x\ y = (\text{case } c\ x\ y \text{ of } Gt \Rightarrow False \mid \_ \Rightarrow True)$$
$$lt\text{-}of\text{-}comp\ c\ x\ y = (\text{case } c\ x\ y \text{ of } Lt \Rightarrow True \mid \_ \Rightarrow False)$$

It was an easy exercise to prove that *is-cmp* c implies that *le-of-comp* c and *lt-of-comp* c satisfy the conditions of Isabelle/HOL's class for linear orders, and vice versa, if $\leq$ and $<$ form a linear order, then also *is-cmp comparator-of*.

We further defined a type class for comparators, *compare*, which demands a constant *compare* :: α *comparator* that satisfies *is-cmp compare*.

In order to define comparators for datatypes, we rely on an auxiliary function that combines a list of elements of type *order* lexicographically.

Definition 1. *We define the function comp-lex* :: *order list* $\Rightarrow$ *order as*

$$comp\text{-}lex\ [] = Eq$$
$$comp\text{-}lex\ (x\ \#\ xs) = (\text{case } x \text{ of } Eq \Rightarrow comp\text{-}lex\ xs \mid \_ \Rightarrow x)$$

Now comparators for lists and trees are easily defined. We just compare the constructors first, and in case of equality recurse and combine the results for each argument via *comp-lex*.

Example 2. Since both lists and trees have one type variable α, the corresponding comparators require a comparator c :: α *comparator* as additional argument. Hence we will define the functions cmp_{list} and cmp_{tree} of types α *comparator* $\Rightarrow$ α *list comparator* and α *comparator* $\Rightarrow$ α *tree comparator*, respectively.

$$cmp_{list} \ c \ Nil \ Nil = comp\text{-}lex \ [\,]$$
$$cmp_{list} \ c \ Nil \ (Cons \ \_\ \_) = Lt$$
$$cmp_{list} \ c \ (Cons \ \_\ \_) \ Nil = Gt \quad\quad\quad (3)$$
$$cmp_{list} \ c \ (Cons \ x \ xs) \ (Cons \ y \ ys) = comp\text{-}lex \ [c \ x \ y, cmp_{list} \ c \ xs \ ys]$$
$$cmp_{tree} \ c \ (Tree \ x \ xs) \ (Tree \ y \ ys) = comp\text{-}lex \ [c \ x \ y, cmp_{list} \ (cmp_{tree} \ c) \ xs \ ys]$$

Both comparators are constructed following a general schema which will be discussed further in Sect. 5, and which produces a comparator cmp_κ of type α_1 *comparator* $\Rightarrow \ldots \Rightarrow \alpha_n$ *comparator* $\Rightarrow (\alpha_1, \ldots, \alpha_n) \ \kappa$ *comparator*.

4   Show

A show function for type α provides a string representation of any given value of that type, i.e., $show :: \alpha \Rightarrow string$. In order to allow for constant time concatenation of results (and thus avoid unnecessary performance regression) the actual transformation into a string is postponed as long as possible. This is achieved by the usual trick of using functions of type $string \Rightarrow string$ (which we will abbreviate to $shows$) instead of plain strings. Then $show$ from above is generalized to $shows :: \alpha \Rightarrow shows$. The original show function is easily recovered by $show \ x = shows \ x \ [\,]$.

In our implementation this is further extended by a *nat* argument representing the "precedence" of the context in which the show function is used, providing more flexibility with respect to parenthesization. For simplicity's sake we omit this detail in the following.

Another quirk that is required by user convenience is a special show function for lists of αs, $shows\text{-}list :: \alpha \ list \Rightarrow shows$. E.g., we usually want lists of characters to be printed as string, i.e., "abc" instead of "$[a, \ b, \ c]$." In addition, show functions are required to satisfy the *show law*:

$$shows \ x \ (ys \ @ \ zs) = shows \ x \ ys \ @ \ zs$$

Together this brings us to the type class *show*:

class *show* =
  fixes $shows :: \alpha \Rightarrow shows$ and $shows\text{-}list :: \alpha \ list \Rightarrow shows$
  assumes $shows \ x \ (ys \ @ \ zs) = shows \ x \ ys \ @ \ zs$ and
    $shows\text{-}list \ xs \ (ys \ @ \ zs) = shows\text{-}list \ xs \ ys \ @ \ zs$

The Show Law. One way of looking at the show law is that show functions do not temper with or depend on output produced so far. To see this, consider the specific instance $shows \ x \ ([\,] @ zs) = shows \ x \ [\,] @ zs$ and observe that this requires $shows$ always to behave as if called with $[\,]$ as second argument.

Our goal is now to automatically derive a show function for a given datatype.

Example 3. Assuming a show function for list elements s, this would look as follows for the list datatype:

$shows_{list}\ s\ Nil = \lambda\text{``}Nil\text{''}$
$shows_{list}\ s\ (Cons\ x\ xs) = \lambda\text{``}(Cons\text{''} \circ \_ \circ s\ x \circ \_ \circ shows_{list}\ s\ xs \circ \lambda\text{``})\text{''}$

Here we use two notational conveniences: $\lambda\text{``}text\text{''}$ is the show function producing the literal string *"text"*; and $\_$ is a show function producing a single space.

For the tree datatype from the introduction it would be:

$shows_{tree}\ s\ (Tree\ x\ ts) =$
$\quad \lambda\text{``}(Tree\text{''} \circ \_ \circ s\ x \circ \_ \circ shows_{list}\ (shows_{tree}\ s)\ ts \circ \lambda\text{``})\text{''}$

The underlying general schema (Sect. 5) produces a show function $shows_\kappa$ of type $(\alpha_1 \Rightarrow shows) \Rightarrow \ldots \Rightarrow (\alpha_n \Rightarrow shows) \Rightarrow (\alpha_1, \ldots, \alpha_n)\ \kappa \Rightarrow shows$.

5   Internal Constructions

In general we have to consider an arbitrary datatype (1) for which we want to define some function

$$f_\kappa :: (\alpha_1 \Rightarrow \sigma_1) \Rightarrow \ldots \Rightarrow (\alpha_n \Rightarrow \sigma_n) \Rightarrow (\alpha_1, \ldots, \alpha_n)\ \kappa \Rightarrow \sigma'$$

that is parameterized by corresponding functions for type parameters and relies on the existence of a function $f_{\kappa'}$ for each datatype κ' that was used in the construction of κ. For comparators and show functions this specializes to

$cmp_\kappa :: (\alpha_1\ comparator) \Rightarrow \ldots \Rightarrow (\alpha_n\ comparator) \Rightarrow (\alpha_1, \ldots, \alpha_n)\ \kappa\ comparator$
$shows_\kappa :: (\alpha_1 \Rightarrow shows) \Rightarrow \ldots \Rightarrow (\alpha_n \Rightarrow shows) \Rightarrow (\alpha_1, \ldots, \alpha_n)\ \kappa \Rightarrow shows$

whose definitions will rely on a comparator (show function) for each of the α_i as well as each κ' used in the definition of κ.

Note that for such κ', the function $f_{\kappa'}$ itself takes arguments for the type parameters of κ'. Thus, for any occurrence $(\tau_1, \ldots, \tau_k)\ \kappa'$ there will be a subterm of the shape $f_{\kappa'}\ g_1 \ldots g_k$, where each g_i depends on the structure of τ_i. For nested recursive datatypes this may result in κ occurring inside a type parameter position of κ', e.g., the rose tree type makes a nested occurrence inside α *tree list*.

Before we discuss this any further, let us have a look at the specification mechanisms of Isabelle/HOL that would in principle support the automatic definition of a function f_κ (like cmp_{tree} or $shows_{tree}$)

One candidate would be Isabelle's function package [5] by Krauss. This would require automatic termination proofs, and recursion through previously defined $f_{\kappa'}$ is only possible after the automatic generation of congruence rules, which both seems at least tedious. Krauss himself remarked that the function package might not be the right solution for our purposes (personal communication).

The other candidate is primitive recursion, which is provided by the datatype package in form of the primrec command. When using primrec, termination is obtained automatically in exchange for certain syntactic restrictions, which we will call *primitive recursive form* in the following. Essentially, we may only perform pattern matching on one argument, and for a left-hand side like g $(C$ $t_1 \ldots t_n)$, the recursive calls must be of the form $maps$ g t_i where $maps$ is a combination of canonical map functions of those types which are involved in nesting. For instance, if g takes lists as argument, then $maps$ is the identity, since the datatype of lists is not nested. If g takes rose trees as argument then $maps$ is the map function on lists, since trees are nested within lists.

Note that neither cmp_{list} and cmp_{tree} from Example 2 nor $shows_{tree}$ from Example 3 are in primitive recursive form. It is well-known how to reduce the pattern matching to only one argument, by moving the pattern matching into a case-expression on the right-hand side. In the case of lists we are done: the defining equations are in primitive recursive form, and from these we can easily derive the equations of Examples 2 and 3. However, in the presence of nesting there is still some work to be done.

For instance, one can apply a nested-to-mutual translation as proposed by Slind and Hurd [11]. We actually applied this definitional principle in our previous version [12,14]. However, it has the disadvantage of not being modular, in the sense that in the presence of nesting we could not reuse existing constants. As an example, in [14] the definition of cmp_{tree} will not contain cmp_{list} itself, but a fresh copy of the definition of cmp_{list}, specialized to lists of trees. And even worse, when proving properties of cmp_{tree}, the tactic has to reprove all properties for the copy of cmp_{list} and cannot reuse properties of cmp_{list}.

In the remainder, we describe another workaround which will establish primitive recursive form w.r.t. primrec, and allow modular proofs. The main problem is that calls like cmp_{list} $(cmp_{tree}$ $c)$ xs and $shows_{list}(shows_{tree}$ $s)$ ts are not of the desired form, as neither cmp_{list} nor $shows_{list}$ is the map function for lists. In general we have to gracefully handle patterns of the shape $f_{\kappa'}$ $(f_\kappa$ $f)$ (where $f_{\kappa'}$ and f_κ might of course take more than one argument function, but such cases can be handled similarly).

The essential idea is now instead of defining f_κ $f_1 \ldots f_n$ $(C_i$ $x_1 \ldots x_n) = \ldots$, to encode the information that is provided by the argument functions f_i already into the type of the first argument C_i $x_1 \ldots x_n$ of type $(\alpha_1, \ldots, \alpha_n)$ κ. This is akin to assuming that the f_i have already been partially applied to the appropriate subterms x_i, thus we call such functions *partially applied* (comparator or show) functions and denote them by prefixing the function name with a p. In the following we depict the type changes in the general case as well as for comparators and show functions:

$$pf_\kappa :: (\sigma_1, \ldots, \sigma_n)\ \kappa \Rightarrow \sigma'$$
$$pcmp_\kappa :: (\alpha_1 \Rightarrow order, \ldots, \alpha_n \Rightarrow order)\ \kappa\ \Rightarrow (\alpha_1, \ldots, \alpha_n)\ \kappa \Rightarrow order$$
$$pshows_\kappa :: (shows, \ldots, shows)\ \kappa \Rightarrow shows$$

Given a partially applied function it is easy to define the originally intended one by using canonical maps:

$$f_\kappa \ f_1 \dots f_n = pf_\kappa \circ map_\kappa \ f_1 \dots f_n$$
$$cmp_\kappa \ c_1 \dots c_n = pcmp_\kappa \circ map_\kappa \ c_1 \dots c_n$$
$$shows_\kappa \ s_1 \dots s_n = pshows_\kappa \circ map_\kappa \ s_1 \dots s_n$$

Now let us turn to the construction of such partially applied functions using the primrec mechanism. To ease matters, we provide some auxiliary Isabelle/ML functions (as opposed to HOL functions that can be reasoned about inside the logic). Please keep in mind that in the following we just describe general schemas of putting together certain terms and *not* recursive Isabelle/HOL functions. They are similar to the interpretation $[\![\cdot]\!]_{\Theta,\Gamma}$ of Slind and Hurd [11], but differ since only the former produce terms which fit the requirements of primrec.

Given a type constructor κ together with a function $f_\kappa :: (\alpha_1, \dots, \alpha_n) \ \kappa \Rightarrow \sigma$, we support the construction of what we call a *map block* for type τ

$$\mathcal{M}_\tau^f = \begin{cases} f_\kappa & \text{if } \tau = (\tau_1, \dots, \tau_n) \ \kappa \\ map_{\kappa'} \ \mathcal{M}_{\tau_1}^f \ \dots \ \mathcal{M}_{\tau_\ell}^f & \text{if } \tau = (\tau_1, \dots, \tau_\ell) \ \kappa' \text{ with } \kappa' \neq \kappa \\ \lambda x. x & \text{otherwise} \end{cases}$$

The purpose of a map block is to relay recursive calls to f_κ through arbitrary layers of type constructors. Note that this matches exactly the requirements of the primrec command.

Given a function $f_{\kappa'}$ for each $\kappa' \neq \kappa$ occurring in τ, we further support the construction of a corresponding *compose block* for type τ

$$\mathcal{C}_\tau^f = \begin{cases} \lambda x. x & \text{if } \tau = (\tau_1, \dots, \tau_n) \ \kappa \\ f_{\kappa'} \circ map_{\kappa'} \ \mathcal{C}_{\tau_1}^f \ \dots \ \mathcal{C}_{\tau_\ell}^f & \text{if } \tau = (\tau_1, \dots, \tau_\ell) \ \kappa' \text{ and } \kappa' \neq \kappa \\ \lambda x. x & \text{otherwise} \end{cases}$$

whose purpose is to apply the $f_{\kappa'}$ functions to (via $\mathcal{M}_\tau^f$) appropriately prepared subterms. In this way we can cleanly separate recursive function calls as accepted by primrec from further processing of the corresponding results (via the $f_{\kappa'}$s).

Compose and map blocks are then combined into $\mathcal{C}_\tau^f \ (\mathcal{M}_\tau^f \ x)$ for a variable x of type τ. We illustrate this general construction in the following example.

Example 4. For $\kappa = tree$ and a variable x of type $\tau = \alpha \ tree \ list$ we obtain

$$\mathcal{C}_\tau^f \ (\mathcal{M}_\tau^f \ x) \ = \ \mathcal{C}_\tau^f \ (map \ f_\kappa \ x) \ = \ (f_{list} \circ map \ (\lambda x. x)) \ (map \ f_\kappa \ x)$$

For comparators with $\tau = (\alpha \Rightarrow order) \ tree \ list$ the last term would be

$$pcmp_{list} \ (map \ pcmp_{tree} \ x) :: \alpha \ tree \ list \Rightarrow order$$

and for show functions with $\tau = shows \ tree \ list$

$$pshows_{list} \ (map \ pshows_{tree} \ x) :: shows$$

which both fit the rules of primrec.

Putting everything together, partial comparators are defined as follows.

$$pcmp_\kappa\ (C_i\ x_1\ldots x_{k_i})\ z = \mathsf{case}\ z\ \mathsf{of}\ C_j\ y_1\ldots y_{k_j}\ \Rightarrow$$

$$\begin{cases} Lt & \text{if } i < j \\ Gt & \text{if } j < i \\ comp\text{-}lex\ [C^{pcmp}_{\tau_{i1}}\ (\mathcal{M}^{pcmp}_{\tau_{i1}}\ x_1)\ y_1,\ldots,C^{pcmp}_{\tau_{ik_i}}(\mathcal{M}^{pcmp}_{\tau_{ik_i}}\ x_{k_i})\ y_{k_i}] & \text{if } i = j \end{cases}$$

Example 5. For our example types this results in the following definitions:

$$pcmp_{list}\ Nil\ z = (\mathsf{case}\ z\ \mathsf{of}\ Nil \Rightarrow comp\text{-}lex\ [\,]\ |\ Cons\ \_\ \_ \Rightarrow Lt)$$
$$pcmp_{list}\ (Cons\ cx\ cxs)\ z = (\mathsf{case}\ z\ \mathsf{of}$$
$$\qquad Nil \Rightarrow Gt\ |\ Cons\ y\ ys \Rightarrow comp\text{-}lex\ [cx\ y, pcmp_{list}\ cxs\ ys])$$
$$pcmp_{tree}\ (Tree\ cx\ cxs)\ z = (\mathsf{case}\ z\ \mathsf{of}$$
$$\qquad Tree\ y\ ys \Rightarrow comp\text{-}lex\ [cx\ y, pcmp_{list}\ (map\ pcmp_{tree}\ cxs)\ ys])$$

For partial show functions the general schema is

$$shows_\kappa\ (C_i\ x_1\ldots x_{k_i}) =$$
$$\lambda``(C_i" \circ \_ \circ C^{pshows}_{\tau_{i1}}\ (\mathcal{M}^{pshows}_{\tau_{i1}}\ x_1) \circ \_ \circ \cdots \circ \_ \circ C^{pshows}_{\tau_{ik_i}}\ (\mathcal{M}^{pshows}_{\tau_{ik_i}}\ x_{k_i}) \circ \lambda``)"$$

Example 6. For the type of rose trees this results in the following definition:

$$pshows_{tree}\ (Tree\ s\ ts) =$$
$$\lambda``(Tree" \circ \_ \circ s \circ \_ \circ pshows_{list}\ (map\ pshows_{tree}\ ts) \circ \lambda``)"$$

which is in the desired primitive recursive form.

It eventually remains to prove the equations of Examples 2 and 3 from these definitions. For this, we mainly demand compositionality of the various map functions, the simplification rules for map functions, the definitions of all participating comparators (or show functions), and the definitions of the partially applied functions. For example, for the comparator of trees we derive the desired equation as follows, where in the step from (7) to (8) we use the compositionality of map_{tree} and map, and we go from (9) to (8) by unfolding both definitions of cmp_{tree} and cmp_{list}.

$$cmp_{tree}\ c\ (Tree\ x\ xs)\ (Tree\ y\ ys) \tag{4}$$
$$= pcmp_{tree}\ (map_{tree}\ c\ (Tree\ x\ xs))\ (Tree\ y\ ys) \tag{5}$$
$$= pcmp_{tree}\ (Tree\ (c\ x)\ (map\ (map_{tree}\ c)\ xs))\ (Tree\ y\ ys) \tag{6}$$
$$= comp\text{-}lex\ [c\ x\ y,\ pcmp_{list}\ (map\ pcmp_{tree}\ (map\ (map_{tree}\ c)\ xs))\ ys] \tag{7}$$
$$= comp\text{-}lex\ [c\ x\ y,\ pcmp_{list}\ (map\ (pcmp_{tree} \circ (map_{tree}\ c))\ xs)\ ys] \tag{8}$$
$$= comp\text{-}lex\ [c\ x\ y,\ cmp_{list}\ (cmp_{tree}\ c)\ xs\ ys] \tag{9}$$

6   Code Equations for Comparators

Recall that our main motivation was to define functions inside the logic which then should become available for code generation. Hence, after having defined comparators as in Example 5, and having proved the equations of Example 2, we just register the latter as code equations. In this way, only comparators will appear in generated code, and the internal construction via the partially applied comparators remains opaque—and for the same reasons, the partially applied show functions will not occur in the generated code.

Still these code equations are not optimal w.r.t. execution time. Especially in languages with eager evaluation, the right-hand side of (3),

$$comp\text{-}lex\ [c\ x\ y, cmp_{list}\ c\ xs\ ys]$$

is problematic. Even if the first comparison $c\ x\ y$ evaluates to Lt or Gt, also the second argument $cmp_{list}\ c\ xs\ ys$ will be evaluated in eager languages.

To avoid this inefficiency, we completely unfold applications of $comp\text{-}lex$ in the right-hand sides of the equations in Example 2 before handing them over to the code generator. To be more precise, unfolding is always performed w.r.t. the following three equations (which are all easily proved):

$$comp\text{-}lex\ [\,] = Eq$$
$$comp\text{-}lex\ [x] = x$$
$$comp\text{-}lex\ (x\ \#\ y\ \#\ xs) = (\text{case}\ x\ \text{of}\ Eq \Rightarrow comp\text{-}lex\ (y\ \#\ xs)\ |\ z \Rightarrow z)$$

The advantage of doing this just for code generation is that we can still use properties of $comp\text{-}lex$ within proofs, e.g., when showing that our comparators really behave like comparators. Moreover, we can keep the canonical structure as described in Example 2 without having to perform lots of case splits.

After the expansion, the right-hand side of the code equation for (3) becomes

$$\text{case}\ c\ x\ y\ \text{of}\ Eq \Rightarrow cmp_{list}\ c\ xs\ ys\ |\ z \Rightarrow z$$

where even in eager languages the recursive call will only be evaluated on demand.

7   Correctness of Generated Functions

Eventually we have to ensure correctness of the generated show functions and comparators. For comparators, this amounts to proving the following soundness theorems for our example types and for the general case, and similar theorems have to be proved regarding the show law.

$$is\text{-}cmp\ c \Longrightarrow is\text{-}cmp\ (cmp_{list}\ c) \qquad is\text{-}cmp\ c \Longrightarrow is\text{-}cmp\ (cmp_{tree}\ c)$$
$$is\text{-}cmp\ c_1 \Longrightarrow \ldots \Longrightarrow is\text{-}cmp\ c_n \Longrightarrow is\text{-}cmp\ (cmp_\kappa\ c_1\ \ldots\ c_n) \tag{10}$$

Although the theorems are clearly sufficient to easily plug together valid comparators, they are not sufficient when proving the soundness theorem for a new datatype which uses nested recursion, such as rose trees. To illustrate the problem, recall the defining equation for cmp_{tree}:

$$cmp_{tree} \; c \; (Tree \; x \; xs) \; (Tree \; y \; ys) = comp\text{-}lex \; [c \; x \; y, cmp_{list} \; (cmp_{tree} \; c) \; xs \; ys]$$

In order to prove the soundness theorem for cmp_{tree}, we clearly need correctness of cmp_{list}. However, since cmp_{list} is invoked on $cmp_{tree} \; c$, the soundness theorem for cmp_{list} can only be applied if we already would have the soundness theorem for cmp_{tree}, and thus the current form of the soundness theorems is not strong enough in the presence of nesting.

As a solution, we always generate pointwise soundness theorems which are based on pointwise properties of a comparator. From the pointwise theorems we can then easily conclude the soundness theorems stated above.

In detail, for symmetry, transitivity, equality, the show law, etc., we define pointwise variants. Here, we only illustrate transitivity. We define transitivity on the level of *order*, and a pointwise variant on the level of comparators. It imposes a stronger variant of transitivity in comparison to (2), which captures all possible combinations of Lt and Eq. This is required, as we want to prove transitivity in a standalone way, without having to refer to symmetry or equality.

definition *trans-order* :: *order* $\Rightarrow$ *order* $\Rightarrow$ *order* $\Rightarrow$ *bool* where
$\quad$ *trans-order* $x \; y \; z \longleftrightarrow$
$\qquad (x \neq Gt \longrightarrow y \neq Gt \longrightarrow z \neq Gt \land ((x = Lt \lor y = Lt) \longrightarrow z = Lt))$

definition *ptrans-comp* :: α *comparator* $\Rightarrow \alpha \Rightarrow$ *bool* where
$\quad$ *ptrans-comp* $c \; x \longleftrightarrow (\forall \; y \; z. \; trans\text{-}order \; (c \; x \; y) \; (c \; y \; z) \; (c \; x \; z))$

The former definition is more low-level, but has the advantage of being smoothly combinable with *comp-lex*, independent of any comparator:

lemma *comp-lex-trans*: assumes *length xs* = *length ys* and *length ys* = *length zs*
and $\forall \; i < length \; zs. \; trans\text{-}order \; (xs \; ! \; i) \; (ys \; ! \; i) \; (zs \; ! \; i)$
shows *trans-order* (*comp-lex xs*) (*comp-lex ys*) (*comp-lex zs*)

In combination with the already proved partial transitivity property of cmp_{list}

$$(\bigwedge x. \; x \in set \; xs \Longrightarrow ptrans\text{-}comp \; c \; x) \Longrightarrow ptrans\text{-}comp \; (cmp_{list} \; c) \; xs \quad (11)$$

we can now prove the partial transitivity property for trees in a modular way.

$$(\bigwedge x. \; x \in set_{tree} \; t \Longrightarrow ptrans\text{-}comp \; c \; x) \Longrightarrow ptrans\text{-}comp \; (cmp_{tree} \; c) \; t$$

We first apply induction on t. So let $t = Tree \; x_1 \; ts_1$ where we can assume the premise and the induction hypothesis.

$$x \in set_{tree} \; (Tree \; x_1 \; ts_1) \Longrightarrow ptrans\text{-}comp \; c \; x \text{ for all } x \quad (12)$$

$$t_1 \in set \; ts_1 \Longrightarrow ptrans\text{-}comp \; (cmp_{tree} \; c) \; t_1 \text{ for all } t_1 \quad (13)$$

We have to prove $ptrans\text{-}comp$ $(cmp_{tree}\ c)$ $(Tree\ x_1\ ts_1)$, i.e.,

$$
\begin{aligned}
&trans\text{-}order\ (cmp_{tree}\ c\ (Tree\ x_1\ ts_1)\ t_2)\ (cmp_{tree}\ c\ t_2\ t_3) \\
&(cmp_{tree}\ c\ (Tree\ x_1\ ts_1)\ t_3)
\end{aligned}
\tag{14}
$$

for all t_2 and t_3. In the general case, at this point we perform a case analysis on both t_2 and t_3, where all of the cases where the three leading constructors are different are easily proved by unfolding the transitivity property followed by simplification. Hence, it remains the interesting case with identical constructors. Let $t_2 = Tree\ x_2\ ts_2$ and $t_3 = Tree\ x_3\ ts_3$. Then, (14) simplifies to

$$
\begin{aligned}
&trans\text{-}order\ (comp\text{-}lex\ [c\ x_1\ x_2,\ cmp_{list}\ (cmp_{tree}\ c)\ ts_1\ ts_2]) \\
&\quad (comp\text{-}lex\ [c\ x_2\ x_3,\ cmp_{list}\ (cmp_{tree}\ c)\ ts_2\ ts_3]) \\
&\quad (comp\text{-}lex\ [c\ x_1\ x_3,\ cmp_{list}\ (cmp_{tree}\ c)\ ts_1\ ts_3])
\end{aligned}
$$

and via theorem $comp\text{-}lex\text{-}trans$, it remains to consider all the comparisons of the arguments of the constructor $Tree$ which leads to the following proof obligations.

$$
ptrans\text{-}comp\ c\ x_1
\tag{15}
$$

$$
ptrans\text{-}comp\ (cmp_{list}\ (cmp_{tree}\ c))\ ts_1
\tag{16}
$$

Here, (15) is immediately solved by (12) and the simplification rules for set. And for (16) we first apply (11), then conclude via the induction hypothesis (13).

The proof for the individual arguments is easily generalized to the generic case and follows a simple schema: whenever we hit some foreign type $(\tau_1, \ldots, \tau_m)\ \kappa$, we use the partial transitivity theorem of cmp_κ and proceed recursively on each τ_i; whenever we hit the comparator under consideration, we apply the induction hypothesis, and whenever we hit a comparator for some type variable, we apply the corresponding premise.

In a similar way, also partial symmetry and equality properties are defined and proved. We separated the three properties and did not define a partial comparator property, as the corresponding proofs are all a little bit different and could not easily be merged into a single one. For example, for transitivity we perform one induction and then do a case analyses on two other elements, whereas for symmetry and equality, a single case analysis suffices.

Having proved the partial properties of a comparator and show function, it is easy to derive the main (global) properties of comparators and show functions, namely soundness theorems in (10) and the show law.

8   Integration into Isabelle/HOL Infrastructure

At this point, we have a machinery to automatically derive various class instances for datatypes. Whereas for hash codes and show functions these mechanisms are immediately applicable, this is not the case for comparators. The reason for the latter is the fact, that comparators are not well supported in the Isabelle distribution, where most algorithms for sorting, search-trees, etc. are defined via class *linorder*, and a combination of $\leq$, $<$, and $=$ is applied. To bridge this gap, we offer three different alternatives.

The first alternative is to bridge everything via *lt-of-comp*, *le-of-comp*, and *comparator-of*. This is done when invoking derive *linorder tree*. This command creates a new class instance for trees, *tree* :: (*linorder*) *linorder*, where the syntax says that if the type parameter α is an instance of *linorder*, then so is α *tree*. Here, $<$ and $\leq$ will be defined as *lt-of-comp* (cmp_{tree} *comparator-of*) and *le-of-comp* (cmp_{tree} *comparator-of*), respectively. With this approach one can easily use all the existing algorithms. For example, we used this approach to generate linear orders for the datatypes of the CAVA LTL model checker [2], without changing a single line in the remaining formalization. However the switch between comparators and orders clearly has a negative impact on efficiency.

The second alternative is to modify the existing algorithms so that they are defined via comparators. Here, we provide an easy solution which performs this change just before code generation. It works as follows. First, we defined a class *compare-order*, which demands that there is a linear order and a comparator *compare*, where the induced orders coincide, i.e., $<$ = *lt-of-comp compare* and $\leq$ = *le-of-comp compare* must hold. Afterwards we provide a method which strengthens the class constraints from *linorder* to *compare-order*, where every two consecutive comparisons are replaced by one comparator invocation with the help of several lemmas of the shape:

(if $x \leq y$ then if $x = y$ then P else Q else R) =
(case *compare* x y of $Eq \Rightarrow P \mid Lt \Rightarrow Q \mid Gt \Rightarrow R$)

For example, the standard code equations to lookup the value of some key in a red-black tree, *rbt-lookup* :: (α, β) *rbt* $\Rightarrow \alpha \Rightarrow \beta$ *option*, are

rbt-lookup Empty k = *None*
rbt-lookup (*Branch c l x y r*) *k* =
(if $k < x$ then *rbt-lookup l k* else if $x < k$ then *rbt-lookup r k* else *Some y*)

but after invoking compare-code (α) *rbt-lookup* they are transformed into:

rbt-lookup Empty k = *None*
rbt-lookup (*Branch c l x y r*) *k* =
(case *compare* k x of $Eq \Rightarrow$ *Some y* $\mid Lt \Rightarrow$ *rbt-lookup l k* $\mid Gt \Rightarrow$ *rbt-lookup r k*)

Note that in the original code equations, α only has to be an instance of *linorder*, whereas the modified version enforces α to be an instance of *compare-order*.

In summary, the second approach is also easily integrated. For example, it suffices to invoke compare-code on all constants *rbt-ins, rbt-lookup, rbt-del, rbt-map-entry, sunion-with, and sinter-with* in order to completely adapt the whole red-black tree implementation to work on comparators. And it suffices to call derive *compare-order list* to make lists an instance of *compare-order*, and similarly for other datatypes. This command internally just combines the soundness lemmas (10) of the comparators to assemble a comparator, and then defines $<$ and $\leq$ via *lt-of-comp* and *le-of-comp*. In this way, we could remove over 600 lines of proofs for manually created linear orders in IsaFoR . Moreover, the change from linear orders to comparators led in theory to a linear speed-up when performing comparisons. To measure the impact in practice, we certified over 4122 termination and complexity proofs that have been produced by various tools during the international termination competition, which the generated code had to validate. Whereas the old code required around 17 min for the certification of all proofs, the new version required less than 4 min.

However, there remains one disadvantage, namely that an existing class instance might interfere with the instance that derive *compare-order* wants to create. For instance, if the ordering on products is defined to be pointwise, then there is no chance to make *prod* an instance of *compare-order*.

Therefore, the third alternative does not require instances of *compare-order*. Instead, one has to copy those functions which are relevant for code generation, manually integrate comparators, and then perform an equivalence proof. Afterwards, one can reuse all of the theorems without much overhead.

As an example, we again consider red-black trees. Here, we manually adapted the lookup function, and the corresponding equivalence proof is straightforward.

primrec *rbt-comp-lookup* :: α *comparator* $\Rightarrow$ (α, β) *rbt* $\Rightarrow$ α $\Rightarrow$ β *option* where
  rbt-comp-lookup c Empty k = None
| *rbt-comp-lookup c (Branch - l x y r) k = (case c k x of*
    Lt $\Rightarrow$ *rbt-comp-lookup c l k*
    | *Gt* $\Rightarrow$ *rbt-comp-lookup c r k*
    | *Eq* $\Rightarrow$ *Some y)*

lemma *rbt-comp-lookup*:
  is-cmp c $\Longrightarrow$ *rbt-comp-lookup c = ord.rbt-lookup (lt-of-comp c)*

Afterwards, a theorem like *map-of-entries*—which required a proof of 66 lines— is easily adapted for comparators via *rbt-comp-lookup* and proved in a single line. Notice that *ord.rbt-sorted* and *ord.rbt-sorted* require the order as a parameter, whereas *rbt-lookup* and *rbt-sorted* implicitly take the order from the type class.

lemma *map-of-entries*: *rbt-sorted t* $\Longrightarrow$ *map-of (entries t) = rbt-lookup t*

lemma *comp-map-of-entries*: *is-cmp c* $\Longrightarrow$ *ord.rbt-sorted (lt-of-comp c) t*
  $\Longrightarrow$ *map-of (entries t) = rbt-comp-lookup c t*
  using *linorder.map-of-entries*[*OF comparator.linorder*] *rbt-comp-lookup* by *metis*

In this way, as a case study we adapted the whole container framework of Lochbihler to use comparators instead of linear orders. Most of the adaptation was straightforward and just required the insertion of suitable equivalence statements like *rbt-comp-lookup*, and the change from *ord.rbt-lookup* to *rbt-comp-lookup*. Moreover, we could remove over 450 lines within the container framework, where manual constructions for orders (now: comparators) and equality-checking have been replaced by one-line invocations of our generators.

9   Conclusion

We presented a mechanism that allows for the automatic derivation of the following operations for arbitrary user-defined datatypes: comparators, show functions, and hash functions. Our work relies on the canonical map functions and corresponding facts that are provided by Isabelle's new datatype package. We further showed how our work can be integrated into existing formalizations, thereby saving lines of code as well as improving the efficiency of generated code.

Acknowledgments. We thank S. Berghofer, J. Blanchette, L. Bulwahn, F. Haftmann, B. Huffman, A. Krauss, P. Lammich, A. Lochbihler, C. Urban, T. Nipkow, D. Traytel, and M. Wenzel for their valuable support w.r.t. motivating our development, information on the old and new datatype packages, and for answering several Isabelle/ML related questions. We thank the anonymous reviewers for their helpful comments. This work was supported by Austrian Science Fund (FWF) projects P27502 and Y757. The authors are listed in alphabetical order regardless of individual contribution or seniority.

References

1. Blanchette, J.C., Hölzl, J., Lochbihler, A., Panny, L., Popescu, A., Traytel, D.: Truly modular (co)datatypes for Isabelle/HOL. In: Klein, G., Gamboa, R. (eds.) ITP 2014. LNCS, vol. 8558, pp. 93–110. Springer, Heidelberg (2014). doi:10.1007/978-3-319-08970-6_7

2. Esparza, J., Lammich, P., Neumann, R., Nipkow, T., Schimpf, A., Smaus, J.-G.: A fully verified executable LTL model checker. In: Sharygina, N., Veith, H. (eds.) CAV 2013. LNCS, vol. 8044, pp. 463–478. Springer, Heidelberg (2013). doi:10.1007/978-3-642-39799-8_31

3. Haftmann, F., Nipkow, T.: Code generation via higher-order rewrite systems. In: Blume, M., Kobayashi, N., Vidal, G. (eds.) FLOPS 2010. LNCS, vol. 6009, pp. 103–117. Springer, Heidelberg (2010). doi:10.1007/978-3-642-12251-4_9

4. Hinze, R., Peyton Jones, S.: Derivable type classes. ENTCS **41**(1), 5–35 (2001). doi:10.1016/S1571-0661(05)80542-0

5. Krauss, A.: Partial and nested recursive function definitions in higher-order logic. JAR **44**(4), 303–336 (2010). doi:10.1007/s10817-009-9157-2

6. Lammich, P., Lochbihler, A.: The Isabelle collections framework. In: Kaufmann, M., Paulson, L.C. (eds.) ITP 2010. LNCS, vol. 6172, pp. 339–354. Springer, Heidelberg (2010). doi:10.1007/978-3-642-14052-5_24

7. Lochbihler, A.: Light-weight containers for Isabelle: efficient, extensible, nestable. In: Blazy, S., Paulin-Mohring, C., Pichardie, D. (eds.) ITP 2013. LNCS, vol. 7998, pp. 116–132. Springer, Heidelberg (2013). doi:10.1007/978-3-642-39634-2_11

8. Magalhães, J.P., Dijkstra, A., Jeuring, J., Löh, A.: A generic deriving mechanism for Haskell. SIGPLAN Not. **45**(11), 37–48 (2010). doi:10.1145/2088456.1863529

9. Nipkow, T., Paulson, L.C., Wenzel, M. (eds.): Isabelle/HOL - A Proof Assistant for Higher-Order Logic. LNCS, vol. 2283. Springer, Heidelberg (2002)

10. Peyton Jones, S.: The Haskell 98 language. JFP **13**(1), 139–144 (2003). doi:10.1017/S0956796803001217

11. Slind, K., Hurd, J.: Applications of polytypism in theorem proving. In: Basin, D., Wolff, B. (eds.) TPHOLs 2003. LNCS, vol. 2758, pp. 103–119. Springer, Heidelberg (2003). doi:10.1007/10930755_7

12. Sternagel, C., Thiemann, R.: Haskell's show-class in Isabelle/HOL. Archive of Formal Proofs, July 2014. http://afp.sf.net/entries/Show.shtml
13. Thiemann, R., Sternagel, C.: Certification of termination proofs using CeTA. In: Berghofer, S., Nipkow, T., Urban, C., Wenzel, M. (eds.) TPHOLs 2009. LNCS, vol. 5674, pp. 452–468. Springer, Heidelberg (2009). doi:10.1007/978-3-642-03359-9_31
14. Thiemann, R.: Generating linear orders for datatypes. Archive of Formal Proofs, August 2012. http://afp.sf.net/entries/Datatype_Order_Generator.shtml
15. Traytel, D., Popescu, A., Blanchette, J.C.: Foundational, compositional (co)datatypes for higher-order logic: category theory applied to theorem proving. In: Proceedings of the 27th LICS, pp. 596–605 (2012). doi:10.1109/LICS.2012.75

Formalising Knot Theory in Isabelle/HOL

T.V.H. Prathamesh[✉]

Department of Mathematics, Indian Institute of Science, Bangalore, India
`prathamesh@math.iisc.ernet.in`

Abstract. This paper describes a formalization of some topics in knot theory. The formalization was carried out in the interactive proof assistant, Isabelle. The concepts that were formalized include definitions of tangles, links, framed links and various forms of equivalences between them. The formalization is based on a formulation of links in terms of tangles. We further construct and prove the invariance of the Bracket polynomial. Bracket polynomial is an invariant of framed links closely linked to the Jones polynomial. This is perhaps the first attempt to formalize any aspect of knot theory in an interactive proof assistant.

Keywords: Formalization of mathematics · Knot theory · Kauffman bracket · Bracket polynomial

1 Introduction

Knot theory refers to a study of mathematical objects which are derived from the intuitive notion of a knotted loop of rope. Its modern day origins lie in Lord Kelvin's theory of vortex atoms. It transcended its origins to become an important area of mathematical research. Its applications extend to various branches of physics, chemistry and biology. Despite the enormous growth in formalization of important results and theories in proof assistants, there has been no formalized theory of knots in an interactive proof assistant to the best of our knowledge.

Knot theory as a discipline is largely centered around study of knots, links and various invariants of knots and links. Knots in the context of the knot theory are simple closed loops in 3-dimensional space. A link refers to a disjoint non-intersecting collection of knots. The constituent knots of a link can be entangled. If a link can be obtained from another link by wiggling and twisting, without involving any cutting and pasting, then the two links are regarded as equivalent. The equivalence class of the (unknotted) circle is called the unknot.

Establishing equivalence of two given knots can be a difficult task. Even though this problem is decidable, the algorithm is extremely complicated and difficult to implement in its generality. There are, however, several algorithms for unknot recognition which have been implemented. Knots and links are often distinguished by the means of various knot and link invariants. Some of the prominent link invariants include Khovanov homology, Alexander polynomial and Jones polynomial.

© Springer International Publishing Switzerland 2015
C. Urban and X. Zhang (Eds.): ITP 2015, LNCS 9236, pp. 438–452, 2015.
DOI: 10.1007/978-3-319-22102-1_29

This paper outlines our formalization of some of the important definitions and results in knot theory in Isabelle/HOL. The choice of interactive proof assistant and the implementation logic was based on the high expressiveness of the logic, presence of the formal proof language (Isar) and reasonably effective automation.

The definitions that have been formalized include links, equivalence of links and framed links. A framed link is a link in which each knot is viewed as being made out of a ribbon instead of a string. Many of the definitions and results contained in our formalization are skipped in this paper for the purpose of brevity. The most significant contribution outlined in this paper is the construction and proof of invariance of the Kauffman bracket or Bracket polynomial, an invariant of framed links which is closely linked to the Jones polynomial.

The formalization described in this paper should be seen a part of an attempt to formalize various concepts and results in knot theory in Isabelle/HOL. Some of the reasons why such a project could be important are listed as follows.

1. It is perhaps the first attempt to formalize results in knot theory in an interactive proof assistant. Deep interconnections of knot theory with various other branches of mathematics, also enable development of various supporting libraries. For instance, this project led to the development of a formalized theory of tensor product of matrices.
2. Given the reliance on computer generated results in knot theory, be it for the purpose of unknot recognition or for the computation of invariants, development of formally verified code and theories should prove to be great importance in improving trust in these results.

The paper is organised along the following lines. In Sect. 2.1, we introduce the reader to the definitions of basics concepts in knot theory that are found in most textbooks. In Sect. 2.2, we introduce tangles which form the basis of our formulation of knots. In Sect. 3, we introduce the formalization of tangles and links in Isabelle/HOL. In Sect. 4.1, we describe how the Kauffman bracket could be defined through tangles. In Sect. 4.2, we describe how the above definition was employed for the purpose of formalization of the Kauffman bracket. Section 5, contains the conclusions and further work.

2   Preliminaries

2.1   Standard Definitions

In this section, we introduce some of the basic concepts in knot theory. We stick to the definitions that are found in most standard textbooks on the subject. For further reading, one could refer to [3]. One may note that the formalization presented in this paper is based on a representation of links through tangles, which is described in Sect. 2.2. The standard definitions are introduced for the sake of completeness.

Definition 1. *A knot K is defined as the image of a smooth, injective map $h : S^1 \to S^3$ so that $h'(\theta) \neq 0$ for all $\theta \in S^1$. An oriented knot refers to a knot* *along with an assigned orientation to the curve. The orientation is denoted by an arrow on the curve.*

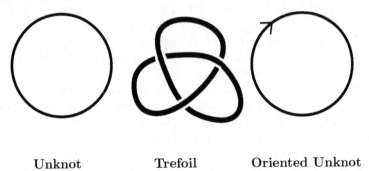

Unknot   **Trefoil**   **Oriented Unknot**

Remark 1. A knot is defined as an embedding in the 3-sphere S^3 and not $\mathbb{R}^3$, because S^3 is compact. The complement of a tubular neighbourhood of a knot is an important object of study from the perspective of 3-manifold topology.

Definition 2. *A link $L \subset S^3$ is a smooth 1-dimensional submanifold of S^3 such that each component of L is a knot and there are only finitely many components. An oriented link refers to a link whose components are oriented knots.*

Two links are considered to the same if there is an *ambient isotopy* between them, which is defined as follows.

Definition 3 (Ambient Isotopy). *Two links L_1 and L_2 in S^3 are said to be ambient isotopic if there exists a smooth map $F : S^3 \times [0,1] \to S^3$ such that*

1. $F|_{S^3 \times \{0\}} = id|_{S^3} : S^3 \to S^3$.
2. $F|_{S^3 \times \{1\}}(L_1) = L_2$.
3. $F|_{S^3 \times \{1\}}$ *is a diffeomorphism* $\forall s \in [0,1]$

Ambient isotopy induces an equivalence relation on links as attested by the following theorem.

Theorem 1. *Ambient isotopy induces an equivalent relation on the set of all links.*

A framed link is a link in which each knot is viewed as being made out of a ribbon instead of a string. One may note that, because of the thickness of a ribbon, two equivalent links may not be equivalent as framed links. However, if two framed links are equivalent, they are equivalent as links without the framing. Framed links are formally defined as follows.

Definition 4 (Framed Link). *A framed m-component link is a collection of m unordered (unoriented) circles smoothly and disjointly embedded in S^3 and such that each component is equipped with a continuous unit normal vector field. Two framed links are equivalent if they are isotopic by an ambient isotopy that preserves the homotopy class of the vector field on each component.*

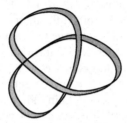

An important invariant of framed links is the Kauffman bracket (also called the Bracket polynomial), which is defined as follows.

Definition 5 (Kauffman Bracket). *[1] Let L be a projection of a framed, unoriented link L. The Kauffman bracket of L, is an element of $\mathbb{Z}[A, A^{-1}]$, with A an indeterminate, computed by the following skein relations*

$$\langle \phi \rangle = 1$$

$$\langle \text{⊗} \rangle = A \langle \text{⊗} \rangle + A^{-1} \langle \text{)(} \rangle$$

$$\langle \text{◎} \rangle = (-A^2 - A^{-2}) \langle \text{○} \rangle$$

where ϕ is the link with no components. The second equation, for example, says that any time you can find three different link projections which look exactly the same except in a small disk, where they look as shown in the equation, then their brackets satisfy this equation. Of course, this means if you happen to know the brackets of the two projections on the right side, this tells you the bracket of the left side. The third equation is interpreted similarly, and gives the effect of removing an unlinked unknot from a link.

Theorem 2. *The Kauffman bracket is an invariant of framed links.*

The Kauffman bracket is closely linked to the Jones polynomial, an invariant of oriented links. If we consider the link L without the framing and induce an orientation on the link L, the Jones polynomial $J(L)$ of this oriented link can be defined in terms of the Kauffman bracket by

$$J(L) = A^{-3\omega(L)} \langle L \rangle \tag{1}$$

where $\omega(L)$ is the writhe of the oriented link diagram. Writhe is an integer valued property of an oriented link diagram that describes the amount of coiling in the 3-dimensional space.

2.2   Tangles and Links

To formalize knot theory using the canonical definitions introduced in the previous sections would become a tedious and unrewarding exercise. This is partly because formalization of even basic results in knot theory using these definitions would require an extremely well developed library of topology and algebraic topology. Moreover proofs of many of the results found in literature are largely guided by pictures, which make it very difficult to formalize.

There are a large variety of syntactic representations of knots and links, which are useful from the perspective of formalization. Closed braids, planar graphs, grid diagrams, DT code and Gauss code are some of the commonly used representations. Given the multiplicity of representations and definitions, fixing a particular way of representing and defining knots and links for the purpose of formalization was among the earliest choices to be confronted. In such a scenario, defining links and framed links in terms of tangles seemed most appropriate for the following reasons

1. Tangles are easy to visualise and defining links through tangles does not lead us too far from an intuitive understanding of links.
2. Framed links can easily be defined using tangle moves.
3. Links are easy to construct from the generating tangles using composition and tensor products.

All the definitions and theorems used in this section can be found in [1].

Definition 6 (Tangle). *A* tangle *is the image of a smooth embedding of a union of circles and intervals into the cylinder $D \times I$, where D is the unit disk in $\mathbb{C}$. The intersection of a tangle with the boundary of the cylinder is required to be transverse, to lie in $X \times (\{0,1\})$, where X is the x-axis in D, and to be exactly the image of the endpoints of the intervals. Tangles are considered up to smooth isotopy of the cylinder leaving $X \times \{0\}$ and $X \times \{1\}$ invariant as sets. The* domain *of a tangle is defined as the number of intersection points of the tangle with $X \times \{0\}$. The domain of a tangle T is denoted by dom(T). The* codomain *of a tangle is defined as the number of intersection points of the tangle with $X \times \{1\}$. The codomain of a tangle T is denoted by codom(T).*

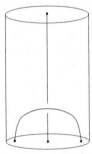

Links are tangles whose codomain as well as domain equal 0. It can be proven that any two links are ambient isotopic if and only if they are isotopic as tangles.

Tangles can thus be treated as a generalisation of links. Tangles also simultaneously serve the purpose of being the building blocks of links, using the operations defined below.

Definition 7 (Tangle Composition). *The composition of two tangles ($\circ$) is a partial function which takes the tangles T_1 and T_2 to the tangle obtained by placing T_1 on top of T_2 and isotoping to ensure that the endpoints match smoothly and then rescaling it appropriately. $T_1 \circ T_2$ is defined only when*

$$\mathrm{dom}(T_1) = \mathrm{codom}(T_2)$$

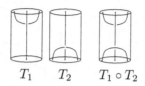

$$T_1 \qquad T_2 \qquad T_1 \circ T_2$$

Definition 8 (Tangle Tensor). *The tensor product of tangles($\otimes$), T_1 and T_2, is obtained by placing T_1 next to T_2 and rescaling it appropriately. To be more precise, T_1 and T_2 are isometrically embedded next to each other in a cylinder with a large enough diameter to accomodate both of them such that the endpoints lie in $X \times \{0, 1\}$. The cylinder is then rescaled to ensure that the diameter is 1.*

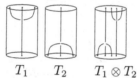

$$T_1 \qquad T_2 \qquad T_1 \otimes T_2$$

Framed links can be similarly generalised as well as built using framed tangles.

Definition 9 (Framed Tangle). *A framed tangle is a tangle where every connected component is equipped with a continuous unit normal vector field. Two framed tangles are equivalent if they are isotopic by an ambient isotopy (of tangles) that preserves the homotopy class of the vector field on each component.*

The following theorem tells us that every tangle can be constructed from a finite set of generating tangles using composition and tensor product and that the equivalence of tangles can be formulated in terms of a finite set of generating relations.

Theorem 3. *Every unoriented, framed tangle is the composition of tensor products of the five tangles*

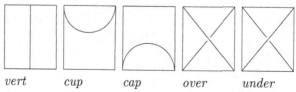

$$vert \qquad cup \qquad cap \qquad over \qquad under$$

and two such products correspond to the same tangle if and only if they can be connected by a sequence of the following moves, where T and S are arbitrary tangles.

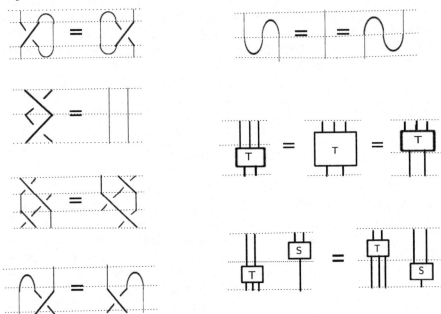

The same is true of oriented, framed tangles, if each generator and each relation above is written with every possible consistent orientation. The same is also true in either case for unframed tangles, if one adds in relation (1) that both equal the identity tangle.

Remark 2. We will refer to a representation of a tangle in terms of the generating tangles, as a *tangle diagram.* Framed tangles upto isotopy can then be treated as the equivalence class of tangle diagrams under the relations above. We will use the term *generating tangles* to refer to the five tangles mentioned in the above theorem - vert, cap, cup, over and under. We will use the term *framed tangle moves* to refer to the moves described above. The corresponding moves for tangles shall be referred to as *tangle moves.*

3   Formalization of Tangles and Links

The Theorem 3 of the previous section enables an algebraic description of tangles. This algebraic description can be used to used to formalize tangles and links in Isabelle/HOL. We begin by defining *generating tangles* as a type.

```
datatype brick = vert
               | cup
               | cap
               | over
               | under
```

We then proceed to define `blocks`, which can be geometrically interpreted as collection of bricks placed next to each other in a horizontal fashion. The type of blocks is defined as follows

> **type_synonym** *block* = *"brick list"*

A `wall` can be interpreted as collection of blocks arranged vertically over each other. It is formally defined as follows

> **datatype** *wall* = *basic block*
> *|prod block   wall*   (**infixr** *"*"* 66)

One may note that every tangle diagram can be represented by a wall. Every wall need not however represent a tangle diagram, because the number of incoming and outgoing strands of constituent blocks need not match appropriately. The following functions are thus used to check if a wall represents a tangle diagram or a link diagram.

primrec *is_tangle_diagram::"wall ⇒   bool"*
where
"is_tangle_diagram (basic x) = True"
*| "is_tangle_diagram (x*xs) = (if is_tangle_diagram xs*
*   then (codomain_block x = domain_wall xs)*
*   else False)"*

definition *is_link_diagram::"wall ⇒   bool"*
where
"is_link_diagram x ≡ (if (is_tangle_diagram x)
*   then*
*   (abs (domain_wall x) + abs(codomain_wall x) = 0)*
*   else False)"*

The `domain` and `codomain` functions mentioned above are used to describe the number of incoming and outgoing strands of a block and a wall respectively.

Defining the tensor product of two tangle diagrams through walls is more intricate process, since we want to ensure that tensor product of two well defined tangle diagrams returns a well defined tangle diagram. The tensor product is defined along the following lines.

primrec *concatenate* :: *"block ⇒ block ⇒ block"* (**infixr** *"⊗" 65*) **where**
concatenates_Nil: "[] ⊗ ys = ys" |
concatenates_Cons: "((x#xs) ⊗ys) = x#(xs⊗ys)"

primrec *make_vert_block:: "nat ⇒ block"*
where
"make_vert_block 0 = []"
| "make_vert_block (Suc n) = vert#(make_vert_block n)"

```
fun tensor::"wall => wall => wall" (infixr "⊗" 65)
where
1:"tensor (basic x) (basic y) = (basic (x ⊗ y))"
|2:"tensor (x*xs) (basic y) = (
        if (codomain_block y = 0)
          then (x ⊗ y)*xs
          else
          (x ⊗ y)
            *(xs⊗(basic (make_vert_block (nat (codomain_block y))))))"
|3:"tensor (basic x) (y*ys) = (
      if (codomain_block x = 0)
        then (x ⊗ y)*ys
        else
        (x ⊗ y)
          *((basic (make_vert_block (nat (codomain_block x))))⊗ ys))"
|4:"tensor (x*xs) (y*ys) = (x ⊗ y)* (xs ⊗ ys)"
```

Tensor product of two walls, when both the walls consist a single block, is equal to placing the blocks next to each other. The following diagram illustrates how the tensor product is defined when one of the walls consists of a single block and the other consists of multiple blocks.

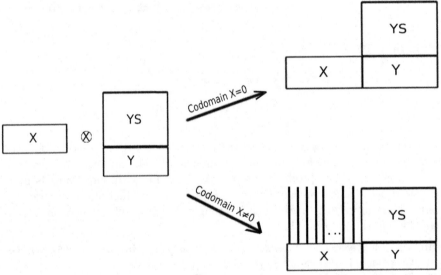

In a similar fashion, tensor product of two walls when the wall on the left hand side consists of more blocks than the wall on the right hand side, are defined. Tensor product of two walls, when both the walls are consituted by multiple blocks is recursively defined in terms of the above steps.

The following theorem tells us that if two walls represent tangle diagrams, then their tensor product is a tangle diagram.

```
theorem tensor_preserves_is_tangle:
  assumes "is_tangle_diagram x"
      and "is_tangle_diagram y"
  shows "is_tangle_diagram (x ⊗ y)"
```

3.1   Tangle Moves and Tangle Equivalence

The tangle moves express a relationship between the walls, and are defined as boolean functions. For instance the following tangle move

is formalized as

definition swing_pos::"wall ⇒ wall ⇒ bool"
where
"swing_pos x y ≡ (x = r_over_braid)∧(y = l_over_braid)"

The relation expressing equivalence of two walls through generating moves is defined by the function linkrel, which is defined as the disjunction of individual tangle moves.

definition linkrel::"wall ⇒wall ⇒bool"
where
"linkrel x y = ((uncross x y) ∨ (pull x y) ∨ (straighten x y)
∨(swing x y)∨(rotate x y) ∨ (compress x y) ∨ (slide x y))"

The function framed_linkrel similarly defines the equivalence of two framed links by omitting appropriate tangle moves. This enables us to define tangle equivalence as follows

inductive Tangle_Equivalence :: "wall ⇒ wall ⇒ bool"   **(infixl** "~" 64)
where
   refl [intro!, Pure.intro!, simp]: " a ~ a"
|equality [Pure.intro]: "linkrel a b ⟹ a ~ b"
|domain_compose:"(domain_wall a = 0)∧(is_tangle_diagram a)
                     ⟹ a ~ ((basic [])∘a)"
|codomain_compose:"(codomain_wall a = 0) ∧ (is_tangle_diagram a)
                     ⟹ a ~ (a ∘ (basic []))"
|compose_eq:"((B::wall) ~ D) ∧ ((A::wall) ~ C)
         ∧(is_tangle_diagram A)∧(is_tangle_diagram B)
         ∧(is_tangle_diagram C)∧(is_tangle_diagram D)
         ∧(domain_wall B)= (codomain_wall A)
         ∧(domain_wall D)= (codomain_wall C)
             ⟹((A::wall) ∘ B) ~ (C ∘ D)"
|trans: "A~B ⟹ B~C ⟹ A ~ C"
|sym:"A~ B ⟹ B ~A"
|tensor_eq:
     "((B::wall) ~ D)
     ∧ ((A::wall) ~ C)
     ∧(is_tangle_diagram A)
     ∧(is_tangle_diagram B)
     ∧(is_tangle_diagram C)
     ∧(is_tangle_diagram D)
     ⟹((A::wall) ⊗ B) ~ (C ⊗ D)"

This definition of `Tangle_Equivalence` states that the wall `xs` is equivalent to the wall `ys` if and only if:-

1. If `xs` and `ys` are same as walls.
2. They are related by a tangle move.
3. If `xs` and `ys` can be expressed as tensor product of two walls representing tangle diagrams, such that the corresponding walls are equivalent.
4. If `xs` and `ys` can be expressed as composition of two walls representing tangle diagrams, such that the corresponding walls are equivalent and the composition is end point matching.
5. If `ys` belongs to the transitive closure of `ys` under tangle equivalence or vice versa.

The equivalence of two framed tangles, in terms of representative walls, is obtained by replacing `linkrel` with `framed_linkrel` in the above definition. These definitions along with the presence of quotient types in Isabelle/HOL, enable us to define both tangle diagrams and framed tangle diagrams as types. Links, framed links, tangles and framed tangles can then be defined as types quotiented under the above equivalence relations.

The formalization described above enables us to prove equivalence of two link diagrams, as illustrated by the following theorem.

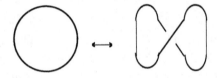

theorem *Example:*
"(basic [cup,cup])∘(basic [vert,over,vert]) ∘ (basic [cap,cap])
          ~ (basic [cup]) ∘ (basic [cap])"

4   Kauffman Bracket

4.1   Kauffman Bracket Through Tangles

The Kauffman bracket of a link, as defined in Sect. 2.1, is fairly cumbersome to formalize in the given form. We take recourse to an alternative method of constructing the Kauffman bracket through tangles, which can be found in [1]. In the form as described in the paper, it is defined as a functor from the category of tangles to the category of linear operators on vector spaces. In essence, it can treated as a map from the set of tangles to the set of matrices over the ring of Laurent polynomials. Consider the map ⟨ ⟩ from the set of tangle diagrams to the set of matrices whose entries are Laurent polynomials such that:-

– **Step 1**: Generators are mapped to the following matrices

$$\langle vert\rangle \quad \longrightarrow \quad \begin{bmatrix} 1 & 0 \\ 0 & 1 \end{bmatrix}$$

$$\langle cup\rangle \quad \longrightarrow \quad \begin{bmatrix} 0 & -A & A^{-1} & 0 \end{bmatrix}^T$$

$$\langle cap\rangle \quad \longrightarrow \quad \begin{bmatrix} 0 & A & -A^{-1} & 0 \end{bmatrix}$$

$$\langle over\rangle \quad \longrightarrow \quad \begin{bmatrix} A & 0 & 0 & 0 \\ 0 & 0 & A^{-1} & 0 \\ 0 & A^{-1} & A - A^{-3} & 0 \\ 0 & 0 & 0 & A \end{bmatrix}$$

$$\langle under\rangle \quad \longrightarrow \quad \begin{bmatrix} A^{-1} & 0 & 0 & 0 \\ 0 & A^{-1} - A^3 & A & 0 \\ 0 & A & 0 & 0 \\ 0 & 0 & 0 & A^{-1} \end{bmatrix}$$

– **Step 2**: It is extended to all tangles in the following fashion.

$$\langle T_1 \otimes T_2\rangle = \langle T_1\rangle \otimes \langle T_2\rangle$$

$$\langle T_1 \circ T_2\rangle = \langle T_1\rangle \circ \langle T_2\rangle$$

The map $\langle\ \rangle$ is a tangle invariant. The proof of invariance involves checking that the map defined above is invariant under tangle moves. If one pays careful attention to the rank of the matrices defined above, one might notice a correspondence between the dimension of the matrix and the codomain and domain of the tangles. The relationship is given by

$$row\ length\ (\langle T\rangle) = 2^{dom(T)}$$

$$column\ length\ (\langle T\rangle) = 2^{codom(T)}$$

As a consequence, it follows that every link is mapped to a 1×1 matrix, whose entry is a Laurent polynomial. This polynomial is a link invariant, which is the same as the *Kauffman Bracket* or the *Bracket Polynomial* [1].

4.2   Formalizing Kauffman Bracket

Formalizing Kauffman bracket on the lines defined above, requires a formalized theory of tensor products of matrices. There was no such formalized theory available in Isabelle/HOL to the best of our knowledge. The tensor product and some of its relevant properties were formalized by building on the existent formalized theory of matrices by Sternagel and Thiemann which can be found in [2]. Two of the important features of this formalization are:-

1. Tensor products can be defined for matrices over any commutative ring with unity.
2. Properties such as associativity and bilinearity of the tensor product have been proved.

Laurent polynomials could be defined by treating Laurent polynomials as rational functions, and by combining aspects of the existent theories on polynomials and fraction fields.

```
type_synonym intpoly = "int poly"
```

```
type_synonym rat_poly = "intpoly fract"
```

In order to define the Kauffman bracket, we begin by mapping bricks to the matrices whose entries are rational functions.

```
primrec brickmat::"brick ⇒ rat_poly mat"
where
"brickmat vert = [[1,0],[0,1]]"
|"brickmat cup  = [[0],[A],[-B],[0]]"
|"brickmat cap  = [[0,-A,B,0]]"
|"brickmat over = [[A,0,0,0],
                    [0,0,B,0],
                    [0,B,A-(B*B*B),0],
                    [0,0,0,A]]"
|"brickmat under  = [[B,0,0,0],
                      [0,B-(A*A*A),A,0],
                      [0,A,0,0],
                      [0,0,0,B]]"
```

The variables A and B used above refer to x and $1/x$. This map is extended to blocks in the following manner.

```
primrec blockmat::"block ⇒ rat_poly mat"
where
"blockmat [] = [[1]]"
|"blockmat (l#ls) = (brickmat l) ⊗ (blockmat ls)"
```

Every wall is mapped to the product of matrices corresponding to its constituent blocks. We refer to the matrix associated to a wall as the Kauffman matrix of the wall.

```
primrec kauff_mat::"wall ⇒ rat_poly mat"
where
"kauff_mat (basic w) = (blockmat w)"
|"kauff_mat (w*ws) = rat_poly.matrix_mult (blockmat w) (kauff_mat ws)"
```

The type rat_poly mat consists of both 'valid' and 'invalid' matrices. The product of two matrices need not be a valid matrix, unless both the matrices are valid and the relevant row lengths and column lengths match. Given that the Kauffman matrix of a wall is obtained by composition of matrices, it is neccessary to ensure that the Kauffman matrix of a wall representing a tangle diagram is a valid matrix.

theorem `effective_matrix_kauff_mat:`
 assumes `"is_tangle_diagram ws"`
 shows `"(rat_poly.row_length (kauff_mat ws)) = 2^(nat (domain_wall ws))"`
 and `"length (kauff_mat ws) = 2^(nat (codomain_wall ws))"`
 and `"mat (rat_poly.row_length (kauff_mat ws)) (length (kauff_mat ws))`
      `(kauff_mat ws) "`

It follows from this result that the Kauffman matrix of a wall representing a link diagram, is a 1×1 matrix. Thus it establishes a correspondence between links and rational functions.

theorem `link_diagram_matrix:`
 assumes `"is_link_diagram ws"`
 shows `"mat 1 1 (kauff_mat ws) "`

The following theorems illustrate that the Kauffman matrix of a tensor product of two tangles is the tensor product of their Kauffman matrices and that the composition of two tangles is the product of their Kauffman matrices.

theorem `tangle_compose_matrix:`
`"((is_tangle_diagram ws1) ∧ (is_tangle_diagram ws2)`
`∧ (domain_wall ws2 = codomain_wall ws1)) ⟹`
`kauff_mat (ws1 ∘ ws2) =`
`rat_poly.matrix_mult (kauff_mat ws1) (kauff_mat ws2)"`

theorem `Tensor_Invariance:`
 `"(is_tangle_diagram ws1) ∧ (is_tangle_diagram ws2)`
 `⟹ (kauff_mat (ws1 ⊗ ws2) = (kauff_mat ws1) ⊗ (kauff_mat ws2))"`

In order to prove that the Kauffman bracket is an invariant of links, it suffices to prove that the map kauff_mat is an invariant of walls under the relation Tangle_Equivalence. This statement is formally expressed as follows.

theorem `"(w1::wall) ~f w2 ⟹ kauff_mat w1 = kauff_mat w2"`

The proof of this theorem consisted of splitting the goal into subgoals using the induction rule, and then checking invariance under various tangle moves.

5 Conclusions and Further Work

In this paper, a formalization of some of the knot theoretic concepts has been presented. These concepts include tangles, links upto ambient isotopy and the Kauffman bracket. The invariance of the Kauffman bracket for framed tangles has been further proved in Isabelle/HOL. The list of results proved in the course of formalization however was not restricted to these results. The total length of the code is over 8000 lines. This also includes development of tensor product for matrices, with proofs of some of the basic results. The proofs can be found online at https://github.com/prathamesh-t/Tangle-Isabelle/

The choice of formalization of knots through tangles has its share of advantages and disadvantages. The correctness of the formalized theory presented here depends on the correctness of the human proofs that establish the equivalence of knot isotopy and tangle moves. The proof can be obtained by morse theory. The proof can also be motivated pictorially. The formal proof of invariance of the Kauffman bracket illustrates the fact that highly nontrivial results can be obtained from this choice of definition. At the same time, the same definitions do make it fairly tricky to prove some of the simpler results such as invariance of the number of components of links.

With respect to our future work, we would like to formalize other knot invariants such as the Alexander polynomial. We would also like to formulate forms of equivalence between links which are important in 3-manifold topology, such as the ribbon equivalence. We intend to extract formally verified code for computation of various invariants including the Kauffman bracket. We are also keen on formalizing more results about rational functions, since many of these results could prove useful in computation of Kauffman bracket. Another possible direction for future research is to formalize results about similar topological objects such as virtual links.

We hope that this work illustrates the provability of results in low dimensional topology and geometry in interactive proof assistants.

Acknowledgments. I would like to thank Siddhartha Gadgil for conceptualising and supervising the project, apart from his many other invaluable suggestions.

References

1. Sawin, S.: Links, quantum groups and TQFTs. Bull. Amer. Math. Soc. (N.S.) **33**, 413–445 (1996)
2. Sternagel, C., Thiemann, R.: Executable Matrix Operations on Matrices of Arbitrary Dimensions, Archive of Formal Proofs (2010). http://afp.sf.net/entries/Matrix.shtml
3. Kauffman, L.H.: On Knots. Princeton University Press, Princeton (1987)

Pattern Matches in HOL:

A New Representation and Improved Code Generation

Thomas Tuerk[1]([🖂]), Magnus O. Myreen[2,3], and Ramana Kumar[3]

[1] Independent Scholar, Brechen, Germany
thomas@tuerk-brechen.de
[2] CSE Department, Chalmers University of Technology, Gothenburg, Sweden
[3] Computer Laboratory, University of Cambridge, Cambridge, UK

Abstract. Pattern matching is ubiquitous in functional programming and also very useful for definitions in higher-order logic. However, it is not directly supported by higher-order logic. Therefore, the parsers of theorem provers like HOL4 and Isabelle/HOL contain a pattern-compilation algorithm. Internally, decision trees based on case constants are used. For non-trivial case expressions, there is a big discrepancy between the user's view and the internal representation.

This paper presents a new general-purpose representation for case expressions that mirrors the input syntax in the internal representation closely. Because of this close connection, the new representation is more intuitive and often much more compact. Complicated parsers and pretty printers are no longer required. Proofs can more closely follow the user's intentions, and code generators can produce better code. Moreover, the new representation is more general than the currently used representation, supporting guards, patterns with multiple occurrences of the same bound variable, unbound variables, arithmetic expressions in patterns, and more. This work has been implemented in the HOL4 theorem prover and integrated into CakeML's proof-producing code generator.

1 Introduction

Pattern matching is ubiquitous in functional programming and in definitions within interactive theorem provers. Through the use of case expressions (a. k. a. match expressions), pattern matching allows for concise and easy to read definitions. For provers based on higher-order logic (HOL), case expressions are not natively part of the logic. To use them, they are processed outside the logic.

The term parsers of all major HOL systems, in particular HOL4 [11], Isabelle/HOL [13], HOL Light [2], and ProofPower, contain an implementation of pattern compilation. This pattern compilation turns case expressions into decision trees consisting of nested applications of *case constants* [1,10], which are defined for each algebraic datatype. A complicated pretty printer prints the resulting decision trees as case expressions. The decision trees can be evaluated efficiently using basic rewriting techniques. They represent complete case splits; no cases overlap and no case is missing. However, the pattern-compilation

© Springer International Publishing Switzerland 2015
C. Urban and X. Zhang (Eds.): ITP 2015, LNCS 9236, pp. 453–468, 2015.
DOI: 10.1007/978-3-319-22102-1_30

implementation in the parser and the complicated pretty printer are a cause for concern in an LCF-style prover. In addition to the inference kernel, these components need to be trusted to some degree. Another disadvantage is that pattern compilation often leads to a huge blow-up in term size. Code extracted from the internal representation is often both hard to read and slow. Moreover, the structure intended by the user is obfuscated by pattern compilation and proofs have to follow the artificial, often very complicated structure of the internal representation.

Our contribution is a new representation for case expressions that avoids these problems. Our representation is able to mirror the user's input syntax faithfully. Therefore parsing and pretty printing are straightforward and the blow-up in term size is avoided. Compared to the decision tree representation, extracted code is of better quality and proofs need to consider fewer cases. Moreover, the new representation supports more advanced pattern-matching features. For example, it supports patterns that include guards, binding a variable multiple times, arithmetic expressions as well as a concept similar to simple view patterns [12].

Related to our work are function definitions packages like the ones implemented in Isabelle/HOL [3], HOL4 [10] and HOL Light [2]. At the top level, they are able to avoid case expressions in the logic completely by using a set of (conditional) equations for function specifications. For example, the length function on lists $(\text{len } l := \text{case } l \text{ of } [] \Rightarrow 0 \mid x :: xs \Rightarrow (\text{len } xs + 1))$ is described by the equations $\text{len } [] = 0$ and $\forall x \, xs. \, \text{len } (x :: xs) = \text{length } xs + 1$. Since arbitrary equations are used instead of case constants, these packages provide all the features offered by our approach like guards, arithmetic expressions or binding a variable multiple times. However, unlike our approach, these packages do not represent case expressions in the logic at all. Instead a set of (conditional) equations is returned. The most striking difference in semantics, compared to case expressions, is that there is no precedence on these equations. This means that overlapping patterns are problematic. One option, implemented by e.g. HOL4 and Isabelle/HOL, is to use pattern compilation to transform the input patterns to a set of non-overlapping patterns. This leads however to the same issues and restrictions described for pattern compilation above. An alternative implemented in e.g. Isabelle/HOL [3] and HOL Light is to prove that overlapping input patterns result in the same value. Enforcing non-overlapping patterns often leads to either very complicated guards or similar blow-ups in the number of cases as compilation to a decision tree.

The work[1] presented here has been implemented in HOL4 [11]. Our representation is very similar to concepts used internally by the HOL Light function definition package[2]. However, we expose our definitions to the user, whereas the function definition package uses it only internally. Since our representation uses Hilbert's choice operator and existential quantification, naive usage is likely to cause problems. In contrast to decision trees, simple rewrite

---

[1] The code can be found under: https://github.com/HOL-Theorem-Prover/HOL/ examples/pattern_matches.

[2] Compare function CASEWISE in define.ml.

techniques are not sufficient. Therefore, a significant part of this work consists
in providing specialised tools for dealing with our case expressions. We provide
parsers and pretty printers[3], and evaluation and simplification tools[4]. There is
support for turning function definitions using our case expression into equations
similar to the ones produced by function definition packages[5]. Furthermore, there
are tools for converting between our case expressions and ones represented by
decision trees. These tools range from untrustworthy parsers and pretty printers
to a pattern-compilation algorithm implemented inside the logic. This allows
the user to choose the right representation for the task at hand. Moreover, our
implementation of a pattern-compilation algorithm[6] lets us leverage some of the
nice properties of decision trees. It is used to check the exhaustiveness of our
case expressions[7] as well as pruning patterns that are made redundant by the
combination of multiple other patterns[8].

Applying code generation to the new representation, we can produce higher-
quality code. We demonstrate the quality improvement with a few examples, and
describe how proof-producing code generation (for CakeML [5]) can be extended
to generate code that mirrors the exact structure of the HOL term and the
concrete syntax provided by the user (Sect. 6).

2   Shortcomings of Decision Tree Representation

All major HOL systems, in particular HOL4 [11], Isabelle/HOL [13], ProofPower
and HOL Light [2], use decision trees based on case constants for representing
case expressions. In this section, we try to illustrate this method and its shortcomings using a number of examples. If not explicitly stated otherwise, we use
HOL4 as the example prover. The implementation in the other systems is very
similar, though.

Basic Example. The classical representation of case expressions is based on case
constants [1,10]. HOL4's datatype definition package produces for each algebraic
datatype definition a case constant that can perform a top-level pattern match on
the constructors of that datatype. HOL4's list datatype (with constructors Nil
and Cons) for example has an associated case constant list_case, which is characterised by the following equations

```
list_case Nil n f = n
list_case (Cons y ys) n f = f y ys
```

---

[3] See patternMatchesSyntax.

[4] See e. g. PMATCH_SIMP_ss or PMATCH_REMOVE_GUARDS_ss.

[5] See PMATCH_LIFT_BOOL_ss.

[6] See e. g. PMATCH_CASE_SPLIT_ss.

[7] See COMPUTE_REDUNDANT_ROWS_INFO_OF_PMATCH and
PMATCH_IS_EXHAUSTIVE_CONSEQ_CONV.

[8] See PMATCH_REMOVE_REDUNDANT_ss.

The datatype definition package also informs the parser and the pretty printer about such case constants. The term `list_case   x   5 (λy ys. y + 3)` is pretty printed as

```
case x of Nil => 5 | Cons y ys => y + 3
```

This representation is also accepted by the parser and parsed to the internal `list_case` representation.

Pattern Match Heuristics. For such simple case expressions, the classical approach works perfectly. However, problems start to appear if the case expressions become even slightly more complex. As an example, let's define b ∨ (a ∧ c) via a case expression:

```
case (a, b, c) of
   (_, T, _) => T
 | (T, _, T) => T
 | (_, _, _) => F
```

For this example, several applications of the case constant for type `bool`, better known as if-then-else, need to be nested. Classically, HOL4 performs the splits from left to right, i.e. in the order a, b, c. This leads to (slightly simplified) the following internal representation and corresponding pretty printed output:

```
if a then (                    case (a,b,c) of
   if b then (                    (T,T,T) => T
      if c then T else T        | (T,T,F) => T
   ) else (                     | (T,F,T) => T
      if c then T else F        | (T,F,F) => F
   )                            | (F,T,_) => T
) else (                        | (F,F,_) => F
   if b then T else F
)
```

Even for this simple example, one can observe a severe blow-up. The number of rows has doubled. One might also notice that the clear structure of the input is lost. Other systems using the classical approach might behave slightly differently in detail but in principle suffer from the same issues. Isabelle/HOL for example also performs pattern compilation always from left to right, but is slightly better at avoiding unnecessary splits.

To combat some of these issues, we extended HOL4's pattern compilation algorithm in early 2013 with state-of-the-art pattern match heuristics[9] presented by Luc Maranget [6]. These heuristics often choose a decent ordering of case splits. Moreover, we also implemented – similar to Isabelle/HOL – the avoidance of some unnecessary splits. With these extensions, the example is compiled to:

---

[9] See `/src/1/PmatchHeuristics.sig` in the HOL4 sources.

```
if b then T else (          case (a,b,c) of
   if c then                    (v,T,v3) => T
      (if a then T else F)   | (T,F,T) => T
   else F                    | (F,F,T) => F
)                            | (v,F,F) => F)
```

However, this improvement has a price. The pattern compilation algorithm in the parser became slightly more complicated and the results are even harder to predict.

Real World Example. The following case expression is taken from Okasaki's book on Functional Datastructures [8]. It is used in a function for balancing red-black trees, which are represented using the constructors Empty, Red and Black.

```
case (a,b) of
   (Red (Red a x b) y c,d) => Red (Black a x b) y (Black c n d)
 | (Red a x (Red b y c),d) => Red (Black a x b) y (Black c n d)
 | (a,Red (Red b y c) z d) => Red (Black a n b) y (Black c z d)
 | (a,Red b y (Red c z d)) => Red (Black a n b) y (Black c z d)
 | other => Black a n b
```

Parsing this term with the classical pattern-compilation settings in HOL4 results in a huge term that pretty prints with 121 cases! Even with our state of the art pattern-match heuristics a term with 57 cases is produced. In this term the right-hand sides of the rows are duplicated a lot. The right-hand side of the last row (Black a n b) alone appears 36 times in the resulting term.

This blowup is intrinsic to the classical approach. Our pattern match heuristics are pretty good at finding a good order in which to perform case splits. For this example, they find an optimal order. There is no term based on case constants that gets away with fewer than 57 cases. However, clever pretty printers might present a smaller looking case expression (see Sect. 6.2).

Also notice that this example relies heavily on the precedence of earlier rows over later ones in the case expressions. If we use – as required by the equations produced by function definition packages – non-overlapping patterns, we get a similar blow-up as when compiling to a decision tree.

3   New Approach

In the previous section we presented the classical approach used currently by all major HOL systems. We showed that the internal representation for this approach often differs significantly from the input. There is often a huge blow-up in size. This leads to less readable and more importantly less efficient code as well as lengthier and more complicated proofs. In the following we will present our new approach and how it overcomes these issues.

3.1   Definition

A row of a case expression consists of a pattern p, a guard g and a right hand side r. We need to model the variables bound by the pattern. Therefore, p, g and r are functions that get the value of the bound variables as their argument. They are of type $\gamma \to \alpha$, $\gamma \to$ bool and $\gamma \to \beta$, respectively. The type of the bound variables γ changes for each row of a case expression. In order to easily fit into HOL's type system, we apply these components to a function PMATCH_ROW which maps them to their intended semantics.

```
PMATCH_ROW p g r := λv.
  if (∃x. (p x = v) ∧ g x) then
    SOME (r (@x. (p x = v) ∧ g x))
  else
    NONE
```

For injective patterns, i. e. the ones normally used, this function models perfectly the standard semantics of case expressions (as e. g. defined in [9]).

It remains to extend this definition of the semantics of a single row to case expressions. Our case expressions try to find the first row that matches a given input value. If no row matches, a functional language like ML would raise a match-exception. Here, we decided not to model this error case explicitly and use ARB (a. k. a. Undef) instead. This constant is used to denote a fixed, but unknown value of an arbitrary type in HOL. Formally, this means that our case-expression constant PMATCH is defined recursively by the following equations:

```
PMATCH v [] := PMATCH_INCOMPLETE := ARB
PMATCH v (r::rs) := case r v of
                      SOME result => result
                    | NONE => PMATCH v rs
```

3.2   Concrete Syntax and Bound Variables

The definitions above let us write case expressions with guards. The body of the list-membership function mem x l, can for example be written as:

```
PMATCH 1 [
  PMATCH_ROW (λ (uv:unit). []) (λuv. T) (λuv. F);
  PMATCH_ROW (λ(y,ys). y::ys) (λ(y,ys). x = y) (λ(y,ys). T);
  PMATCH_ROW (λ(_0,ys). _0::ys) (λ(_0,ys). T) (λ(_0,ys). mem x ys)
]
```

This syntax closely mirrors the user's intention inside HOL. However, it is rather lengthy and hard to read and write. Therefore, we implemented a pretty printer and a parser for such expressions, enabling the following syntax:

```
CASE 1 OF [
  ||. [] ~> F;
  || (y,ys). y:: ys when (x = y) ~> T;
  || ys. _:: ys ~> mem x ys
]
```

For rows, we write bound variables only once instead of repeating them for pattern, guard and right-hand side. Moreover, there is support for wildcard syntax. Finally, we provide the CASE . OF . notation for PMATCH and reuse standard list syntax for the list of rows. Thus, in contrast to the classical approach the parser and pretty printer are straightforward.

3.3 Advanced Features

Our representation provides more expressive case expressions than the classical approach. We don't enforce syntactic restrictions like using only datatype constructors or binding variables only once in a pattern. Fine control over the bound variables in a pattern allows inclusion of free variables, which act like constants. Finally, there are guards.

These features can be used to very succinctly and clearly express complicated definitions that could not be handled with the classical approach. Division with remainder can for example be defined by:

```
my_divmod n c :=
  CASE n OF [
    || (q, r). q * c + r when r < c ~> (q,r)
  ]
```

The new case expressions are not even limited to injective patterns. They can for example be used to perform case splits on sets.

```
CASE n OF [
  ||. {} ~> NONE;
  ||(x, s). x INSERT s when ~(x IN s) ~> SOME (x, s)
]
```

3.4 Congruence Rules

Case expressions are frequently used to define recursive functions. In order to prove the well-foundedness of recursive definitions, HOL systems use a termination condition extraction mechanism, which is configured via congruence rules[10]. We provide such congruence rules for HOL4.

$$
\forall v\ v'\ \text{rows rows}'.\ (
$$
$$
(v = v') \wedge (r\ v' = r'\ v') \wedge
$$
$$
(\text{PMATCH } v'\ \text{rows} = \text{PMATCH } v'\ \text{rows}')) \Longrightarrow
$$
$$
(\text{PMATCH } v\ (r :: \text{rows}) =
$$
$$
\text{PMATCH } v'\ (r' :: \text{rows}'))
$$

$$
\forall p\ p'\ g\ g'\ r\ r'\ v\ v'.\ (
$$
$$
(p = p') \wedge (v = v') \wedge
$$
$$
(\forall x.\ (v = (p\ x)) \Rightarrow (g\ x = g'\ x)) \wedge
$$
$$
(\forall x.\ (v = (p\ x) \wedge g\ x) \Rightarrow
$$
$$
(r\ x = r'\ x))) \Longrightarrow
$$
$$
(\text{PMATCH\_ROW } p\ g\ r\ v =
$$
$$
\text{PMATCH\_ROW } p'\ g'\ r'\ v')
$$

These rules lead to very similar termination conditions as produced by the congruence rules for the classical decision trees. Therefore they work well with existing automatic well-foundedness checkers.

---

[10] See e.g. HOL4's *Description Manual* Sect. 4.5.2 or Isabelle/HOL's manual *Defining Recursive Functions in Isabelle/HOL* Sect. 10.1.

Remark. The observant reader might wonder, why `PMATCH_ROW` uses 3 functions as arguments instead of just one function returning a triple. This would simplify parsing and pretty printing, but cause severe problems for recursive definitions using `PMATCH`. HOL4's machinery would not be able to use the resulting congruence rules, since their application would require higher-order matching. We expect that Isabelle/HOL would be fine with rules that require higher-order matching, but have not tested this.

4   Evaluation and Simplification

If one naively expands the definition of `PMATCH`, one easily ends up with huge terms containing Hilbert's choice operator and existential quantifiers. To avoid this, we developed specialised tools for HOL4 to evaluate and simplify our case expressions. As a running example consider

```
CASE (SOME x, ys) OF [
  || y. (NONE, y::_)   ~> y;
  || (x',y). (x', y::_)  ~> (THE x')+y;
  || x. (SOME x, _)   ~> x;
  ||. (_, _)  ~> 0
]
```

Pruning Rows. For each row of a `PMATCH` expression, we check, whether its pattern and guard match the input value. If we can show that a row does not match, it can be dropped. If it matches, all following rows can be dropped. We don't need a decision for each row. If it is unknown whether a row matches, the row can just be kept. Finally, if the first remaining row matches, we can evaluate the whole case expression. Applying this method to the running example results in

```
CASE (SOME x, ys) OF [
  || (x',y). (x', y::_)  ~> (THE x')+y;
  || x. (SOME x, _)   ~> x
]
```

Partial Evaluation. In order to partially evaluate `PMATCH` expressions, we try to split the involved patterns into more primitive ones. For this we split tuples and group corresponding tuple elements in multiple rows into so-called *columns*. In the running example the first column contains the input `SOME x` and the patterns `x'` (where `x'` is bound) and `SOME x` (where `x` is bound). The second column contains `ys` as input and `y::_` and `_` as patterns.

If the input value of a column consists of the application of an injective function, e. g. a datatype constructor, and all patterns of this column contain either applications of the same injective function or bound variables, this column can be partially evaluated. We can remove the function application and just keep the arguments of the function in new columns. For rows containing bound variables, we substitute that variable with the input value and fill the new columns with fresh bound variables. In our running example, we can simplify the first column with this method. This partial evaluation leads to:

```
CASE (x, ys) OF [
   || y. (x'', y::_) ~> x'' + y;
   || x. (x, _) ~> x
]
```

Now, the first column consists of only variables and can be removed:

```
CASE ys OF [
   || y. y::_ ~> x + y;
   ||. _ ~> x
]
```

The semantic justification for this partial evaluation is straightforward. Essentially we are employing the same rules used by classical pattern compilation algorithms (compare e. g. Chap. 5.2 in [9]). However, implementing it smoothly in HOL4 is a bit fiddly. It involves searching for a suitable column to simplify and instantiating general theorems in non-straightforward ways.

Integration with Simplifier. Pruning rows and partial evaluation are the most important conversions for PMATCH-based case expressions. Other useful conversions use simple syntactic checks to remove redundant or subsumed rows. Additionally, we implemented conversions that do low-level maintenance work on the datastructure. For example, there are conversions to ensure that no unused bound variables are present, that the variable names in the pattern, guard and right-hand side of each row coincide and that each row has the same number of columns. All these conversions are combined in a single conversion called PMATCH_SIMP_CONV. We also provide integration with the simplifier in form of a simpset-fragment called PMATCH_SIMP_ss.

The presented conversions might look straightforward. However, the implementation is surprisingly fiddly. For example, one needs to be careful about not accidentally destroying the names of bound variables. The implementation of PMATCH_SIMP_CONV consists of about 1100 lines of ML.

5   Pattern Compilation

We provide several methods based on existing pretty printing and parsing techniques to translate between case expressions represented as decision trees and via PMATCH. The equivalence of their results can be proved automatically via repeated case splits and evaluation.

More interestingly, we implemented a highly flexible pattern-compilation algorithm for our new representation. As stated above, our simplification tools for PMATCH are inspired by pattern compilation. Thus, the remainder is simple: we provide some heuristics to compute a case-split theorem. Pattern compilation consists of choosing a case split, simplifying the result and iterating.

5.1 Constructor Families

We implemented the heuristics for finding case splits in form of a library called constrFamiliesLib. This library maintains lists of ML functions that construct case splits. An example of such a function is a literal-case function that performs a case distinction based on nested applications of if-then-else when a column consists only of bound variables and constants. However, the main source of case splits found by the library are *constructor families*.

A constructor family is a list of functions (constructors) together with a case constant and a flag indicating whether the list of constructors is exhaustive. Moreover, it contains theorems stating that these constructors have the desired properties, i. e. they state that all the constructors are injective and pairwise distinct and that the case constant performs a case split for the given list of constructors. If a column of the input PMATCH expression contains only bound variables and applications of constructor functions of a certain constructor family, a case-split theorem based on the case constant of that constructor family is returned.

constrFamiliesLib accesses HOL4's TypeBase database and therefore automatically contains a constructor family for each algebraic datatype. This default constructor family uses the classical datatype constructors and case constant. One can easily define additional constructor families and use them for pattern compilation as well as simplifying PMATCH expressions. These families can use constructor functions that are not classical datatype constructors. Such constructor families provide a different view on the datatype. They lead to a feature similar to the original views in Haskell [12]. This is perhaps best illustrated by a few examples.

List Example. One can for example declare [] and SNOC (appending an element at the end of a list) together with list_REVCASE as a constructor family for lists, where list_REVCASE is defined by

```
list_REVCASE l c_nil c_snoc =
  if l = [] then c_nil else (c_snoc (LAST l) (BUTLAST l))
```

With this declaration in place, we get list_REVCASE l 0 (λ x xs. x) automatically from compiling

```
CASE l OF [
  || . []   ~> 0;
  || (x, xs). SNOC x xs ~> x
]
```

5.2 Exhaustiveness Check/Redundancy Elimination

The case-split heuristics of our pattern compilation algorithm can be used to compute for a given PMATCH expression an exhaustive list of patterns with the following property: a pattern in the list is either subsumed by a pattern of the original PMATCH expression or does not overlap with it. There are no partial overlaps. Moreover, subsumption can be checked easily via first-order matching.

We can use such an exhaustive list of patterns to implement an exhaustiveness check[11]. We prune all patterns from the list that are subsumed by a pattern and guard in the original PMATCH expression. Pruning the list with respect to a pattern p and guard g, consists of adding the negation of g to all patterns in the list that are matched by p. Then we remove patterns whose guard became false. If after pruning with all original patterns and guards, the resulting list is empty, the original pattern match is exhaustive. Otherwise, we computed a list of patterns and guards that don't overlap with the original patterns and when added to the original case expression make it exhaustive.

The same algorithm also leads to a powerful redundancy-detection algorithm[12]. To check whether a row is redundant, we prune the exhaustive list of patterns with the patterns and guards of the rows above it. Then we check whether any pattern in the remaining list overlaps with the pattern of the row in question.

6   Improved Code Generation

In this section we turn our attention to code generation. It has become increasingly common to generate code from function definitions in theorem provers. Code generators for HOL operate by traversing the internal structure of HOL terms and producing corresponding code in a suitable functional programming language, e. g. SML, OCaml or Scala.

As discussed in Sect. 2, the classical per-datatype case constants can produce harmful duplication and a significant blow-up in the size of the HOL terms. Code generators that walk the internal structure of the terms are likely not to realise that there is duplication in various subterms and therefore to produce code that is unnecessarily verbose and somewhat unexpected considering what the user gave as input in their definition.

Our PMATCH-based case expressions avoid accidental duplication and instead very carefully represent the user's desired format of case expressions in the logic. As a result, even naive term-traversing code generators can produce high-quality code from HOL when PMATCH-based case expressions are used.

In what follows, we first illustrate the quality difference between code generated from the classical approach versus from PMATCH-based case expressions. Then, we will explain how our proof-producing code generator for CakeML has been extended to handle PMATCH. Typically, code generators do not provide any formal guarantee that the semantics of the generated code matches the semantics of the HOL definitions given as input, but code generation for CakeML is exceptional in that it does produce proofs. We have found that PMATCH is beneficial not only to the generated code itself but also when producing proofs about the semantics of the generated code.

---

[11] See PMATCH_IS_EXHAUSTIVE_CONSEQ_CONV.

[12] See PMATCH_REMOVE_REDUNDANT_ss.

6.1   The Quality of Generated Code

Simple code generators, which traverse the syntax of the HOL terms, produce code that is very similar to the internal representation. Take for instance a variation, using a catch-all pattern, of the basic example in Sect. 2. When compiled classically, this case expression results in a term that repeats the 5 on the right-hand side. The generated code also repeats the 5. We show the user input on the left and the output of code generation[13] on the right.

```
case x of                    case v5 of
   Cons y Nil => y + 3          Nil => 5
   | _ => 5                     | Cons v4 v3 => case v3 of
                                      [] => v4 + 3
                                      | Cons v2 v1 => 5
```

If we instead input the example above as a PMATCH-based case expression, we retain the intended structure: the result does not repeat the 5 and the Cons y Nil row stays on top. It is easy for a code generator to follow the structure of a PMATCH term, so the generated code reflects the user's input.

```
CASE x OF [                   case v2 of
   || y. Cons y Nil ~> y + 3 ;    Cons v1 Nil => v1 + 3
   ||. _ ~> 5                     | _ => 5
]
```

In this simple example, the duplication of the 5 is not too bad. However, for more serious examples like the red-black tree balancing function (in Sect. 2) the difference in code quality is significant. The code generated from the classical version is 90 lines long and unreadable, while the code generated from the PMATCH-based case expression is almost identical to the input expression, i.e. readable, unsurprising and only 8 lines long.

The red-black tree example is not the only source of motivation for producing better code. Our formalisation of the HOL Light kernel, which we have proved sound [4], contains several functions with tricky case expressions. For the HOL Light kernel, we want to carry the soundness proof over to the generated implementation, so it is important that the code generator can also produce proofs about the semantics of its output.

The helper function raconv (used in deciding alpha-convertibility) has the most complex case expression in the HOL Light kernel:

```
raconv env tm1 tm2 =
  case (tm1,tm2) of
    (Var _ _, Var _ _) => alphavars env tm1 tm2
  | (Const _ _, Const _ _) => (tm1 = tm2)
  | (Comb s1 t1, Comb s2 t2) => raconv env s1 s2 ∧ raconv env t1 t2
  | (Abs v1 t1, Abs v2 t2) =>
    (case (v1,v2) of
       (Var n1 ty1, Var n2 ty2) => (ty1 = ty2) ∧
                                    raconv ((v1,v2)::env) t1 t2
     | _ => F)
  | _ => F
```

---

[13] Our code generator renames variables. Here x has become v5, for example.

For this and other examples, a significant case explosion happens when parsed using the classical approach. The generated code is verbose, and the performance of our verified CakeML implementation [4] suffers as a result. By using `PMATCH`-based case expressions as input to the code generator, we retain the original structure and avoid the explosion.

6.2 Why Good Input Case Expressions Matter

For generating high-quality code, there is an alternative to rephrasing the input: post-processing the generated code. Such post-processing is (in a simple form) implemented as part of the case-expression pretty printers and the code-generation facilities of all major HOL systems. For translating decision trees into `PMATCH` expressions, (see Sect. 5) we implemented a similar, but more powerful post-processor, which combines rows by reordering them and introducing wildcards.

As discussed in Sect. 2, the 5 input cases of the red-black tree example produce 57 cases when printed naively in HOL4. After re-sorting and collapsing rows, our post-processor reduces this to 8 cases. Isabelle/HOL's pretty printer and code generator produce 41 cases. What's worse, these figures depend on the exact form of the internal decision tree. For another valid decision tree our post-processor produced e. g. 25 cases. So, post-processing can improve the result, but the results are still significantly worse than good user input.

For CakeML, there is the added difficulty that we need to *verify* the post-processing optimisation phase. Formally verifying optimisations over a language which includes closure values is very time consuming, as we have found when working on the CakeML compiler [5]. The reason is that optimisations alter the code, and, through closure values, code can appear in the values of the language. As a result, every optimisation requires a longwinded value-code relation for its correctness theorem.

Reasoning about and optimising `PMATCH`-based case expressions is much simpler. Moreover, the `PMATCH`-based approach allows manual fine-tuning of the exact form of the case expression in the logic, before the automation for code generation takes over. In general this leads to better results.

6.3 Proof-Producing Code Generation for CakeML

It is straightforward to write a code generator that walks a `PMATCH` term and produces a corresponding case expression in a functional programming language like CakeML. For CakeML, we additionally need to derive a (certificate) theorem which shows that the semantics of the generated CakeML code matches the semantics of the input `PMATCH` term. In this section, we explain how the proof-producing code generator of Myreen and Owens [7] has been extended to handle `PMATCH`-based case expressions.

The proof-producing code generator has been described previously [7]; due to space restrictions, we will not present a detailed description here. Since the approach is compositional, it is sufficient for the purposes of this paper to focus on the form of the HOL theorems that are produced during code generation. These theorems relate generated code (deep embeddings) to input terms (shallow embeddings) via the CakeML semantics. They are of the the following form.

```
assumptions ⟹ Eval deep_embedding env (inv shallow_embedding)
```

Here `Eval` is an interface to the CakeML semantics, and the `env` argument is the semantic environment. The assumptions are used to collect constraints on the environment. The refinement invariant `inv` describes how a HOL4 value is implemented by a CakeML value. For example, for lists of Booleans, the appropriate refinement invariant would relate the HOL value `Cons F Nil` to the value `Conv "Cons" [Litv (Bool F); Conv "Nil" []]` in the semantics of CakeML.

The code-generation algorithm traverses a given shallow embedding bottom-up. To each subterm se, it applies a theorem of the form `... ⟹ Eval ... env (inv se)`, where `inv` is the refinement invariant appropriate for the type of `se`. Assumptions that relate shallow and deep embeddings are discharged via recursive calls. Other assumptions are either collected or discharged directly. The by-product of this forward proof is a deep embedding constructed in the first argument of `Eval`.

In order to support `PMATCH`, we need to provide theorems of the following form to this algorithm:

```
... ⟹ Eval (...) env (inv (PMATCH xv rows))
```

For an empty set of rows, the CakeML semantics of case expressions raises a `Bind` exception, whereas `PMATCH` results in `PMATCH_INCOMPLETE`. There is no connection between these two outcomes. Therefore, the following theorem intentionally uses the assumption false (`F`) to mark that one should never end up in this case.

```
F ⟹ Eval env (Mat x []) (b (PMATCH xv []))
```

The case of non-empty pattern lists is more interesting. The theorem is long and complicated, so we explain its parts in turn. First, let us look at the conclusion, i.e. lines 11 and 12 below. The conclusion allows us to add a pattern row, `PMATCH_ROW`, to the shallowly embedded `PMATCH` term and, at the same time, a row is added to the deep embedding: `Mat x ((p,e)::ys)`.

```
1     ALL_DISTINCT (pat_bindings p []) ∧
2     (∀v1 v2. (pat v1 = pat v2) ∧ v1 = v2) ∧
3     Eval env x (a xv) ∧
4     (p1 xv ⟹ Eval env (Mat x ys) (b (PMATCH xv yrs))) ∧
5     EvalPatRel env a p pat ∧
6     (∀env2 vars.
7         EvalPatBind env a p pat vars env2 ∧ p2 vars ⟹
8         Eval env2 e (b (res vars))) ∧
9     (∀vars. (pat vars = xv) ⟹ p2 vars) ∧
10    ((∀vars. ¬(pat vars = xv)) ⟹ p1 xv) ⟹
11    Eval env (Mat x ((p,e)::ys))
12        (b (PMATCH xv ((PMATCH_ROW pat (K T) res)::yrs)))
```

Now let us look at the assumptions on the theorem and how they are discharged by the code generator when the theorem is used. The subterm evaluations are on lines 4 and 6–8. The code generator derives theorems of these forms by recursively calling its syntax-traversing function. As mentioned above, each such translation comes with assumptions and these assumptions are captured by variables p1 and p2. When the theorem above is used lines 9 and 10 will be left as assumptions, but the internal assumptions, p1 and p2, are passed in these lines to higher levels (see [7] for details).

The other lines 1, 2 and 5 are simple assumptions that are discharged by evaluation and an automatic tactic. Line 1 states that all the variables in the pattern have distinct names. PMATCH allows multiple binds to the same variable, but CakeML's pattern matching semantics does not allow this. Line 2 states that the pattern function in HOL is injective; and line 5 states that the CakeML pattern p corresponds to the pattern function pat in the current CakeML environment env and based on refinement invariant a for the input type.

The CakeML code generator can only generate code for PMATCH-based case expressions when there is an equivalent pattern expression in CakeML. This means, for instance, that one cannot generate code for case expressions with multiple binds to a variable, those that use non-constructor based patterns, or those that use guards. PMATCH-based case expressions that do not fall into this subset can usually be translated by removing these features first. We provide automated tools which work for most situations[14], although using this feature-removing automation can, in the worst case, lead to significant changes in structure of the terms, even replacing them with bulky decision trees similar to those of the classical approach.

We have used this PMATCH-based translation to produce high-quality CakeML code for all of the case expressions in the HOL Light kernel.

7 Summary

This paper presents a new representation, PMATCH, for case expressions in higher-order logic which faithfully captures the structure of the user's input. Because pattern-matching structure is retained, proofs over PMATCH expressions are simpler, and code generated from PMATCH expressions is better. Moreover, PMATCH is more general than currently-used representations: it supports guards, views, unbound variables, arithmetic expressions in patterns and even non-injective functions in patterns.

In addition to the new representation itself, we provide tools for working with PMATCH expressions in HOL4. Our tools include a parser and pretty printer, conversions for simplification and evaluation, and a pattern-compilation algorithm inside the logic. This pattern compilation is used to check the exhaustiveness of lists of patterns as well as for implementing powerful techniques for eliminating redundant rows. Furthermore, we have extended CakeML's proof-producing code generator to translate PMATCH expressions into high-quality CakeML code.

At present, our tools are already more powerful and convenient than the existing support for case expressions in the major HOL systems. In the future we plan

---

[14] The exceptions are non-constructor patterns that are not part of a constructor family.

to extend them further. In particular, we plan to improve the support for advanced patterns like arithmetic expressions.

Acknowledgements. The second author was partially supported by the Royal Society UK and the Swedish Research Council.

References

1. Augustsson, L.: Compiling pattern matching. In: FPCA, pp. 368–381 (1985). http://dx.doi.org/10.1007/3-540-15975-4_48
2. Harrison, J.: HOL light: an overview. In: Berghofer, S., Nipkow, T., Urban, C., Wenzel, M. (eds.) TPHOLs 2009. LNCS, vol. 5674, pp. 60–66. Springer, Heidelberg (2009). http://dx.doi.org/10.1007/978-3-642-03359-9_4
3. Krauss, A.: Partial recursive functions in higher-order logic. In: Furbach, U., Shankar, N. (eds.) IJCAR 2006. LNCS (LNAI), vol. 4130, pp. 589–603. Springer, Heidelberg (2006)
4. Kumar, R., Arthan, R., Myreen, M.O., Owens, S.: HOL with definitions: semantics, soundness, and a verified implementation. In: Klein, G., Gamboa, R. (eds.) ITP 2014. LNCS, vol. 8558, pp. 308–324. Springer, Heidelberg (2014). http://dx.doi.org/10.1007/978-3-319-08970-6_20
5. Kumar, R., Myreen, M.O., Norrish, M., Owens, S.: CakeML: a verified implementation of ML. In: Jagannathan, S., Sewell, P. (eds.) POPL, pp. 179–192. ACM (2014)
6. Maranget, L.: Compiling pattern matching to good decision trees, September 2008
7. Myreen, M.O., Owens, S.: Proof-producing translation of higher-order logic into pure and stateful ML. J. Funct. Program. **24**(2–3), 284–315 (2014)
8. Okasaki, C.: Purely Functional Data Structures. Cambridge University Press, Cambridge (1998)
9. Peyton Jones, S.L.: The Implementation of Functional Programming Languages (Prentice-Hall International Series in Computer Science). Prentice-Hall Inc., Upper Saddle River (1987)
10. Slind, K.: Function definition in higher-order logic. In: von Wright, J., Harrison, J., Grundy, J. (eds.) TPHOLs 1996. LNCS, vol. 1125, pp. 381–397. Springer, Heidelberg (1996)
11. Slind, K., Norrish, M.: A brief overview of HOL4. In: Mohamed, O.A., Muñoz, C., Tahar, S. (eds.) TPHOLs 2008. LNCS, vol. 5170, pp. 28–32. Springer, Heidelberg (2008)
12. Wadler, P.: Views: A way for pattern matching to cohabit with data abstraction. In: POPL, pp. 307–313. ACM (1987)
13. Wenzel, M., Paulson, L.C., Nipkow, T.: The isabelle framework. In: Mohamed, O.A., Muñoz, C., Tahar, S. (eds.) TPHOLs 2008. LNCS, vol. 5170, pp. 33–38. Springer, Heidelberg (2008). http://dx.doi.org/10.1007/978-3-540-71067-7_7

Author Index

Printed in the United States
By Bookmasters